MANY URBANISMS

MANY URBANISMS

Divergent Trajectories of Global City Building

MARTIN J. MURRAY

Columbia University Press

New York

publication supported by a grant from
The Community Foundation *for* Greater New Haven
as part of the Urban Haven Project

Columbia University Press
Publishers Since 1893
New York Chichester, West Sussex
cup.columbia.edu

Library of Congress Cataloging-in-Publication Data
Names: Murray, Martin J., author.
Title: Many urbanisms : divergent trajectories of global city building / Martin J. Murray.
Description: New York : Columbia University Press, [2021] | Includes bibliographical references and index.
Identifiers: LCCN 2021022066 (print) | LCCN 2021022067 (ebook) |
ISBN 9780231204064 (hardback) | ISBN 9780231204071 (trade paperback) |
ISBN 9780231555357 (ebook)
Subjects: LCSH: Cities and towns—Study and teaching. | Urbanization. | Megacities.
Classification: LCC HT109 .M87 2021 (print) | LCC HT109 (ebook) |
DDC 307.762—dc23
LC record available at https://lccn.loc.gov/2021022066
LC ebook record available at https://lccn.loc.gov/2021022067

Cover design: Julia Kushnirsky
Cover images: © istockphoto

CONTENTS

Acknowledgments *vii*
Preface *xi*

Introduction: Rethinking Global Urbanism at the Start of the Twenty-First Century 1

PART I: CONVENTIONAL URBAN THEORY AT A CROSSROADS

1. The Narrow Preoccupations of Conventional Urban Studies 23

2. The Universalizing Pretensions of Mainstream Urban Studies: Generic Cities and the Convergence Thesis 45

PART II: TRAJECTORIES OF GLOBAL URBANISM AT THE START OF THE TWENTY-FIRST CENTURY: A FIRST APPROXIMATION

3. Globalizing Cities with World-Class Aspirations: The Emergence of the Postindustrial Tourist-Entertainment City 67

4. Struggling Postindustrial Cities in Decline 91

5. Sprawling Megacities of Hypergrowth: The Unplanned Urbanism of the Twenty-First Century 126

6. Building Cities on a Grand Scale: The Instant Urbanism of the Twenty-First Century 166

PART III: THE FUTURE OF URBANISM

7. Conclusion: Urban Futures 199

....

Notes 211

Bibliography 293

Index 351

ACKNOWLEDGMENTS

This book originated out of a persistent, nagging unease with mainstream currents in urban scholarship that all-too-often cavalierly dismiss some cities as exemplars of "failed urbanism" and others as backward and incapable of "catching up" with "good cities." I have spent a great deal of my time as an urban scholar inspecting closed-down factories and abandoned homes, poking around collapsing buildings peopled with squatters, exploring irregular settlements that house the shelterless poor, and accompanying public police and private security operatives on night-time excursions in the "bad parts of town." I have come to regard these out-of-the-way places and the activities that take place around them as part and parcel of contemporary urbanism, no less theoretically significant and personally fascinating than glitzy shopping malls, gated residential estates, New Urbanist enclaves, gleaming skyscrapers, and gentrified tourist meccas. For me, there is a bit of theoretical arrogance in establishing abstract standards for an idealized vision of cities without slums, without squatters, without irregular work, without crumbling infrastructure, without shuttered storefronts, and without abandoned houses. Blight and ruin are just as much a part of urbanism as growth and development. With all of this messiness in my mind, I set out in this book to rescue all those "troublesome and unpleasant things" (deindustrialization and shrinkage, slums and informalities, collapse and ruination) about cities and frame them as objects of inquiry in their own right.

Scholarly research and writing never takes in completely detached and isolated settings. In writing, we all accumulate huge debts to fellow scholars who have influenced our ways of thinking, and to colleagues, friends, and family who have contributed in ways that often go overlooked and unacknowledged. This book took shape out of an extended engagement with debates and controversies in urban studies. I found myself dissatisfied with various efforts to forge a single, one-size-fits-all urban theory to account for urbanization everywhere. I also became unconvinced that cites are unique, one-of-a-kind entities. I have found these two extremes—a rigorous scientific approach that adopts an a priori agreed-upon set of analytic concepts to study cities, on the one hand, and an "anything-goes" celebration of the radical uniqueness of cities, on the other—to be particularly unhelpful for trying to make sense of global urbanism today.

In looking for patterns that have come into being in the early twenty-first century, I arrived at the conclusion that it was possible to classify city building into four distinct types. Of course, it would be naïve and not wise on my part to suggest that any one city embodied the characteristics that I attribute to this fourfold classification scheme. I am not talking about actually existing cities, but I am interested in investigating how city building as an abstract process has crystalized into distinct streams or trajectories. City building is not a random, happenstance process, but a patterned process that produces different kinds of cities that coalesce around shared characteristics.

Like my companion book, *The Urbanism of Exception*, *Many Cities: Divergent Trajectories of Global Urbanism* is a work of synthesis. I have sought to weave together the research and writing of other urban scholars in order to advance a rather broad argument that captures what I regard as a decisive turn in global urbanism at the beginning of the twenty-first century. The kinds of city building that dominated from the middle of the nineteenth to the late twentieth century have disappeared. But how do we understand the seismic shifts that characterize city building in the twenty-first century? *Many Cities* offers a critique both of those efforts to establish a homogeneous set of principles to define urbanism and of research and writing that never moves beyond the differences and particularities of cities.

Over many years, I have engaged in countless conversations about the ideas that form the foundation for this book. I am particularly grateful to Anne Pitcher, Garth Myers, David Bieri, Gavin Shatkin, Filip De Boeck, Maria Arquero de Alcarón, Bob Beauregard, David Wilson, Jason Hackworth, Phil Harrison, Cecil Madell, Lisa Weinstein, Mona Fawaz, Josh Akers, Eric Seymour, and R. J. Kocialniak for engaging with me in impromptu and informal conversations about themes in this book. While I have not seen him in decades, Mike Davis and his writing (*City of Quartz*, *Dead Cities*, *Ecology of Fear*, *Evil Paradises*) remain a major inspiration for me. Taubman College of Architecture and Urban Planning and the Department of African-American and African Studies, University of Michigan, provided supportive platforms that provided me with the intellectual space and freedom to think and write. In Taubman College, Dean Jonathan Massey and Associate Chair Geoff Thun offered encouragement and support.

It has been a pleasure and a privilege to work with Eric Schwartz, editorial director, and Lowell Frye, editorial assistant, at Columbia University Press. Both Eric and Lowell embraced this project from the start and quickly moved it through the approval process on its way to becoming a book. I especially appreciate their professionalism and thoughtfulness. I would also like to thank the entire support staff at Columbia University Press, particularly Marielle Poss, director of Editing, Design, and Production. I would like to acknowledge the great work of Ben Kolstad, editorial services manager at KnowledgeWorks Global, and his team for their excellent copyediting work. I am especially grateful for the two anonymous readers who read and commented on the manuscript. Their insights were particularly helpful when I undertook revisions.

Writing books is always a time-consuming process that requires dedication, commitment, and self-absorbed retreat. I often become distracted and lost in thought. I apologize for sometimes appearing (and even being) selfish. As always, I am most appreciative of the support, encouragement, and love that I always get from my best friend and partner, Anne Pitcher. While I do not always acknowledge it at the time, I always test my ideas in conversations with her. If she raises doubts about my ideas, I always revise them. I would like to acknowledge how much I get from interacting with Jeremy (now a professor), Andrew (almost finishing his Ph.D.), and Alida (off to law school). My parents, Bob and Margaret, are no longer with us. My four brothers—Mark (twin), Dennis (middle), Greg (next to last), and Thomas (always the little brother to the rest of us)—are as irreverent and self-centered as ever. They all believe that professors do not really "work." I dedicate this book to my larger family network, both here and gone.

I should note that some parts of chapter 6 in this book have already been published under the title "Cities on a Grand Scale: Instant urbanism at the Start of the 21st Century," in *The Routledge Handbook on Spaces of Urban Politics*, ed. Kevin Ward, Andrew E. G. Jonas, Byron Miller, David Wilson (New York: Routledge, 2018), 184–196. Whatever parts I have borrowed from this book chapter are reprinted here with permission.

PREFACE

Global urbanism at the start of the twenty-first century has reached an important watershed: for the first time in human history, the urban population has passed over the symbolic threshold where the majority of people now live in cities. At a time when the unprecedented scale of worldwide urbanization has marked the dawn of what in academic and journalistic discourse has been referred to as the new Urban Age, what it means to project the future shape of cities-yet-to-come—and to talk of sustainable urban development and to provide for human security in fragile urban environments—has assumed enormous significance.[1] As a world-historical process, urbanization has lurched inexorably forward in space and time, "explod[ing] inherited morphologies of urbanism at all spatial scales," creating "new, rescaled formations of urbanized territorial organization," and intensifying "socio-spatial independencies across places, territories and scales." At the start of the twenty-first century, as Neil Brenner and Christian Schmid have persuasively argued, "the resultant, unevenly woven urban fabric" has acquired

> extremely complex, polycentric forms that no longer remotely approximate the concentric rings and linear density gradients associated with the relatively bounded industrial city of the nineteenth century, the metropolitan forms of urban development that were consolidated during the opening decades of the twentieth century or, for that matter, the tendentially decentralizing, nationalized urban systems that crystallized across the global North under Fordist-Keynesian capitalism.[2]

The overwhelming bulk of urban population growth in the foreseeable future is expected to take place not only in the Global South's sprawling megacity regions but also in its relatively small and medium-sized cities. This shift in the scale and scope of worldwide urbanization has meant that the prominent "leading cities" of North America and Europe have become "increasingly anomalous embodiment[s]" of contemporary urbanism.[3] Scholars have estimated that 93 percent of population growth in cities will occur in the so-called developing nations in the foreseeable future, with 80 percent of that growth taking place in Asia and Africa. By the end of the first decade of the twenty-first century, there were more than four hundred cities with a population of at least one million and nineteen cities with a population over ten million. In 1975, just three cities could be classified as "megacities" with populations of ten million or more, only one of them in a "less developed" country. As of 2017, there were thirty-seven megacities. The United Nations estimates that by 2025 there will be close to fifty megacities, with the overwhelming majority located in so-called less developed countries. Other projections suggest even larger numbers, forecasting that Asia alone will have at least thirty megacities by 2025, including Mumbai (2015 population of 20.8 million), Shanghai (35.5 million), Delhi (21.8 million), Tokyo (38.8 million), and Seoul (25.6 million).[4]

As their sheer numbers have proliferated, megacities of hypergrowth have displayed a clear break from prior patterns of urbanization and urban spatial form on a global scale. Scholars, policy makers, and journalists have used a wide variety of analytic constructs to characterize and categorize rapidly growing cities in the world's poorer countries. The most notable terms include "emerging," "developing," "underdeveloped," "less developed," "backward," "modernizing," the "Third World," and the "Global South."[5] Each of these categories has its own history and connotations and, as such, carries (often unacknowledged) ideological biases and normative baggage. Each category offers what amounts to a one-dimensional shorthand that sometimes obscures as much as it illuminates about structural inequalities, persistent poverty, and the division between haves and have-nots on a global scale. Yet there is no question that the process of urbanization is taking place at an accelerated rate in fast-growing cities in relatively poor countries. For the foreseeable future, the great majority of new urban residents on a global scale will be poor, will lack regular wage-paying work, and will have to cobble together livelihoods under impoverished conditions in resource-depleted cities where public assistance and municipal services are nonexistent for the majority of residents.[6]

Yet the actual picture is much more complicated than can be seen simply by pointing to the dominant patterns and general trends that characterize global urbanism at the start of the twenty-first century. Focusing exclusively on the accelerated pace of urban population growth on a world scale can sometimes conceal as much as it reveals. While the overall trajectory of urban transformation points toward larger and larger cities, contemporary processes of urbanization have resulted in divergent patterns of global urbanism. The distinctiveness of these patterns suggests that it is impossible to account for all contemporary urbanization with a single, universalizing, one-size-fits-all theoretical framework.[7]

These dissimilarities are indeed stark. Well-known cities like London, New York, Chicago, Tokyo, Paris, Barcelona, and other "globalizing" (or already globalized) cities with world-class aspirations remain in the spotlight as exemplars of healthy, sustainable, and competitive urbanism. Yet authoritative estimates suggest that one in six cities worldwide experienced substantial loss of population even before the 2007 American subprime mortgage crisis and the onset of the late-2008 global economic slowdown. In the United States, for instance, the 2006 census estimates revealed that sixteen of the twenty largest cities in the 1950s had contracted in population size, often by huge amounts.[8]

A substantial body of scholarly literature has drawn attention to this growing subset of cities in the United States, Europe, and elsewhere that are confronting sustained socioeconomic decline and the challenges associated with deindustrialization, shrinking populations, property abandonment, and the downsizing of municipal services. What is clear is that this phenomenon of "shrinking cities" transcends national borders. Urban contraction resulting from capital flight and disinvestment, downward shifts in employment opportunities, and population loss is a worldwide dilemma that cuts across continents, regions, and even localities. Cities that have grown slowly or suffered population shrinkage can be found everywhere.[9] Many older industrial cities in core areas of the world economy continue to face population declines that are taking place in some instances on an unprecedented scale. Over the past half-century, 370 cities throughout the world with populations over one hundred thousand have shrunk by at least 10 percent.[10] Metropolitan regions across the United States, Canada, Europe, and Japan have projected double-digit declines in population in the coming decades.[11] Suffering from disinvestment, the hemorrhage of stable employment, and significant population loss, these distressed postindustrial cities in decline are "loser cities" that stand in stark contrast to the massive city-building projects of the Asia Pacific Rim, the "spectacular urbanism" of the Persian Gulf, and the "instant cities" of southern China.[12]

Then there are those struggling cities with burgeoning populations, inadequate and overstretched infrastructure, and virtually nonexistent regulatory mechanisms—unplanned and unmanageable cities like Kinshasa, Lagos, Karachi, Dhaka, Lima, Manila, Cairo, and Caracas, to name a few—which, as Jennifer Robinson has suggested, "do not register on intellectual maps that chart the rise and fall of world and global cities."[13] Lacking the observable qualities of genuine "city-ness," these bloated metropolitan conurbations do not seem to function in recognizable ways.[14]

What has emerged at the start of the twenty-first century is a kind of asymmetrical urbanism, in which shifting patterns of highly uneven urban development have gone hand in hand with expanding gaps between wealth and poverty not only among cities but also within cities.[15] In the contemporary age of globalization, cities around the world have become enmeshed in an uneven spatial geography where the so-called healthy, vibrant cities with world-class aspirations are emerging as key nodal points in the global economy, while the loser cities that are unable to compete are bypassed and left behind. The megacities of hypergrowth are growing so fast that opportunities for wage-paid employment have failed to keep pace with the expanding numbers of work seekers, existing infrastructure is so overburdened that it has virtually collapsed, and municipal services cannot adequately provide for long-term residents or new arrivals. Adding even more confusion to this diverse mixture of urban typologies is the steady accretion of master-planned, holistically designed cities built entirely from scratch—the instant cities that originated in the Persian Gulf, popped up along the Asia Pacific Rim, and eventually spread to the Indian subcontinent, elsewhere in the Middle East, Africa, and beyond.[16]

Urbanization is a complex, multifaceted, and sometimes contradictory global process that proceeds along multiple pathways without a privileged, common end point. To think in terms of alternative trajectories is to challenge the claims to a singular urbanization process that takes place through recognizable stages along a predetermined linear pathway.[17] It is necessary to engage critically with new thinking about global urbanism and urbanization processes, to deviate from the conventional practice of evaluating "Third World" cities from the fixed (and privileged) perspective of "First World" cities, and to dispense with the rigid demarcation between (exploiting) core zones and (exploited, dependent) peripheral regions of the world economy, as if these distinctions somehow represented a permanent and unchanging geographical fix.[18]

URBAN FUTURES

This wholesale shift in the center of gravity of global urbanization away from the historic Euro-American core of the world capitalist economy has provoked both renewed popular curiosity about cities around the world and scholarly concern about what the patterns and rhythms of global urbanism portend for the future. As a general rule, cities concentrate poverty and deprivation, but they also represent perhaps the best hope of escaping these circumstances. Beyond the demographic hyperbole and stereotyped imaginings of urban dystopia about rogue "cities gone wild" (chaotic and feral) that has accompanied the explosive expansion of megacities of hypergrowth, it has also been widely acknowledged in both popular culture and scholarly writing that cities are the key strategic sites for global flows of finance, trade, and information, which are the indisputable lifeblood of the contemporary capitalist world economy.[19]

Well-known "global city" theorists like Saskia Sassen have described a spatial ecology of global centrality and marginality that has essentially taken shape as a ranked hierarchy of rival "alpha cities" in fierce competition to become leading command-and-control centers in the world economy.[20] Her analysis can be interpreted in one of two ways—either as a Darwinian claim about intense competitive rivalries and the eventual survival of the fittest, or as a Durkheimian argument about the evolving functional specialization of cities arranged unevenly in what might be called the new international division of labor. Whether or not one cares to subscribe to such formulaic ecological characterizations of globalization, one key theme remains: despite all the scholarly discourse about deterritorialization and the borderless world, cities and their place within a world of cities continue to matter, and to matter a great deal.[21]

If London, Paris, Chicago, and New York at the start of the twentieth century assumed dominant roles as the premier crucibles and living laboratories for world-historic experiments with urban modernity, the hyper-urbanism of the new millennium—characterized by explosive growth in the sheer size and vast geographic scale of cities—has altered the terrain upon which to make sense of the experiences and expectations of contemporary urbanity.[22] In the face of the irreversible trend toward the urbanization of the globe, the fate of humanity has become ever more inextricably tied to the socioeconomic (and political) life of cities, to the variegated sites of urban space and administrative structures, and to the networks and flows of capital, migration, and trade that not only tie global-city regions together but also link them in wider urban fields.[23] As Trevor Hogan and Julian Potter have suggested, the coming of supersized megacities "seems inseparable from the ambivalent and transient experience of modernity—the ideals of liberty, individuality, property, accelerating progress," and, conversely, the grim realities of grinding poverty, immobility, anonymity, and social marginalization for those left out of, and hence excluded from, the mainstream of urban life. The growth of megacities has amplified the dysfunctionalities and social evils associated with late-nineteenth- and early-twentieth-century modern industrial cities—the exemplars of urban modernity at the time.[24] The dynamism of the modern metropolis as the engine of capitalist growth and expansion was not without its seamy underside of crime and vice, environmental degradation, and horrific deprivation. The mirror image of the capitalist modernity associated with affluent neighborhoods and upscale housing can be found in such squalid building typologies as overcrowded tenements and insalubrious slums.[25] These urban dysfunctionalities never disappeared. They have mutated, metastasized, and reappeared under different guises (and have assumed different meanings) as shantytowns, ghettoes, informal settlements, and squatter camps in today's sprawling megacities of hypergrowth.[26]

Those ordinary people who lay claim to a "right to the city" continue to struggle with a long litany of challenges, including securing gainful employment, obtaining access to decent and affordable housing, and compensating—through their own initiative and sacrifice—for inadequate infrastructure and the lack of inclusive governance. Confronted with entrenched patterns of social inequality and the absence of meaningful channels for upward mobility, marginalized city dwellers often come face-to-face with a life of absolute impoverishment and immiseration.[27]

While the growing importance of megacities (and global-city regions) in the world economy is widely acknowledged, there remains considerable disagreement over how to interpret the explosive growth of urbanism on a global scale.[28] Urban scholars and policy makers together oscillate between starkly opposing points of view. These prognostications range from unblinking optimism to skeptical pessimism. Does the unprecedented growth of the world's cities offer limitless opportunities for a better life, or does it portend a looming disaster of monumental proportionsjust waiting to happen. ? On the one side, a great deal of scholarly writing and popular journalism has coalesced around a confident, optimistic discourse that envisions cities as inviting arenas of livability, opportunity, and transformative potential.[29] The well-entrenched fantasy projection of a global evolutionary trajectory toward urban growth and trickle-down prosperity has remained an astonishingly tenacious belief in mainstream developmentalist circles. Edward Glaeser, in particular, is an enthusiastic advocate of the inherent value of urban living, arguing that cities are durable engines of innovation and wellsprings of material wealth that offer unprecedented opportunities for sustained upward mobility. In short, these scholars and journalists argue that cities are humanity's greatest creation and the best hope for the future.[30]

In contrast, an opposing dystopian discourse has crystalized around the view that unregulated population growth and uncontrolled spatial expansion forfeit whatever advantages cities seem to offer those urban residents who cannot compete for scarce resources. There is no shortage of disturbing images of rapidly growing cities around the world as vast dumping grounds or repositories for the jobless poor.[31] Many urban scholars have warned that the inability of large numbers of struggling cities to absorb work seekers into the mainstream of urban life creates the conditions for a coming catastrophe.[32] In the megacities of hypergrowth, unplanned and chaotic urbanization has continued to take a huge toll on the quality of the material environment and on human health, contributing to ecological, socioeconomic, and political instability. As Mike Davis has provocatively asked, "does ruthless Darwinian competition, as increasing numbers of poor people compete for the same informal scraps, ensure self-consuming communal violence as yet the highest form of urban involution?"[33] An estimated one-third of nearly three billion urban dwellers today live in what are typically called "slums," defined as informal settlements where people cannot secure necessities like clean water, adequate housing, and proper services. As a result, an estimated 1.6 million urban residents die each year for lack of such basic amenities as clean water and proper sanitation.[34] Without a doubt, growing urban inequalities and diminishing opportunities for upward mobility have greatly reinforced patterns of social polarization, exclusion, and marginalization in cities around the world.[35]

One enduring image has portrayed the contemporary city as a social space for the circulation of civic freedoms and the practices of political autonomy—that is, an evolving terrain that provides the public setting for the monopolization, contestation, and negotiation of power, politics, and economic exchange.[36] Whether large or small, cities are contested sites where the politics and practices of citizenship are played out, often in dramatic fashion. The entanglements of rights, entitlements, and claims to belonging careen back and forth between what Egin Isin has called the fault line between *urbs* and *civitas*—that is, the gap between the experience of the city on the one side, and its enormous productive capacities for human progress and its promise of inclusive participation in civil and democratic polity on the other.[37] Economic interdependence goes hand in hand with legal rights of private ownership and formal political-legal autonomy. Instrumental market rationality and the impersonal cash nexus intricately bind buyers and sellers together in ever-expanding circles of exchange and circulation. Yet engagement in "increasingly abstract forms of association of which money is the paradigmatic form" subjects urban dwellers to a market logic that both offers opportunities for enrichment and operates as a powerful force for entrapment.[38]

THE NARROW FOCUS OF CONVENTIONAL URBAN STUDIES

Despite their claims to universal applicability, the dominant theories and methodical approaches used to study cities—constituting what Ananya Roy has called the "universalizing knowledge-space" of mainstream urban theory—have remained largely tied to the experience of a handful of leading world-class cities of Europe and North America. These cities serve as the wellsprings for ideas about urban modernity and the privileged source for theorizing about urbanism and urbanization.[39] All too often, these vibrant, healthy, "developed" cities located in the historic core areas of the capitalist world economy form the paradigmatic reference point and the discriminating lens through which to view diverse global urban experiences.[40] These leading world-class cities function as the standard-bearers for what a "good city" should be and the benchmark for defining what less successful cities should emulate.[41] As a general rule, scholarly inquiry has remained largely fixated on ranking cities in relation to ordered hierarchies and categorizing cities according to levels of development. Grounded in the principles of classical modernization theory, these mainstream approaches to urban studies favor teleological narratives of evolutionary progress in which cities develop incrementally through distinct stages toward a shared goal of sustainable, healthy urbanism.[42]

The excessive focus on identifying the precise defining characteristics of what constitutes a "globalizing city" with world-class aspirations—and the attendant fixation on the rank-ordering of cities in accordance with their position in an arranged hierarchy of relative achievements—has tended to flatten out and homogenize our understanding of the contemporary experience of urban modernity.[43] The conventional urban studies literature has been largely unable to break away from the long-standing regulative fiction that the process of urbanization is a more or less singular pathway that unfolds in fairly distinct stages, with cities distributed along a continuum that stretches from the "fully developed" (mature) metropolis at one pole to "less developed" cities that lack all the important features of modern urbanity at the other pole, with a lot of "developing" cities somewhere in between, striving to catch up by ascending through the hierarchy of achievement.[44]

Rethinking conventional urban theory requires shifting our focus away from the strong fixation on hierarchies, rank orders, and "success stories" and adopting instead a vantage point that starts with the diversity, heterogeneity, and unevenness of urban experiences on a global scale. Examining diverse urban experiences in a world of cities enables us to draw upon new approaches to theorizing that are not already tied to such paradigmatic constructions as global cities, world-class cities, creative cities, smart cities, and competitive cities—to name just a few.[45]

Properly understood, the point of departure for understanding contemporary global urbanism—the emergent cities of the twenty-first century—is recognizing the diverse trajectories of urban transformation, the heterogeneity of spatial forms, and the asymmetrical patterns of agglomeration and density in contrast with dissolution and dispersal. Dispensing with conventional approaches to studying cities rooted in the "modernization" myth—that is, a priori conceptions of normal and expected routes to a perceived end point called "development"—enables us to explore growth and decline without becoming trapped in a framework that looks for deviations from an a priori norm. This teleological approach relies upon such well-worn binary oppositions as "developed" versus "developing," and "First World" versus "Third World," which indelibly mark cities as "ahead" or "behind" in the (fast or slow, accelerated or stalled) drive toward modernity. Breaking away from conventional approaches to thinking about global urbanism requires an analytic sensitivity to gaps and lags, to bypassing and leapfrogging, and to accelerating and slowing down.[46] Grasping these multiple trajectories of urban transformation at the start of the twenty-first century—what Patsy Healey has called the "recognition of contingency and complexity"—requires that we abandon the preconceived notions that govern our understanding of cities and urbanization.[47]

Processes of growth and development (sometimes fast-paced and sometimes agonizingly slow) are inextricably intertwined with stagnation and decline. As a general rule, conventional urban studies literature has continued to rely on abstract concepts, analytic categories, and descriptive markers that were first developed and gradually refined over time to account for the meteoric rise in the nineteenth and twentieth centuries of the great modern industrial metropolises that constituted the core of the capitalist world economy. Yet with the emergence of diverse patterns of urbanization, conventional models used to explain the evolving dynamics of growth and development of the modern metropolis that characterized the Fordist era of industrial capitalism have become outmoded.[48]

Framed in bold strokes, what characterizes global urbanism at the start of the twenty-first century is the remaking of cities in ways that sidestep and skip over the patterns of urban growth and development that characterized the late-nineteenth-century era when the great modern industrial metropolises of North America and Europe assumed dominant positions in top-down management of the capitalist world economy. At present, the patterns of global urbanism do not conform to orthodox one-size-fits-all models or rigid paradigms that rank cities by what they have and what they lack. The production of a multiplicity of different cities that respond to different pressures and logics makes the call for a more thorough, cosmopolitan reimagining of urban theorizing on a global scale even more urgent.[49]

Rigid adherence to the universalizing claims of an all-encompassing urban theory—with its roots in the experience of globalizing cities in North America and Europe—neglects the heterogeneity, the diversity, and the shifting geographies of global urbanization. As Helga Leitner and Eric Sheppard have persuasively argued, it is necessary to reject the pretension that a single, omnibus theory can sufficiently account "for the variegated nature of urbanization and cities across the world."[50] In a similar vein, Jamie Peck has contended that reconfiguring urban theory must "occur across scales, positioning the urban scale itself, and working to locate cities not just within lateral grids of difference, in the 'planar' dimension, but in relational and conjunctural terms as well."[51] Focusing attention on difference, not as a deviation from an expected pattern of development but as the fundamental constituent feature of global urbanism, provides an alternative point of departure for understanding urbanization on a world scale.[52]

Efforts to (re)theorize urban studies from the perspective of the Global South have called for a paradigm shift in our understanding of urbanization and urbanism.[53] "Seeing from the South," as Vanessa Watson has suggested, provides an alternative vantage point from which to unsettle and dislodge "taken-for-granted assumptions" about what it means to talk about urban modernity and the supposed destination of urban development.[54] Challenging the core regulative principles of those mainstream urban theories that treat Northern urbanization as the normative benchmark from which to evaluate progress toward urban modernity allows us to contest "the dogma of a universal and teleological model of urbanization."[55]

In contrast to mainstream approaches to urban studies, the alternative conceptualization of "ordinary cities" has steered urban thinking away from the inordinate fixation on globalizing cities with world-class aspirations as the wellspring of ideas about "good urbanism," proposing instead the idea that cities everywhere should be drawn into a wider theoretical engagement in the production of new geographies of theory making.[56] This proposed movement toward a postcolonial comparative urbanism opens up new room for thinking across sites and across diverse historical circumstances in order to understand the nature of the interrelations and connections between and among cities located outside the core areas of the world economy.[57]

OPEN-ENDED EXPLORATIONS IN SEARCH OF ADEQUATE THEORETICAL FRAMEWORKS

My interests in exploring the divergent trajectories of global urbanism at the start of the twenty-first century are as much speculative and conjectural as they are analytical and diagnostic.[58] This book is about "imagining cities in advance of their arrival" as much as deciphering cities as they exist at present.[59] At the risk of oversimplification, cities are complex assemblages of material objects arranged in various patterns of density, serving multiple functions and purposes, and held loosely together via infrastructural networks. City building is always unfinished and incomplete. The balance between material form and density, as well as between proximity and distance, constantly evolves. As Saskia Sassen has suggested, "in this incompleteness lies the possibility of making."[60] City building is not a random, haphazard process without purpose and deliberation, but neither is it a uniform process converging around a common spatial-temporal destination or end point.[61]

This book is animated by three questions. First, how do we grasp the rapidly evolving realities of global urbanism at the start of the twenty-first century—and the multiple urban worlds that take tangible shape in different places—as city-building practices follow multiple trajectories across geographical space and historical time? At root, teasing out and identifying these patterns involves an exercise in classification. Second, how is it possible to theorize about global urbanism in ways that recognize the diversity, distinctiveness, and historical specificity of cities, while at the same time acknowledging that the contemporary world of cities is interconnected and subject to "widely circulating practices of urbanism" that often produce similar patterns and outcomes?[62] In other words, urbanism does not conform to inviolable "laws of motion," yet it is not the outcome of purely random happenstance. Third, how do we distance ourselves from universalizing theoretical approaches that rely upon a priori (and ex cathedra) conceptual frameworks to understand the putatively singular, overarching logic of urbanization, while at the same time avoiding the radical relativism (what Robert Beauregard has called "radical uniqueness") that treats each city as an irreducible special case with a peculiarity and uniqueness all its own?[63] Addressing these questions requires that we excavate the main underlying assumptions and core regulative principles that have for quite some time framed and guided mainstream urban studies as a distinct field of inquiry.[64]

By the end of the twentieth century, the unfolding of new, haphazard patterns of urban growth and development that gave rise to vast, spatially fragmented, and distended polycentric conurbations without obvious or fixed boundaries suddenly unsettled and destabilized the taken-for-granted conventional distinction between the city (with its high-density downtown core) and the suburb (with its low-density building typologies at the urban edge).[65] The appearance of these amorphous, horizontally expansive, post-urban spatial configurations on a boundless scale—what urban theorists have variously called the extended metropolis, the 100-mile city, the limitless city, sprawl city, exopolis, postmetropolis, metrourbia, or the city turned inside out—has thrown into doubt the concept of the city as a coherent category of analysis.[66]

New spatial patterns of urbanization have effectively undermined the conventional understanding of cities as discrete, spatially bounded places that have an integrity all their own. What might be called extended urbanization—an idea that falls into line with Neil Brenner and Christian Schmid's notion of "planetary urbanization"—reflects multifaceted urbanizing impulses that have pushed their tentacles outward into exurban territories (hinterlands) in haphazard ways that defy simple classification.[67] As a dynamic process of spatial production, urbanization has broken down barriers and leapfrogged over obstacles, dissolving and absorbing whatever stands in the way. The shape-shifting nature of urbanizing processes that produce new spatial forms and novel kinds of territorial urbanization "proceeds through the appropriation and deployment of multiple logics of spatial production and articulation."[68] Dispersed settlement patterns, ribbons of densifying growth, and elongated corridors of circulation and transportation have produced "intensely operationalized landscapes" that in no way resemble what conventional urban studies has long understood as the city.[69]

As a universally recognized and identifiable object of inquiry, the notion of the city is, upon close inspection, ephemeral and unstable. Looking at cities in the limited sense of administrative (political) units with arbitrarily imposed boundaries yields very little by way of understanding the underlying dynamics and processes of urban transformation. Cities consist of multiple, contested territories that are embedded in complex local histories.[70] The disruptive force of unregulated outward sprawl, the development of regional patterns of haphazard growth, the proliferation of edge cities that undermine the idea of monocentric urbanism (defined by density, proximity, and agglomeration), and emergent configurations of extensive urbanization, or what might be called "continuous settlement," imply that it is time to rethink what we mean by the city, and by urbanism more generally.[71] Cities have no a priori or fixed ontological status but are socially produced and continuously transformed by a multiplicity of forces and imperatives operating at every geographic scale. In short, epistemologically speaking, cities as objects of inquiry are as much arbitrary social constructions as actual places with an ontological status of their own.[72]

The aim of this book is to recast our thinking about trajectories of global urbanism at the start of the twenty-first century. By adopting a broad spatial and temporal framework that unsettles conventional approaches to urban theorizing, I hope to identify patterns of regular recurrence, historical evolution, and genuine novelty in contemporary processes of globalizing urbanization. The driving force behind these efforts to rethink mainstream urban studies is to challenge those theoretical approaches that aspire to universal validity and general explanations without respect for time and place. All such universalizing theories, albeit in different ways, fail to acknowledge the temporal and spatial specificity of place and context as the substructure and building blocks for knowing.[73] The persistence of such epistemological categories as First World and Third World, developed and developing, are unhelpful binary distinctions that paralyze our imagination, obscuring our ability to see beyond the developmental paradigm of linear pathways leading to the end goal of the "modern metropolis." Similarly, vague terms like "failed," "distorted," or "stalled" urbanism suggest that some cities lag behind and need to catch up with some idealized image of what a "good city" ought to be.[74]

Rethinking urban theory enables us to conceive of cities beyond the West not as derivative copies, imitations, or counterfeits of the real urbanism of the Global North but as hydra-headed, polymorphous, and mutating ensembles of physical objects and social practices.[75] Mainstream urban studies typically frame overburdened cities of the (so-called) Global South as an assortment of empirically interesting yet anomalous cases, that amount to not much more than reservoirs of stylized facts or storehouses of anecdotal information that yield poignant stories of desperation and victimization or heroic manifestations of sheer grit, but little else.[76] They are "objects to be theorized," but "they are never able to theorize back"; that is, they are not able to produce ideas that contribute to building new theoretical understandings of global urbanism.[77] In such an analytic division of the urban world, cities beyond the West—those "off the map" of structural relevance—are looked upon as "the intractable, the mute, the abject, or the other-worldly," that is, pale replicas of an imagined genuine urbanism, and exemplars of a stalled, truncated, or even counterfeit modernity.[78]

GRAND NARRATIVES OF PROGRESS THROUGH EMULATION

The dominant current in the applied, policy-focused literature still follows the conventional narrative of development as the singular goal of policy implementation. The assumption is that if policy recommendations are implemented correctly and effectively, then poor cities will, over time, experience progress toward sustainable development. This viewpoint lends itself to the notion that current inequalities and hardships are in some ways an acceptable state of temporary limbo, constituting a kind of unfortunate but necessary rite of passage that will eventually resolve itself. All newcomers to cities have at some time experienced periods of suffering and socioeconomic hardship, but they were eventually able to work their way out of poverty and ascend the ladder of upward mobility.[79]

It is my contention that urban transformation is neither linear (following a recognized and predetermined direction) nor cumulative (additive and not reversible). There are no singular, fixed end points or directional beacons to guide urban transformation. This nonteleological approach rejects the assumption of requisite sequences, or identifiable stages or phases, of urban growth and development that define the trajectories of urban transformation.[80] Yet while urban transformation is multidirectional and rhizomatic, it is not directionless and inexplicably chaotic. There are identifiable patterns and shared features. Connections, networks, and flows inextricably bind cities to one another in ways that make it impossible to treat them as singular, bounded, and isolated units of analysis. This connective tissue operates in asymmetric ways that privilege certain places while punishing others. The "unruly materiality of the urban" has inspired the proliferation of imaginative theoretical projects. Urban theorizing "takes place in the midst of fields of politics, power and practice" that arise and erupt from the diversity of urban experiences.[81]

The collapse of the grand narrative of urbanization (and the march toward progress), which tried to subject disparate urban realities and experiences worldwide to the reductive model of the modern metropolis in all its pristine glory, has opened creative theoretical space that has enabled us to embrace urbanism as a multitude of conditions that do not conform to a single universal model of transformation.[82] In searching for new ways of understanding global urbanism, I contend that it is necessary to maintain a largely inductive, open-ended approach to analytic revision and conceptual experimentation. The deductive, universalizing ambitions of grand theorizing run the risk of not only foreclosing conceptual innovation but also subsuming difference and variation under a single analytic framework. In contrast, a call for theoretical openness requires, as Jennifer Robinson has argued, "thinking with variation and repetition, rather than trying to 'control for difference.'"[83] The shift in the geographic center of global urbanization, coupled with the diversity in the existing patterns of urban transformation, has revealed gaps and fissures in mainstream urban scholarship. Universalizing theorizations of urbanization focus on establishing the core features that characterize the presumed essential nature of cities. Yet the elastic, flexible, and malleable boundaries of cities and urbanization more generally have undermined theoretical and methodological efforts to identify a coherent object of inquiry. Making sense of the diverse trajectories of global urbanism requires abandoning the quest for a privileged starting point and a single analytic framework to account for the range of different urban outcomes.[84]

Precisely what the process of urbanization means in regard to ideas about urbanism and cities has become increasingly blurred because of complex and hybrid patterns of land use and human settlement. "The vast urban agglomerations taking shape across the world," as Saskia Sassen has suggested, "are often seen as lacking the features, quality, and sense of what we think of as urbanity [and city-ness]."[85] Depending on specific geographic, climatic, economic, and cultural conditions, numerous and often radically conflicting outcomes accompany urban transformation and development: the hyperdense megalopolis of high-rise vertical urbanism inexplicably coexists with extensive horizontal growth and seemingly endless sprawl; traditional street life of small-scale "walkable" urbanism exists side by side with high-speed traffic corridors and freeways; the material durability of architecture and the built environment is overlaid by the immaterial ephemerality of lived daily experience ("soft urbanism"); a feverish proliferation of enclave spectacles is incongruously juxtaposed against sites of dereliction and ruin; the fast-track urbanism of the Asia Pacific Rim and the Persian Gulf coexists with deindustrializing cities faced with shrinking populations along with the slow strangulation and gradual disappearance of smaller cities and towns in relatively rich countries with high levels of industrial development.[86]

As the crowning achievement of the industrial age, the planned urbanism of high modernism has lost its special place as the agreed-upon model for the future of cities in the face of the massive expansion of informal housing settlements and self-built shelter as the dominant modes of city building on a global scale.[87] As opposed to thinkers from the nineteenth century onward who looked upon the emergence of the modern metropolis as an up-to-date cosmopolitan outpost standing against the traditionalism and parochialism of rural backwardness, the idea of urbanism today "no longer indexes a normative cultural concept"—expressed in lively debates over the civil realm of social life, public space and the public sphere, architectural spectacle, the emblematic figure of the flaneur, along with the blasé attitudes, and anonymity—but "represents a cosmos of extremely varied notions determined by geographic, cultural, and individual preferences." If we want to grasp the complexity of urbanism at the start of the twenty-first century, "we have to capture it in all its disguises, gradations, and transformations occurring simultaneously on a global scale."[88]

As a means of understanding the everyday experience of ordinary people outside the mainstream of urban life, the once reigning paradigm of modernity has proved to be an inadequate analytic tool. To accept and even embrace the merits of indecipherability, ambiguity, and illegibility requires a rethinking of the basic concepts and analytic constructs that are embedded in the modernist paradigm of rational order and spatial stability.[89] Following Bonaventura de Sousa Santos, making sense of ordinary urbanism in struggling cities requires extensive epistemological shifts in the kind of knowledge production that can build the field of critical urban studies.[90]

This book challenges long-entrenched ideas about how we conceive of the meaning and significance of the informal settlements that provide housing and livelihoods for growing numbers of otherwise shelterless urban poor who inhabit struggling cities of hypergrowth around the world. For the most part, the mainstream scholarly literature has viewed these irregular settlements through the prisms of modernization and development theories, the breakdown of sociospatial order, the rise of endemic crime and violence, or debates about the cultures of poverty. Yet these shantytowns have proved to be both more durable and more multifaceted than any of these perspectives anticipated.[91] Far from being accidental outcomes (or unfortunate outliers) that emerged in the shadows of more dynamic formal economies and public regulatory regimes, unplanned, unregulated, and irregular housing arrangements have become permanent and integral features of urban life on a global scale.[92]

MANY URBANISMS

INTRODUCTION

Rethinking Global Urbanism at the Start of the Twenty-First Century

At the start of the twenty-first century, conventional approaches to understanding urban transformation on a global scale have become unsettled.[1] Operating with inherited paradigmatic frameworks and outdated analytic categories, mainstream urban studies scholarship seems incapable of accounting for not only the unprecedented pace and scale of global urbanization but also the rich diversity of urban experiences. The bewildering range of transformation and expansion has so fundamentally reshaped urban life that contemporary cities around the world bear little resemblance to the much-celebrated "modern industrial metropolis" of the nineteenth and twentieth centuries. Perhaps ironically, the persistent, almost obsessive fixation on identifying the distinguishing marks—visibility, recognition, and differentiation—that characterize "globalizing" cities with world-class aspirations comes at a time when the promise of a better future (via incorporation into the mainstream of urban life) appears further out of reach than ever for city dwellers around the world. As a discursive category and mobilizing myth, the global city / world city concept "conjures up imaginaries of high modernity, mega-development, twenty-first century urbanity, [and] progressive urban futures in the new millennium." Yet skeptics have claimed that the global city / world city hypothesis "is often used not so much as an analytical tool but as a 'status' yardstick to measure cities in terms of their global economic linkages, to locate their place in a hierarchy of nested cities, and to assess their potential" to reach the pinnacle of achievement and success.[2]

Yet there is a kind of cavalier pretention and almost deliberate blindness that anoints globalizing cities with world-class aspirations with the coveted status of "successful" urbanism. Celebratory narratives of competitive advance and accomplishment overlook those relations of domination and exploitation that accompany the attainment of world-class status in the world of cities. The unstated assumption is that other cities—noncompetitive, struggling, and backward—need to adopt "international best practices" that originate from the experiences of healthy, competitive, world-class cities in order to move upward in the rank order of successful urbanism. Urban transformation is a messy and contradictory process: growth and decline operate in tandem, and planned interventions (whether private or public) often fail to attain their aspirational goals.[3]

CONTRASTING URBANISMS: SHIFTING GEOGRAPHIES AND DIVERSE TYPOLOGIES

At the start of the twenty-first century, globally competitive, vibrant, and healthy cities (the so-called First World Cities) seem more the exception than the general rule.[4] Despite the hollow rhetoric of eventual recovery, struggling postindustrial cities suffering from widespread abandonment and neglect, sharp demographic decline, high unemployment rates, and low skill levels face so many structural constraints it is difficult to imagine regeneration strategies that can successfully catalyze sustained growth outside a few isolated pockets.[5] Similarly, the unprecedented hypergrowth of the sprawling megacities of the (so-called) Global South has not only fundamentally altered both the shape and the sheer expanse of worldwide urbanism but also undermined the capacity of formal regulatory regimes to plan and manage the use of urban space.[6] As new arrivals in unprecedented numbers pour into these struggling cities, these expectant work seekers come face-to-face with the grim realities of limited opportunities for stable wage-paid employment, the availability of decent housing, or access to social services and workable infrastructure. Finally, the "instant cities" of the Asia Pacific Rim, the massive city-building projects of coastal China, and the spectacular urbanism of the Persian Gulf have radically recalibrated the speed at which city building has unfolded on a global scale.[7] These efforts at "fast-track urbanism" have introduced an entirely new tabula rasa approach to city building in ways that have displaced the incremental gradualism of earlier centuries.[8]

One problem with universalizing theories of urban transformation is their continued reliance upon conceptual frameworks, analytic categories, and fixed distinctions that were developed under specific historical and spatial conditions; they may have been appropriate for understanding urban transformation at an earlier epoch or in a different time frame but no longer carry the same theoretical relevance under current circumstances. Unquestioned ideas about the universal and essential characteristics of the city and urbanization have remained deeply entrenched in urban thought, especially in fields related to policy recommendations and interventions. To radically reframe the universalizing pretensions of mainstream urban theory means to begin with the assertion that urbanization is, as Neil Brenner has cogently argued, "an uneven, variegated process, one that is necessarily articulated through contextually specific patterns and [divergent] pathways."[9]

All in all, the sheer complexity of global urbanism at the start of the twenty-first century has called into question the continued relevance of dominant European- and North American–centered modes of theorizing about urban transformation and urban life.[10] Over the past several decades, a significant body of scholarly research and writing has challenged conventional analytic understandings of a single, homogeneous settlement type (the city) as a universal, all-encompassing category or a stable, coherent object of inquiry consisting of comparable and measurable properties.[11] This rising tide of critical scholarship has cast into doubt the once-unifying ideal of universal stages of urban growth that marked a singular, linear pathway toward the ultimate goal or end point of urban development. What has animated this critical reappraisal has been the emergence since the end of the twentieth century of entirely new, unexpected, and deeply entrenched patterns of urban transformation.[12]

CHALLENGING MAINSTREAM URBAN THEORY: TRANSCENDING INHERENT DUALISMS

World-class cities assume an exalted place as models or paradigms, generating key theoretical insights into understanding urbanism and useful policy recommendations about how to become competitive, sustainable, resilient, and creative. In contrast, struggling cities—mired in "backwardness" and "underdevelopment"—are classified as problematic exemplars of "incomplete," "wayward," or "failed" urbanism, requiring diagnosis and reform. Topographical representations of this sort often become trapped in a crude geographic divide between Western modernity and the underdeveloped non-West, as a way of distinguishing between cities of the developed and the developing worlds.[13] Pushing against the "regulating fiction" of the First World global city as the wellspring for theorizing about urban transformation and the primary source for new ideas about international "best practices," Jennifer Robinson has called for the development of a robust urban theory that can overcome its "asymmetrical ignorance."[14]

Historically speaking, theorizing about cities has drawn its inspiration from ideas that "draw strong links between the urban and the modern."[15] Over time, mainstream urban studies literature has continued to rely on analytic concepts and binary oppositions that were first developed and gradually refined to account for the origins and evolution of the modern industrial metropolis that came into being during the Industrial Age starting in the mid- to late-nineteenth century. The core epistemological currents that have animated mainstream urban theory "drew on specific (Western) versions of urban modernity" to identify those universalizing (and homogenizing) impulses that were seen to define the pathways of urbanism everywhere.[16]

Rethinking mainstream urban theory means moving beyond the universalizing and essentialist presuppositions that dominate conventional approaches to understanding urban transformation on a global scale. Challenging these basic axiomatic principles enables us to overcome the narrow fixation in mainstream urban studies that privileges leading globalizing cities of North America and Europe as key locales of study and sources of general knowledge.[17] Rethinking conventional urban theory calls for a more nuanced and open-ended agenda that both challenges the ontological and epistemological presuppositions of mainstream research but also offers alternative approaches for understanding the trajectories of global urbanism as objects of inquiry in their own right.[18]

As Neil Brenner and Christian Schmid have cogently argued, "inherited analytical vocabularies and cartographic methods do not adequately capture the changing nature of urbanization processes, and their intensely variegated expressions, across the contemporary world."[19] The difficulties that urban scholars often encounter in theoretically specifying the variety, diversity, and heterogeneity of urban experiences stem from two analytic flaws. The first follows from an undue reliance upon ideal-typical models of urban transformation and development derived from identifying the salient features of key globalizing cities of Europe and North America. The methodological problem with the ideal-typical approach is that it tends to generalize from a specific set of characteristics, relying on this narrow set of attributes as the standard yardstick or indicator of "normal," or expected, urbanization everywhere.[20] Viewed through this paradigmatic frame of reference, struggling cities (whether growing too fast or not enough) always appear as the outcome of flawed, distorted, or unfinished patterns of urbanization that lack particular attributes of genuine "city-ness."[21]

This tendency to define struggling cities by what they lack easily falls prey to an evolutionary way of thinking that characterizes them as "not yet" developed and in need of remedial intervention. By treating distressed cities as deviations that lag behind or stray away from an allegedly normal path of urbanization, they are too easily reduced to structural irrelevance as anomalous outliers or else subjected to a kind of moralizing discourse that prescribes remedies to overcome their inadequacies and backwardness.[22] In the current phase of globalization, this almost canonical fixation on achieving the coveted status of "world class" hinders the capacity of mainstream approaches to account for the genuine diversity that characterizes the varieties of contemporary urban experiences, thereby precluding a more nuanced reading of global urbanization as a complex, contradictory, and multilayered process of cross-fertilization, mimicry, and outright plagiarism.[23]

The second flaw arises from thinking of cities in terms of vertically arranged hierarchies. This mode of classification reinforces the tendency in conventional urban studies to rank cities according to ideal typifications of alleged success and failure on a range of dimensions. Global cities, world-class cities, competitive cities, and creative cities are those that have achieved the enviable status of exemplary role models, and paradigmatic embodiments of success. In contrast, loser cities—sometimes called "forgotten cities"—are those that suffer from a host of deficiencies and (to add insult to injury) have failed to address the sources of distress.[24] Not only are these analytic distinctions biased and misleading, but they also have contributed to a bifurcation in the field of urban studies between the (exalted) work of theoretical reflection and the (remedial) policy practice of development.[25] Urban theory functions as a privileged scholarly enterprise, broadly focused on and produced in the West, that generates models and policy recommendations. In contrast, development studies as a distinctive field of inquiry is a policy-oriented approach that focuses on the diagnosis and treatment of problems of Third World megacities. This theoretical asymmetry rests on a presumed eschatology—the assumption that leading world-class cities in North America and Europe are the models to be emulated and that urban development along a preassigned path will inevitably converge at a common end point called development.[26]

MULTIPLE PATHWAYS OF URBAN TRANSFORMATION

The kind of urban transformation taking place at the start of the twenty-first century is a rather confusing story of stark contrasts. The unrelenting pressures of globalization have rescaled global finance and unleashed new patterns of worldwide production and trade. The wholesale restructuring of the capitalist world economy that began in the 1970s and has continued through the opening decades of the twenty-first century has fundamentally reconfigured the relationships between cities on a global scale and the networks and circuits within which they are embedded.[27] The overarching shift from so-called Fordist to post-Fordist global production in the capitalist world economy put in motion the creation of vast networks of factories, semiautonomous affiliates, new distribution techniques, and technological capabilities in the fields of communications and informatics that resulted in a radically reordered world of cities.[28]

With the turn toward entrepreneurialism and neoliberal modes of urban governance, new strategies of city building on a global scale have reshaped urban landscapes in fundamental ways and, in so doing, have redefined the rights of citizenship by reconfiguring the terms of access to urban space, social services, and infrastructures.[29] The emergence of new patterns of social inequalities both within and among cities has brought about new patterns of displacement and dispossession that have culminated in new forms of segregation and spatial exclusion. The combination of these shifting global networks and deepening interdependence is commonly collapsed in the term "globalization."[30] The multiform impulses of globalization have triggered a massive transformation of cities on a world scale. As a result, global cities, along with aspiring world-class cities, have taken an exalted place at the upper echelons of the ranked hierarchy of cities, whereas the sprawling megacities of hypergrowth—along with those deindustrializing cities that have experienced abandonment and decline—have fallen by the wayside in the competitive race for recognition and status, unceremoniously dismissed as either seemingly unmanageable polyglots or hopeless losers.[31]

Understanding the multiple pathways of urban transformation at the start of the twenty-first century poses serious challenges for scholars interested in theorizing about urban futurology and for planners engaged in the practice of making these better places to live. In the conventional urban studies literature, debate and discussion over the last three decades have largely revolved around three analytic conceptions: the "world-city hypothesis,"[32] the "global city" thesis,[33] and "world city networks."[34] Whatever disagreements might exist among these perspectives, they share the view that networks, circuits, and flows link different cities together and these asymmetrical connections play the dominant role in shaping global urbanism in the world economy at a time marked by the current phase of globalization.[35] The tendency in conventional urban studies literature to privilege these conceptual frameworks—what Tim Bunnell and Anant Maringanti have called "metrocentricity"—limits the scope of theoretical development and empirical inquiry.[36] This methodological bias toward privileging globally connected and networked cities and their role as powerful organizing nodes of the world economy has effectively pushed aside alternative research agendas.[37]

Despite numerous critiques from a variety of angles, these mainstream theories remain appealing to scholars and policy makers eager to understand the relationship between cities and the global economy.[38]

Nevertheless, an insurgent countercurrent of urban scholarship has called for the rethinking of conventional urban studies in ways that capture a wider and fuller range of diversity, heterogeneity, and complexity of cities on a world scale. The rising chorus of scholarly voices calling for a rethinking of conventional urban studies is impossible to ignore.[39] Recent calls for "provincializing global urbanism" (Ananya Roy and Helga Leitner),[40] for engaging with "urban theory beyond the West" (Tim Edensor and Mark Jayne),[41] for the recovery of "ordinary cities" (Jennifer Robinson, Ash Amin, and Stephen Graham),[42] for discovering "new geographies of theory" (Ananya Roy),[43] and for "radical openness" (Jennifer Wolch)[44] indicate a growing interest in reconceptualizing the field of urban studies in ways that move beyond the universalizing and teleological impulses of the prevailing theories of urban growth and transformation. City building around the world seems to defy such presumptions that urban transformation conforms to a single set of normative propositions and expectations. These efforts at imposing "grand narratives" of urban transformation on the complex realities of city building reflect a kind of neo-Hegelian mindset, with its totalizing Zeitgeist framework and linear teleology. Rather than conforming to a singular logic of urban transformation, global urbanism seems to have diverged along multiple pathways. Urban studies thus needs a rethinking that embraces a kind of theoretical openness and flexibility, seeking through comparison and contrast to account for the historical-spatial specificity of those "ordinary cities" that, as Jennifer Robinson has put it, are "off the map"—relegated to the residual category of structural irrelevance.[45] Engaging horizontally with a world of diverse, distinctive cities—instead of remaining fixated with the ranked hierarchies of global cities—can open a broader conversation about the meaning of transnational urbanism in the twenty-first century.[46]

The distinctive feature of the current phase of globalization is the integration of so-called national economies through the expansion, diffusion, and networking of capital flows and trade, information, and technologies on a global scale.[47] Beginning in the 1970s and accelerating in subsequent decades, fundamental shifts in the technology and organization of industrialized production led to the socio-spatial unraveling of the Fordist-era dynamics of large-scale and spatially concentrated industrial complexes—organized for mass production and mass consumption—and their replacement with lean, flexible systems of production of the post-Fordist era. This global dispersion has called into existence new networks and connections that appear in the form of vast commodity chains that stretch across the globe. In the current post-Fordist global era, leading branches of industry, such as high-tech information systems, aerospace, electronics, automobiles, and financial services, "are globally organized and selectively localized, depending on the relative global and regional competitive advantage that each place offers."[48] This "glocalization" of industrial production has brought about an unprecedented expansion in both the scale and scope of metropolitan regions, thereby rendering increasingly anachronistic monocentric models of inner-city urban cores surrounded by dependent suburbs, not only in the United States but also in Western Europe and Latin America.[49] The demise of the modern metropolis has gone hand in hand with the rise of sprawling, polynucleated metropolitan regions (variously referred to as the postmetropolis, the exopolis, the limitless city, and the 100-mile city) where the once reigning binary form of high-density central city and low-density suburbs has exploded into "a regional carpet of fragmented communities, zones and spaces."[50]

At the time when development was largely defined in terms of modernization theory and "the way forward lay in mimicking or even surpassing Western models of performance," aspiring cities in the (so-called) developing world sought to refashion themselves as industrializing manufacturing hubs in the pursuit of economic growth. While it is evident that the universalizing developmentalist goals associated with modernization theory can no longer function as coherent models for understanding the crosscurrents of global urbanism, new theoretical orientations and frameworks have yet to develop beyond their infancy and to emerge as a fully formed alternative. In short, once regnant frameworks are dying, but alternative approaches have not yet been born.[51]

RETHINKING CONVENTIONAL URBAN THEORY

Despite the diversity and heterogeneity of urban experiences, mainstream urban studies literature has generally failed to pay attention to the pace and tempo of urbanization around the globe or to the relationship between fast-track cities and incremental urbanism, on one side, and sprawling megacities of hypergrowth and postindustrial shrinking cities collapsing in on themselves, on the other.[52] Existing mainstream theories of urban transformation provide little ground for understanding struggling cities in distress other than to suggest that they are either anomalies or deviations from an expected preexisting North American (Western) pattern of competitive and successful achievement.[53] The tendency to view distressed cities (whether expanding too fast or declining too much) purely in terms of their particularity and radically unique exceptionality (with all of its "exotic" overtones) or, equally important, as failed, incomplete, or deficient examples of the "normal" expectations of urban development effectively diminishes their significance as real objects of inquiry in their own right.[54]

As Ananya Roy has cogently argued, dislocating the hegemony of leading world-class cities of Euro-America as the center of theoretical production requires more than simply expanding the geographic scope of research to study noncompetitive cities (those struggling with a range of social and structural inequities) as strange, anomalous, deviant, and esoteric empirical cases. In the "mattering map" of mainstream urban theory, global/globalizing cities have "hogged the explanatory limelight," to borrow a phrase from Jamie Peck, relegating struggling cities in poor countries to the sidelines of structural importance, recognized solely as "(lagging) exceptions to a modernist-metropolitan" norm of First World urbanism. "In this imperialist mode of theorizing," those distressed, struggling cities lacking efficient management systems, up-to-date infrastructure, and stable middle-class employment opportunities have become "little more than leftover categories," classified under the sign of "those empty and chaotic concepts like 'third world city,' which were (ill-)defined largely in terms of presumed developmental and cultural deficits."[55]

This way of seeing merely "keeps alive the neo-orientalist tendencies" that interpret Third World cities through the dystopian lens of chaotic unmanageability, mired in the hopelessness of problems without solutions. Moving the center of theory making away from an exclusive reliance on the leading "primate" cities of North America and Europe involves considering the distinctive experiences of cities elsewhere as capable of generating productive theoretical frameworks. Rethinking our conventional understanding of global urbanism depends as much on incorporating insights derived from a critical examination of Lagos, Manila, and Kinshasa as it does on ideas emanating from Chicago, Los Angeles, and New York.[56]

As Jennifer Robinson has argued, it is time to write ordinary cities (those that fall outside the rubric of globalizing cities with world-class aspirations) into urban theory, beginning by moving beyond conventional categories (First World versus Third World, Western versus non-Western), overused labels (global cities, world-class cities), and ranked hierarchies (developed and developing, leading and lagging) that, by definitional fiat, foreclose other possible approaches or angles of vision. These binary characterizations only contribute to a conceptual blindness that obscures our ability to think outside the constraints of linear time. Rethinking conventional urban studies requires us to build alternative analytic frameworks that take seriously the complexity and heterogeneity of global urbanism.[57] Bringing ordinary cities back into the mainstream of urban studies enables us to appreciate the diverse and multiple trajectories of urban growth and development, the endless possibilities for leaping ahead or falling backward, and the hybrid mixing that characterize global urbanism in the twenty-first century.[58]

The steady stream of such new image categories as ordinary cities, shrinking cities, postindustrial cities, instant cities, megacities, cosmopolitan cities, creative cities, slumdog cities, unequal cities, shadow cities, invisible cities, rebel cities, wounded cities, torn cities, divided cities, rogue cities, mongrel cities, feral cities, pirate towns, the postmodern metropolis, subaltern urbanism, neoliberal urbanism, postcolonial urbanism, hybrid urbanism, makeshift urbanism, subaltern urbanism, and so forth, reflects a growing unease in urban studies scholarship with the current state of theory.[59] These efforts to take seriously different kinds of urbanity provide ample testimony for the need to broaden our fields of inquiry and expand our horizons if we wish to account for the diversity and complexity of urban forms and experiences today.[60]

Developing new approaches in urban studies means breaking away from theorizations that subsume cities under the singular logic of globalization or position cities in terms of their functional roles within such all-encompassing scripts as the "urban age" or "planetary urbanization"—evocative phrases that have opened new fields of inquiry but, at the end of the day, may obscure as much as they illuminate.[61] Theorizing about global urbanism at the start of the twenty-first century requires a new sensitivity to difference, the recognition of multiple and asymmetrical trajectories of urban transformation, and the maintenance of a balance between the search for general patterns and the appreciation of local particularities and diversity.[62]

DIVERGENT TRAJECTORIES OF URBAN TRANSFORMATION: A FIRST APPROXIMATION

At the risk of oversimplification, it is possible to identify four distinct trajectories of urban transformation around which city building has coalesced at the start of the twenty-first century. First, those vibrant, healthy, developed First World cities that the mainstream urban studies literature has long relied upon as the wellspring of ideas about urban modernity represent prototypical exemplars of globalizing cities with world-class aspirations. In contrast, struggling postindustrial cities in decline (sometimes called shrinking cities), are the noncompetitive "losers" that have failed to make the make the giant leap forward into the glittering world of culture-led redevelopment by replacing dependence on manufacturing and industrial production with "soft" services—information technologies, tourism, and entertainment. Third, sprawling megacities of hypergrowth—not so long ago dismissively pigeonholed under the narrow epistemological category of Third World cities—have become vast depositories of impoverished work seekers who are unable to find formal wage employment and whose everyday survival depends upon irregular work, informal trade, and self-built housing. Finally, master-planned, holistically designed "instant cities" that have exploded along the Asia Pacific Rim, the Persian Gulf, and elsewhere represent exemplary expressions of fast-track urbanism, or what Shiuh-Shen Chien and Max Woodworth have called the "urban speed machine."[63]

These divergent trajectories of urban transformation must be understood as sui generis expressions of global urbanism with particular properties of their own. They do not constitute aberrations, exceptions, or outliers that have strayed from some overarching, universalizing master narrative of urbanization—deviations from some theoretically specified normal, expected pathway of urban transformation heading toward a shared end point called sustainable development. These four trajectories of global urbanism have an integrity all their own; each is an object of inquiry in its own right.[64]

No actually existing city embodies completely the logic of assemblage that gives rise to these four divergent trajectories. Classifying urban transformation into four distinct pathways is a methodological exercise somewhat akin to identifying Weberian-inspired ideal types. These typological distinctions must be seen as helpful heuristic devices, and not as closed categories. These ideal types are abstract stylized categories generated by piecing together features derived from a variety of urban experiences. Each of the four city types is open to the possibilities of interruption, uneven temporalities, and spatial asymmetries.[65] These four types do not exhaust the range of permutations and hybrid variations that characterize the world of cities, but they do point to divergent pathways of urban transformation that are often overlooked, misunderstood, or undertheorized in the conventional urban studies literature.

The complexity of these four city types calls not for the deductive testing of preexisting theories but for inductive theory building in order to generate new understandings of global urbanism. To engage with these multiple pathways of global urbanism requires us to rethink the concept of urban development, not as a universal, one-size-first-all solution to the challenges of urban transformation but as a locally determined, historically specific toolbox of strategies designed to promote the standard of living and socioeconomic health of particular cities in historically specific conditions.[66]

TABLE 1 Paradigmatic Representation of Ideal Types of City Building at the Start of the Twenty-First Century

COMPETITIVE WORLD-CLASS CITIES	POSTINDUSTRIAL CITIES IN DECLINE
Functional Characteristics	*Functional Characteristics*
From industrial to postindustrial	Once industrial, deindustrializing
Tourist-entertainment nexus	
Financial services	
Command-and-control centers	Junk businesses
Highest and best use of land	Vacant land and abandoned properties
Specialized functions of landed property	Variegated uses
Integrated infrastructure	Disintegrated/haphazard infrastructure
Competitive markets for landed property	Nonfunctioning property markets
"Clean" enterprises (hi-tech; design)	Detritus of industrial past: brownfields ruin and abandonment
Dense and vertical	Dispersed, undercrowded
Concentrated investment	Disinvestment
Metaphorical Expressions	*Metaphorical Expressions*
Globalizing cities with world-class aspirations	"Loser cities"
Creative/sustainable/smart	Unsustainable Shrinking
Morphological Features	*Morphological Features*
Creative destruction of built environment	Hollowed out; perforated
Symmetrical	Asymmetrical
Coherent form	Incoherent form
Aesthetics	*Aesthetics*
Sustainable, resilient	Unsustainable, at risk

Live/work/play	Ugly
Governance	*Governance*
Managed	Unmanageable, beyond management
Strong service delivery	Shrinking services, underserviced
Predictable and reliable	Unpredictable and unreliable
MEGACITIES OF HYPERGROWTH	**FAST-TRACK URBANISM**
Functional Characteristics	*Functional Characteristics*
Never industrial, marginally industrial	Postindustrial Entrepots, gateway cities
Overcrowded space	Controlled space
Unregulated land use	Master-planned, holistically designed
Unmanaged hypergrowth	Managed growth
Concentrated population	Dispersed population
Metaphorical Expressions	*Metaphorical Expressions*
Cities off the map	Instant cities
Ordinary cities	Cities in a box
Morphological Features	*Morphological Features*
Sprawling	Rational, coherent
Spontaneous, haphazard	Planned from above
Runaway urbanism	Dubai-ization
Opaque boundaries	Rigid boundaries
Formless	Distinct enclosed spatial form
Unpredictable	Overly predictable
Aesthetics	*Aesthetics*
Congested, unsanitary	Clean
Disorderly	Orderly
Governance	*Governance*
Unplanned	Overplanned
Unmanaged, unregulated	Top-down management

The first city type comprises those vibrant, competitive, and healthy cities that have made the successful transition from reliance upon industrial production and mass manufacturing as their primary engines of economic growth to the adoption of service-oriented culture and creative industries as the lynchpins for sustained economic prosperity.[67] This shift to the postindustrial information age, which some scholars have called "cognitive-cultural capitalism," has marked a major turning point in thinking about cities and global urbanism.[68] These prosperous cities, with First World amenities and up-to-date services, have attracted the lion's share of scholarly attention over the past several decades.[69] As the historical epoch of Fordist capitalism of mass manufacturing in the urban industrial age faded away, those postindustrial cities that have successfully capitalized on newly developed clean industries, emerging as major financial-service centers relying on cutting-edge information technologies, have come to play a dominant role in the world economy. These vibrant postindustrial cities—the new age boomtowns of postmodern urbanism—represent the prototypical exemplars of globalizing cities with world-class aspirations.[70]

With the success of the new informational city of high finance, information technology, start-up entrepreneurship, innovation hubs, and digital media, replacing the paradigmatic industrial city of mass production with creative service economies and high-end consumption, has meant that this new prototype of success has become the barometer for debates about sustainability, livability, and smart growth.[71] Not only do these cities that boast of quality-of-life attributes, state-of-the-art aesthetics, and experience economies become symbolic markers of achievement, they have also become aspirational models for urban growth coalitions seeking to elevate their status in the hierarchical rankings of competitive urbanism.[72] These vibrant postindustrial cities not only create the privileged platform for urban theorizing about cosmopolitanism, modernity, and global citizenship; they also constitute the touchstone for fresh new ideas about what the "good city" should be.[73]

At the other end of the spectrum is a growing number of previously developed postindustrial cities trapped in an accelerating downward spiral of decline and ruin. In one of six cities worldwide, the trend is not steady growth and expanding population but rather job loss and demographic shrinkage.[74] Urban contraction resulting from disinvestment, employment decline, and population loss is a worldwide phenomenon. While those globalizing cities with world-class aspirations are relatively prosperous and are able stay on a growth path, struggling cities in distress fail to attract people and investment and, as a result, experience stagnation and fiscal austerity instead. These declining cities are characterized first and foremost as embattled places that suffer from abandonment, neglect, and ruination.[75]

The associated web of problems includes pronounced demographic decline, high unemployment rates, low skill levels among working-age populations, underresourced city departments and disappearing municipal services, outdated and crumbling infrastructure, and high levels of urban dysfunctionality. As a general rule, the steady accretion of struggling cities in distress is typically associated with the rise and decline of the long twentieth-century era of Fordist industrialization. The conjoined processes of deindustrialization and job loss associated with accelerating global competition from the 1960s onward lie at the root of the stagnation and malaise of these struggling cities.[76] In the language of conventional urban studies, these are declining cities, shrinking cities, or abandoned cities—the failures in the celebratory world of aspiring world-class cites. While the vibrant competitive cities have successfully reinvented themselves as postindustrial creative hubs, noncompetitive cities have failed to make the giant leap forward into the glittering world of culture-led redevelopment and to restore the prosperity they once enjoyed during the industrial age of high modernism.[77] These postindustrial cities in decline are the outcome of social forces that have produced ruin and abandonment.[78]

The urban future—as Ananya Roy, Edgar Pieterse, Mike Davis, and numerous others have stressed—lies elsewhere, in cities outside the historic heartlands of global capitalism such as Shanghai, São Paulo, Cairo, Mumbai, Mexico City, Rio de Janeiro, Dakar, Manila, Lagos, and Kinshasa.[79] Unlike the once-industrial cities that have fallen into disrepair, these sprawling megacities of hypergrowth have experienced massive population growth without a commensurate expansion of wage-paid work and affordable housing fit for human habitation.[80] Stretching outward, these ballooning cities seem to defy the normal expectations or standards of what a modern metropolis should be. Seen through the lens of conventional urban studies, they appear as chaotic and hopelessly unmanageable, lagging behind, trapped in a downward spiral of uncertainty and desperation, where efforts to impose order, stability, and coherence on the urban form typically end in frustration or outright failure.[81]

The dynamic and unprecedented growth of megacities (and megaregions) outside the North American and European core areas of the world economy has unsettled conventional understandings of the pace and scale at which cities evolve and develop over time. Taken together, the seemingly unfettered spatial expansion, the sheer scale of growing population size, the dense patterns of human settlement, the absence of any unifying logic or centripetal force exerting a gravitational pull around a central core, the haphazard urban form, the breakdown of the functional utility of land use, the great disparities of wealth and asymmetrical access to requisite resources, and high levels of risk for ordinary urban dwellers have undermined what it means to talk about orderly urbanism, or even conceive of an evolving metropolitan form by reference to stages of urban growth.[82]

The idea that megacities of hypergrowth portend the future of urbanism has become a powerful trope animating thinking about urbanization as a global process. Seeing megacities as exemplary expressions of the urban future "expands the boundaries of urban modernity to include places once thought to be outside of history or lagging behind it." While this framing correctly rejects the teleological "assumption that time always moves in progressive directions," this understanding unfortunately "removes from the view the fact that the images of disorder are actually signs of an uneven geography of wealth and power produced within, not outside or prior to, histories of global capitalism."[83]

These sprawling megacities of hypergrowth appear as anomalous exceptions to expected patterns of orderly urbanization. With massive informal settlements, few opportunities for regular wage-paid work, broken-down infrastructure, and environmental degradation, these unplanned and unregulated urbanizing agglomerations have produced new spatial configurations that have never existed before in such exaggerated and extreme form.[84]

The real challenge for urban theory and planning practice is not merely to describe the failure of these megacities of hypergrowth to follow in the footsteps of the great cities of Western modernity, but to seek a deeper comprehension of why these large sprawling megalopolises have evolved in the complex ways they have. What is needed is a much more nuanced and rounded theoretical appreciation of urban transformation that not only acknowledges the spatial unevenness of the urbanization process on a global scale but also treats poor urban residents of squatter camps and informal settlements as active agents in constructing meaningful lives for themselves, rather than simply passive victims of inexorable structural processes beyond their control.[85]

As a general rule, conventional urban studies literature has failed to provide a full picture of the complicated dynamics of urban life in the megacities of hypergrowth. These cities cannot be understood apart from the pressures of globalization, which have exposed them to the unpredictable vagaries of the world market and subjected them to triumphant neoliberal ideologies—with their stress on downsizing and privatization of public services. What needs to be grasped both theoretically and empirically is that these large megalopolises operate in accordance with distinctive logics, norms, and values that produce highly complex social worlds that cannot be simply dismissed as hopelessly unmanageable and chaotic.[86]

Finally, the instant cities of the Asia Pacific Rim, the Persian Gulf, and elsewhere are expressions of fast-track urbanism, in which new cities appear as if out of nowhere and constructed de novo at hyperspeed. These master-planned, holistically designed cities built from scratch are testaments to the power of city builders to dispense with incrementally rebuilding existing urban landscapes and instead to embark on their own, starting over with a blank slate.[87] Dubai and other cities of the Persian Gulf may have served as an originating paradigm for this type of city building, but experiments with new types of compressed cities have appeared everywhere.[88]

What propels fast-track urbanism is the wish image of contemporary global capitalism for instant gratification.[89] The new instant cities mimic what Frederic Jameson has called "postmodern hyperspace"—that stretching of space and time to accommodate the multinational global space of late capitalism: fluid networks of capital, up-to-date telecommunications, generic airports, undifferentiated hotels, equally fungible office parks, and globalized commodities and their logos and franchises.[90] Iconic, excessive, and extravagant are the repeatable mantras that accompany city-building efforts—crucial pillars in the entrepreneurial narrative that seeks to assert the coming of age of instant cities as recognized players in the global marketplace of finance, trade, and tourism. The real estate developers behind this fast-track urbanism have adopted a high-modernist agenda for city building, including a belief in technical prowess, abstract rationality, social engineering, and market efficiency.[91] Master-planned, holistically designed, and privately managed cities exemplify a novel kind of radical hypermodernity in which the worthiness of place is determined almost exclusively by the ability to extract value from real estate. Regulation by market competition leads almost inevitably to segregation of spaces on the grounds of the ability to pay.[92]

Like postindustrial cities in decline and the sprawling megacities of hypergrowth, the instant cities of the Asia Pacific Rim and the Persian Gulf are often seen as anomalous outliers that deviate from the conventional understanding of city building as a slow, incremental process that unfolds in fits and starts over historical time. These master-planned, holistically designed instant cities appear to come from nowhere, without any intermediate steps, built entirely from scratch in accordance with preconceived guidelines. The city-building processes that characterize instant cities defy the long-standing image of the layered city that arises, in fits and starts, out of the ruins of the old.[93] Instant cities represent a new model of city building in which the incrementalism of the past no longer holds sway, and where building with a tabula rasa approach has taken on a momentum all its own. The instant cities model has deliberately broken any lingering ties to historicism, an outlook which was central to earlier classical traditions that sought to link urban spaces with the historical past.[94]

The meteoric rise of instant cities that originated in the Persian Gulf and the Asia Pacific Rim has spread haphazardly across the globe. In the nineteenth and twentieth centuries, industrialization, modernism, and modernity defined the development and nature of city life around the world. At the dawn of the twenty-first century, new concepts like globalization, postindustrial urbanism, and tourist-entertainment machines have become signifiers of the profound shift defining the character of world-class cities and the lives of their inhabitants. At the macro level, this shift affects not only the world-class aspirations of particular cities but the evolving role of globalizing cities within the world economy.[95] The unprecedented scale and scope of city building in the Asia Pacific Rim and the Persian Gulf challenge the conventional image of the modern metropolis of Western modernity—and even the very idea of the city.[96] These instant cities deviate from the conventional understanding of urban growth and development as a gradual process in which creative destruction of the built environment, incremental rebuilding, and constant erasure and reinscription produce an ever-evolving urban form.[97]

For the most part, scholarly attention directed at these four types of cities has operated within separate epistemic communities, each with its own objects of research, vocabulary, and matters of concern. The resulting silo effect has reinforced the development of disjointed fields of inquiry, ensuring that scholars do not engage in dialogue but instead talk past one another.

ALTERNATIVE ANALYTIC FRAMEWORKS

Using the analytic framework of global urbanism as a theoretical lens enables us to explore diverse pathways of urban transformation at the start of the twenty-first century. This line of reasoning coincides with the research agenda associated with the planetary urbanization paradigm. This perspective seeks "to decenter inherited approaches to the urban question that begin with the idea of 'the' city as a bounded spatial unit and then look 'inwards' towards its neighborhoods, built environments and social fabric (as, for example, in classic Chicago School models), or 'outwards' towards the metropolis, the region, the territory, and the world."[98] By inverting the analytic starting point, it is possible to radically reframe the question of urbanization away from the search for a single, universal, standardized measure of success and failure toward a focus on the uneven, geographically variegated, and contradictory urbanizing processes that produce different types of cities.[99] Forecasts for the urban future project the continuation of current trends toward differentiation and polarization in which some competitive, "healthy" cities can expect to remain vibrant and to prosper and thrive, while those uncompetitive, "unhealthy" cities will in all likelihood suffer further stagnation and decline. These multiple trajectories are not antithetical to each other but, on the contrary, are complementary and inseparable. These four subsets of cities loosely resemble ideal types but do not conform to this methodological approach in the conventional sense. They must be understood relationally—that is, not as independent from each other but as mutually constituted through webs of interrelationships, networks, and connections.[100]

The purpose of this classificatory exercise is not to superimpose new labels on particular types (or subsets) of cities but rather to introduce new ways of thinking about global urbanism at the start of the twenty-first century. The goal is to uncover how spatial restructuring on a global scale has produced highly unequal and uneven urban spaces around the world, each with an integrity and logic of its own. These efforts at rethinking global urbanism echo Jennifer Robinson's call for a "revitalized and experimental international comparativism that will enable urban studies to stretch its resources for theory-building across the world of cities."[101] This shift away from universalizing theorizations that produce ranked hierarchies of cities toward a relational understanding of a world of cities enables us to expose deeper connections between and among cities.[102] Identifying these subsets of city types is a useful exercise for heuristic purposes. These four types of cities indicate identifiable patterns of urban transformation but do not suggest a clear teleological direction. They do not represent hermetically sealed silos but simply provide a useful classificatory scheme for recognizing differences that are intersecting, overlapping, and constantly mutating.[103] Thus, it is not possible to unambiguously classify all cities as belonging exclusively in one of these four boxes. Seen from another angle, cities are not essential entities distinguished by a fixed set of characteristics; rather, they are kaleidoscopic spaces with overlapping attributes.[104]

As a general rule, conventional urban studies literature has tended to operate epistemologically in hermetically sealed ways that reinforce the unspoken assumption that globalizing cities with world-class aspirations, postindustrial cities under stress, the slum cities of the global South, and the instant cities built from scratch have nothing in common with one another. Only rarely have scholars brought these distinct city types into shared conversation.[105] The silo effect of separate literatures makes it difficult to address the common threads of urban transformation on a global scale.

Properly understood, these contrasting urban development scenarios are not exceptions; instead, they exemplify the uneven spatial geographies of global capitalism at a time of late modernity.[106] Globalizing cities with world-class aspirations operate in the same competitive field as postindustrial cities in decline.

Similarly, the megacities of hypergrowth have become the repositories, or vast catchment zones, for surplus humanity—those vast legions of work seekers who cannot be absorbed into the stable world of wage-paid employment. Last, the proliferation of instant cities represents the dreamscape of late modernity in which starting over and building afresh offer a way to escape the difficult task of revitalizing existing urban landscapes.[107]

Rethinking conventional urban theory challenges the universalizing impulses of grand narratives that project a single, linear, privileged path of urbanization toward a shared goal of developed maturity.[108] Calling these commonly used dualisms and typologies into question encourages us to destabilize the idea of a shared end point and to introduce instead an openness to multiple pathways and alternative routes with no preconceived terminus. In challenging the axiomatic presumptions of universality, teleology, and linearity, efforts at rethinking urban studies focus instead on the historical specificity of urban transformation, multiple trajectories of urban transformation, and the diversity of outcomes. A revised and enlarged urban theory involves thinking about global urbanism in terms of leapfrogging and mixing, circumventing and combining. Dispensing with the expectation of linear progression of development along a single pathway enables us to locate innovative city-building practices that bypass or sidestep what convention urban studies has suggested are necessary stages of urban transformation. Similarly, the blending of practices and ideas enables cities to adopt unusual and unexpected hybrid forms.[109]

PART I

Conventional Urban Theory at a Crossroads

CHAPTER 1

THE NARROW PREOCCUPATIONS OF CONVENTIONAL URBAN STUDIES

As a general rule, scholarly research and writing on urban development and city life have drawn inspiration from the great transformations that occurred in such metropolitan centers as London, Chicago, New York, Paris, Tokyo, and Los Angeles. In the conventional understanding, these theoretically privileged cities "act as the template against which all other cities are judged."[1] Mainstream urban theories rest on a set of ideas, propositions, and prescriptions grounded in the presumption that market liberalism and entrepreneurial "good governance" policies will enable cites that "lag behind" to eventually "catch up" by emulating those globalizing cities with world-class aspirations.[2]

Conventional approaches to urban studies literature, as Jennifer Robinson has persuasively argued, have been divided between Western and other cities—that is, between "celebrations of 'modernity,' " on the one hand, and the "promotion of urban development," on the other.[3] For the most part, the key assumptions and analytical constructs that guide mainstream urban theories are generated out of research and writing on a handful of leading cities of North America and Western Europe—that is, those developed, vibrant, and healthy cities that are globally connected in terms of finance, trade, and investment. These globalizing cities with world-class aspirations serve as the wellspring of ideas about what makes a city competitive, creative, and sustainable.

In contrast, when distressed cities in poor countries—with their broken-down infrastructure, unregulated street life, and poor-quality housing—become visible in mainstream urban theory, they are usually assembled under the sign of underdevelopment, exemplars of failed, stalled, or distorted urbanism.[4] In short, they are outside of mainstream theory, typically portrayed as anomalous and inscrutable places of underdevelopment and backwardness—exceptional places where comprehensive regulatory regimes, land-use planning, and good democratic governance happen only under exceptional circumstances.[5] They are the megacities of hypergrowth, literally bursting at the seams, overwhelmed by "poverty, disease, violence, and toxicity."[6] These precarious cities provide the disturbing setting for an expanding planet of slums, where the radical decoupling of urbanization from industrialization has gone hand in hand with the "late capitalist triage of [surplus] humanity."[7]

The mainstream urban studies literature renders these megacities of hypergrowth as a kind of terra incognita without distinctive form, without clear-cut boundaries, and without the regulatory capacity to forge order out of chaos. To the extent that informal economic activities and self-built housing are the driving forces behind the creation of new territories, networks, and relations outside the control and regulation of state authority, these sprawling megacities of hypergrowth remain elusive and almost incomprehensible. Dismissed as unable to generate theoretically relevant insights for urban studies, these struggling cities under distress become the site for exotic ethnographic inquiry that reveals the gritty underside of global urbanism.[8] As a general rule, these cities are interpreted through the lens of developmentalism, a largely prescriptive approach—as Jennifer Robinson has put it—that broadly understands these places to be sorely lacking in the normal (and expected) qualities of good "city-ness." These distressed cities are in need of remedial intervention in order "to improve capacities of governance, service provision, and productivity."[9]Mainstream urban scholarship typically focuses on such questions as global preeminence, world-class rankings, culture-led redevelopment, starchitecture and the Bilbao effect, entrepreneurial modes of governance, creativity, sustainability and livability, experience economies, and flagship tourist and entertainment megaprojects. While answers to these questions yield considerable understanding of urbanization on a global scale, they do not exhaust the range of possible ways of making sense of different types of urban experiences that fall outside the purview of globalizing cities with world-class aspirations.[10]

THE INHERENT DANGERS OF FACILE GENERALIZATIONS

Particular cities have historically been portrayed as exemplary expressions of the leading edge of the urbanism of their era, acting as metonyms, prototypes, and paradigms for the urban dynamics of the time.[11] In a well-known essay, Walter

Benjamin identified Paris as the "capital of the 19th Century."[12] At the turn of the twentieth century, scholars and writers singled out New York as the quintessential "modern metropolis."[13] In the 1920s and 1930s, urban sociologists affiliated with the "Chicago School" identified Chicago as an exemplary laboratory in which to investigate the generic claim that cities were social organisms with distinct parts held together by internal processes.[14] In the 1990s, in describing what he regarded as the emergent "postmetropolis," Edward Soja famously remarked that "it all comes together in Los Angeles."[15]

If the prototype for urban development is derived primarily from the histories and experiences of a handful of leading globalizing cities with world-class aspirations, then the "structurally irrelevant," economically marginal, or globally inconsequential cities of what used to be called the Third World appear to present a kind of curious anomaly or outlier.[16] Because they do not possess in equal measure the redeeming qualities of genuine "city-ness" that characterize those globalizing cities that aspire to world-class status, they are classified as struggling cities in the developing (and still underdeveloped) world.[17] As a result, they are regarded as imperfect, stunted, and underdeveloped versions of what they ought to be (and may someday become)—that is, vibrant, well-functioning, healthy clones following in the footsteps of the much-celebrated globalizing cities.[18]

What is needed is a rethinking in urban studies away from ready-made, deductive theories that seek to generalize across multiple cases and a movement instead toward a kind of genuine theoretical openness and flexibility that seeks, through comparison and contrast, to account for the historical-spatial specificity of particular cities.[19] The fascination with grand, totalizing theories whose universalizing claims are applied to cities everywhere tends to conflate divergent processes of urban growth and development and to flatten and homogenize the rich diversity of what constitutes the urban experience. This way of thinking "emphasizes the commonalities across all types of cities" and treats differences as more or less cosmetic variations on "systematic regularities that are susceptible to high levels of theoretical generalization."[20]

Rethinking conventional approaches to urban studies calls for seeing urbanization through the lens of a world of diverse, distinctive cities, or what have been called ordinary cities—that is, those cities that seem to resist inclusion in a priori analytic classifications and ranked hierarchies—as distinctive objects of inquiry in their own right, thereby enabling us to see them as historically specific entities rather than as embodiments of abstract models.[21] Broadening the field of inquiry in this way provides a foundation for a much deeper, more nuanced understanding of the complexity of urbanization as it has occurred in different areas of the world. Only by moving away from conceptual frameworks that seek to pigeonhole cities into preestablished categories and classification schemes is it possible to open new ways of understanding cities and city-making processes from a variety of different perspectives.[22]

GLOBAL CITIES / WORLD CITIES PARADIGM

Over the past several decades, the pioneering work of a number of urban theorists has drawn attention to the pivotal role that the current wave of globalization has playing in fostering intensified competition among cities that aspire to "world-class" status.[23] In the main, much of this scholarly writing about urban rivalries has crystallized around a research agenda that has been labeled the "global cities" paradigm. The discursive power of the global cities hypothesis depends upon identifying certain key attributes of cities that, taken together, function as the driving force behind the globalizing impulse. By framing their research agenda around identifying which cities possess these indispensable attributes (and in what quantities), scholars operating in these paradigmatic frameworks have—perhaps inadvertently—triggered a competitive urge among local growth coalitions that, in the rhetoric of city boosterism, seek to find ways to advance in the ranked hierarchy of aspiring world-class cities.[24]

Scholars operating in the global cities paradigmatic framework have sought to reassess the strategic importance of a handful of leading cities as key command-and-control centers within the interlocking globalizing dynamics of finance and trade. A central axiom in this innovative theoretical framework is the contention that, for giant transnational corporate enterprises, the spatial dispersal of production and manufacturing operations over great distances requires a parallel territorial concentration of command-and-control functions in key cities located at the apex of the global urban hierarchy. As locally based and integrated organizational hubs for global business and financial functions, these global cities have attracted a disproportionate share of corporate headquarters and giant financial institutions, along with high-end, advanced producer services and business-related activities (telecommunications, conferencing facilities, advertising, media, design and cultural industries, transport, and real estate developments).[25]

Taken as a whole, this global cities model—with its particular stress on the evolving networks of interconnected urban centers—has sparked a great deal of substantive research that has advanced our understanding of the place and function of certain strategically located cities in the spatial geography of the world economy. By situating large, metropolitan regions within a common globalizing framework, this approach has opened up genuine possibilities for fruitful comparisons that promise to yield new insights into the changing roles of cities in the contemporary world economy.[26]

Yet despite its considerable strengths as an orienting framework for empirically grounded research, the global cities approach is not without its theoretical limitations as an overarching paradigm for understanding urbanization as a global process. While some scholars have questioned the validity of its empirical claims, others have challenged the global cities paradigm on strictly theoretical grounds.[27] By identifying certain key urban centers as material manifestations of the structural processes of globalization, the global cities approach has incorporated certain functionalist and economistic biases into its theoretical reasoning. This way of theoretically specifying its object of inquiry within a limited scope of concerns has inadvertently contributed to a narrowing of the field of vision for urban studies. By privileging the functional roles and specializations of large metropolitan centers in the global marketplace and categorizing cities into a ranked hierarchy roughly in accordance with the economic power they command, the global cities approach limits, in an a priori way, the kinds of questions that can be legitimately addressed within its theoretical framework.[28]

In large measure, research and writing on global cities and their aspiring clones have become a hegemonic knowledge project, generating ideas, concepts, and policy recommendations directed at providing insights and lessons for struggling cities in distress to follow.[29] In both the scholarly literature and popular accounts, global cities combine the status of both field site for research and laboratory for experimental innovation. In the emergent world of international best practices, policy consultants and other professional experts have profiled, analyzed, and dissected global cities, looking for their secrets and local recipes for economic success. The local attractions and relative strengths of cities—due to their peculiar (historically specific) characteristics and intangible but distinctive qualities—"easily get lost in the scramble to [decipher and] learn by rote the lessons of the global city." The hegemonic force of the urban pedagogy that separates cities into (globally connected) winners and (locally constricted) losers—either/or characterizations that figure so prominently in the impressionistic rankings of popular urban studies—seems to be an integral part of "the knowledge project of neoliberal globalization."[30] A kind of cargo cult theory has emerged as a social force all its own, a "normative therapeutic" that has crystalized around the mantra "build it like this and wealth will certainly come." The shifting back and forth between theoretical analysis and policy prescriptions has contributed to what Rob Shields has called "the symmetry of urban scholarship with the fad for city rankings in [popular] journalism."[31]

As is the case with most if not all overarching paradigms, there is an inherent danger in seeking to explain far too much from far too small a sample of cities. As a heuristic device, the global cities framework has become so generalized and pervasive in the scholarly literature that the difference between global cities and those relatively vibrant, healthy cities (or just big cities) has become somewhat blurred.[32] Taken to its extremes, the global cities paradigm has largely succumbed to a categorizing imperative, reducing scholarly research to something akin to a taxonomic exercise in which cities are catalogued, labeled, and assigned a place in a ranked hierarchy according to the relative socioeconomic power they command in the world economy.[33] One main complaint leveled against the conjoined global cities / world cities paradigms is that their narrow focus on structured hierarchies and rank orders acknowledges cities only as nodes within globalized networks of financial and business-oriented transactions.[34] Restricting our gaze to "structural positioning in transnational economies" means that what we learn about cities comes "only by inference from their ranking in interurban relations and flows."[35] Despite its obvious insights, the global cities approach has tended to map its own historically specific geographies of ranked hierarchies onto that of mainstream development theories, prescribing what unhealthy, stagnating, loser cities must do in order to rise in the rank order of what it means to be world class.[36] To the extent that it focuses almost exclusively on transnational business and financial networks that link leading metropolitan centers in a ranked global hierarchy, the global cities approach tends to ignore those cities that are economically marginal and insignificant as hubs for global production and finance. As a consequence, these cities virtually disappear, falling off the map (figuratively speaking) of global importance.[37]

POLARIZING TENDENCIES IN URBAN STUDIES

As Austin Zeiderman has observed, the field of urban studies has become entangled in a complicated web of "intense theoretical self-reflection."[38] Faced with the undeniable realities of seismic shifts in processes of urbanization on a world scale, urban scholars have grappled with how to reformulate the conceptual underpinnings of existing theories in order to make sense of what we mean by cities and urbanism more generally. The nature of urbanization as a multilayered and spatially uneven process has come under scrutiny, as has the most appropriate scalar vantage point from which to understand what is happening and why. The steady proliferation of such terms as "global urbanism," "worlding cities," "planetary urbanization," "the urbanism of exception," and "cities in a world of cities" have entered into our conceptual vocabulary as testament to the theoretical uncertainty that has engulfed urban studies.[39]

To a certain degree, efforts to expand the spatial scale of inquiry has followed in the footsteps of scholarly currents that began in the late twentieth century with a turn toward investigating transnationalism and globalization. In urban studies, recent attention to the scale of the global and the planetary has gone hand in hand with increased skepticism of the existing ontological and epistemological foundations of conventional understandings of cities and urbanism. On the one hand, objections to the underlying spatial assumptions that have long framed the field of urban studies (namely, looking at isolated cities in single nation-states) have led to efforts to "theorize urbanization as an extended, 'planetary' process that transcends the boundaries of any one particular city." This approach, which Zeiderman has subsumed under the rubric of "global urban theory," has effectively opened up conceptual space for a more worldly, globally focused urban studies. On the other hand, postcolonial critiques of the narrow geopolitics of knowledge production "have made evident the need to displace the Euro-American locus of urban theory and to advance more cosmopolitan perspectives on contemporary urban life in its multiple and varied forms."[40] In mounting critiques of mainstream urban theories that extrapolate from the experiences of cities located in the Global North, theorists influenced by postcolonial thinking have questioned the epistemological validity of all universalizing claims to a general theory of urbanization on a global scale.[41]

At a time when cities are increasingly shaped by a multiplicity of transnational connections and globalizing processes, the challenge is for urban theory to balance a recognition of the interconnectedness and commonalities that come from widely circulating practices of city building on a world scale with an appreciation of the distinctiveness of urban experiences. As a general rule, the field of urban studies has not effectively engaged with exactly how one can go about writing and "theorizing across the very evident differences and similarities which exist among cities."[42]

Scholarly efforts to make sense of cities and urbanization have tended to gravitate toward two opposing poles. On the one side, one scholarly current has insisted that the starting point for a proper understanding of urbanization on a global scale is an agreed-upon set of general principles and a "coherent concept of the city as an object of theoretical inquiry."[43] The goal of these universalizing theorizations of urbanization is to rescue urban theory from eclecticism and particularism by identifying the essential characteristics (or intrinsic features) of the city as a distinctive entitywith sui generis properties of its own. Gesturing toward a hegemonic metanarrative that subsumes all cities under its terms of engagement, these universalizing theories seek to establish a common analytic framework that can account for variations in the empirical makeup of cities that arise from different contextual circumstances.[44]

Writing against what they have characterized as an iconoclastic "new particularism," Allen Scott and Michael Storper, for example, argue in favor of "a meaningful concept of urbanization with generalizable insights about the logic and dynamics of cities."[45] As key advocates for the inherent value of conceptual abstraction and theoretical rigor, Scott and Storper go to great lengths to challenge what they consider the misplaced claim that "cities are so big, so complicated and so lacking in easily identifiable boundaries that any attempt to define their essential characteristics is doomed to failure." The new particularism is a source of rich "empirical detail and descriptive color" but, at the end of the day, is totally inadequate because it overlooks the "systematic regularities in urban life that are susceptible to high levels of theoretical generalization."[46] To counteract this tendency to treat every city (because of its "bewildering degree of individuality") as a sui generis special case, they argue instead that cities can be understood in terms of a single, all-encompassing theoretical framework that rests on an understanding of the economies of agglomeration and the urban land nexus.[47]

Universalizing approaches to understanding the nature of cities seek to monopolize the knowledge terrain with the assertion that their perspective is privileged because of its unique capacity to uncover the essence of urbanization. For Scott and Storper, "at least some of the cacophony" that exists in scholarly research and writing on cities "can in part be traced back to the failure of scholars to be clear about these matters of definition and demarcation." The crucial task for urban theory is to uncover and delineate "the inner logic of urbanization" through a rigorous examination of extant regular patterns and common features.[48]

In seeking to establish the grounds for a unifying and comprehensive theory of the "general nature of urbanization as a particular mode of spatial integration and interaction," Scott and Stoper have argued against what they regard as those antitheoretical currents that fall into a kind of urban exceptionalism in which "every individual city is an irreducible special case."[49] This tension between identifying recurrent patterns and "commonalities across all types of cities," on the one hand, and the stress on particularities and ontological differences among cities, on the other, brings into the open a great deal of the messiness and inconsistency in urban theorizing today.[50]

As a general rule, the field of urban studies has long been torn between seeking general lawlike regularities and repeatable occurrences as the foundation for theory building, at one end of the spectrum, and focusing exclusively on peculiarities and idiosyncratic specificities in the evolution of cites, at the other end. Debates in urban studies mimic these longstanding philosophical disputes concerning knowledge-production. Urban theorists sympathetic to the nomothetic position have sought to identify common properties and regular patterns to explain the evolving trajectory of urbanism on a global scale. In contrast, scholars inclined to the idiographic perspective have looked upon cities as more or less unique entities that have to be understood on their own terms rather than as instances of a type or class.

These decidedly different epistemological starting points trace their origins to the philosophical distinction between nomothetic and idiographic modes of knowledge production. In the late nineteenth century, the German philosopher Wilhelm Windelband coined the terms "nomothetic" and "idiographic" to refer to different approaches to evidence-based knowledge.[51] This distinction parallels the search for general properties in the natural sciences, on the one hand, and the specification of individual features in the humanities, on the other. Nomothetic approaches have their roots in positivist methodologies guiding social science inquiry and efforts to pattern fields like geography and urban studies after the natural sciences. Scholars tied to this tradition seek to reveal general patterns with universal scope. Nomothetic knowledge aims at finding shared properties that are common to a class of similar entities, with the ultimate goal of deriving universalizing theories or lawlike generalities to account for these observed regularities. In contrast, idiographic approaches can trace their origins to humanist aesthetic sensibilities. Scholars operating in this vein stress diversity, difference, and the uniqueness of each instance under investigation. Idiographic knowledge focuses on providing in-depth descriptive accounts of particular entities by stressing their unique (temporally circumscribed and spatially demarcated) characteristics. The nomothetic-idiographic divide is sometimes referred to as the search for general theories, with "covering laws" that blanket multiple cases, versus individual particular knowledge.[52]

In many ways, the Scott-Storper endorsement of a universalizing urban theory, positioned in opposition to what they regard as undue stress on the singular and distinctive features of cities, mimics the classical distinction between nomothetic and idiographic approaches. In mounting their defense of a general theoretical framework, Scott and Storper caution against "the unfortunate influence of post-structuralist philosophy in urban studies" (with its "semantically inflated jargon") that "effectively dissolves the city away as a structured socio-geographic entity" and "encourages in turn a rampant eclecticism so that the city as such tends to shift persistently out of focus."[53] Taking a step further, they point to what they consider "the overblown interpretative schemas that post-structuralism licenses and their tendency to crowd out analytically-oriented forms of social (and especially economic) enquiry in favour of a conceptually barren search for difference, particularity and localism."[54]

Rooted in the tradition of scientific naturalism, universalizing theories of urbanization rest on a priori conceptual abstraction which, in turn, requires hard-and-fast distinctions between "essential and accidental, [and] intrinsic and contextual" characteristics of the city.[55] Universalizing theorizations offer a certain parsimonious elegance, conceptual coherence, and analytic clarity for understanding urbanization as a process. While commendable in its quest for finding the explanatory power of an all-encompassing general (and unified) urban theory, universalizing approaches reduce complexity and difference to secondary importance as "little more than variations on a universal form," substituting simplified formulas for the appreciation of diversity, and subsuming "the complex and heterogeneous urbanisms" that have evolved around the world "into an already existing universalizing analysis of urbanization."[56]

To argue for generic, one-size-fits-all theories of urban transformation is to fall into the trap of adopting universalizing models that, by definition, treat difference as deviation from the expected patterns of urbanization. The diversity of cities around the world presents a challenge to those urban theories that typically draw the only source of their inspiration from single contexts, individual cases, or paradigmatic instances.[57] What we gain from adopting a unified urban theory in cutting through "complexity to reveal underlying patterns (i.e. simplification) and separat[ing] causality from change," we lose by way of rigid adherence to "an overly tidy sense of what constitutes a scientific approach."[58]

DECONSTRUCTING ILL-INFORMED AND ILL-FITTING UNIVERSALISMS

By seeking to establish, once and for all, a unified general urban theory rooted in a foundational understanding of agglomeration and clustering, Scott and Storper choose to ignore (unwittingly, perhaps) the inherently unequal structures of power and the injustices associated with cities.[59] But as Oli Mould persuasively points out, cities are not ontologically distinct from the social and political relationships in which they are embedded. Not only does the conceptual framework championed by Scott and Storper "depoliticize the urban," it also uncritically adopts a "somewhat unidirectional developmentalist discourse" that treats "underdevelopment, stagnation, and decline" as anomalies and deviations that fall outside the fixation on "growth-focused reasoning" and its teleological underpinnings. To reduce our understanding of the urban to the identification of essential characteristics runs the risk of "rendering the city as a logical, linear, sequential and (more damagingly) exclusively economic technological system." The narrowness of this thinking has produced a theoretical orientation that is at once too instrumental, too deterministic, and too economistic.[60]

At the other end of the spectrum, an emergent current in the urban studies literature has tended to eschew a priori theoretical abstraction altogether and to dispense with the committed search for general patterns and commonalities across cases, preferring instead to focus on the particularities of individual urban experiences. Sometimes associated with assemblage theory, actor-network theory, and/or postcolonial approaches to comparative urbanism, this tendency toward seeing cities as one-of-a-kind entities with properties, characteristics, and experiences all their own lays particular stress on the exceptionality, peculiarity, and uniqueness of each city. This orientation toward treating cities as sui generis has attracted a great number of adherents, particularly those influenced by ideas that originated in cultural and postcolonial studies. Scholarly research and writing that tend toward eclectic understandings of urban life recoil against theoretical approaches that position all cities within a single script—a framing device that, by definition, downplays variation and difference as theoretically inconsequential.[61]

To argue for the incommensurability of different kinds of cities within the field of urban theory is to fall under the spell of what Robert Beauregard has called "radical uniqueness"—the a priori contention that cities are so idiosyncratic and exceptional that theorizing about similarities and differences between them is meaningless. There is much more at stake than simply identifying the peculiarities of cities. By endorsing the unmediated uniqueness of all cities, urban studies—as a theoretical endeavor that aims at identifying common patterns by means of a comparative perspective—would come to a halt. "With comparison impossible," Beauregard has argued, "thinking beyond the single case would be useless."[62] At root, scholarly inquiry that too readily endorses the radical uniqueness of cities exhibits an antitheoretical bias amounting to a kind of agnostic relativism—an "anything goes" approach to inquiry. To assert that cities are so complex, so multidimensional, and so context-specific that they defy simple classification is tantamount to suggesting that differences so overwhelm similarities that seeking to identify common trends, general patterns, and shared experiences is more or less a meaningless exercise. As Beauregard has argued, "Only city-specific theory would remain, that is, theory focused on the conditions and dynamics within particular places."[63] Advocating the exceptionality of individual cities undermines our capacity to make generalizations and to create typologies—features that lie at the core of theoretical reasoning.[64]

These polarizing tendencies obscure two fundamental stylized facts about the divergent trajectories of global urbanism. On the one hand, the evolving patterns of global urbanism have not converged around a single mode of urbanization; city building has proceeded in different directions and at varying speeds of transformation. In short, the historical specificity of divergent trajectories of urbanization cannot be dismissed as deviations from an expected norm but instead require explanation in their own right. Speaking in terms of multiple, alternative trajectories represents an effort to rescue the discussion of urban transformation from the linear reductionisms that defined conventional modernization theory and its developmentalist corollary.[65] On the other hand, urbanization does conform to distinct and recognizable patterns. More specifically, one can argue with a degree of certainty that city building falls into distinct clusters or types. This classificatory scheme rests on ideal types. These do not signify hermetically sealed containers. Ideal types arise from extrapolating common features from actual cities, but no actual city can be unambiguously classified into one and only one box. City types overlap, blend, and intersect. No single theory can account for the variegated nature of cities and the multiple pathways of urbanization around the world. Different analytic frameworks and theoretical orientations "gain traction by their capacity to develop compelling explanations of urban processes and spaces suited to particular spatial-temporal contexts."[66]

PLANETARY URBANIZATION

Recent controversies over the usefulness of the "planetary urbanization" paradigm have brought to the forefront long-standing debates pitting abstract theorizing against concrete empirical specification—that is, a stress on finding general patterns versus offering "thick descriptions" of local particularities. By drawing attention to how multidimensional and multiscalar processes of urbanization have dissolved and blurred the spatial distinctiveness of dense urban cores and surrounding nonurban peripheries, the planetary urbanization paradigm has not only problematized the urban as a "bounded unit of territory or a particular place-defined object" but also highlighted the "dynamic interdependency of the increasingly continuous web of dense agglomerations with their 'operational landscapes' of production, extraction, and circulation to sustain them." As Neil Brenner and Christian Schmid have provocatively argued, there is "no longer any *outside* to the urban world."[67]

The idea of planetary urbanization is premised on the claim that multiscalar urbanizing processes have extended their tentacles in all directions and into once untouched arenas, enclosing, swallowing, and dissolving everything in their path.[68] The dynamic processes of urbanization do more than simply spread outward; they also become "the vortex for sucking in everything the planet offers: its land and wealth, its capital and power, its culture and people—its dispensable labour-power," while simultaneously expelling its unwanted detritus, leaving behind a useless residue of disposable objects and people. In short, urbanization produces expansion and growth, on the one hand, and wreckage and abandonment, on the other.[69]

Viewing urbanizing processes through the lens of planetary urbanization has triggered a rethinking of the ontological stability of the idea of the city. In referring to what they call "methodological cityism," Hillary Angelo and David Wachsmuth have opposed what they regard as the unwarranted "analytical privileging, isolation and perhaps naturalization of the city in studies of urban processes."[70] By foregrounding the geographical embeddedness of cities in interdependent networks instead of their alleged separateness, the planetary urbanization approach allows for a much more inclusive conceptualization of the materialities of urbanization and the intertwined linkages among cities. Instead of treating the city and its Other (suburbs, countryside, rural hinterland, exurban frontier, and wilderness) as two independent and conflicting social worlds, it is more useful to think of their relationality and connectivity.[71]

As should be expected, the planetary urbanization approach has attracted its share of unconvinced skeptics. These skeptical voices have raised concerns about the extent to which planetary urbanization translates "across geographic scales and location"; and they call instead for empirically grounded approaches to understanding the historical specificity of local conditions.[72] Despite their dissimilar vantage points, critics of the planetary urbanization thesis discern what they regard as its totalizing impulses that are routinely manifested in indifference to questions of gender, sexuality, race, and the postcolonial condition.[73] Along with similar universalizing theories, planetary urbanization embodies, as Brenner summarizes the views of its detractors, "the latest in a long line of Eurocentric, masculinist, and heteronormative epistemologies that perpetuate the hubristic 'god-trick' of denying their own embeddedness within the social processes they aspire to understand."[74]

Broadly speaking, it is possible to identify a common thread that runs through the critiques of planetary urbanization. "Whether focusing on empirical, epistemological or theoretical grounds," critics tend to juxtapose a "manner of difference against a mode of abstraction," arguing that the planetary urbanization framework—"as an abstract theory of large-scale social processes"—"is inadequate or incomplete to the extent that it fails to accommodate [and overlooks all manner and sundry of everyday] embodied and place-based [manifestations] of social difference, either in its production (as a result of situated knowledge), or its application (in a three-dimensional and contradictory world)."[75] This inattention to difference amounts to an epistemological blindness, which results, in turn, in the marginalization and occlusion of alternative viewpoints, social practices, and experiences.[76]

Despite the passage of time, the lingering influence of the nomothetic-idiographic distinction, pitting a "universalizing" epistemological orientation against a "particularizing" one, has not faded away. This differentiation is a rather crusty old philosophical demarcation in social science inquiry, dividing scholars wishing to emulate approaches to causal analysis found in the natural sciences from those grounded in the ethnographic traditions associated with "local knowledge." While framed without reference to earlier philosophical controversies, recent critical commentaries that direct their opprobrium toward the planetary urbanization paradigm bear a striking resemblance to the nomothetic versus idiographic dispute, with its roots in the search for general patterns in contrast to the celebration of difference, singularity, and historical specificity.[77]

Much of this controversy amounts to little more than a tempest in a teapot. There is a place for grand theorizing in the planetary urbanization vein as much as there needs to be grounded research focusing on local conditions and particular circumstances. Critics, skeptics, and detractors alike have attributed an unwarranted coherence and integrity to the planetary urbanization thesis that simply does not exist. The charge that planetary urbanization is premised on a totalizing discourse is unwarranted.[78] At root, the planetary urbanization perspective functions more as a metatheoretical orienting framework than as a comprehensive "theory of everything."[79] It does not rest on rigid epistemological principles, a fixed definition of essential properties, or a locked conceptual toolbox. Rather than endorsing a single, universalizing epistemology of the urban that can "encompass, subsume, or supersede other perspectives in urban studies," advocates for the planetary urbanization approach subscribe to "the need for plural, heterodox" approaches to knowledge production.[80] As Christian Schmid has argued, the planetary urbanization framework is "grounded upon a basic hypothesis: that the contemporary urbanizing world cannot be adequately understood without systematically revising inherited concepts and representations of the urban."[81] By offering a metatheoretical provocation instead of a closed circle or school of thought, it offers a flexible set of propositions through which to unpack urbanization as a multiscalar, variegated, spatially (and substantively) uneven, volatile, unfinished, and contradictory process, one that is "necessarily articulated through contextually specific patterns and pathways."[82]

COMPARATIVE URBANISM

If we are to avoid the twin dangers of grand overarching universalism on one side and narrow particularism/exceptionalism on the other, it is necessary to rethink comparative approaches to the study of cities.[83] To properly engage with the diverse trajectories of global urbanism requires a theoretical flexibility and inventiveness that only comes with a willingness to experiment with different approaches that may dislodge and reinvent concepts that "drive and shape conversations about the nature of the urban."[84]

Over the past several years, the reanimation of debates about comparative urbanism—from epistemological, theoretical, and methodological as well as empirical perspectives—has opened up new vantage points from which to make sense of global urbanism at the start of the twenty-first century. The resurgent interest in comparative approaches to urbanism on a global scale goes far beyond a mere revival of older and well-established comparative traditions of twentieth-century social sciences. This new attention to comparative urbanism calls for more experimental approaches to understanding cities in two fundamental ways: (1) apparently different and seemingly unrelated cases are given at least as much attention as those that are similar and highly interconnected; and (2) the ordinary and the mundane are taken as equally important as the powerful and the paradigmatic. Rethinking both the conceptual geographies of urban theory and the geographies of urban theory production, these new approaches to comparative urbanism also seek to destabilize long established categories such as First World and Third World cities, developed and developing cities, global cities and world cities. While these dichotomies were themselves the conceptual products of earlier efforts at urban theorizing and implicit comparative reasoning, they have lost some of their power as analytic instruments.[85]

As a distinct mode of inquiry, comparative urbanism "aims at developing knowledge, understanding, and generalization at a level between what is true of all cities and what is true of one city at a given point in time."[86] Only by comparing and contrasting urban experiences worldwide, as Dennis Rogers has argued, "can we expect to gain insights into more general urban trends and dynamics."[87] It goes without saying that every city is different in historically specific ways—this is the distinctive nature of place. What requires our attention is how distinctive cities with historically specific characteristics can share common features with otherwise very different cities. Clearly, as Jan Nijman has argued, "difference and similarity are two faces of the same coin, each other's opposites, mutually exclusive, and reciprocally exhaustive." Cities are both similar and different. The aim of comparative urbanism, then, is to systematically study similarity and difference among cities or urban processes.[88]

This turn toward comparative urbanism offers a way "to redress the uneven and restricted geographic foundations of inherited [theoretical] approaches to understanding urbanism," and it provides a framework for "building theoretical insight from a diversity of specific urban outcomes, processes, and contexts."[89] A relational perspective enables us to appreciate that cities are not singular entities existing in pristine isolation. Their interconnectedness shapes their place in regional and global hierarchies. Divergent trajectories of global urbanism cannot be understood a priori without empirical investigation. This endeavor is a theory-building undertaking rather than a theory-confirming exercise. A sustained and vigorous vigilance for similarities and differences enables us to expand and revise existing theories and yet always be aware of their intrinsic limitations, rather than seeking to confirm them with deliberately selected examples.[90]

For Beauregard, a variety of "nested categories"—real estate capitalism, the built environment, infrastructure, the growth-versus-decline couplet, gentrification, distinct spatial typologies (enclave, ghetto, citadel)—enable us to explore cities and then interpret commonalities and differences. Thinking about relationships among categories, for example, enables us to recognize the difference between resurgent cities experiencing regeneration after a temporary period of stagnation and declining cities faced with continued depletion of jobs, residents, and tax revenues.[91]

Along with analytic concepts, categories are the foundation of theoretical reasoning. They enable us to group together empirical matters of fact and to distinguish and juxtapose differences. This active role of theoretical reasoning allows us to move beyond single cases by teasing out similarities and differences. Thinking in terms of categories and classification schemes is the antithesis of radical uniqueness because it allows us to identify commonalities and shared features across cases.[92]

LINGERING ATTACHMENT TO MODERNIZATION AND DEVELOPMENT

Following James Fraser, one can ask: what are the underlying socio-spatial assumptions about the urbanizing world that treat some cities as exemplars of modernity, innovation, and creativity, while others are cast as backward or behind and, worse yet, largely irrelevant places without redeemable qualities? A growing body of scholarly thinking has challenged the geopolitical assumptions involved in mapping cities onto a hierarchically arranged political-economic continuum that ranks them in ordered hierarchies, because it tends to perpetuate a bifurcated mode of understanding cities in which some are regarded as modern and up-to-date and others as lagging behind and hence in need of development. The point of departure for mainstream theorizing about globalization is the assertion that the increasing mobility of capital, commodities, and ideas has broken down spatial barriers.[93] The corrosive effects of globalization have led to the dissolution of administrative barriers and juridical boundaries. In place of an imagined world of bounded places (and strong attachment to place), we are faced with a world of border crossings and virtually unimpeded flows across space.[94] By conceiving of all cities as operating on the same trajectory, theorizing about urban space easily falls into a reductionist exercise of creating ever more exacting measures to gauge the competitive rivalry between places in terms of their aspirational success in achieving the status of world class or, to put it another way, in terms of their relative " globalness," "globality," or "centrality."[95]Framed this way, mainstream theorizing about globalization bears a striking resemblance to the classic theories of modernization—what Doreen Massey has called "the old story of modernity," with its linear conception of unfolding chronological time and its singular trajectory of progress. "Once again," she argues, theorizing in this way "convenes spatial difference into temporal sequence." With a preestablished end point, this approach "denies [by definitional fiat] the possibility of multiple trajectories" of urban transformation.[96]

The prescriptive power of the idea of urban development as both a goal and an evaluative standard has long held aloft the promise of a better future while simultaneously claiming validity for the entire world—without questioning the acceptability of its underlying assumptions and premises. The core propositions of classic theories of modernization presupposed that deviation from a predestined development trajectory results in sustained backwardness and underdevelopment.[97]

The once reigning myth of modernization and its goal of progress have faded. Just as proponents of the concept of multiple modernities (or varieties of modernity) have challenged Eurocentric idealizations of the past and have sought to capture diverse and polycentric perspectives of the future, so too a renewed openness to theoretical innovations can begin to explore the diversity of urban outcomes, multiple pathways of urbanization, and the shifting modalities of urban life at the start of the twenty-first century.[98]

Rethinking mainstream urban theory allows us to challenge essentialist understandings of cities as bounded entities that take a universal form and instead to conceive of them as hybrid ensembles of historically malleable socio-spatial relations.[99] The inherited analytic lens of urban modernity has restricted the capacity of theorizing about global urbanism to dispense with the regulative fiction of linear pathways toward a shared future defined by the goal of development.[100] This tendency to think about cities in terms of competitive and noncompetitive, winners and losers, creative and unimaginative, robust and stagnant is a difficult habit to break. These differences are typically embedded in moralizing discourses about what the "good city" should be and the celebratory language of sustainability, smartness, and eco-friendliness.[101]

Jennifer Robinson argues that theoretical insights about cities cannot be derived from the experiences of a handful of "successful" cities alone and that a new approach to understanding global urbanism should widen the terrain in order to gain insights from a broad range of different circumstances and diverse settings. For this reason, she has envisaged a critical urban theory that does not rely on predetermined categories that assign cities to places in a ranked hierarchy but instead builds on a "cosmopolitan comparativism" that places all cities equally within the same analytical field of investigation.[102] Within this heterogeneous field, the differences across and within cities must be thought of as diverse experiences rather than as exemplars of a hierarchical division.[103] In order to learn from different circumstances, Robinson has argued, it is not vibrant global cities or struggling Third World cities that should be central for scholarly analysis and policy recommendations, but what she has termed "ordinary cities" in all their complexity, diversity, and peculiarity.[104]

LIMITATIONS OF CONVENTIONAL URBAN STUDIES

Despite claims of inclusiveness and universality, conventional theories of urbanization have largely ignored struggling cities at the margins of success or else treated them as expressions of failure, with the result that these places are underemphasized and undertheorized.[105] Certainly, in the vast spectrum that characterizes the world of cities, there are large numbers of noncompetitive, "loser" cities that do not fare very well in the rank order of integration into global circuits of power. The challenge for urban studies scholarship is to bring those ordinary cities (to borrow a term from Jennifer Robinson) back into the scholarly dialogue as objects of inquiry in their own right.[106]

Only by establishing some distance from conventional understandings of urban transformation derived from the experiences of the leading cities of North America and Europe is it possible to recognize that the complex dynamics of urbanization on a global scale cannot be neatly fitted into the categories and analytic frameworks that have animated mainstream urban studies for quite some time.[107] In seeking to grasp the complexity of the urban experience at the start of the twenty-first century, it is necessary to connect ongoing processes of urbanization with the current phase of globalization. It goes without saying that no city is untouched by the forces of globalization, but the nature of this connection depends upon the specific conduits of transmission and the networks that mediate them. In thinking about these relationships, we should avoid the egregious mistake of conceiving of the global and the local in ways that parallel such binary oppositions as the universal and the particular, the macro-structural and the micro-situational, and the permanent and the contingent. Global networks are maintained, adjusted, secured, and configured in local settings. Seen in this light, making sense of urbanization on a global scale revolves less around conceiving of a one-way transmission belt that implants "a generalized *global* onto every local surface," than on remaining ever vigilant to multiple pathways and back-and-forth multiscalar interactions. On one hand, the disruptive force of globalization can result in the substantial reconfiguration ("reterritorialization") and even the wholesale disappearance or annihilation ("deterritorialization") of localities. The power of finance, commodities, and ideas can easily overwhelm whatever pristine and unprepared localness stands in the way. On the other hand, particular localities can sometimes acquire their place-specific power precisely as a consequence of their strategic alignments with global networks. Rather than thinking of the local as a given terrain existing before and outside the global, it is more fruitful to see these realms as relationally connected and jointly produced through their mutual interaction.[108]

Cities have always been structurally enmeshed in global economic and geopolitical networks, so the challenge for us is to distinguish the present moment of globalization from long-term trends in urban transformation. The current phase of global economic restructuring has brought about a wholesale transformation in the transnational division of labor that, in turn, has put into motion new urban dynamics on a world scale. This process of globalization has not only triggered fundamental shifts in the worldwide flows of capital, commodities, labor, and information but also redefined the functional roles of cities in terms of their transnational linkages and their position in the international division of labor.[109] On the one hand, the globalizing cities with world-class aspirations have sought to position themselves in the upper echelons of the world urban hierarchy of achievement. In seeking to move upward in this rank order, city boosters in these aspirant world-class cities have tried to exploit whatever competitive advantage they may have in such fields as tourism, arts and culture, and financial and producer services. On the other hand, struggling cities with little or nothing to offer in the global marketplace have faced the dire prospect of disappearing into ruin and decay, thereby "falling off the world map," at least in terms of connections to the world economy and subjective perception.[110]

The ongoing geopolitical restructuring of the world economy unleashed by globalization has gone hand in hand with long-term structural trends in worldwide patterns of urbanization. Over the past half-century, the scale of urban population growth around the world has been unprecedented. As Mike Davis has pointed out, cities have absorbed nearly two-thirds of the global population explosion since 1950, and the present urban population (3.2 billion) is larger than the total population of the world in 1960.[111] Cities in the impoverished regions of the Global South have expanded at much higher rates than cities in the relatively wealthy regions of North America and Europe, in some cases eclipsing even the most apocalyptic forecasts of (neo)-Malthusian overpopulation. Contrary to the hopeful visions of scholars and planners who imagine livable, resilient, and sustainable cities, the capacity of many of the world's megacities (and other large urban agglomerations) to accommodate their burgeoning populations has lagged far behind the pace of urbanization.[112]

In mainstream urban studies literature, urbanization has conventionally been understood as a by-product of economic development. But this analytic framework—rooted as it is in neoclassical dual-economy theorizations of supply and demand—fails to account for the phenomenon of "urbanization without growth" observed beginning in the late twentieth century in the marginal regions of the world economy.[113] Perhaps most alarming about current patterns of rapid urbanization is that the population of the world's poorest cities has continued to expand unabated, in spite of stagnant or declining real wages, rising prices, and skyrocketing urban unemployment. Urbanization without accompanying economic growth has produced a kind of perverse urbanism or surplus urbanism, in which the radical decoupling of population expansion from industrialization has severed whatever previous linkages existed between expectant new arrivals and opportunities for regular wage-paid employment.[114] The exponential growth of overcrowded and impoverished slums, shantytowns, and informal settlements that have metastasized around many of the world's largest cities provides ample visible evidence for what some researchers have called runaway urbanism or excessive urbanization.[115] Recent trends in rapid urbanization have prompted some scholars to speak of the most rapidly expanding megacities of hypergrowth as the quintessential "shock cities" of overurbanization at the start of the twenty-first century.[116]

The globalization of economic transactions via transnational capital investment and trade has fostered intense rivalries among cities seeking competitive advantage in attracting outside capital investment in the built environment. In trying to position themselves favorably in the world economy, city officials have often plowed public resources into signature tourist attractions and showcase entertainment extravaganzas, rather than into projects aimed at sustainability, affordable housing, or poverty alleviation.[117] In ordinary cities struggling with a range of dysfunctionalities, easy access to the global flow of ideas, information, and images has profoundly reshaped the identities of well-to-do urban residents, as their sense of belonging and social allegiances becomes less rooted in specific localities and more tied to global consumer culture.[118] With the increasing privatization of basic social services and the overall neglect of physical infrastructure, the privileged middle classes with global connections have increasingly cloistered themselves into the fortified enclaves of Western modernity that have materialized in cities all over the world. As the propertied, the privileged, and the powerful have retreated behind the protective shield of gated residential communities, enclosed shopping malls, and other barricaded sites of luxury, the urban poor are often left to their own devices, fending for themselves in urban environments that have deteriorated almost beyond repair.[119]

The enhanced mobility of capital and trade liberalization have created new configurations of power that have both opened up opportunities to get ahead and imposed constraints on what is possible. At the same time that the forces of globalization have fostered increased functional interdependence and integration into the world economy, struggling cities in distress have experienced a partial retreat into localism, where specialized and particularistic interests have crystallized into distinctive communities, each with its own mode of urban living. Ranging from impromptu and spontaneous to semipermanent and structured, these social groups go their own way, often bypassing official channels and fashioning their own rules of the game, in ways that are "relatively autonomous of the dynamics prevailing in the rest of the [global] urban system."[120] Paradoxically, as the physical and social spaces of the urban landscape have become more fragmented, disconnected, and disjointed, social collectivities of all kinds—whether grounded in religious affinities, shared neighborhood affiliations, ethnic identifications, "homeboy" or hometown associations, youth culture, criminal gangs, or other kinds of socio-cultural attachments—have formed, evolved, and metastasized, offering everything from solace and emotional support, mutual aid, protection, and occasional reward to leisure activities and access to opportunities.[121]

The centrifugal forces of globalization and localization have magnified tendencies toward spatial fragmentation and social polarization on a world scale. In circumstances where urban entrepreneurialism and business enterprise have become the new dogma of development discourse, cities around the world have experienced a ruthless division of urban space in which cocooned islands of extreme wealth and social power are interspersed with places of deprivation, marginalization, and decline.[122] A great deal of the scholarly literature on urbanization begins—implicitly or explicitly—with normative prescriptions of how cities should ideally function when formal management mechanisms are in place. Framed in this way, sprawling cities of hypergrowth typically appear as exemplary expressions of failed, distorted, or stalled urbanism, lacking the basic requirements and attributes of genuine urbanity that mark the "normal," expected urbanization of healthy cities elsewhere. A more fruitful approach starts with the premise that struggling cities in distress are, as AbdouMaliq Simone has put it, "works in progress," at one and the same time driven forward by the inventiveness of ordinary people themselves and held in place by inertia and slowness in adapting to changing circumstances.[123] The challenge for urban theory is to critically assess how struggling cities in distress actually work under straitened circumstances. This stress on provisionality and improvisational "getting by" enables us to conceive of struggling cities under stress as unfinished and contingent sites where the possibilities of urban becoming are not already predetermined. In cities everywhere, there is the semblance of vitality and resourcefulness, with urban residents typically relying on their own ingenuity and inventiveness to stitch together, reconfigure, and make sense of their daily lives.[124] Yet there is also the appearance of entropy and stasis, in which capacities for innovative change are stalled by ruin, choked with waste, and clogged with useless objects out of place, where "enormous creative energies have been ignored, squandered, and left unused."[125]

It is undoubtedly true that struggling cities around the world are, for the most part, distressed places in need of good governance, effective and efficient management, refurbished infrastructure, greater popular participation in local decision making, sustainable livelihoods, and expanded opportunities for socioeconomic advancement. Yet they are more than simply manifestations of failed urbanism. We need to reach outside this diagnostic mindset with its normative injunctions in order to discover and appreciate the historical specificity of ordinary cities around the world—how they make themselves at the same time as they are made.[126]

CHAPTER 2

THE UNIVERSALIZING PRETENSIONS OF MAINSTREAM URBAN STUDIES

Generic Cities and the Convergence Thesis

Efforts to understand the dynamics of global urbanism at the start of the twenty-first century have for far too long been weakened by the tendency in conventional urban studies to seek all-encompassing, one-size-fits-all explanations for urban transformation on a global scale.[1] This "convergence thesis" argues that the powerful homogenizing pressures of globalization at the time of late modernity have produced similar spatial and socioeconomic outcomes in globalizing cities with world-class aspiration around the world.[2] Because of the seemingly pervasive and irresistible power of globalization, distance and geography no longer matter in ways they once may have, resulting in a kind of generic "sameness" and homogeneity in urban spatial form and urban processes.[3] One indication of this trend toward seeing through a universalizing lens is the burgeoning literature on gentrification, gated enclaves, and suburban sprawl that begins with the axiomatic presumption that global expressions of these spatial patterns are variations on a common theme.[4]

Contemporary mainstream urban studies literature is often overwhelmed by dismal stories of the power of globalization and neoliberalism to dissolve differences between cities. Much of this urban scholarship suffers from a kind of theoretical determinism that views cities and urban spaces as helpless, receptive pawns that offer little resistance or opposition when subjected to global forces in the age of neoliberalism. Countless case studies have detailed the serial reproduction of look-alike building typologies, neoliberal regulatory regimes, gentrifying tendencies, and similar design stylistics that all seem to originate in the leading cities of the Global North. These case studies often warn of a looming urban dystopia in which cities have become virtually indistinguishable.[5]

In stressing the homogenizing impulses of globalization and the normalizing logic of neoliberalism, the urban convergence thesis postulates that globalizing cities are becoming more alike in certain key respects, regardless of their geographic location and their historical specificity. The key elements of this convergence include such characteristic features as suburbanization and sprawl, gated residential communities, entrepreneurial governance and growth-machine politics, gentrification, social inequality and segregation, and spatial polarization. The steady accretion of these shared features has eroded local specificity and erased differences between cities.[6]

Prevailing mainstream perspectives that focus on the impact of globalization on urban form suggest that globalizing cities with world-class aspirations are experiencing an inexorable process of Westernization or Americanization.[7] In writing about cities in Southeast Asia, Howard Dick and Peter Rim, for example, have observed what they regard as a distinct Americanization of city life, visible in such features as gated communities, cocooned shopping malls and entertainment complexes, high-rise office buildings, freeways, and edge cities. For Dick and Rimmer, this process of convergence means that cities in Southeast Asia can no longer be seen as a discrete category of the Third World city. "Any attempt to explain either the historical or contemporary urbanization of southeast Asia as a unique phenomenon is therefore doomed to absurdity," they argue.[8] The logic of globalization has made the paradigm of the Third World city obsolete and hence irrelevant and meaningless.[9] As a consequence, "there should now be a single urban discourse" indifferent to regional difference.[10]

A great deal of scholarly writing suggests that globalizing cities are increasingly converging around generic (that is, similarly designed and formatted) redevelopment schemes. Constructing huge megaprojects and inserting them into existing metropolitan fabrics has become a standardized city-building strategy for real estate developers seeking to emulate what a world-class city should look like. In his distinctly provocative way, the architect Rem Koolhaas has taken this line of reasoning further than anyone else. "Is the contemporary city like an airport—'all the same'?" he asks.[11] Like airports, which are modern in exactly the same way, the generic city is a city without an identity—no past, no future, no distinction, and no character.[12] Eventually, Paris will turn into Las Vegas and London will become Atlanta.[13] The convergence thesis projects the future of urbanism as endless repetition of equally fungible features that can be found in cities almost everywhere. Koolhaas stresses patterns of urban sprawl as an essential characteristic of the urban future in which density is artificially created in the form of urban simulacra: shopping malls, theme parks, and museum environments.[14]

The concept of convergence is a kind of euphemism. It is not only abstract, but equivocal and Delphic. Urban scholars have used it to signify the unassailable inevitability of the homogenizing impulses of the contemporary age of globalization. The rhetorical deployment of convergence embodies a kind of absolutism that suggests no alternative.[15] Evolutionary grand narratives insist that cities everywhere should eventually "modernize" and converge around a common set of characteristics.[16] One strong current in the scholarly literature on convergence argues that large-scale urban redevelopment projects are part and parcel of a top-down global strategy of metropolitan gentrification. Neil Smith, for example, has argued that the process of gentrification, "which initially emerged as a sporadic, quaint, and local anomaly in the housing markets of some *command–center* cities," has blossomed into a far-reaching urban revitalization strategy on a global scale, "densely connected into the circuits of global capital and cultural circulation."[17] In a similar vein, Rowland Atkinson and Gary Bridge have argued that the gentrifying impulse has become such a powerful global force that it constitutes "the new urban colonialism."[18]

Scholars who perceive a convergence of urban form around a Western or American model warn against such deleterious consequences as environmental degradation, spatial polarization, growing social inequalities, and class segregation.[19] As a general rule, scholarly accounts of globalization and urban reconfiguration start with a picture of an urban fabric whose patterns of everyday life and institutional rules are disturbed by the sudden imposition of global networks. These networks penetrate urban life and restructure economic relations, introducing new types of employment and levels of income commensurate with the command-and-control sites of global cities in the core areas of the world economy, resulting in new levels of social differentiation between those who become part of the networks and those who are left out. The winners in this grand narrative flock to secluded, gated residential communities and gentrified neighborhoods, and they spend their leisure time and discretionary incomes in enclave spaces that replicate similar ones in other globalizing cities. This insertion into the global space of flows produces new levels of inequality and polarization in employment, incomes, and use of the built environment. These structural tendencies toward polarization threaten to evolve into a potentially explosive situation of marginalization and social exclusion.[20]

THE CREATIVITY FIX: POLICY CONVERGENCE AROUND "CREATIVE CITIES"

Prevailing mainstream urban studies and policy-making perspectives have typically taken the view that in the competitive race to acquire world-class status, cities need to emulate what seems to have produced internationally acclaimed success elsewhere.[21] As a general rule, the successful urbanism of such world-class cities as New York, London, and Amsterdam is trumpeted as a normative ideal, something that struggling cities should strive to model themselves after.[22] For those that operate with a hierarchical view of global urbanism, what distinguishes top-tier world-class cities from those that seek to advance in the rank order is their capacity to compete successfully for new investment and well-paying employment opportunities.[23]

In the orthodox mainstream view, successful postindustrial cities at the dawn of the new age of cognitive capitalism are those that converge around the development of knowledge-intensive economic activities, creative industries, and innovation clusters.[24] One particularly fashionable urban redevelopment script suggests that a successful remedy for remaining or becoming competitive is to attract the kinds of knowledge-based creative industries and innovative grassroots entrepreneurialism that promote "smart" growth.[25] In its simplest formulation, the main idea behind these policy recommendations is that the advent of a new phase of knowledge-based cognitive capitalism has brought about a fundamental shift in the requirements for competitive success on the global stage. The emergence of what might be termed the cognitive-cultural economy, especially those arenas "that are explicitly geared to cultural production," is dependent upon radically new forms of technological innovation and "productive organization with a strong focus on flexible labor processes."[26] This new conventional wisdom suggests that creating the proper conditions for the capture of innovation and creativity is the key to successful urban redevelopment.[27]

Over the past several decades, the conjoined discourses of creativity and the creative city have figured prominently in mainstream urban policy debates, where the strategic agenda has stressed cobbling together new growth paths for postindustrial cities.[28] As a rhetorical call to action and an alleged wholesale paradigm shift, the creative cities mantra suggests that the reversal of urban fortunes is a matter of finding the right recipe (or toolkit) for success to put deindustrializing cities back on track.[29] As Jamie Peck has argued, the most conspicuously successful innovation in the recent history of urban policy making is the widespread adoption of mainstream policy recommendations designed to stimulate the creative growth of struggling urban economies transitioning from an ossified industrial past to a vibrant postindustrial future—usually by way of market-friendly interventions in the socio-cultural sphere, as a way of attracting and retaining innovative clusters and talented individuals.[30] This creative cities model has replaced the once dominant approach that called for the indiscriminate pursuit of economic growth at virtually any cost. For at least the past decade, both scholarly urban studies literature and public policy perspectives have stressed the benefits of the creative cities agenda, in which sponsorship of the arts, culture, and knowledge industries plays an increasingly important role in urban competitiveness. This urban redevelopment agenda has often adopted the well-worn terminology of the cognitive-knowledge economy, such as dynamic clusters, proximity and agglomeration, and adaptive innovation, as a way of promoting place-based strategies (such as innovation districts) in order to trigger socioeconomic revitalization in postindustrial cities.[31]

What indicates the success of these policy recommendations is that their reach seems to have become nearly ubiquitous, even if the geographic spread of these ideas amounts to not much more than vague mimicry or pale imitation.[32] Catering to the mercurial tastes and preferences of the "creative class" has become, almost simultaneously, a favored redevelopment strategy for aspiring world-class cities, an urgent imperative for competitiveness, and a well-worn cliché of contemporary urban policy making.[33] Like sustainability and resilience, the creative cities label has become one of those key traveling ideas that offer a normative policy framework for aspiring world-class cities.[34] Global in scale and scope, the burgeoning market for creativity policies extends from the very top to the very bottom of the ranked hierarchy of cities, animated by a pervasive sense of competitive urgency to "get ahead" in order to avoid falling behind.[35]

In an era of knowledge-driven growth, the influential creative cities thesis rests on the presumption that the good fortunes of aspiring world-class cities are increasingly tethered to the preferences, tastes, and lifestyle choices of the creative class—a hypermobile group of talent-rich and entrepreneurially inclined individuals who can become the wellspring of economic prosperity for distressed cities. Primarily associated with the proselytizing efforts of Richard Florida, the creative-class thesis holds that a uniquely mobile class of talented individuals has become the principal carrier (and catalyst) for innovative capacity and that the task for aspiring world-class cities is to cater to this elite group with various incentives and inducements, including the sponsorship of a vibrant urban cultural life, rich in luxury amenities, arts and entertainment venues, attractive downtown housing opportunities, and trendy lifestyle options.[36] In the normative projection of successful urban restructuring, the creative class takes its rightful place as the model subject of progressive urbanism and the "natural force for rapid development."[37]

The creative cities approach represents a new stage in the evolution of the entrepreneurial city.[38] What distinguishes this iteration from earlier manifestations is its heavy reliance on Richard Florida's thesis about the positive relationship between the creative class and economic growth. Since the 2002 publication of his wildly popular but deeply contentious book, *The Rise of the Creative Class*, Florida's ideas have been broadly assimilated into the soft (immaterial) infrastructure of urban entrepreneurialism, spreading initially across the United States and later to aspirant world-class cities around the world.[39] The creativity fix was especially especially popular in slow-growth postindustrial cities, where local growth coalitions and city officials hoped that a creative-city revitalization strategy might be powerful enough to reverse decades of relative decline and socioeconomic stagnation.[40] Mesmerized by the creativity fix, growth coalitions, loosely consisting of real estate developers, commercial property owners, professional image makers, city planners, and municipal authorities, often come together to orchestrate a plethora of urban promotional activities and planning strategies designed to cater to "a distinct set of urban motifs presumably commensurate with creative class lifestyles, cultural practices, and consumption habits."[41]

With the spread of the creative cities strategy, the heightened propensity for aspiring world-class cities to borrow (and even plagiarize) what appear to be cutting-edge policy ideas and organizational technologies from one another reflects the growing global significance of reflexive networks of fast-track policy circuits.[42] The conduits for rolling out solutions to reaching maximum competitiveness include a plethora of international policy conferences and fact-finding study trips, linked with marketing agencies, promotional companies, and consulting firms that engage in the highly lucrative business of selling policy products to aspiring world-class cities.[43] Packaged as highly stylized and readily consumable models, creative cities strategies are tailored to move quickly in the rarefied atmosphere of fast policy transfer. "Framed as enabling technologies," they are reduced to easily digestible components to "facilitate their portability from place to place and their adaptability to a range of local conditions." The widespread popularity of the creative cities approach marks the ascendency of conditions fostering "*endemic policy mobility*, the characteristics of which include growing deference to 'international best practice' models, near continuous learning from like-minded (and ideologically aligned) others, and the blurring of jurisdictional boundaries."[44] These strategic policy initiatives "share remarkably similar analyses, conclusions, and policy ambitions."[45]

Prevailing mainstream urban development models have focused attention on the aggressive pursuit of investment opportunities and property-led economic growth, city marketing and "hard branding," supply-side inducements (such as tax abatements and investment credits), neighborhood gentrification, market-driven culture and entertainment districts, and luxury retail revitalization.[46] The established revitalization strategy of deliberately manipulating business climates in order to attract mobile investment and nurture talented individuals has effectively been overlaid with, and retrofitted around, this new vision of innovation and creative growth. Mainstream urban policy makers present this "new iteration of the urban development game" as a kind of zero-sum exercise in which there are winners and losers.[47] According to the prevailing normative script that has accompanied the new creative cities approach, "a refusal to play practically invites competitive failure."[48]

Without a doubt, the creative cities discourse has become a durable urban policy position for consultants who market ideas worldwide.[49] Despite resting on a shaky evidence base and circular causal reasoning, the creative class / creative cities thesis has proved to be remarkably seductive, not only to scholars in urban, cultural, and innovation studies but also to desperate city officials looking for a quick fix and opportunistic policy advocates who travel around the world. For all intents and purposes, the "creativity fix" has become the new universal formula for urban competitiveness, just as its accompanying policy routines—city marketing, place promotion, and hipster (New Age) downtown gentrification—have become decidedly formulaic and somewhat blandly generic. Many aspiring world-class cities appear to pursue the same policy formula comprising one or more of the following elements: nascent creative and knowledge industries, innovation clusters, high-paying occupations filled by the creative class, signature architecture and flagship megaprojects, and luxury sites for the expression of leisure-oriented consumer identities. Creative industry quarters, art galleries and museums recycled from old industrial buildings, monumental architecture, iconic public art installations, convivial post-public spaces for social gathering, luxury downtown residential apartments, and multistoried shopping gallerias have become the stock-in-trade features that mark the success of revitalized competitive cities. So many struggling cities in search of competitive advantage are trying to be distinctive and to stand out from the crowded field, but too many are seeking to do so by simply copying each other.[50]

Like innovation, creativity is a nebulous, malleable term that allows for multiple interpretations. Despite its weak explanatory power, the creativity thesis continues to have an enduring afterlife in the whirlwind world of urban policy. While exaggerated pronouncements about the creative city and its favored inhabitants have certainly benefited from clever promotional campaigns, their evident allure and alleged salience have little to do with their intrinsic explanatory power of the prescriptive model of creative growth—or even indeed with the inventiveness of the associated marketing efforts. Rather, the creative cities mantra has traveled so far and so fast because its advocates have artfully grafted it onto the current political-economic terrain of entrepreneurial urbanism and actually existing neoliberalism.[51] The creativity script is a place-making strategy that encodes an engaging image of successful transformation based on a set of principles that combine the contemporary fixation on culture-led development with neoliberal economic imperatives. The progressive rhetoric—notably, its explicit embrace of social diversity and tolerance—is dwarfed by the overriding imperatives of the entrepreneurial city.[52]

The discourses and practices of creative cities policy-making recommendations dovetail neatly with the prevailing order of neoliberal urbanism, where the dominance of real estate capitalism and market-led urban redevelopment, the polarization of labor and housing markets, the wholesale retrenchment of public services and social programs, and accelerating inter-city competition for jobs, investment, and assets have gone virtually unchallenged.[53] The creative cities approach represents a "soft" policy fix for municipal officials who seek to bolster the competitive advantage of their struggling cities. While it depends on at least some modest discretionary public spending on cultural assets, the creative cities concept reproduces the dominance of supply-side market competition as the primary driving force behind urban policy. This prescriptive remedy for struggling cities has simply repackaged and strengthened the kind of orthodox growth-oriented, market-reinforcing, and gentrification-friendly policies, cleverly reclaiming what amounts to a conventional property-led redevelopment paradigm under the fashionable creativity rubric.[54]

The creative cities approach as a policy formula for urban planning practice does not represent simply a technical exercise but an ideological intervention into urban life.[55] At the end of the day, the discourse of the creative city amounts to little more than a rhetorical device that can "placate the hearts and minds" of local city officials, encouraging them to believe "that they are actually doing something whilst doing hardly anything at all."[56] The creative cities approach has served as an "intellectual technology" aiming at the invention of new macro-scale social agents (that is, the creative class) who are expected to perform the role of heroic saviors in fostering urban regeneration.[57] The creativity fix often appears as a standard one-size-fits-all recipe that can be applied anywhere.[58] One weakness is that it reinforces the tendency toward reductionist and simplistic understandings of the complex processes of urban regeneration, thereby directing policy makers and planners "to opportunistic rather than strategic thinking." Framing urban revitalization through the narrow lens of a single creativity formula can easily overlook or ignore deeper structural problems facing struggling cities related to their place in the uneven flow of capital around the globe.[59]

Following the injunction derived from Pierre Bourdieu, the principal task for an in-depth investigation of a social object like the creative cities approach is not to examine the strategic policy initiatives as such but to examine "the social space from which [such an initiative] derives its distinctive, differential and relational properties."[60] In practice, the creativity fix is part of a broader and deeper shift toward new kinds of entrepreneurial urban management used to boost the tarnished image of ailing cities and persuade highly mobile global capital and professional and service classes that putting down roots in the city is a reasonable strategy for them.[61] Critics of the creative cities approach point to what Mary Donegan and Nichola Lowe have called the dark side—that is, the inability of self-styled creative cities to address disturbing questions of income inequality and the possibility that importing large numbers of talented individuals will exacerbate disparities between different people and neighborhoods, leading to creative gentrification and displacement.[62] The creativity fix is a two-edged sword that typically combines regeneration with dislocation. For the most part, skeptics argue that the creative city approach overlooks "the stark inequalities which characterize life for countless urban dwellers."[63]

DEFICIENCIES OF THE CONVERGENCE THESIS

The convergence thesis suffers from some inherent flaws that weaken its capacity to account for the full spectrum of urban transformations at the start of the twenty-first century. One major problem with prevailing perspectives on convergence is that they put undue stress on observed similarities in urban trends and thereby gloss over important sources of divergence rooted in local cultural, geographic, and institutional dynamics. Research and writing on convergence are rooted in a fundamentally misleading conceptualization of urbanization as a universal phenomenon with local specificities. Framed this way, different urban constellations of disparate features "are forced onto the procrustean bed of a universal model" of convergence, thereby erasing, as if by methodological fiat, fundamental differences that diverge from the alleged common characteristics. The convergence perspective begins with a more or less essentialist understanding of cities in which they are regarded as manifestations of an a priori universal form rather than treated as variegated ensembles of historically changing socio-spatial processes and relations.[64]

Despite the entrenched popularity of the global cities / world cities framework as an analytic point of departure for examining emergent patterns of global urbanism, a rising chorus of dissident voices suggests that the idea of convergence overstates the power of social actors and institutional arrangements operating at a global level and underestimates local agency and contingency.[65] The excessive focus on convergence is problematic for two basic reasons. First, in starting with the assumption that Western patterns of urban transformation represent a universal template and then seeking to determine whether urban forms in other cities of the world "fit" (or do not fit) this mold, theories of convergence "inherently privilege similarity over difference," thereby either downplaying variation or treating it as an anomaly or deviation.[66]

A second problem with the convergence thesis is that it overstates the power of structural forces operating at a global level and downplays the capacities of local actors under place-specific circumstances to shape outcomes to their desired ends.[67] In assuming a one-way transition of ideas from North to South, convergence frameworks fail to identify the dynamics of localism in shaping urban development. This approach largely disregards the possibility of real estate promoters, property developers, policy makers, planners, and other powerful social actors shaping city-building projects to suit their own interests.[68] Hank Savitch and Paul Kantor, for example, argue that city builders strategically shift their behavior to take advantage of opportunities in the global marketplace. Cities are constrained by the dynamics of the global economy, but they are not its prisoners or passive victims. Local growth coalitions have more maneuverability in some cities than in others in the endless game of development. Local governance and innovative planning initiatives can combine with economic fortune and urban policies to provide resources that expand or contract the scope for choice.[69]

A great deal of scholarly research has revealed the diversity of urban experiences in the face of globalization. Geographic circumstances and historical context certainly influence the reception of (and reaction to) the pressures of reaction to globalization, moderating its deleterious homogenizing effects.[70] Some scholars have emphasized the specific historical trajectories of cities that produce distinct urban geographies.[71] George Lin, for example, makes an important distinction between convergence in urban processes and divergence in urban forms.[72] Similarly, Lawrence Ma and Fulong Wu contend that the convergence thesis does not allow for the possibility that similar surface features of urban spatial form may be created by different processes in different places, arguing that the pressures of globalization can be mediated by local forces and processes embedded in local culture, history, or economic and political systems.[73]

The uncritical adoption of a Westernization or convergence analytic framework can conceal as much as it reveals. The excessive focus on convergence reflects a central contradiction in mainstream urban theories in the contemporary era of globalization. The widespread and undiscerning use of concepts and categories derived from the experience of globalizing cities with world-class aspirations—what Gavin Shatkin has called a "tyranny of terminology"—has taken hold in the conventional scholarly literature on cities and globalization. The labeling of diverse phenomena in cities around the world with terms originating and theorized with reference to a handful of leading Western cities—categories like edge cities, suburbanism and peri-urban sprawl, Disneyfication, gated residential communities, and gentrification—have tended to obfuscate our understanding of the variety and variability of global patterns of urbanization, thereby impairing our capacity to understand urban change and its causes and to derive useful implications for urban planning and local policy. Examining city-building efforts through the lenses of local actors and competing interests reveals crucial local differences in urban development processes that are shielded from view if we restrict our analytic frame of reference to Westernization models or convergence approaches.[74] Such ideas as the generic city were put forward as a way of grasping the future of global urbanism,[75] but their overextended use reifies the unfounded premise that globalization equals homogenization. The notion of the generic city celebrates the global proliferation of a homogenized Eurocentric urban modernity, but it does not take us very far in framing a comparative approach to global urbanism.[76]

Framing globalizing cities through the abstract analytical device of globalization rather than through an appreciation of the myriad, messy, and contradictory conditions of everyday life has a tendency to reinforce the one-sided view that power rests with the global in the ongoing struggle to overcome the obstacles embedded in the local. From this perspective, it is easy to see how scholars have seen global forces as the triggering mechanism for such processes as gentrification and global competition as the driving force behind real estate speculation.[77] Approaches that stress the corrosive power of global forces correctly recognize how traveling ideas like international best practices are indeed consequential because of their widespread appeal as tools for city builders. Yet these approaches risk reinforcing a rigid dichotomy in which an abstract, remote, and seemingly all-powerful global is positioned against a concrete and authentic local that is incapable of mounting any sustained challenge to those outside forces. There is no doubt that global forces shape city-building processes, yet historically specific conditions on the ground also influence eventual outcomes.[78]

A number of scholars have called for a more nuanced understanding of the tension between the global and the local, and for a view of urban transformation as not simply imposed from above and outside, but rather an inherently negotiated process.[79] Richard Child Hill, for example, has argued that structures of global economic and political power constitute "nested hierarchies" in which "parts and wholes are not subordinated to one another." In this view, cities "both facilitate the globalization process" and at the same time "follow their own relatively autonomous trajectories."[80] Similarly, scholars such as Abidin Kusno, Nezr AlSayyad, and others have examined what they call an emergent "third space"—a hybrid, fluid space of messy mixtures between the local and the global, as people in specific localities reshape cities according to local social, cultural, and political exigencies.[81] The common thread that connects these conceptual frameworks is a deliberate effort to restore social agency to urban analysis and refute perspectives that depict local residents as "impotent, passive and guileless"—powerless spectators who observe physical, spatial, and socio-political changes that they neither control nor understand."[82]

The convergence thesis puts undue stress on diffusion, replication, and serial repetition at the expense of appropriation, translation, and hybridization.[83] Put another way, this perspective overlooks the inherent tension in globalization between mimicry and adaptation. While new city-building efforts often borrow, copy, and even plagiarize ideas and practices from global cities, the interaction between global and local actors in different institutional settings invariably produces divergent outcomes.[84] The encounter with globalization often results in hybrid mixtures that bear little resemblance to their original source of inspiration. Looking at the trajectories of global urbanism through this alternative lens of place-specific forces reveals differences in urban development processes that are shielded from view if we employ Westernization or convergence as our only analytic framework.[85]

SOUTHERN THEORY: A WAY FORWARD, OR A CUL-DE-SAC?

For at least the past decade, the scholarly field of urban studies has undergone a protracted phase of rich experimentation and open-ended inquiry, generating a proliferation of paradigmatic frameworks, new conceptual configurations, and different methodological approaches.[86] Inspired by efforts to make sense of the shifting geographies of global urbanization, this rethinking in urban studies has sought to develop place-sensitive understandings of the contradictory dynamics of urban transformation that are more aligned with the diversity of cities and regions. What has animated this rethinking is the growing recognition that the conceptual tools inherited from mainstream approaches to urban studies are ill equipped to unpack and make sense of the changing contours of global urbanism.[87]

Over the past two decades, scholars rooted in a wide range of different intellectual traditions have sought to disrupt and overturn the prevailing orthodoxies that have revolved around the privileged position of the global cities / world cities paradigms and their fixation with uncritically emulating their alleged success at climbing the ranked hierarchies of urban relevance. In drawing attention to ordinary cities, Ash Amin and Stephen Graham, for example, cast doubt on the totalizing claims inherent in mainstream theories and called instead for a nonessentialist understanding of the "multiplex city," where "diverse ranges of relational webs coalesce, interconnect and fragment."[88] The aim of these dissenting voices has been to establish analytic distance from those mainstream theoretical and methodological traditions that have sought to identify, classify, and hence systematize (inter)urban differences, and to connect these to universalizing explanatory hierarchies within a unified field of urban studies.[89] By advocating a universalizing platform for theorizing, mainstream urban thinking has systematically neglected the diversity and shifting geographies of global urbanization that cannot be contained within conventional analytic frameworks.[90] Critics of conventional urban theorizing have advocated the deconstruction (and abandonment) of universalistic or global models. More specifically, they have called for "for a reinvigoration of comparative urban studies; for the provincialization of urban theory; for the recognition of divergent circumstances, localized complexity and unpatterned diversity; for explanatory circumspection and humility; and, in some cases, for a [wholesale] retreat from generalized urban theories themselves." These efforts at deconstructing dominant paradigms in mainstream urban studies have successfully exposed the blind spots of conventional approaches to urban studies but have not yet developed a coherent alternative perspective.[91]

One strong current, linked to postcolonial theory and subaltern urbanism, has suggested that conventional Western analytic frameworks originating in the Global North have little of value to contribute to understanding cities of the Global South.[92] The growing recognition of the limited explanatory capacity of mainstream urban theories—with their core ideas originating in the Global North, their standardized urban models of urban form and structure, and their totalizing visions of a "convergent, modernizing urban future"—has given rise to calls for the development of a distinctly "Southern theory," "theory from the South," or "urban theory beyond the West."[93] This vantage point of seeing from the South reflects a grave discomfort and skepticism with the continuing dominance of mainstream urban studies literature—loosely (and somewhat imprecisely) referred to as Northern or Western urban theory—to shape the terms of debate and to supply the main categories of analysis through which to understand the trajectories of global urbanism at the start of the twenty-first century.[94] With its roots in a postcolonial perspective, this Southern critique cautions against the uncritical application of Northern theory, with its universalizing pretensions and its functionalist bias, to cities everywhere. In particular, those who advocate this Southern turn stress the limited capacity of mainstream urban studies to appreciate the complexity and diversity of cities beyond the West outside of a conceptual framework that conceives of them as hopelessly damaged places, lagging behind, and needing to catch up through imitation and replication.[95] One contribution of this emergent Southern theory is to rescue modernity—or, more precisely, what amount to "multiple, alternative modernities"—from the linear reductionisms of the modernization theory that dominated scholarly thinking in the 1950s and 1960s.[96]

The embryonic Southern turn in critical urban studies has framed megacities of the Global South as incubators of future urbanism and the leading edge of processes of globalizing modernity.[97] Operating on the premise that Eurocentric perspectives on the city are outdated and anachronistic, this emergent Southern theory suggests that it is time to turn to a wider range of urban experiences from which to draw analytic insights with the aim of generating a new theoretical understanding of global urbanism.[98] Correlatively, the emergence of a theory from the South reflects what some scholars have called a postcolonial turn in many social science and humanities disciplines and corresponds with a move away from the analysis of cities rooted in approaches derived solely from political economy.[99]

In exploring the dynamics of global urbanism at the start of the twenty-first century, what is to be gained from taking seriously the new ideas and new experiences of urban living that characterize cities of the Global South? One key proposition in the current formulation is that cities of the South represent new spaces of experimentation that prefigure the emerging future of cities of the West (or the North).[100] In the urban studies literature, the calls to move beyond conceiving of global urbanism in terms or ranked hierarchies along a continuum of success and failure have produced a new sensitivity to a flattened playing field consisting of a world of cities. In planning theory, one perplexing question arises: what alternative urban policies and practices might follow from the adoption of a Southern perspective?[101]

It is undeniable that urban theory-making has become a more diversified and heterodox undertaking, originating "from a greater variety of epistemological perspectives, geographical and disciplinary contexts and underpinned by a wider range of methodological approaches."[102] Nevertheless, the all-too-frequent tendency toward overly generalized critiques, misrepresentation of key arguments and claims, and the silencing of alternative perspectives has at times polarized debate in unhelpful ways. The inattention to nuance can allow legitimate questions and skepticism to mutate into overstretched truisms in ways that can stifle genuine dialogue.[103] Repeated time and again, well-meaning skepticism can turn into oversimplification, "polemical shorthand and rhetorical stretching," hyperbolic caricature, and inadvertent distortion.[104]

The global/world cities and Southern turn perspectives might appear at first glance to be mutually incompatible approaches to understanding the shifting contours of global urbanism. Yet each has something useful to offer.[105] On the one side, the strength and appeal of the global/world cities perspective (framed in terms of the pressures of globalization) are that it provides a coherent and theoretically grounded account of the dramatic process of spatial polarization at work in the world of cities today.[106] This literature has drawn attention to the ways the powerful pressures of globalization at a time of late modernity have reinforced existing divisions but also created new kinds of spatial polarization between elite enclaves catering for the affluent and abandoned landscapes of deprivation and impoverishment.[107] At a time when the modernist vision of holistic city building has collapsed, the steady accretion of new satellite cities and luxurious urban enclaves in cities around the world has led to a kind of splintering urbanism in which powerful urban elites have commandeered premium infrastructures for themselves.[108]

On the other hand, the vantage point of seeing from the South enables us to destabilize existing mainstream urban theories by questioning the rigid underlying premises guiding scholarly research and writing. One primary aim behind these calls for a theory from the South is to dislodge deeply entrenched claims of universal applicability embedded in mainstream theories of modernization and conventional development studies and, in so doing, to challenge underlying spatial assumptions about the urban world that render some cities exemplars of modernity and innovation while casting others aside as lagging behind and, worse yet, insignificant and forgotten places.[109] In response to postcolonial sensibilities, urban scholars have expressed a new eagerness to abandon Western-centric assumptions about cities and to treat the heterogeneity of urban practices, identities, and processes as objects of inquiry in their own right.[110] Taking this idea a step further, scholars operating from the perspective of Southern theory have argued that mainstream urban theories derived from the experience of Western cities are not sufficiently powerful to account for the dynamics of what Susan Parnell and Sophie Oldfield have called "southern urbanism."[111]

The body of scholarly work loosely called Southern theory has persuasively argued that scholars should avoid evaluating cities around the world by a single measuring stick or set of criteria, a manoeuver that inevitably cast most cities of the Global South into the residual category of failed (or not-so-real) urbanism.

Instead, one must be more attentive to the ways in which differences matter and to the ways in which exploring cities of the Global South can reshape our understanding of the great diversity of urban experiences in less hierarchical ways.[112] The postcolonial challenge to mainstream urban theories offers a persuasive critique of one-size-fits–all models of urban transformation and the kinds of imitative urbanism that suggests a one-sided transfer (or diffusion) of ideas from the Global North to the Global South.[113] The celebration of difference and diversity, rooted in the postmodern critique of modernist thinking, offers a way out of what Lila Leontidou has characterized as "absurd caricatures of cities not conforming to 'expected' patterns."[114]

With its roots in postcolonial theory and subaltern urbanism, this seeing-from-the-South perspective has focused attention on differences and dissimilarities of urban experiences outside the West. Among other things, this alternative angle of vision represents an effort to construct an analytic orientation that can "speak back to putative 'centres' of geography in transformative ways."[115] The influence of Southern theory has led many scholars to argue that cities of the Global South reflect a distinct political, social, and aesthetic logic, grounded in informality and improvisation as the dominant modes of urban life, and that mainstream urban studies and planning theory literatures do not adequately account for the everyday realities of cities: the intense spatial juxtapositions of land use and people; the volatile mix of socioeconomic and cultural discourses that animate politics; and the patterns of political contestation rooted in historical, social, and cultural particularities.[116] In contrast to mainstream theories of global convergence, the emergent scholarly literature on subaltern urbanism—a perspective that has been a major influence in efforts to rethink urban studies—has focused on the "practices and technologies that bring an increasing heterogeneity of calculations, livelihoods, and organizational logics into a relationship with each other."[117]

DEFICIENCIES OF SOUTHERN THEORY AS AN ALTERNATIVE PARADIGM

Over the past decade, cities of the Global South have increasingly been understood as exemplary expressions of emergent forms, trajectories, and processes of twenty-first-century urbanism.[118] This shift in focus has been a welcome rejoinder to the "the asymmetric predominance [and enduring presence] of North American and European cities as paradigmatic models or prototypes for conventional urban research and debate."[119] This Southern turn has opened mainstream urban theorizing and debate "to more sustained engagement with experiences and issues from a wider world of cities beyond those of the Global North."[120] Yet it is important to question what kinds of theoretical frameworks and geographical imaginations have been mapped onto cities of the Global South in these efforts to challenge the dominance of Western theory.[121] Unless the actual historical specificities and grounded realities are used to disrupt and reframe existing theoretical perspectives and Northern-centric models, there is a risk of remaining trapped at the abstract level of critique without providing an alternative theoretical approach.[122]

Despite Ananya Roy's insistence that "the centre of theory-making must move to the Global South," a great deal of care must be taken to avoid positioning cities of the Global South as the new paradigmatic exemplars for understanding twenty-first-century urbanism. Substituting Southern cities for Northern ones risks replicating the existing overemphasis on superlatives and universalizing categories that characterize mainstream urban theorizing with its roots in the experiences of North American and European cities. The value of the Southern turn rests with generating new theoretical dialogue and opening up new channels of urban research and policy formation within a wider world of cities.[123]

Though provocative, calls to decolonize and de-Westernize urban theory have not moved very far beyond a broad-based critique of existing analytic frameworks and dominant ideas. Calls to "provincialize global urbanism" are thick with normative projections but thin with alternative theoretical formulations.[124] Surely, the binary registers of North versus South, or West versus the Rest, are useful metaphorical tools. While they work effectively as a defamiliarizing tactic, their deployment as analytic constructs runs the risk of a sloppy reverse essentialism in which Europe and its traditions are treated as a monolithic entity and the Global South functions as an undifferentiated Other.[125] These critical observations do not invalidate the distinctions, but they do caution against an essentialist reading of hard-and-fast binary oppositions. These opposing poles need to be understood as an ongoing construction that is critically relevant to understanding the trajectories of global urbanism at the start of the twenty-first century, rather than a distinction that marks a prori differences.[126]

One inherent problem with such metaphorical constructions is that these terms almost always lack precision and clarity. To replace one privileged vantage point with another seems merely to reproduce the same problems with theorizing about cities under a different guise. Rethinking urban theory is "never simply a matter of empirical expansion or spatial extension ('adding in' cities of the South, or Southern urbanism as counterpoint)" to the dominant narrative of conventional urban theorizing.[127] In the emergent scholarly literature on Southern theory, there are at least three different meanings attached to the idea of the Global South.[128] The first is hemispherical—that is, a geographic designation of a particular location. Yet to conceive of the South as a strictly geographical designation means to impose a locational-spatial straightjacket on what cannot be other than a relational category—one that refers to distinct social relations and social processes, not to a specific location.

The second meaning attached to the notion of the Global South derives from the connotation of developmental backwardness, a circumstance of arrested development lagging behind the North. In some ways, this characterization simply substitutes polite terms for what amounts to the reproduction of the old-fashioned distinctions (which have now fallen out of favor) between the developed First World and the underdeveloped Third World. While no one claims that this division of the world of cities into two distinct camps represents a permanent condition, the broad categories of North and South are largely synonymous with the conventional binaries of rich and poor, developed and developing.[129] Dichotomies of this sort reproduce generalizations about rich/developed cities of the North, on one hand, and poor/developing cities of the Global South, on the other, that cannot be sustained under careful empirical scrutiny. This kind of analytic framing unfortunately has produced another essentialist reading of cities rooted in geographical location. This simple dichotomy overlooks the variety and complexity of asymmetrical patterns of urban transformation that produce both wealth and poverty at the same time, regardless of geographical location.[130]

Lastly, the third meaning attached to the notion of the Global South is rooted loosely in the idea of the postcolonial, a relational term connoting asymmetrical relations of power and exploitation that cannot exist without the North as it mirror opposite. In this regard, as Srinivas Aravamudam has suggested, the " 'North' and 'South' are a direct analogy of what was explored as 'West' and 'East' in Edward Said's rendition of Orientalism as a dualist ontology of scapegoating and demonization."[131]

The analytic distinction between cities of the Global North and those of the Global South is inherently unstable, porous, mutable, and often indecipherable. As Jean and John Comaroff have put it, "It is not difficult to show that there is much South in the North, much North in the South, and more of both to come in the future."[132] Despite the fact that the term "cities of the Global South" has replaced "the Third World" as a more or less popular usage, the label itself is inherently slippery, inchoate, and unfixed. The idea of the Global South describes less a geographic place than a polythetic category with limited analytical reach.[133]

Seen in the most positive light, such slogans as "theory from the South" or "seeing from the South" indicate a sensitivity to the relationality of urban transformation on a global scale. Yet the idea of Southern theory suffers from undue geographic vagueness and analytic imprecision. The Global South is not a homogeneous space or a stable ontological category (or a condition of being) symbolizing subalternity, informality, and indeterminacy, but it is a gesture toward acknowledging the asymmetrical and highly differentiated character of global urbanism.[134]

While Southern theory provides a useful platform for reorienting thinking about global urbanism, it has limited value as an analytic category that can serve explanatory purposes. Just as critics have suggested that the concept of neoliberalism has become an "overstretched and ill-defined signifier" through overuse, it might be similarly argued that the Global South as a blanket term has become an "unstable and ambiguous label," all too often used in ambiguous and contradictory ways.[135] The proponents of Southern theory have reinforced the emergent skepticism with the methodological appropriateness of building theory via "cross-case extrapolation and pan-urban abstraction."[136] Calls for seeing from the South and building Southern theory have contributed more to exposing and undermining of the universalizing pretensions and essentialist currents of mainstream urban theories than to building alternative analytic frameworks, or to the aspirational goal of what Jamie Peck has called "urban-theoretical renewal and reconstruction."[137]

Without a doubt, rethinking urban theory involves the rejection of unwarranted generalizations, sweeping universals, and unhelpful abstractions. Yet at the same time, one must recognize the need to remain skeptical of what Jamie Peck has called "spatialized logics of inversion." There is an inherent danger in reading or treating the southern city as a "kind of inversion of a dominant other, as a mirror image or 'theoretical negative' defined in opposition to a received (and relatively singular and static) 'Northern' or 'Western' urbanism."[138] In seeking to compensate for the dismissive tone of mainstream urban theories, it is not helpful for proponents of Southern theory to engage in what Ann Varley has called "inverted valorization," such that "the attributes of the purportedly 'negative' pole are instead valued positively."[139] To simply juxtapose a newly discovered Southern theory (with its stress on vibrancy, indeterminacy, and radical openness) against an outmoded Northern theory (with its unwarranted universalizing pretenses and nomothetic-deductive orientation) "run[s] the risk of generating new essentialisms, possibly even a slide into new forms of urban-geographical determinism."[140]

Thinking in this way can easily "result in a privileging of the principles of diversity, singularity, and difference against those of commonality, uniformity, and connection," instead of "efforts to hold these in relational tension."[141] As Ann Varley has persuasively argued, the challenge for retheorizing the field of urban studies is to recognize diversity and the lack of convergence "without inadvertently falling back into the trap of assuming incommensurable difference across a North–South divide."[142] If cities of the Global North and those of the Global South are treated as hermetically sealed mirror opposites, then it is impossible to grasp the intersecting and cross-fertilizing currents that blur and dissolve such hard-and-fast distinctions.[143]

PART II

Trajectories of Global Urbanism at the Start of the 21st Century

A First Approximation

CHAPTER 3

GLOBALIZING CITIES WITH WORLD-CLASS ASPIRATIONS

The Emergence of the Postindustrial Tourist-Entertainment City

Cities are no longer just built; they are imaged. City designers, like others who observe the metropolis, image and re-image cities through the calculated use of media.

Sam Bass Warner and Lawrence Vale[1]

In the face of the unrelenting pressures of globalization, city builders in aspiring world-class cities have undertaken large-scale publicly and privately financed downtown redevelopment programs in order to enhance their competitive advantage in the rank order of the global cities hierarchy. These experiments with urban regeneration have taken place within the framework of the global restructuring of the world economy. Faced with the declining significance of industry, manufacturing, and related activities as the main wellsprings for sustained economic vitality, new urban growth machines (which consist of loose alliances of city boosters, large-scale business enterprises, and key civil leaders) have increasingly looked not to the production of goods but to the provision of leisure and consumption opportunities as the main catalysts for reanimating their historic downtown cores and surrounding inner-city neighborhoods. Economically robust downtowns, leisure-and-tourist zones, upscale shopping malls, cultural precincts, convention centers, gentrified housing, signature architecture, mixed-use "live-work-and-play" quarters, and extravagant megaprojects are the glitzy new face of aspiring world-class cities in the new millennium.[2] As Allen Cunningham has suggested, urban space is everywhere becoming a "commodity working within the logic of the market economy."[3] The processes of spatial restructuring that lie at the heart of urban rejuvenation reconfigure urban spaces as a composite assemblage of place identities arranged to attract consumers.[4]

Although urban theorists have always recognized the importance of cities as cultural spaces of consumption, theorizing about the modern metropolis has typically taken the production of manufactured goods and services as the key to understanding the dynamics of urban growth and development.[5] Yet as the new postindustrial "informational city" of high finance, producer services, information technology, and digital media has replaced the prototypical modern industrial city of mass production, city boosters have turned their attention to considerations of "smart growth," sustainability, livability, quality-of-life attributes, aesthetics, and "experience economies." As the economic center of gravity has shifted away from the historical epoch of industrial (Fordist) capitalism, the culture and creative industries have assumed a greater role in reenergizing urban fortunes. These culture-led strategies for downtown revitalization have focused on entertainment, tourism, and leisure services as the lynchpins for restoring urban prosperity.[6] This wholesale shift in service-based efforts at downtown revitalization has spawned the development of an entirely new vocabulary to describe these trends: the "tourist city,"[7] "fantasy city,"[8] "city of spectacle,"[9] "city of illusion,"[10] and the "city as entertainment machine."[11]

PLACE MARKETING AND THE ENTREPRENEURIAL CITY

It is perhaps best to consider the entrepreneurial city as an imaginary city, constituted through a plethora of images and representations.

Tim Hall and Phil Hubbard[12]

During the period of rapid urbanization in the late nineteenth century, the key task confronting municipal officials was to develop strategies for managing the politics of city building for a growing populace by providing the kinds of public services essential to health, safety, and civic education. A century later, municipalities have set their sights on very different goals. Starting in the late twentieth century, city regimes began to devote enormous creative energies and financial resources not simply to the basic and conventional municipal functions of public administration but also to the task of making cities, in the words of Dennis Judd and Susan Fainstein, "places to play."[13] This goal of fashioning the "tourist-entertainment city" has involved reshaping urban space, more often than not in partnership with private real estate developers, in ways designed to appeal primarily to affluent residents and visitors. This imperative has attracted interest in even the poorest, most blighted "loser" cities such as Detroit, Gary (Indiana), Flint, and Newark.[14] In short, as Sharon Zukin has observed, "culture is more and more the business of cities."[15] This urban obsession with glamor and spectacle amounts to a preoccupation with "a politics of bread and circuses."[16]

Spurred on by the desire to maintain their competitive edge in the global marketplace, city builders in aspiring world-class cities have jettisoned demand-oriented provision of public goods for collective consumption and instead embraced supply-side, market-driven approaches to urban regeneration.[17] Economic competitiveness, responsive and flexible governance, and the expanded participation of elite business coalitions in strategic thinking about urban revitalization have become the new mantras in the turn toward the "entrepreneurial city."[18] Market-led approaches to urban regeneration have stressed the role of municipal officials as facilitators in cobbling together various collaborative arrangements with large-scale real estate developers, public-private partnerships, and other strategic alliances of key corporate stakeholders. The adoption of entrepreneurial modes of urban governance has spawned the implementation of such policy goals as "lean management," deregulation, and privatization and has triggered talk of downsizing, rightsizing, off-loading, unbundling, and outsourcing.[19] This shift toward flexible and proactive entrepreneurial approaches to urban management has meant that planning initiatives have become more fragmented and selective. Comprehensive, citywide, holistic planning schemes that characterized modernist approaches to city building have been supplanted by piecemeal approaches to urban regeneration that usually consist of a patchwork of targeted (place-specific), stakeholder-led projects.[20]

As a general rule, the goals of urban entrepreneurialism have fostered an ethos whereby growth coalitions are compelled to direct their attention to the creation of wealth for property investors rather than its distribution in public services.[21] This wholesale shift from redistributive modes of urban governance (or what David Harvey has called "managerialism") to pro-growth, entrepreneurial approaches has meant finding ways to attract investment from footloose capital into large-scale redevelopment projects.[22] In the mantra of competitive cities, the ultimate goal of urban entrepreneurialism and the politics of growth at any cost is to use property-led regeneration strategies to become more integrated into global networks of cities, and hence to climb the ranked hierarchy of aspiring world-class cities.[23]

The turn toward urban entrepreneurialism is underpinned by an ideology of private enterprise that emphasizes the leading role of corporate business in constructing prestige flagship projects as a key vehicle for reviving urban fortunes through culture-led downtown regeneration.[24] Large-scale real estate developers have become social and spatial engineers, assuming expanded responsibilities for planning the built environment, producing and redefining the public interest, and determining the social uses of privatized space.[25] Planned unit developments (PUDs) have, in effect, institutionalized the practice of privatizing the planning process. Incorporated into local zoning ordinances, PUDs have given real estate developers the extraordinary latitude to come forward with their own proposals for lot layouts, building densities, and street patterns on large parcels of land.[26]

New urban growth coalitions have stressed public-private partnerships, place marketing, and culture-led development as the key components of competition-based and market-led strategies to jump-start downtown revitalization.[27] This "fight for the global catwalk" (to borrow a useful metaphor from Monica Degen) has typically involved the sponsorship of flagship property-led redevelopment projects where "cities compete with each other by parading made-up images of different areas of the city which advertise these spaces as favorable and attractive to business and leisure."[28] As is often the case, public-private coalitions hire private planning consultants in order to shape the future of cities. They often claim

FIGURE 3.1 Tourist Entertainment Cities: New York's Times Square. Times Square is one of the largest tourist attractions and economic hubs of New York City. Sergey Zolkin

that the planning procedures they institute typically incorporate consensus-based, collaborative, and inclusionary techniques, rather than elite-centered and expert-driven models derived from past practices.[29]

For aspiring world-class cities, the neoliberal principles of entrepreneurial urbanism demand not only the erasure of the outdated built environment produced to sustain earlier methods of industrial production but also the dismantling of older regulatory mechanisms for the public management of the city invested in civic authorities. The processes of creative destruction of urban space have opened up speculative development through the installation of "privatized, customized, and networked urban infrastructures intended to (re)position cities within supranational capital flows" and fostered urban regeneration through place-making projects in order to make cities more competitive and marketable.[30] The new urban politics consists of using location marketing and privatized regulatory regimes to promote the wholesale restructuring of urban landscapes.[31] Facing stiff competition from other globalizing cities, urban management has become an endless real estate placement game in which city builders focus on fashioning favorable locations that can attract corporate headquarters, entrepreneurial start-ups, and tourist experiences, thereby transforming historic downtown cores into aesthetically pleasing landscapes of consumption, mixed-use shopping precincts, and themed entertainment destinations.[32]

FROM BUSY WORKPLACE TO LEISURELY PLAYGROUND: THE CITY AS TOURIST-ENTERTAINMENT MACHINE

Regeneration strategies that focus on entertainment and cultural consumption depend on selling the city to business travelers, tourists, and local leisure seekers.[33] Place making lays particular stress on how various place-marketing strategies have produced a city of surface, calculated for a primarily visual and aesthetic effect.[34] These efforts to sell the city have inspired their own specialized terminology: imagineering (the expression coined by the Disney Corporation to refer to the engineering of imaginary places),[35] Disneyfication,[36] theming,[37] staging,[38] and branding (and even hard-branding).[39] City imaging, place marketing, and the packaging of urban life into fragmented, commodified units for sale to a bourgeoning tourist and business services industry have taken root in cities aspiring to achieve world-class status.[40]

In their efforts to sell the city, contemporary urban design and architectural practices lay particular stress on the all-important centrality of the visible image and the aestheticization of urban spaces, thereby effectively transforming the cityscape into a scenographic landscape of visual consumption, an object to be gazed upon as pure spectacle.[41] The plundering of architectural ruins, the restoration of abandoned buildings, the reappropriation of disused ornamental styles, and the repackaging and recycling of what were once underutilized or neglected locations combine to condense the historic fabric of the city "into a displaced, artificial condition, a form of *tableau* configured and framed without a context—or within a context that no longer matters."[42] The incorporation of revitalized and revalued urban spaces into "theatrically staged compositions" transforms the city into "gentrified, historicized, commodified, and privatized landscapes"—that is, an entertainment theme park. The City of Spectacle, "with its endless flux of combinatorial forms" that are decoratively dispersed across the "broken surface" of the cityscape, as Christine Boyer has suggested, reduces the city to "the play of pure imagery."[43]

Giving precedence to appearance over substance and establishing the primacy of the façade in the creation of urban disguises effectively reduce the effect of architecture to two dimensions, where the construction of simulated environments lacks both depth and substance.[44] Framed this way, architecture in the city appears as an autonomous artistic expression that has been effectively separated from any consideration of historic context or existing urban inequalities.[45] By reducing the city to its surface patina, aestheticization promotes a kind of consensus around the status quo, thereby distracting attention from urban difference, and hence from the real social and economic injustice that exists within the city.[46] The modernist penchant for monumentality in city building communicated the ideals of the permanence, authority, and power of the capitalist order. In contrast, the postmodern architecture of historicized pastiche creates ephemeral and transitory places that lay particular stress on the playful, circumstantial, and ludic qualities of urban life.[47]

Property-led urban regeneration schemes have become a major reason for the homogenizing appearance of aspiring world-class cities. While they vary in the minute details of architectural styling and ornamental finish, they are generically similar in conceptual design and practical execution.[48] The net result is the appearance of what Guy Julier has termed "designscapes": distinctive ensembles of office buildings, retail space, condominium towers, cultural amenities, renovated spaces, green landscaping, and street furniture that are cobbled together in assembled packages that vary a little but not a lot.[49] Place marketing in the entrepreneurial city focuses on the construction of such entertainment destinations as new "cosmopolitan city-center lifestyles" as a way of signaling distinction and authenticity.[50] The success of property-led redevelopment undertakings in Bilbao, Barcelona, the London Docklands, and La Défense in Paris quickly became the most seductive of all "traveling ideas," producing what amounted to the serial reproduction of designscapes imitated elsewhere.[51]

Large-scale urban development projects have increasingly become important sources of branded and imagined place identity.[52] The Salford Quays on the Manchester Ship Canal was one of the earliest examples of the place-making strategy constructed around similar design motifs and look-alike building typologies. Initially developed in 1982 through public-private partnerships on the site of underutilized docklands, this megaproject has expanded to include mixed-use amenities like apartment buildings, office blocks, luxury hotels, and retail space, linked together with the Imperial War Museum North (designed by Daniel Libeskind) and the Lowry arts complex (a landmark cultural venue designed by James Stirling and Michael Wilford). Other examples include the South Bank and Paddington Basin redevelopments (London), Espace Léopold and the European Union District in Brussels, the refurbished financial district in the Dublin docklands, Potsdamer Platz and the science-technology-university complex at Adlershof ("Eagle's Court") in Berlin, the Kop van Zuid in Rotterdam, the Euralille complex in Lille, Donau City in Vienna, Gunwharf redevelopment (Portsmouth), HafenCity (Hamburg), Brindleyplace (Birmingham), the Ørestad project in Copenhagen, Rotermann Quarter (Tallinn, Estonia), Melrose Arch (Johannesburg), the CityLife project and Fiera Milano Rho exhibition complex in Milan, and the 1998 World Expo site in Lisbon.[53]

For the most part, these are exemplary expressions of what Leslie Sklair has termed "scripted spaces"—stage sets for the propagation and conduct of the culture ideology of consumerism in the era of late capitalism.[54] Just as the Paris arcades signified the triumph of consumer capitalism at the end of the nineteenth century, these "brandscapes," as Anna Klingmann has so persuasively argued, are the visible expression of contemporary hyper-capitalism standing at the conjuncture of economic globalization and corporate interests. These deliberately designed building typologies "constitute the physical manifestations of synthetically conceived identities transposed onto synthetically conceived places, demarcating culturally independent sites where corporate value systems materialize into physical territories." As material expressions of deeper social relations, brandscapes "have become key elements in linking identity, culture, and place."[55]

Subjecting urban landscapes to the logic of commercialization has resulted in the disappearance of conventional public spaces and their replacement with new, privately managed places with altered forms and functions. Thus, the market for urban spaces develops in tandem with a new "industry of place." The raw materials for building these new themed or imagined places are images of existing cities and their recognizable buildings, monumental architecture, construction sites, well-known streets, and world-famous squares. Sometimes city builders have imitated entire city complexes. A new industry has developed, euphemistically known as cultural mining—a kind of cultural archeology that seeks out spectacular details but, above all, urban images.[56] The spatial products that result from such cultural mining of visual images are not really places in the conventional sense, but are dressed up as places. Although these spatial products assume a real-life physical appearance, they remain independent of their original inspiration. They are the embodiment of the postmodern notion that in the contemporary modern world everything is everywhere simultaneously.[57]

Equally important, these themed entertainment destinations are the precursors of a new type of imagined community, one that is primarily disposed to falling under the spell of sensory sensations, impressions, or experience.[58] In this postmodernist city building, the heightened importance of sounds, smells, atmosphere, aura, and related sensations that animate the senses have replaced the modernist principles of functionality, efficiency, and utility. These themed entertainment destinations depend upon the disappearance of the original. By stressing stageability and performance, they have effectively blurred the conventional distinction between everyday life and art. These themed entertainment destinations resemble three-dimensional urban stage sets. These placeless places—what Edward Relph called "the non-place urban realm"—are not rooted in any actual place but are imaginary spaces that scenographically reproduce cityscapes of either real or invented cities.[59]

Theming space is linked to the wholesale rejection of the modernist approach to urban form, with its stress on technological rationality, functional differentiation, and streamlined efficiency.[60] Such well-known postmodernist architects as Robert Venturi and Jon Jerde have proudly proclaimed that there is no harm in giving people what they want.[61] Postmodernist architecture cultivates the idea of fragmented urban space consisting of varied sections that overlap and intersect. The haphazard and unpredictable way these spaces are utilized, often temporarily, emerges from this cobbled-together collage. Dispensing with the modernist ideal of establishing coherent and holistic urban landscapes, postmodernist design has turned its attention to the local scale. This narrow focus has resulted in the creation of a series of highly specialized spaces that typically assume an eclectic character. These postmodernist spaces are usually replicated many times over and, taken as a whole, resemble a maniac's diary indiscriminately filled with disconnected entries.[62] Contemporary architecture planning has increasingly directed attention not to the creation of new architectural or spatial values but to replicating existing patterns, generating urban reproductions, one after another.

Seen in this light, theming represents a shortcut, or a kind of advertising ploy, that leads to the experience of spatial hyperreality, with the underlying aim of promoting the image of a city as an exceptional place.[63]

STAGING THE TOURIST-ENTERTAINMENT CITY: PLACE MARKETING AND THE CULTURAL POLITICS OF URBAN REINVENTION

City builders in aspiring world-class cities have undertaken culture-led regeneration efforts in the belief that sponsorship of such high-profile, flagship megaprojects as sports and entertainment sites, arts festivals, museum quarters, and historical heritage sites enhances the city's image while serving as a catalyst for successful downtown urban renewal.[64] This culture-led approach to urban regeneration reflects what some have called the "new conventional wisdom" in urban policy.[65] In a crowded marketplace marked by intense competition to attract corporate investors, business travelers, and tourists, the stress on place promotion has become a central element of civic boosterism.[66]

The economic fortunes of postindustrial cities with world-class aspirations have come to depend largely on energizing a new growth-machine politics around the promotion of cultural industries and attractions, their competitive advantage derived not from manufacturing and industrial production but from service industries, cultural amenities, and place identities.[67] Culture-led regeneration strategies typically revolve around the designation of distinct cultural quarters where an assemblage of venues such as art galleries, museums, performing arts centers, and other public exhibitions functions as the catalyst for building tourist-entertainment industries.[68] Cities compete by reinventing themselves as distinctive places of consumption in which to satisfy new upscale demands for commercialized leisure, recreational activities, and other sensory experiences. City boosters have often turned to the marketing of tradition through sponsorship of arts and music festivals, theatrical performances, annual celebrations, fashion events, parades, and other large public gatherings.[69] In order to sell cities to middle-class consumers, city builders have focused on the construction of subsidized convention centers and sports arenas, the fashioning of arts-and-entertainment districts, and the regeneration of historic "old towns."[70] As Loretta Lees has argued, such redevelopment efforts rely in part on rhetorical tropes firmly embedded in a form of liberal romanticism and associated beliefs about the connections between sociocultural diversity, street-level vitality, and aesthetically pleasing urban space.[71] Aggressive marketing and promotional packaging often includes a stress on clean-and-green lifestyles, the branding of new localities as must-see destinations, the abundance of high and popular arts, the "villaging" of city centers to evoke lost or mythical qualities of public life, historicized quarters, and the lure of spectacular buildings.[72]

FROM FUNCTIONAL BUILDING TYPOLOGIES TO SYMBOLIC ARCHITECTURE

As a general rule, the driving force behind modernist architecture was captured by the dictum "form follows function." With what might be called the postmodern turn, architecture has replaced the goals of efficiency, rationality, and functionality with pastiche, playfulness, and ornamentalism.[73] These changes in architectural style can be traced to broader socioeconomic shifts, notably the desire for differentiation at a time of accelerated global competition.[74] Branding, place promotion, and "experience management" have emerged at the forefront of contemporary architectural practice.[75]

In their pursuit of visibility and distinctiveness, city boosters have often turned to trophy buildings and signature architecture as high-profile launching pads for urban regeneration.[76] What distinguishes these showcase projects from conventional building typologies is that they typically attract attention by virtue of their aesthetic qualities rather than their functional utility. The signature buildings themselves—rather than the practical purposes they serve and the activities contained within them—become the central focal point, creating the aesthetic experience.[77] City watchers have attributed the extraordinary urban renaissance of Bilbao, resurrected from the archetype of a declining industrial port city into the new mecca of urbanism, to the stunning success of the Guggenheim Museum (designed by the noted celebrity starchitect Frank Gehry) in attracting international tourists.[78] The Guggenheim Bilbao epitomizes the powerful intersection of three ingredients—an emblematic icon, a global trademark, and a signature architect—that converge to produce a "brand" with global name recognition.[79] This use of signature architecture to enhance the appeal of cities, sometimes referred to as the Guggenheim/Bilbao effect, has spread to cities aspiring to world-class status. City builders elsewhere have sought, with some success, to mimic the Bilbao experience as a model for urban regeneration, hoping to use avant-garde, look-at-me architecture as a means to jump-start urban revitalization.[80]

Culture-led regeneration strategies have focused on the construction of such niche zones as fashion districts, quaint urban villages, or museum quarters as key components of the experience economy.[81] In what might be called the "valorization of milieu," these zones depend upon the agglomeration effects of clustering museums, restaurants, art galleries, and other themed entertainment sites within close proximity.[82] Subjected to the pressures of maintaining financial solvency, museums have shifted their focus from mono-functional places specializing in the collection and conservation of cultural artifacts to becoming multidimensional tourist destinations that put a premium on an entertaining experience.[83] The traditional core activities of established museums—the preservation, archiving, restoration, and display of assembled collections of artifacts and the pursuit of scholarly research—have been replaced by an emphasis on serving as multiuse entertainment sites or cultural supermarkets that combine spectacle with consumption.[84] No longer serving a primarily educational function, museums have become something akin to cultural amusement parks for housing temporary, traveling exhibitions and sponsoring themed blockbuster events.[85]

In the contemporary age of globalization, the city of spectacle has developed at a pace that has easily outdistanced the efforts of architects and planners to shape them.[86] There is no longer sufficient time to plan cities as coherent wholes. As a result, the role of architecture in the city has fundamentally changed from that of designing functional buildings to that of creating aesthetically pleasing commodified experiences, "tied first and foremost to speculation in future identity and to real estate values." This shift in the practice of city building comes with "an atmosphere of higher risk," coupled with the prospect of higher reward. As postindustrial cities from Bilbao to Rotterdam can attest, real estate developers—operating in conjunction with municipal officials—"deal with the risk posed by the pace and unpredictability" of the contemporary climate of competitive urbanism not by conservatively hedging their bets or by improving existing conditions but by creating enclaved spaces that aim to be self-contained worlds-in-themselves. These self-enclosed enclaves, often referred to in official planning parlance as "overlay zones," generally fall into two distinct categories. The first type is embodied in such places of singular identity as gated residential communities, private golf estates, or retirement villages with a full range of services and amenities. The second type is characterized by an exaggerated hyper-identity—what Roger Sherman has called "a kitchen-sink urbanism descendant from the theme park." Without a doubt, the results of this brave new urbanism have been unsatisfying and often unsettling. Nevertheless, their undeniable popular and economic success has transformed the culture and space of urban life. These enclaves reflect the fact that, in the contemporary world of subcultures and self-interest, "the inflation of identity and 'tactics of negotiation' matter more to the urban game than location or context." What matters most is the clever use of aesthetic design and stylistics, "cunningly conceived as a pheromone that will attract a wide range of potential audiences."[87]

The serial reproduction of tourist spectacles, cultural theme parks, business improvement districts, enterprise zones, waterfront developments, and gentrified housing schemes—all of which are scattered across the urban landscape—is not simply an aggregated outcome of spontaneous local pressures; it reflects the powerful disciplinary effects of worldwide market competition pitting each aspiring world-class city against the others. The desire to replicate the Bilbao effect has lured countless city builders in depressed postindustrial cites to seek to mimic the success brought about by the fusion of a signature building (the Guggenheim Museum), a starchitect (Frank Gehry), and a global trademark.[88]

The success of the regeneration strategy at Bilbao in capturing the tourist imagination prompted city boosters far and wide to seek to emulate its formula. As a result, "every city," Sharon Zukin sarcastically observed, "wants a 'McGuggenheim,' "[89] The unanticipated consequence of this crass imitation is that the more cities compete to be different, the more they seem to look strikingly similar, each outfitted with its own sculptural flagship building and generic mixed-use regeneration scheme.[90]

The current frenzy for constructing spectacular architecture goes well beyond functional utility to creating some sort of signature trophy building.[91] For aspiring world-class cities, no building type quite embodies the exaggerated importance of symbolic meaning in architecture as much as the high-rise skyscraper.[92] A high-rise office tower is more than a marvel of engineering prowess; it is also a symbol of urban, national, and corporate identity—a claim to modernity.[93] City boosters use spectacular architecture as a way to announce their hoped-for emergence on the world stage.[94] As exemplars of the principle that height is a symbol of power, unusually tall buildings like the Burj Khalifa skyscraper in Dubai, the Pentronas Towers in Kuala Lumpur, and Taipei 101 in Taiwan, along with other equally spectacular architectural monuments, have greatly enhanced the cachet of aspiring world-class cities seeking to secure their place among the leading metropolitan centers.[95]

URBAN REVITALIZATION IN POSTINDUSTRIAL URBAN LANDSCAPES: REVALORIZING URBAN WASTELANDS

In responding to the mounting pressure of interurban competition on a global scale, city builders in aspiring world-class cities have often initiated large-scale redevelopment projects designed to revitalize neglected neighborhoods or distressed industrial districts that had fallen into disrepair.[96] These place-making strategies have targeted such derelict sites as abandoned rail yards, redundant docklands, or underutilized warehouse districts. These efforts at revalorization typically conform to some combination of a gentrifying impulse, where the ruined built environment is demolished to clear the path for entirely new buildings with new functions, and a cultural heritage impulse, where the restoration of historic districts seeks to retain a visual connection to the past through the preservation of physical traces of what came before.[97]

For deindustrializing cities, the goals of urban entrepreneurialism have often centered on filling in derelict industrial wastelands with new postindustrial service industries focusing on innovation, tourism, and leisure.[98] The ruins of the modern industrial city often provide real estate developers with the raw materials for constructing the flashy new city of spectacle that arises, like the Phoenix, from the ashes of the old. Derelict and abandoned remnants of the industrial past—docklands, port facilities, factories, warehouses, power plants, rail yards, and other disused debris—form the basic building blocks for the manufacture of cultural heritage sites.[99] These efforts to revalorize and re-signify the vernacular historic landscape amount to what Monica Degen has called the "designer heritage aesthetic."[100] The restoration of old historic buildings and quaint narrow streetscapes provides physical memory markers of what the city used to be. By gesturing toward authenticity, rebuilding strategies that convert derelict sites of ruin into packaged tourist destinations animate a sense of nostalgia for a carefully manicured past. The scenographic arrangement of historic building sites, pedestrianized walkways, and vernacular street furniture is a particular "compositional form which explicitly relies on a series of familiar, non-disturbing and comfortable views from [the] architectural past."[101]

Waterfront redevelopment provides an exemplary expression of the shift from outmoded industrial uses of derelict urban space to the new postindustrial tourist-entertainment city.[102] The mixed-use redevelopment of outmoded and often contaminated waterfronts represents a challenging component of urban revitalization and also a significant real estate opportunity for corporate capital.[103] As Sharon Zukin has suggested, the corporate blueprint for success that drives these themed entertainment destinations is typically restricted to constructing a narrowly focused microcosm of the past or a bleached-out principle that defines these tourist destinations is "simply a visual *theme*" without uncomfortable reminders or troubling histories.[104] Whereas the Guggenheim Museum in Bilbao, the Burj Khalifa in Dubai, and other monumental architectural spectaculars decontextualize the future, festival marketplaces like Faneuil Hall Marketplace in Boston, Harborplace in Baltimore, and South Street Seaport in

New York decontextualize the past, replacing rich texture with a smooth and soothing patina.[105] Festival marketplaces typically masquerade as cultural heritage sites, yet without the commitment to historical accuracy required of museums and archaeological sites. Reusing industrial ruins enables real estate developers to create "instant *faux* tradition" that can be immediately transformed into a marketable commodity.[106]

The speculative nature of real estate investments in place promotion has produced a risk-averse logic of entrepreneurial caution. These risk-aversion strategies have spawned, in turn, the serial reproduction of tested formulas for success. Such familiar megaprojects as the redevelopment of disused waterfronts, the retrofitting of old warehouse districts, and the construction of "historic districts" follow familiar patterns that have been reproduced in aspiring world-class cities around the world.[107] In calling attention to a cultural and leisure precinct called Southbank located on the riverfront opposite the central city grid of Melbourne, Quintin Stevens and Kim Dovey offer a predictable repertoire of elements: "the river's edge is lined with scheduled entertainment venues and saturated with choreographed street theatre, public artworks and illusory soundscapes, intended to attract a well-heeled clientele and to frame leisure within a context of consumption."[108] In their efforts to generate culture-led redevelopment, real estate developers have tended to homogenize places "with an endless repetition of standard devices," ranging from recognizable building types to predictable advertising slogans.[109] The serial reproduction of standardized formulas and programmatic ingredients means that places become more like franchises that look and feel the same everywhere.[110] Produced under social conditions that resemble the manufacture of consumer products, architecture—and the built environment more generally—has become a cultural commodity that conforms to "the same patterns of both standardization and market differentiation."[111] One may ask: how can architecture create a distinct place identity and respect the differences of places if the built form is merely replicated in multiple sites?[112]

THE SPECULATIVE NATURE OF LARGE-SCALE URBAN REVITALIZATION PROJECTS

Referring to the ever-increasing phalanx of postindustrial cities in the British Isles that have undertaken exclusionary, profit-driven, waterside regeneration projects, the psycho-geographer Iain Sinclair once wrote "any puddle will do"—observing that no matter how bitter the climate or inauspicious the view, waterfront projects never seem to go out of fashion in postindustrial landscapes. The redevelopment of the Speirs Wharf Canalside cultural quarter in the postindustrial brownfield landscape of North-West Glasgow is a stunning example of speculative investment in an emergent tourist-entertainment city. The canalside master plan for the cultural quarter is a neoliberal mixture of soft policy options that borrow liberally from "creative city" (Charles Landry), "creative class" (Richard Florida), "new urbanism" stylistics (Andres Duany and Elizabeth Plater-Zyberk), and "smart growth" principles. The master plan is specifically framed in the language of austerity, suggesting that only a significant infusion of capital can be the catalyst for urban regeneration in a time of economic crisis. Ultimately, waterfront redevelopment—which links historic preservation and cultural heritage sites with entrepreneurialism—revolves around the extraction of value from land and property and an increase in the socio-spatial tax base, making the smart growth and new urbanism formulas a strategy for profit-making investment, despite their green pretensions and rhetoric of community participation.[113] Undergirded by a buoyant ideology of growth, waterfront redevelopment schemes often fall short of expectations.[114]

The ongoing remaking of underutilized downtown urban cores, abandoned and derelict waterfronts, and blighted historic buildings through urban redevelopment megaprojects is part and parcel of the urbanization of neo-liberalism.[115] Megaprojects are the quintessential exemplars of urban entrepreneurial strategies. They represent the means through which globalization becomes urbanized, or the places where the forces of globalization become rooted.[116] The primary objectives of megaprojects are to foster entrepreneurial goals through large-scale growth-oriented strategies.[117] Contemporary megaprojects are the primary embodiments of the new postindustrial, consumption-oriented image of the city.[118] Megaprojects range from new transportation linkages to business districts (La Défense, Paris; Docklands, London) and commercial centers (Potsdamer Platz, Berlin).[119] Megaprojects like the redevelopment of rail stations, highways, and airports are key examples of planned, large-scale strategic interventions into the contemporary urban fabric aimed at better connecting and revitalizing key inner-city locales.[120]

As a general rule, large-scale urban redevelopment projects depend upon practices of exception, whereby real estate developers in partnership with compliant municipal administrations are able to bypass statutory regulations and the regulatory agencies that oversee them and to gain relaxations of existing planning standards, including zoning variances and interim development controls. In the typical case, large-scale redevelopment projects rely upon the creation of public-private partnerships and independent agencies with special or exceptional powers of intervention and decision making to clear away obstacles and blunt any public outcry.[121] The establishment of these regimes of exception is premised on the ability to override existing bureaucratic jurisdictions, to reassign administrative responsibilities and practices, and to concentrate decision making in the hands of compliant agencies, boards, and commissions.[122] These exceptionality measures foster special governance regimes that "lie at the margins" of conventional statutory planning and involve a "significant redistribution of policy-making powers, competencies, and responsibilities" to "quasi-private and highly autonomous organizations" operating outside the direct oversight of public authorities.[123]

Critics have pointed to the downside of large-scale megaprojects.[124] Many of these prestigious state-driven infrastructure projects, such as the Big Dig in Boston, the Los Angeles subway line, and the Denver International Airport, are legendary for their cost overruns.[125] The proponents of such megaprojects were successful in gaining approval because they used what Bent Flyvbjerg has called the "Machiavellian formula": "a fantasy world of underestimated costs, overestimated revenues, undervalued environmental impacts and overvalued regional development effects."[126] Based on a study of several hundred projects in more than twenty countries, Flyvbjerg has questioned the "professional expertise of engineers, economists, planners and administrators" because they consistently misrepresent available information and deliberately disregard or downplay risks.[127] Because of this lack of accountability, Flyvbjerg has referred to the continuing practice of investing in such risky undertakings as the "megaproject paradox."[128]

What is striking about many of these flagship megaprojects is their physical similarity and their convergence around their market orientation and the dominance of corporate enterprise.[129] Oversaturated markets react with downward pressure on profits. Overbuilding leads to underutilization.[130]

Enmeshed in the cultural and economic logics of late capitalism, enchanted sites of postindustrial revitalization are trapped in the ceaseless, whirlwind cycles of competitive renewal to attract investment and visitors in order to offset the falling rate of profit. Yet carefully designed tourist bubbles almost unavoidably experience a diminution (sometimes gradual and sometimes abrupt) of their appeal to uniqueness—"hence, their ability to enchant—as their success becomes compromised by a combination of consumer fatigue and indifference, and competitor imitation and innovation." Waterfront redevelopment projects and other tourist-entertainment sites are compelled to engage in never-ending cycles of refashioning via fresh capital investment in order to sustain and enhance their continuing appeal to potential consumers.[131]

TOURIST URBANISM

The tourist Venice is *Venice: the gondolas, the sunsets, the changing light, Florian's, Quadri's, Torcello, Harry's Bar, Murano, Burano, the pigeons, the glass beads, the vaporetto. Venice is a folding picture-post-card of itself.*

Mary McCarthy, *Venice Observed*[132]

People taking business trips flew into cities and out again and didn't see the countryside at all. They didn't see the cities, for that matter. Their concern was how to pretend they had never left home.

Anne Tyler, *The Accidental Tourist*[133]

One of the most striking new urban forms that came into existence during the last few decades of the twentieth century is tourism urbanization—that is, the deliberate restructuring of the spatial landscapes of cities and towns to facilitate the short-term visits of large numbers of people so they can consume some of the many pleasurable goods and services on sale.[134] Tourist urbanism has blossomed into a truly global industry dominated by transnational hotel firms, entertainment corporations, casino resorts, and professional sports franchises.[135] Indeed, local urban residents can become tourists in their own cities. The enhancement and rejuvenation of the architectural environment, in combination with the transformation of social and cultural milieux, create a coherent space of consumption that signals both the appropriation of unused spaces for capital accumulation and a particular interpretation of what constitutes fashionable symbolic and cultural capital.[136] These entertainment destinations are theme-o-centric; that is, everything from specific sites to the city (and even nature) itself conforms to a scripted theme, typically drawn from history, popular entertainment, or current fad. These places construct new kinds of public life by transforming the sights and sounds of everyday life into a phantasmagoric spectacle.[137]

The success and appeal of tourism are due to the ability of its promoters to make excursions to exotic, faraway places both popularly accessible and expressive of social distinction and cultural difference. The paradox of seeming both obtainable and exclusive is a central engine in the making of mass consumer culture of which tourism, vacations, and travel are an integral part.[138] Tourism advertising forms part of a hermeneutic circle—a closed semiotic system that links the representations of a tourist destination to the actual tourist experience by creating a flexible set of expectations that the tourist industry is designed to accommodate.[139] Even as themed entertainment destinations model themselves on the tourist experience, tourism itself recodes space as time. Tourist entrepreneurs typically promise travelers idyllic escape from the routines of daily life to destinations where time stands still or the past lives on, uncontaminated by modernity.[140]

As an image-making process that combines both globalization and localization, tourist urbanism is an exercise in global place making that involves a mixture of homogenizing and particularizing influences where the contradictory forces of sameness and uniformity come up against difference and hybridity.[141] In other words, tourist-entertainment cities gravitate between the extremes of exploiting minute spatial differentiations in order to sell the uniqueness of place, on the one hand, and the production of "recursive and serial monotony" in the form of ambience, styles, and narratives in heritage sites, hotels, and retail outlets, on the other.[142] Corporate enterprises that promote tourist urbanism are simultaneously placeless and place-saturated.[143] On the one side, they accentuate the place themes in tourist-entertainment sites by using iconic images and symbols—like Planet Hollywood, McDonald's, and other almost universally recognized name brands—that connect the specific location with the globally familiar. On the other, the business of tourist urbanism is place specific, with "place as its raw material."[144] To the extent that it is institutionalized and commercialized, the practice of tourist urbanism standardizes the experience of collecting cultural signifiers in a restricted period of time.[145] Unlike other commodities, the tourist experience is spatially fixed and is consumed on location. The stress on place distinctiveness, local particularities, uniqueness, difference, variety, and diversity requires that the construction of tourist-entertainment sites involves a historically specific appropriation and reworking of global influences to fit the local context.[146]

As global place makers, tourist and entertainment corporations seek to transfer the logic of commodity production to the production of tourist places.[147] This reconfiguration of localities or sites into marketable objects invariably replaces a commitment to genuine authenticity with an appeal to a staged authenticity in which local history and cultural heritage become manufactured or simulated for tourist consumption.[148] The creation of the urban tourist experience typically involves the postmodern blurring of leisure boundaries where travel to exotic locations blends cultural heritage sites with retail shopping and other modes of consumption.[149] The practice of tourist urbanism reaches its low point with the global production of non-places (Disneyland, Disney World) and non-things (mass-manufactured souvenirs and the serial reproduction of kitsch objects).[150]

In seeking to carve out niche markets that cater for unique experiences, city boosters around the world have set their sights on the promotion of distinct kinds of entertainment, ranging from mega–sports events to annual festivals and parades.[151] The hosting of sports mega-events as launching pads, or platforms, for sustained economic growth has taken hold in cities with world-class aspirations.[152] While city boosters have enthusiastically championed these high-profile, short-term sports mega-events, critics have charged that these one-time-only occurrences are "politically and economically high-risk ventures" that rarely produce the results that their proponents promise at the outset.[153]

Tapping into such diverse themes as sexual identities and pleasure travel, tourism boosters in aspiring world-class cities have gone to great lengths to fashion specific places as one-of-a-kind entertainment destinations.[154] Place production often includes territorially bounded quarters ("ethnoscapes" like the proverbial Chinatown or Little Italy), which feature ethnic cuisine, specialized street markets, and feigned gestures toward cultural heritage.[155] Entertainment-oriented urban regeneration is linked to broader questions concerned with the political economy of tourism, the transformation of public spaces into privatized places of consumption, and the creation of thematically packaged "total environments" separated from the existing urban fabric.[156] Yet the urban tourism industry is notoriously fickle—price sensitive to good and bad news, changing fads, and shifting trends in consumer preference. Like any other economic activity based on supply and demand, urban tourism can easily become a victim of short-term thinking where oversaturated markets become economically unsustainable.[157]

NEW MODES OF URBAN GOVERNANCE UNDER THE SIGN OF NEOLIBERALISM

The wholesale turn toward privatization, deregulation, and urban entrepreneurialism—processes for which "neoliberal city" serves as a convenient shorthand—figures prominently in the scholarly discourse on downtown revitalization in globalizing postindustrial cities.[158] In responding to the challenge of neoliberal restructuring, municipal authorities in aspiring world-class cities have adopted new regulatory regimes that have removed institutional constraints, legal barriers, and administrative impediments that stood in the path of implementing market-led solutions to the restoration of urban prosperity. The thinking behind neoliberal governance is that downsizing, outsourcing, and rightsizing enables market competition to efficiently distribute municipal services to where they are most effective in catalyzing urban revitalization.[159]

Yet as an epistemological category, the term "neoliberalism" (along with all of its surrogate formulations) often functions in urban studies literature as little more than a somewhat crude and blunt instrument used to cast aspersions on everything from the deleterious effects of privatization to the dangers of urban entrepreneurialism. While the term has heuristic value in helping to frame our thinking about shifting modes of urban governance from a Keynesian-welfarist model to a decidedly entrepreneurial approach, one must take care not to overextend its use into a one-dimensional, one-sided analytic device that seeks to explain too much and, as a consequence, accounts for too little.[160] As Jamie Peck has insightfully suggested, the "incautious deployment" of neoliberalism has far too often "given rise to theoretical overreach." Taking this line of reasoning a step further, he points to "the dangers of invoking 'neoliberalism' as an underspecified *zeitgeist* signifier, the explanatory significance of which sometimes does not extend beyond gestural invocations of the 'background scenery' of socio-institutional transformation, drawn only in broad brushstrokes."[161]

Neoliberal policies and practices work on the basic principle that private enterprises operating through market competition can more efficiently and effectively manage urban space than conventional allocation mechanisms of public authorities.[162] The shift to neoliberal modes of urban governance does not necessarily mean a decrease in the amount of open, accessible places for social congregation and chance encounter, yet it does mean a change in how these post-public spaces are produced and regulated.[163] A great deal of scholarly attention has been directed at understanding how municipalities enter into officially sanctioned contracts—especially the increasingly popular type of hybrid ownership model called public-private partnerships—to provide incentives for the privately owned enterprises to create or manage public spaces.[164] As a general rule, state agencies and private companies have initiated new ways of cooperating in the provision of municipal services and infrastructure. The various mechanisms include contracts and concessions, build-operate-and-transfer (BOTs) arrangements, public-private joint ventures, and informal and voluntary methods of cooperation. Through deregulation, state agencies have enabled private companies to compete in the commercial marketplace with public agencies and state enterprises. Municipalities have used corporatization to require state-owned public enterprises to compete in the capitalist marketplace with private firms and to cover their costs and manage their operations more efficiently. Moreover, municipalities have permitted and even encouraged businesses, community groups, cooperatives, private voluntary associations, small enterprises, and other nongovernmental organizations (NGOs) to provide social services.[165]

Under circumstances of lean government, where municipal governance has become an enabling agency for sponsoring private enterprise, real estate developers have acquired, in many instances, virtual carte blanche to implement large-scale megaprojects that effectively reshape urban landscapes. This practice of "unsolicited urbanism"—a catchy term coined by Chris Gibson—refers to the mechanisms through in which property developers are able to manipulate, circumvent, and otherwise subvert formal regulatory guidelines in order to implement large-scale real estate development projects.[166] Unsolicited urbanism connects property, power, and money in an unholy alliance. As Dallas Rogers and Chris Gibson suggest, unsolicited urbanism not only legitimates private development monopolies over specific sites in ways that are inconsistent with the neoliberal tenets of market competition but also normalizes secretive decision-making processes that often take place outside the glare of public scrutiny, engagement, and access. This practice—which might be called "planning-as-deal-making"—allows large-scale private property developers to overcome procedural barriers embedded in preexisting statutory planning frameworks and to legitimate secretive proposals to "unlock" and otherwise appropriate valuable land and useful assets.[167] For Rogers and Gibson, unsolicited urbanism takes planning as deal-making beyond the conventional neoliberal recipe and into uncharted territory. "By creating the conditions for monopoly deals over specific high-value sites," the practice of unsolicited urbanism eliminates altogether the requisite public tender/bidding processes and "accompanying market competition."[168]

In the post-liberal city, new modes of urban governance—a vast array of pseudo-public authorities, public-private partnerships, redevelopment agencies, special-purpose commissions, and unelected entities operating outside of public oversight—have become the primary mechanisms for soliciting, approving, and implementing large-scale private real estate projects. City-building projects have increasingly sought approval "through the logic of exception, whereby large-scale real estate developers demand exemptions from existing regulatory frameworks, including tax allowances, relaxation of land-use regulations, zoning variances, and accelerated approval processes."[169] The cobbling together of new technical-regulatory procedures marks a shift from standardized rule-bound frameworks with their single-purpose mandates to a flexible, fluid system of exception-granting ad hoc deal-making.[170] Rule through exception has become the norm.[171] These accumulated powers of exception—the unimpeded capacity to bypass and rescind the normal palette of rules and regulations—provides powerful real estate developers with wide discretion to "unbundle and rebundle" existing regulatory regimes to suit their own purposes by bending public authorities to their way of seeing the city.[172] The emergence of these new modes of concessionary urbanism enables coalitions of real estate developers, global financial interests, and high-ranking government officials, working with engineering and financing consultants, to tap into floating circuits of global finance and "to access, redevelop and extract value from otherwise unattainable city sites and assets."[173]

POST-PUBLIC SPACE IN THE POSTINDUSTRIAL METROPOLIS

Collectively, the serial reproduction of such privately imagineered designer landscapes as enclosed shopping malls, corporate plazas, arcades, gallerias, reinvented Main Streets (with their quaint storefronts and curbside restaurants), and other contrived or themed stage sets create an illusion of open public space from which the risks and uncertainties of everyday life are carefully eliminated. Often created as facsimiles of some distant place or time—past or future—these themed entertainment destinations are corporate productions associated with the tourism and entertainment industries.[174] The packaging of historic cultural-entertainment districts, using museums, sports complexes, and convention centers, has emerged as the new formula for pumping life into the downtown centers of old industrial cities.[175] The steady accretion of such pseudo-public spaces has not only become the centerpiece of place-marketing efforts to sell the city in the image-conscious world of global neoliberalism but has also gone hand in hand with new mechanisms of social exclusion and marginalization of unwanted Others. While appearing to be open and inviting, these post-public spaces have become contested sites of competition, conflict, and antagonism.[176] In trying to account for the exclusionary logics that regulate access to (and hence the use of) these hybrid spaces, an entirely new vocabulary—domestication, purification, pacification, and the like—has come into play.[177] Whatever the nuanced difference among these terms, they support a grand narrative about the gradual erosion of the more authentic forms of public life that historically defined modern cities.[178] These terms blend together with a range of concepts like securitization, ordering, taming, sanitizing, homogenizing, commercializing, and controlling to describe the radical devaluation of public space (and the withering away of the public realm) as a powerful social and political ideal in the modern city.[179]

On balance, the disciplining of urban space gravitates between hard and soft approaches.[180] On the one side, there are those who stress the growing militarization of urban space—that is, the expanded and visible use of physical force and intimidation (zero-tolerance policing) to pacify the cityscape.[181] Seen in this light, an emergent revanchist urbanism represents the seamy side of the new tourist-entertainment cities.[182] This kind of dominating power is typically equated with the marked presence of physical barriers, proactive (and visible) policing, and invasive surveillance technologies.[183] As John Allen has persuasively argued, the exercise of power in these post-public or privatized public spaces depends ultimately on some form of domination, where the choice over who gains access is constrained by certain kinds of watchful authority or discriminating rules of entry.[184] Put more broadly, an emergent kind of "fortress urbanism" has come to dominate contemporary city-building efforts, where urban landscapes are increasingly divided into citadel office complexes, security zones, and bunkerlike enclaves.[185]

In general, such functionally specialized enclaves as gated residential communities, city improvement districts, and business precincts depend upon narrowly defined (and fixed) principles governing which people are the legitimate "authorized users" and which are not. Those who enter these carefully monitored (and highly regulated) places are under no illusions about who is wanted or who is not. In order to maintain tight control over access and to monitor use, functionally specialized enclaves typically rely on various types of filtered exclusion, where the exercise of power works through an assemblage of walls, gates, barriers, electronic monitoring systems, and hard policing so that only the "right types" of authorized visitors are permitted to enter.[186]

On the other side, municipal authorities have introduced more subtle, less visible, yet equally effective methods of managing and regulating the use of urban space. Such post-public themed entertainment sites as festival marketplaces, enclosed shopping malls, and glitzy "shoppertainment" extravaganzas depend for their commercial success upon the staged performance of a certain type of accessibility, inclusiveness, and openness. The scripted nature of post-public accessible spaces—often a stage set for the performance art of seeing and being seen—enables anonymous strangers to socially congregate and mingle freely, yet they "unknowingly remain subject to a form of control that is regularized, predictable, and far from chaotic."[187] Unlike monofunctional enclaves where access is carefully monitored, themed tourist-entertainment sites provide an experience of authentic social mixing that is not illusory but real.[188] There is a genuine invitation to mingle, circulate, and inhabit these inviting spaces, not as passive consumers of ersatz spectacle and counterfeit imagery but as active participants in the experience itself.[189]

Closure and exclusion in these multifunctional sites rely upon complex modes of power that are less obtrusive and more subtle than those hard mechanisms of spatial control that monitor and regulate the use of space in functionally specialized enclaves.[190] There is more to the exercise of domination in commercial post-public places than the exclusionary logics of physical barriers and visible displays of authority. Closure in the kinds of post-public spaces that rely upon a kind of spatial openness is not just about domination but also about seduction. The power of seduction works "through the experience of the space itself," through its ambient, sensory qualities of sights, sounds, and smells. Ambient power derives from the character of these post-public spaces, where the spatial layout, aesthetically pleasing architecture, and controlled experience produce "a particular atmosphere, a specific mood, a certain feeling" that together induce conformity with expected behavior and action.[191] Somewhat akin to "pacification by cappuccino" (an evocative metaphor coined by Sharon Zukin), ambient power works through the flexible appeal of seduction rather than brute force—that is, through persuasion rather than strict regulation.[192] The sensory qualities of these post-public places create a virtual "economy of effect" where the actual experience, rather than commodified objects per se, becomes that which is consumed.[193] This mesmerizing logic of seduction effectively channels options in manageable directions, restricts choices, and curtails possibilities, thereby "enticing visitors to circulate and interact in ways that they would not otherwise have chosen."[194]

Control in these types of post-public accessible space works through what can be called the subtle "domestication of space."[195] Closure is achieved through a seductive type of inclusion rather than enforced exclusion. In making post-public spaces attractive to "acceptable users" but not to others, architects have softened their features, eliminating hard edges and erasing uncertainty, and designed out indeterminate places, opening them to more sedate and relaxed forms of recreation and leisure, while simultaneously monitoring these sites through a range of largely surreptitious security measures. For example, in his study of Bryant Park (New York), David Madden has suggested that the reorientation of conventional public space—what he terms "publicity without democracy"—has effectively decoupled places of social congregation from discourses of democratization, citizenship, and self-development and connected them ever more firmly to consumption, commerce, and social surveillance.[196]

The registers of power that operate in these post-public places are more subtle, but no less insidious, than the hard-edged policing and management of fortified enclaves.[197] The conjoined tactics of displacement and exclusion are always small-scale and incremental, silent and stealthy. At the end of the day, these softer, indirect, and seemingly benign techniques of social sorting and gentrification by stealth accomplish their goals of social cleansing just as effectively as large-scale strategies of urban removal and spatial clearances.[198]

CHAPTER 4

STRUGGLING POSTINDUSTRIAL CITIES IN DECLINE

Shrinking cities have been shrinking for as much time as they have been growing.

Keller Easterling[1]

Shrinking Cities represent a fin-de siècle *realization that modernity's optimistic engagement with urban decline, as a reversible and episodic misfortune preying on good-planning deprived cities, was after all, a chimera.*

Ivonne Audirac[2]

The current hypermodern age of globalization has gone hand in hand with the emergence of so-called competitive cities—those thriving, healthy metropolises with world-class aspirations that have effectively sidestepped the deleterious effects of job loss related to deindustrialization and the fiscal challenges brought about by declining public expenditures for municipal services, a shifting tax base, and austerity budgets.[3] In negotiating the opportunities that accompanied global restructuring of production and finance, competitive cities have largely turned to market-driven and culture-led regeneration strategies in order to compete in the global marketplace. In jockeying for position on the world stage, new growth coalitions are constantly on the lookout for new ways of promoting their distinctiveness to lure investment, create jobs, expand their tax base, and in the end, restore and enhance local prosperity.[4] The most successfully competitive cities—celebrated in the scholarly literature as global cities or world-class cities—have reached the apex of this ranked hierarchy of thriving metropolises where they peer down from the commanding heights of the world economy.[5]

Generally speaking, what underlies this focus on competitive cities is a normative impulse.[6] Those cities deemed to have achieved the status of competitiveness establish the standards and benchmarks that other, less successful cities strive to emulate.[7] These second-tier cities—which might be called emerging market cities, or globalizing cities with world-class aspirations—become trapped in mimicry if not outright plagiarism, seeking to replicate the highest-value opportunities and sustainable competitive advantage in order to propel themselves upward in the ranked hierarchy.[8]

The scholarly attention directed at aspiring world-class cities has spun off a large subset of secondary literature that highlights such ideas as creative cities (Landry and Bianchini, Florida), entrepreneurial cities (Glaeser and Kerr), and smart cities (Vanolo).[9] For the most part, the writing in these fields has assumed a decidedly prescriptive policy orientation, offering recommendations for achieving some sort of comparative advantage, thereby gaining a competitive edge.[10] For example, research and writing on urban entrepreneurship and the entrepreneurial city have largely focused on the implementation of market-led redevelopment strategies, flexibilization and deregulation of public regulatory frameworks that might hinder free-market values, and privatization of municipal services.[11] In contrast, critical responses to various kinds of policy advice based on entrepreneurial strategies are often framed around the triumph of neoliberal urbanism and the emergence of the neoliberal city.[12]

The flipside of the highly successful competitive city is the "noncompetitive city"—a pejorative term that suggests some degree of failure of vision, administrative incompetence, or worse. Whereas competitive cities with world-class aspirations are able to effectively "fight for the global catwalk," noncompetitive cities languish in a liminal state of permanent crisis. Characterized by stagnating growth, declining investment, and job loss, these cities exemplify the consequences of failed urbanism: suffering from neglect and disrepair, passed over, bypassed, and abandoned.[13] Without a theoretical understanding of decline and shrinkage, growth becomes the default conceptual framework for understanding urban transformation. Fashioned in this way, urban decline becomes an aberration, an outlying condition, and a deviation from the norm. These struggling cities in decline are increasingly rendered redundant and superfluous by virtue of their structural irrelevance in relation to "glocal" networks of production and consumption. The postindustrial cities of the American Rust Belt (Detroit, Flint, Toledo, Cleveland, Youngstown, Gary, Buffalo), the "shrinking cities" of post-socialist Central and Eastern Europe (Halle/Leipzig, Ivanovo [Russia]), and "shock cities" destroyed by war (Sarajevo, Aleppo) and social catastrophe (New Orleans, Houston, Chernobyl/Pripyat) epitomize this type of distressed urbanism. As a result of increasing competition on an ever-widening global scale, some cities are losing out and falling behind.[14] Postindustrial cities in decline seem to languish in suspended animation, neither vibrant and alive nor dead and motionless, embedded in a kind of stalled or interrupted time, trapped in the stillness of entropy and ruin.[15]

Primarily located in what were once the industrial heartlands of the core areas of the capitalist world economy and the post-socialist bloc of Central and Eastern Europe, depressed postindustrial cities in decline (often referred to as "shrinking cities") are characterized by factory closures and mass layoffs, obsolete and decaying infrastructure, vacant and deteriorating housing, abandoned buildings, unusually high concentrations of unemployed and underemployed people, the very poor and homeless (often immigrant and elderly populations), and heavily polluted environmental wastelands.[16]

In mainstream urban theory, struggling postindustrial cities in decline occupy a subordinate position within a spatial taxonomy defined primarily by binary oppositions. In the world of cities, there are winners and losers, competitive and noncompetitive ones, and growing and stagnating ones.[17] Postindustrial cities in decline are always situated on the losing side of these polar extremes—denigrated and stigmatized. The subtext of these binary modes of classification is that decline and abandonment are both unwanted and inferior outcomes, conditions to be avoided in order to prevent ruin and rejection.[18]

Struggling postindustrial cities in decline might be considered part of the "wrong story," as Keller Easterling has put it—"things that are not supposed to happen."[19] Conventional theories of urban growth have long regarded shrinkage or decline as a temporary setback in the linear trajectory of urban transformation, where recovery is only a matter of getting the proper market-led strategies in alignment with enabling municipal policies.[20] Revisiting this faulty premise allows us to consider shrinkage and decline as long-term, structural components of urban transformation. Rethinking conventional theories of urban transformation requires us to do away with the presumption that growth is the normal expected pattern of development while decline is an unexpected aberration or deviation from the norm.[21]

Whereas aspiring world-class cities are buoyed with fanciful visions of progress and future success, postindustrial cities in decline are accompanied by dystopic narratives of profound loss interspersed with bittersweet memories.[22] Littered with neglected infrastructure, abandoned properties, and underutilized buildings that no longer serve their originally intended primary purpose, postindustrial cities in decline reflect the lost aura of modernity that failed to deliver on its promise to produce steady growth and prosperity or, to paraphrase Rem Koolhaas, to transform quantity into quality.[23]

Conventional urban planning practice has remained largely fixated on finding the magic elixir that can regenerate, rejuvenate, and resuscitate these dead and dying places. There is no lack of solutions calling for fresh capital investment, business-friendly tax credits and rezoning, and other market-driven panaceas for jump-starting economic growth in these abandoned, declining, and stagnant cities. The romantic attachment in the scholarly literature to cultural-led redevelopment, to creative cities, and to competitive cities positioned to rise in the rank order of aspiring world-class cities only contributes to the misery of those loser cities that are marked as embarrassing failures, basket cases, and eyesores.[24]

The recent upsurge in attention directed at shrinking cities (a rather imprecise term that has nevertheless captured the imagination of urban scholars) has catalyzed a discussion that runs counter to the groundswell of celebratory rhetoric around the themes of globalization, gentrification, urban lifestyles, cultural-led rejuvenation, tourist and entertainment cities, and the creative class.[25] Scholars and policy makers invented the term "urban shrinkage" to indicate the processes of physical abandonment and economic decline in many once-industrial metropolitan regions negatively affected by disinvestment and the resulting job losses and socioeconomic decline that accompanied the transition to post-Fordism. Urban shrinkage can be characterized as a multidimensional phenomenon encompassing metropolitan regions, cities, and even parts of cities that have experienced a dramatic decline in the economic, demographic, and social foundations that historically contributed to their overall health and prosperity.[26] The causes for this urban decline are many and complex: unfettered suburban sprawl, socioeconomic transformation from traditional industrial-based economies to postindustrial new urban economies, economic restructuring from state economies to market economies, aging populations and demographic shifts, and the like.[27] At the end of the day, as Ivonne Audirac has cogently suggested, shrinking cities represent a *fin de siècle* realization that the optimistic engagement of global modernity with urban decline, framed as a reversible and episodic misfortune preying on distressed cities deprived of good planning practices, was after all just a chimera or a delusional pipe dream.[28]

THE SHRINKAGE SYNDROME: POSTINDUSTRIAL CITIES IN DECLINE

Once the object of attention for only a relatively small group of architects and urban design experts located primarily in Europe, cities and towns that have experienced significant socioeconomic deterioration brought about by plant closures, shutdowns of mining facilities, business bankruptcies, mortgage and tax foreclosures, and sustained population losses have become the focal point of much public discussion and debate over the past two decades or so. Large cities and small towns are not immune to the unrelenting pressures of urban shrinkage.[29] The socioeconomic decline of the coal-mining and mill towns (Scranton, Wilkes-Barre, Lancaster/York, Lewisburg) of the industrial heartland of central Pennsylvania exemplifies this hidden story of shrinkage.[30] Urban shrinkage has appeared in unexpected places, and with greater frequency than once imagined possible. Traveling exhibitions, popular commentaries, television reporting, newspaper accounts, journalistic exposés, film festivals, art installations, policy documents, and documentary cinema—such as *Sketches of Kaitan City* (Kazuyoshi Kumakiri, 2010), *Hula Girls* (Sand-Il Lee, 2006), *Requiem for Detroit?* (Julien Temple, 2010), *Detropia* (Heidi Ewing and Rachel Grady, 2012), *Sometimes City* (Tom Jarmusch, 2011), *Cleveland: Confronting Decline in an American City* (Lincoln Institute of Land Policy, 2016), and *Pruitt-Igoe Myth* (Chad Freidrichs, 2011)—reached wide popular audiences, triggering curiosity and interest that spilled beyond the somewhat narrow confines of academia.[31]

This pervasive shrinking-city syndrome marks the end of the modern industrial era when large-scale manufacturing and production were clustered in the vicinity of large and medium-sized cities in the core areas of the capitalist world economy. Although specific circumstances may differ from one metropolitan region to another, the multidimensional causes are deeply embedded in the processes of globalization that accelerated beginning in the late twentieth century. In the current age of hypermobility of capital, production and circulation processes are spatially dispersed to new manufacturing sites around the world, yet globally integrated through transport, logistics, telecommunications, and finance.[32] Struggling cities in Europe, North America, and other core areas of the world economy have been forced to cope with the effects of long-term socioeconomic restructuring brought about by deindustrialization and job loss. Faced with dwindling corporate investments in production facilities, cities whose prosperity and relative well-being were largely dependent upon manufacturing industries

FIGURE 4.1 Postindustrial Cities in Decline: Brightmoor Neighborhood, Northwest Detroit, Michigan. Abandoned streetscape and vacant land. Photograph by Martin Murray

have come face-to-face with the devastating consequences of deindustrialization and resulting job loss.[33] As homeowners experience foreclosure and business owners go bankrupt, tax revenues decline. As a result, city departments fail to maintain municipal services.[34] The resulting austerity urbanism offers little room for maneuver as public funding disappears for all sorts of social programs and recovery projects.[35]

The increasing mobility of capital set in motion a fierce competition among cities, each seeking to attract new investments (or at least retain old investments) or to maintain or rise in the ranked hierarchy of global success.[36] As a general rule, the unrelenting pressures of global competition have tended to sidestep once robust industrial cities. These cities have either successfully negotiated the transition to postindustrial dependence on services, tourism-entertainment, and finance, or they have struggled to remain healthy and vibrant in the face of innumerable challenges.[37]

These struggling postindustrial cities in decline seem to present an intractable conundrum for mainstream urban theory and planning practice.[38] Conventional theories of urban growth presume that the right mixture of ingredients (fresh capital, location, market competition, good governance, flexible regulatory frameworks, and generous tax subsidies) represents the proper approach to revitalization, regeneration, and restarting the tried-and-true engines of future prosperity. Distressed cities suffering from abandonment and neglect seem to defy conventional expectations about patterns of human settlement and normal (or expected) pathways of urban growth and development. Their presence has repeatedly frustrated the desire for certainty inherent in modernist urban planning, governance, and development.[39] They have typically been viewed through a moralizing lens that suggests that they lack the visionary leadership with wisdom and imagination, the proper regulatory and institutional frameworks, and an open-ended entrepreneurial playing field upon which to base rational, technical solutions to restoring prosperity. These prescriptive pronouncements are typically framed in ways that amount to something akin to "blaming the victim," with austerity budgets, cutbacks in social services, and the supposedly dispassionate rule of the market disproportionately disadvantaging the poor and vulnerable.[40]

The rise of the competitive, entrepreneurial city has resulted in a new social geography of exclusion—"black holes of social exclusion throughout the planet"—with growing inequalities between competitive cities that have integrated into the global networks and uncompetitive cities that have not succeeded in carving out a place in these networks.[41] Knowledge-based industries, innovation, networking, and creativity appear to be the key features determining economic performance and spatial differentiation.[42] Some cities attract investments in knowledge-based and service industries and the kinds of skilled and qualified workers affiliated with them, while others lose their economic base, shedding jobs and population. These polarization processes occur at every scale—globally, regionally, and even locally within the cities themselves.[43]

GENERAL PATTERNS OF STAGNATION AND CONTRACTION

Throughout the twentieth century, thousands of new cities came into existence across the United States. Yet at the same time, the opposite was occurring. A smaller but sizable number of cities and towns were dying, shutting down their municipal governments, relinquishing their local autonomy, and returning to legal and administrative dependence on counties. Between 2000 and 2010, for instance, 138 large and midsize cities across the United States lost population due to a variety of causes.[44] During the same decade, at least 130 local governments in cities, towns, and villages dissolved as a result of financial stress—nearly as many as incorporated during the same period.[45]

While the origins of urban malaise can be traced to the last decades of the twentieth century, the global financial crisis of 2007–2008 brought into sharp relief the surprisingly large numbers of cities and metropolitan regions in the United States, Japan, and Western and Eastern Europe faced with disinvestment, population loss, high unemployment, and fiscal insolvency.[46] Yet for the most part, conventional urban studies literature and planning theory have found it nearly impossible to budge from the position that decline is merely the absence of growth, with market-led strategies of regeneration representing perhaps the best and only remedy for stagnation.[47]

As the interest in urban decline has expanded in both scholarly and policy circles, urban shrinkage has acquired a new meaning, connoting a variety of urban afflictions and encompassing cities of both the Global North and Global South. Recent scholarly research and writing have suggested that postindustrial cities in decline—commonly referred to as the "shrinking cities syndrome"—represent a previously unknown type of city, one that is qualitatively and quantitatively different from previous encounters with urban decline.[48] For some, shrinking cities are like the proverbial canary in the coal mine—the early warning signs of the impending global urban crises of modernization, suburbanization, and metropolitanization that have enveloped the industrialized world.[49] According to Philipp Oswalt, for example, shrinking cities at the start of the twenty-first century are the harbingers of the end of two centuries of industrialization fueled by fossil energy; a crisis of cheap mobility, aging populations, and new waves of technological restructuring; and a crisis of social polarization, with the worsening of social divisions between growing places, connected to global circuits of capital and information, and shrinking places almost totally dependent on local resources.[50] As a consequence, the processes of urban shrinkage have become so ubiquitous that theorizing about decline has become as important as thinking about growth.[51] The proliferating language of postindustrial renaissance, image making, and civic boosterism provides ample testimony that, for cities that have experienced deindustrialization and disinvestment, the image of the transformation of declining industrial wastelands ("ruinscapes") into robust postindustrial landscapes that are "ripe for investment, places of profit, and sources of benefit to capital" has squeezed out alternative visions for a better future grounded in conceptions of equity and justice and ideas about "sharing economies."[52]

As a field of inquiry, conventional urban planning practice has little to say about what distressed cities should become following extensive population and employment loss, disinvestment, and abandonment, or how planners should operate when the prospects for redevelopment are virtually nonexistent, at least in the foreseeable future.[53] The doctrinaire (and dogmatic) fixation on growth has remained deeply embedded in conventional planning culture. Faced with shrinking financial resources and dwindling populations, urban planners have typically reverted to the conventional fallback position of encouraging market-led growth and economic revitalization.[54] Despite clear evidence to the contrary, spatial planning policies have often revolved around a kind of hopeful optimism, operationally akin to denial, that looks upon shrinkage as a temporary cyclical phenomenon that will eventually right itself rather than a permanent, irreversible condition.[55]

GENEALOGIES: HISTORICAL AND CONCEPTUAL ORIGINS

While its precise origins remain clouded in ambiguity, the term "shrinking cities" appears to have been first used in a systematic way more than three decades ago by Hartmut Häußermann and Walter Siebel in their extended examination of the negative repercussions associated with deindustrialization in the Ruhr Valley.[56] During the nineteenth century, the marriage of coal and steel transformed the entire region. Despite the boom years after World War II, the coal and steel industries underwent significant decline starting in the 1970s, if not before, as the result of a deepening structural crisis of intensified global competition, technological change, lower productivity, idle production facilities, and high unemployment.[57] The idea of shrinking cities was derived from the German term *schrumpfende Städte*. In their book *Neue Urbanität*, Häußermann and Siebel investigated processes of urban shrinkage in relation to the balance between growth and stagnation.[58] They identified what they considered the two main causes for urban shrinkage: first, the centrifugal force of suburbanization—the decline of inner-city areas in favor of growing areas in the periphery; and second, deindustrialization and capital flight that triggered the erosion of the industrial foundations of urban prosperity.[59]

Although it may have been used earlier, the term became embedded in a distinctly German discourse about the relationship among growing, stagnating, and declining cities.[60] From the start, the shrinking-cities discourse gathered different kinds of urban decline under a single term.[61] In the 1990s, scholars employed the concept of shrinking cities to describe the specific characteristics of urban transformation during the post-socialist transition in Eastern Europe, and eastern Germany (the former German Democratic Republic) in particular, where the promise of market-driven growth, including privatization and downsizing, led to significant numbers of plant closures, population loss, austerity budgets, and the erosion of public services.[62]

The expanded use of the term "shrinking cities" is often associated with the German *Stadtumbau Ost* (Regeneration East) Program, which was designed to confront the challenges of cities undergoing considerable population losses in eastern Germany following reunification. To draw attention to struggling cities, urban policy makers and planners in Germany searched for a less provocative and more neutral-sounding term, trying to avoid the pejorative connotations of such value-laden concepts as "decline," "abandonment," "blight," or "ruin." The term "shrinking cities" (*schrumpfende Städte* in German) quickly became one of those traveling ideas applied to an ever-widening circle of circumstances.[63]

The term achieved worldwide popularity with the attention-getting launch of a traveling exhibition starting in 2002 called the Shrinking Cities Project sponsored by the *Kulturstiftung des Bundes* (German Federal Cultural Foundation) and financed largely by the German federal government. Carried out from 2002 to 2005 under the curatorship of Philipp Oswalt (Berlin), in cooperation with the Leipzig Gallery of Contemporary Art, the Bauhaus Dessau

FIGURE 4.2 Post Industrial Cities in Decline: Abandoned Packard Plant, East Side of Detroit. The Packard plant, one of the premier automobile production facilities in Detroit in the 1920s, is a collapsing hulk and industrial ruin today. Martin Murray Photograph

Foundation, and the architectural magazine *Archplus*, this traveling exhibition was linked to a large-scale research undertaking focused on four metropolitan regions that epitomized urban shrinkage. What became known as the Shrinking Cities International (SCI) project brought together four interdisciplinary teams to study shrinking cities around the world: Detroit; Manchester and Liverpool; Ivanovo (Russia); and Halle/Leipzig (eastern Germany). In addition to cities in the United States, Great Britain, Russia, Japan, and Germany, the touring exhibition visited a number of other countries particularly suffering from the impact of massive urban shrinkage.[64]

The overall effect of the Shrinking Cities Project was to turn the German term *schrumpfende Städte* into a global theme. Through the use of art installations, film, and photographic images of deserted and derelict industrial sites, abandoned and decaying housing estates, and neglected public spaces of depopulated cities, the project conveyed a distressing impression of utter urban devastation. The first phase of the project, presented in a number of books, aimed to broaden the debate about shrinking cities in eastern Germany to a global audience.[65] At each of the four research sites, a local group consisting of an interdisciplinary team of urban geographers, cultural experts, architects, journalists, and artists joined together to investigate the processes of shrinkage by mapping the causes and consequences of socioeconomic and demographical decline. Each case represented a specific kind of shrinkage: Detroit, shrinkage triggered by suburbanization; Manchester/Liverpool, shrinkage caused by deindustrialization; Ivanovo, shrinkage associated with post-socialism; and Halle/Leipzig, shrinkage brought about by a a combination of causes.[66]

The second phase of the project involved a call for deliberate interventions to address urban shrinkage, especially in Germany, other than just the demolition of abandoned and derelict buildings. By the mid-2000s, the concept of shrinking cities gained traction among other urban theorists and planning practitioners, including those at the Institute of Urban and Regional Development at the University of California, Berkeley, who established a Shrinking Cities Group.[67] Formed in 2009, the Shrinking Cities International Research Network (SCIRN) established a loose alliance of urban scholars whose collaborative work aimed to understand different types of city shrinkage and the role that different analytic approaches, strategies, and planning policies have played in the regeneration of these cities. The core message of the shrinking cities movement focused on four interrelated themes: (1) urban shrinkage is not a new phenomenon; (2) urban shrinkage in the current historical conjuncture is the outcome of multidimensional processes and is one spatial manifestation of globalization; (3) it is necessary to rethink and redefine orthodox understandings of urban decline (typically characterized as the absence of growth); and (4) it is possible to imagine urban repair and revitalization without dependence on remedial actions designed to stimulate growth.[68]

Despite these efforts to soften the term, "shrinkage" has still managed to carry a lingering stigma that suggests some sort of aberration, abnormality, or deficiency. Unlike an idea like "creative destruction," which implies a necessary step in order to enable progress, shrinkage suggests stagnation and entropy or a reversal and (in metaphorical terms) a moving backward in time.[69] In countering the negative stereotypes associated with "struggling cities in decline," policy makers and urban scholars in the United States pushed the discourse in the opposite direction, adopting the term "legacy cities," a deceptive and unhelpful term grounded in a kind of restorative nostalgia that provides little leverage to unpack the dynamics of deindustrialization in the U.S. Rust Belt and beyond.[70]

SCOPE AND BREADTH

Urban shrinkage on a global scale is certainly not a new or particularly recent phenomenon.[71] Journalists, scholars, and policy makers have long documented urban decline and contraction resulting from population loss and deindustrialization, leading in the worst-case scenarios to the eventual abandonment of vast areas of commercial and industrial undertakings and the neglect of entire residential neighborhoods.[72] However, even if urban decline is not new, it has spread, predominantly in industrialized countries but also all over the world. In Europe, the process of urban shrinkage has expanded at a rapid pace.[73] According to authoritative estimates, out of 220 large and medium-sized European cities, 57 percent lost population in the period from 1996 to 2001, as did 54 percent of the larger metropolitan regions.[74]

The burdens of decline are not limited to only a handful of unfortunate cities that can be labeled as special cases, deviant outliers that stray from the expected trajectory of urban growth and development. On the contrary, urban contraction has increasingly become a fairly normal pathway of urban transformation. Paradoxical as it may seem in a period of rapid urban growth around the world, it has been reported that about 370 cities with more than one hundred thousand residents have either temporarily or lastingly undergone population losses of more than 10 percent in the past fifty years. Between 1990 and 2000, more than one in four cities around the world experienced some population loss.[75] This situation is especially pronounced in Europe, where nearly one-third of all cities with more than two hundred thousand inhabitants have undergone at least one decade of population decline in the past forty-five years. According to a comprehensive survey undertaken by Manuel Wolff and Thorsten Wiechmann, 20 percent of European cities experienced shrinkage between 1990 and 2010.[76] Other researchers have suggested that by the early twenty-first century about 40 percent of all European cities were losing population.[77] In large measure, the negative growth trend is associated with cities in North America and Europe, where the number of shrinking cities has increased at a faster rate in the past fifty years than the number of expanding cities. In the United States alone, thirty-nine large or medium-sized cities have endured some population loss, while forty-nine cities in the United Kingdom, forty-eight in Germany, and thirty-four in Italy shrank in population between 1990 and 2000.[78]

According to authoritative estimates, one in six cities worldwide experienced substantial population loss and functional decline even before the 2007–2008 subprime mortgage crisis in the United States and the resulting global economic slowdown and recession.[79] In the United States, for instance, the 2006 census estimates revealed that sixteen of the twenty largest cities in the 1950s experienced significant population shrinkage by the start of the twenty-first century.[80] Thus, once heavily industrial cities like Buffalo, Cleveland, Detroit, Pittsburgh, Flint, and St. Louis have lost more than half their population from their peak numbers in the 1950s.[81] Between 1950 and 1980, Detroit lost half a million people, while Cleveland and Philadelphia each lost roughly three hundred thousand.[82] From a peak number of 676,000 in 1950, the population of the Pittsburgh inner-city area fell almost 50 percent over the following four decades. At its worst, the city lost almost ten thousand people a year.[83] Similarly, Baltimore and Philadelphia have shrunk by nearly a third. Cities outside the Northeast and Midwest—Birmingham, Memphis, Norfolk, Richmond, and pre-Katrina New Orleans—have also experienced significant population losses. Smaller industrial cities—Ashland (Kentucky), Camden (New Jersey), East St. Louis (Illinois), Flint (Michigan), Reading (Pennsylvania), Youngstown (Ohio), and Wheeling (West Virginia) have shrunk in similar fashion.[84] Ohio alone has so many shrinking cities—every major city except Columbus—that one observer has called it "the failed state of Ohio."[85] On a broader scale, *Forbes* drew attention to urban decline in 2008 by publishing lists of "fastest-dying cities" and "fastest-dying towns."[86]

As these cities have lost population, declining demand for commercial properties and residential dwellings has created a new urban landscape dominated by vacant lots and abandoned buildings. As younger workers with skills and education have departed, they have left behind aging populations with few marketable skills that would allow them to find work elsewhere. The residents who have

FIGURE 4.3 Postindustrial Cities in Decline: Riverbend Neighborhood, Detroit. Aerial view of housing vacancy and blighted landscape, Riverbend neighborhood, with factory and the downtown skyline in the background. Olaia Chivite Amigo, Drone Aerial Photograph

remained have become poorer, lacking skills, labor-force attachment, or mobility to compete in the regional or national labor market. Ironically perhaps, while the downtown cores and inner-ring suburbs have lost both residents and jobs, the surrounding metropolitan regions (particularly the outer-ring suburbs) have remained fairly static or increased in size.[87] To cite one case out of many, Philadelphia lost half a million residents between 1950 and 1990, while the surrounding metropolitan region gained more than one million during the same period.[88] Put in a wider framework, growth and decline operate in tandem, and the resulting patterns of urban expansion and contraction cannot be reduced to simple causes and predictable outcomes.[89]

According to authoritative estimates, between 1950 and 2000 more than 350 large cities around the world experienced, at least temporarily, significant losses in population. In the 1990s alone, more than a quarter of the world's largest cities experienced shrinkage.[90] A German government–sponsored project reported that since the mid-twentieth century, more than 450 cities of more than one hundred thousand lost at least a tenth of their population, including fifty-nine in the United States.[91] The system shock caused by the post-socialist transition after 1989 resulted in three out of four cities in Central and Eastern Europe experiencing population shrinkage.[92] The principal sites of shrinking cities in Europe are the post-socialist countries, especially Latvia, Lithuania, Estonia, Bulgaria, Romania, Hungary, Slovakia, the Czech Republic, Poland (particularly, the large-scale coal- and steel-producing agglomerations of Upper Silesia), Russia, Ukraine (particularly the Donetsk Basin region), and eastern Germany.[93] Some cities in northern Europe have continued to shrink in population and socioeconomic relevance, particularly port cities with obsolete shipping facilities, mining towns, and cities built around specific industries like steel production.[94] In Western Europe, the contrast between growing/stable cities and declining/suffering ones is more marked, particularly in Great Britain, Germany, and even France.[95] Southern Europe is also characterized by wide gaps, for example between fast-growing cities in Spain and Portugal and stagnant or shrinking ones in Italy.[96]

It is now widely acknowledged that declining population in European cities has become what some scholars have called the new normal. According to some estimates, almost 42 percent of all large European cities experienced shrinkage during the 1990s.[97] Postindustrial restructuring in what was then West Germany began in earnest in the 1960s and 1970s, leading to the widespread shrinkage of urban cores.[98] The entire coal-mining, steel-making region of the Ruhr Valley (including such notable cities as Dortmund, Essen, Bochum, and Duisburg), an area roughly comparable to the U.S. Rust Belt, experienced considerable contraction and decline.[99] By the mid-1980s, shrinking cities in West Germany had become distinct from growing ones.[100] After German reunification in 1990, the situation became considerably more pronounced. Many cities of the former German Democratic Republic (East Germany) experienced rapid deindustrialization accompanied by population losses averaging around 6 percent between 1991 and 2002. In some medium-sized towns, the population declined between 20 and 30 percent (for example, 29 percent in Hoyerswerda, 21 percent in Schwerin and Halle), indicating the huge geographic variations characteristic of shrinking processes.[101] High rates of unemployment, over 20 percent in many cities in eastern Germany, resulted in a massive movement of population to cities in the western regions of Germany.[102] As many knowledgeable observers have noted, shrinking in eastern Germany is not a temporary phenomenon awaiting readjustment but signals a structural condition of potentially lengthy duration. Population declined even further between 1999 and 2020, taking place at a rate of around 15 percent in the worst affected regions of Ostthüringen, eastern Thuringia, Nordthüringen, and Altmark.[103]

Old industrial cities in France and Great Britain—especially those tied to mining operations, textile mills, and steel factories—have experienced demographic loss and socioeconomic restructuring as capital investment and financing, in response to the pressures of globalization, have moved from heavy industry to tourist, entertainment, and information-technology activities.[104] From Nord-Pas-de-Calais to the Ardennes, from Lorraine to the belt of industrial centers of the Massif Central, from the former industrial cities of Scotland (Greater Glasgow and the Western Isles) to the Black Country, the process of deindustrialization took its toll. Flows of capital and people have increasing migrated away from older industrialized regions toward large cities and urban centers more actively engaged with technologically advanced enterprises.[105]

City shrinkage is not a random process, but instead assumes distinct forms in different locations.[106] While the relationship between the deindustrializing impulses associated with globalization and urban decline is fairly obvious, the distinct patterns and locations of urban shrinkage have manifested themselves differently in accordance with national, regional and local contexts. In Great Britain, for example, those locations most affected by shrinkage are mainly large metropolitan regions and/or industrial areas clustered around the middle and north of the country.[107] In Scotland, shrinking cities can be found in the central belt and also, as in Wales, in certain peripheral areas, including the offshore islands. Shrinkage has significantly affected Scotland, especially the Greater Glasgow metropolitan region and the Western Isles. In contrast, the south and southeast of England show little sign of shrinkage.[108] Shrinking cities in France are primarily located in the center of the country, unlike the east-west divide that characterizes shrinkage in Germany.[109] Fertility rates in France have remained relatively robust by European standards, prompting Emmanuèle Cunningham Sabot and Sylvie Fol to suggest that urban shrinkage has been a "silent process."[110] In highlighting what has happened at the local level, Cunningham Sabot and Fol distinguish three types of shrinking cities in France. The first type is primarily related to the long-term effects of deindustrialization, particularly in the declining industrial regions of Lorraine, Nord-Pas-de Calais, and Haute-Normandie. The second type consists of small towns away from major road and railway infrastructures, particularly those in the regions of Campagne-Ardennes, Midi-Pyrénées, Limousin, and Bourgogne. The third type concerns city centers and inner-city areas shrinking as a result of suburbanization and metropolitan sprawl. These processes are particularly evident in the fast-growing regions in the south of France, such as Marseille, Avignon, Toulin, and Perpignan.[111]

Among British cities, Liverpool and Manchester represent perhaps the most extreme examples of urban shrinkage. Starting with the economic crisis of the 1970s that was triggered in large part by deindustrialization, both cities underwent significant population loss, leaving declining inner-city neighborhoods, vacant land, and derelict buildings.[112] Despite efforts at urban regeneration over the past two decades, Liverpool has continued to shrink, whereas the story of Manchester combines successful revitalization of key sites with long-term stagnation and neglect in surrounding areas.[113]

It is sometimes overlooked that the worst urban decay and abandonment have occurred not in big cities but in small ones like Hartlepool (a former shipbuilding and steel town on the northeast coast of England), Hull (northeast England), Middlesbrough (often ranked near the top of the "worst places to live"), and Wolverhampton (west of Birmingham).[114] Urban policy experts urged public officials to abandon failing cities and towns across the north of England, such as Rochdale, Bolton, Blackburn, Grimsley, and Burnley, and to concentrate instead on helping local residents find jobs elsewhere.[115] Staff writers contributing to the *Economist* ignited a controversy in late 2013 when they openly called on public officials to recognize that they were battling against unyielding forces of decline and to abandon these places rather than continue to pour in financing of social benefits and regeneration projects not likely to reverse the downward spiral of ruination.[116]

Urban shrinkage is not confined to the cities of North America and Europe. In Japan, urban populations (outside Tokyo) are aging and shrinking, and the effect is particularly pronounced in certain locations.[117] Kitakyushu, located at the northern tip of Japan's southern island Kyushu, became one of the largest industrial areas in the country after its first iron-and-steel mill opened in 1901. Yet from the 1970s onward, it slipped into decline, and in the decade after 2006 its population decreased by 72,000.[118] Uneven patterns of "de-growth" have been particularly apparent in the Osaka-Kyoto-Kobe region (*Keihanshin*), where the combination of deindustrialization and declining fertility rates (coupled with the ultra-aging of the population) has resulted in urban shrinkage.[119] These post-growth dynamics have also spread to South Korea, where the combination of suburban sprawl and low fertility rates has drawn population from cities like Seoul.[120] While it may seem paradoxical, cities in India, including Delhi, Kolkata, and Mumbai, have experienced population shrinkage in their historical cores of the cities in the last decade.[121]

Similarly, small cities and towns in Mexico and resource-depleted mining towns in Australia, Canada, and South Africa have struggled to retain residents and bolster their local economic fortunes.[122] However, the most striking urban shrinkage in Australia has occurred in smaller towns, or clusters of small towns, in semirural regions that have suffered from prolonged droughts. Urban shrinkage has also taken place in small and medium-sized cities linked with mining and manufacturing whose cycles of growth and decline have been directly associated with fluctuations in the global marketplace for manufactured goods and minerals.[123]

The lion's share of attention devoted to postindustrial cities in decline, however, has focused on exemplary expressions of spectacular failure in such large deindustrializing cities and metropolitan regions as Detroit, Cleveland, and the cities of the Ruhr Valley. Often overlooked in the hyperbolic posturing over ruin and abandonment is that small former manufacturing cities in New England, the mid-Atlantic states, and the Great Lakes region have experienced slow and gradual decline for well over half a century.[124] Similarly, many small cities and towns in Europe have experienced stagnation and decline.[125] Without marketable resources and underserved by transportation infrastructure, these smaller cities are unable to compete against larger cities with employment opportunities and access to rail and road linkages. Located on the margins in relation to thriving cities, they have difficulty replacing manufacturing companies that have pulled out. The effects of post-Fordist restructuring have proved far more devastating in small cities, where continued economic vitality was dependent on a single industry, or even one large manufacturing company, such as IBM in Binghamton (New York) or Corning Glass in Corning (New York). Similarly, the demographic shrinkage of nearly one-third of French urban areas has primarily affected smaller cities. Located in the central part of the country, far from the coasts and the borders, these small cities do not have the benefits of major transportation infrastructure and metropolitan networks. With few economic development opportunities, they have been left behind in the competitive struggle fostered by globalization.[126]

UNPACKING THE USE AND MEANING OF SHRINKAGE

While urban shrinkage has increasingly shaped policy discourses in Europe, Japan, and to a lesser extent in the U.S. Rust Belt, the precise meaning of this term has remained elusive.[127] Hence, it is necessary to unpack the concept of shrinking cities, trace its origins and development over time, and eventually clarify the confusion over the multiple interpretations ranging across various spatial scales. It is also important to situate how the concept of shrinkage has ranged between its narrow use as an analytic tool to explain existing or emergent conditions on the ground and its much looser deployment as a descriptive umbrella term to denote the scope and scale of negative externalities associated with the downward spiral of decline brought about by some combination of deindustrialization, job loss, and a decaying built environment.[128]

Yet despite its fashionable usage in scholarly and planning discourse, the term is not supported by a clear definition, a robust theoretical framework, and rigorous empirical analysis. Shrinkage assumes different meanings when applied to different spatial scales. With a few exceptions, the concept of shrinkage has conventionally been applied to the middle-range level of urban agglomeration, focusing on metropolitan patterns of decline and abandonment rather than the micro scale of neighborhoods, on one side, or the macro scale of city-regions, on the other.[129] That almost one-third of shrinking cities are relatively small can perhaps explain why urban shrinkage has not received as much attention in the scholarly literature as might be expected.[130]

At the start, scholars and policy makers used the term "shrinking cities" to describe the processes of decline that many post-socialist East German and Eastern European cities experienced after the collapse of the Soviet Union.[131] In time, it became the preferred nomenclature to describe a particular kind of urban experience worldwide.[132] The term gained a wider scholarly audience when it was applied more broadly to making sense of postindustrial cities in distress in Europe and the United States, where post-Fordist deindustrialization processes starting in the 1970s led to the restructuring of the economic foundations for urban prosperity, with far-reaching negative social consequences.[133]

In the current age of globalization, the worldwide spatial restructuring of industrial production put into motion intense competition between deindustrializing cities and city-regions seeking to attract knowledge- and information-based business activities or tourist-entertainment amenities as a way to restore prosperity. Those cities and city-regions that were unsuccessful in this global competition suffered from abandonment and decline.[134] Cities and city-regions in northern Europe, Japan, and the United States suffered from a combination of deindustrialization and the effects of the so-called second demographic transition, characterized by falling fertility rates, aging populations, and declining average household size.[135] Thus, when considered as a worldwide phenomenon, processes of urban shrinkage combine a broad range of features in varying combinations.[136] Depending on where and under what circumstances, the articulation between global conditions and local contexts has given rise to specific types of shrinking cities and different approaches to understanding urban shrinkage.[137] Both the causes and consequences of urban shrinkage are so varied that it is virtually impossible to identify a universal explanation or single pathway. Urban shrinkage is a multidimensional process that operates at multiple scales, consisting of cities, sections of cities, or entire metropolitan areas that have experienced dramatic (and seemingly irreversible) decline in their socio-economic base and in the prospects for their recovery in a form that resembles what came before.[138]

THE MYTHICAL CONSTRUCTION OF SHRINKING CITIES

Despite widespread acknowledgment that urban shrinkage is an actual condition with real-life consequences, scholars and policy makers remain undecided about how to characterize decline and abandonment or how to respond to the challenges of "growth reversal."[139] Is urban shrinkage a paradox, an anomaly, an exception, or the expected outcome of uneven development? Addressing these questions is complicated by the overabundance of opinion and rhetoric that careens back and forth between denial and obfuscation, Pollyannaish optimism and dystopian dread.[140]

In both scholarly literature and popular accounts, certain deeply embedded myths typically accompany alarmist and dystopian stories of urban decline and abandonment. The first is the consoling myth of a lost idyllic past—a linear narrative that recounts a tragic fall from grace that begins from a historical highpoint where confidence in continuous growth and at least a modicum of social harmony prevailed and proceeds through a downward spiral of decline toward ruination. Yet growth and contraction are themselves highly fractured and contentious processes, inundated with social conflict and disruption, in which winners are rewarded and losers are punished.[141]

The second myth constructs decline as if it were more or less evenly spread over the metropolitan landscape, with blight and ruin engulfing everything in their path. Yet urban decline is never homogeneous or uniform. Decline is a multidimensional process with asymmetric and uneven outcomes.[142] For some, urban neglect and abandonment constitute the death knell for their way of life. For others, it is an opportunity to "take advantage of the disadvantage."[143] Opportunities for exploitation arise under the precarious conditions resulting from distress and vulnerability. A third mythical construction rests on a normative logic that suggests that decline is the absence of growth—an expression of market failure and the breakdown of the expected relationship between supply and demand. Yet decline is not the inverse of growth; they are not opposing forces but complementary processes that always work in tandem. City building is a relational process in which expansion and contraction are integrally connected.[144] A fourth mythical construction treats shrinkage as somehow an unexpected deviation from an expected, normal trajectory of growth. Yet if urban shrinkage is regarded as an exceptional outlier, then it becomes nearly impossible to explain. Decline is not an anomaly or an aberration; it is part of the inherent unevenness of capitalist investment in urban space.[145]

Urban decline and abandonment represent a challenge to capitalist narratives of unbounded progress and punctuated growth—both achieved through "creative destruction" and market competition. Ruin and decay suggest that private-property regimes and the market for landed property—regarded as deeply ingrained in the social fabric and hence the principal mode of urban transformation—may be temporary, contingent, and even fragile.[146] Cities have always been haunted by the prospect of ruin. The fanciful projection of ruined cities represents a counterimage to the dominant progressive narrative of industrial modernity.[147]

SHRINKAGE IDENTITY

When cities suffer from substantial population loss, high vacancy rates, abandonment, and neglect, they share common elements in what can be characterized as a "shrinkage identity."[148] While the linkage between shrinkage and crisis is almost always mentioned in the scholarly literature, this relationship is rarely explored from a conceptual and interpretive point of view. Urban decline depends upon the intensity and impact of these processes. As Robert Beauregard has suggested, "just because a city has fewer residents and fewer jobs does not mean that it is experiencing decline; the issue is the composition of these changes, their pace and the resultant distribution of costs and benefits."[149] Whatever the local circumstances, the one common denominator of urban contraction is that shrinking cities display symptoms of breakdown and crisis.[150] This combination of contraction and crisis can be traced in large measure to the negative impact of globalization at the time of late modernity.[151]

To be sure, the scholarly literature on the characteristics, the causes and consequences, and the modes of governance that have accompanied shrinking processes has grown considerably over the past several decades. A great deal of this discussion originated with efforts to understand the post-socialist transformation of cities and metropolitan regions of eastern Germany after reunification in 1990. Studies of shrinking cities in the former German Democratic Republic framed urban decline as the outcome of two concurrent and intertwined processes—deindustrialization and sustained population loss—with wide-ranging effects on the functioning and governance of local communities and often of whole metropolitan regions.[152] The physical manifestations of this decline were considerable: abandoned factories, decaying neighborhoods with underoccupied buildings, soft real estate markets, chronic property abandonment combined with empty wastelands, high crime, and a huge oversupply of dwellings and resulting residential vacancies.[153] Because scholars from what might be called the German

School view urban shrinkage as an enduring if not permanent condition, they have argued for a shift in planning practices away from conventional paradigms focusing on fostering growth toward more appropriate strategies based on right-sizing and downsizing. Among other elements, this shift in policy thinking calls for intensified residential property demolition and the repurposing of the built environment to adjust to demographic and economic change.[154]

Unlike growing metropolitan areas in the United States, Rust Belt cities in the Midwest around the Great Lakes and along the Atlantic seaboard display patterns of urban decline similar to those identified with shrinking cities like Dresden, Halle, and Leipzig in eastern Germany.[155] Deindustrialization, accelerated job loss, and out-migration have conspired to produce cities and city-regions in distress. As in many post-socialist cities of Central and Eastern Europe, the impact of deindustrialization in U.S. Rust Belt cities has triggered a number of ripple effects in which the out-migration of capital, jobs, and people has undermined the local tax base, thereby reducing the level and quality of public services and amenities.[156]

Another, often overlooked, kind of urban shrinkage is associated with resource-depleted cities. Natural resources must be sought where they can be found. Path dependency on resource extractive industries makes cities vulnerable in the extreme. Cities that owe their existence to the availability of minerals, lumber, fossil fuels, and even fishing are often subjected to recurrent boom-and-bust cyclesof resource-extractive industries. For example, the abrupt closing of mining operations in the towns of Taeback, Jeongsun, Samcheok, and Youngwol in the Taeback Mountains of Korea resulted in considerable urban shrinkage.[157] Boomtowns like the logging city of Yichun (northeast China) and Wansheng District (Chongqing, China) represent on the tip of the iceberg. State officials have found that fifty of the estimated 390 mining towns in China have already run dry, leaving three million miners jobless and affecting about ten million more people.[158] Similarly, resource-exhausted cities in North America, Russia, Eastern Europe, Australia, and elsewhere have faced comparable problems associated with plant closings, joblessness, and population loss.[159]

CAUSES AND CONSEQUENCES OF URBAN SHRINKAGE

Understanding urban decline must start from the premise that processes of deindustrialization and resulting job loss, dwindling population, abandonment of the built environment, housing vacancies, and fiscal austerity are not uniform, inevitable, or separate from political decision making and popular contestation. These processes of deterioration are place-specific. Though similarities can be identified among postindustrial cities in decline from the U.S. Rust Belt to the Ruhr Valley, the sociohistorical specificities with which these processes of abandonment and ruin play out have fundamentally transformed the lives and livelihoods of millions of people in ways that reflect the nuanced history, uneven spatial relations, nature of declining industries, and local politics of each location.[160]

Despite recognition of the highly uneven patterns of decline, there has been little discussion and debate about the heterogeneous structural causes of shrinking cities.[161] While wholesale restructuring of the capitalist world economy beginning in the 1970s marked the turning point toward the exacerbation of uneven urban development, scholarly research and writing have largely focused on deindustrialization, population loss, and property abandonment as the most important factors distinguishing growing cities from shrinking ones. This superficial overdetermination, however, masks the root causes of uneven urban development on a global scale.[162]

Whether proximate or otherwise, the causes of urban shrinkage are as varied as they are numerous.[163] Suburban sprawl, deindustrialization and disinvestment, out-migration, population loss, and abandonment, among other conditions, have resulted in highly uneven and complicated patterns of growth and shrinkage in metropolitan areas.[164] While the most simplified symptom of urban shrinkage is considerable and constant loss of population, the more far-reaching process of urban contraction and its causes and consequences are multifaceted and diverse. It is difficult if not impossible to disentangle the long-term structural conditions associated with urban shrinkage from the more immediate proximate causes with local origins. The reasons behind urban shrinkage cannot be reduced to a single determinant, or even a cluster of determinants, but must be understood as a complex body of interrelated processes, causal relationships, and feedback loops. Global forces often combine with local conditions to trigger urban shrinkage. Causes and effects reinforce each other in a never-ending cycle. At the end of the day, it is not really possible to identify a grand explanatory heuristic (or single causal chain) that can account for contraction and decline. Looking for a universal, one-size-fits-all theoretical explanation for shrinkage is an unattainable undertaking.[165]

Urban shrinkage is a multidimensional process that results from the interplay of different macro-processes that manifest themselves in various combinations at the local scale. One major cause of urban shrinkage can be traced to the deleterious effects of deindustrialization. Without compensatory capital investments in

FIGURE 4.4 Postindustrial Cities in Decline: Abandoned Residential Neighborhood, Riverbend, Detroit. North of Jefferson-Chalmers, the Riverbend district is one of the most blighted and vacant residential neighborhoods in Detroit. Olaia Chivite Amigo Drone Aerial Photograph

employment-generating activities, plant closures contribute to severe decreases in the number of available jobs and a rise in unemployment. Urban population shrinkage generally comes with a selective migration process, with the most qualified, the most skilled, and the youngest departing first. Those least capable of leaving are also the most vulnerable populations (the poorest, the oldest, and people with fewer skills). These people are the most adversely affected by the socioeconomic effects of unemployment and deterioration in their standard of living, and they are left in a liminal state of "captivity." This process of selective out-migration reinforces the downward spiral of declining cities and regions.[166] Within shrinking cities, low-income neighborhoods are generally the first to be adversely affected by decline. At the same time, socio-spatial inequalities tend to increase on a metropolitan regional scale. Urban decline results in diminishing financial resources, which typically spills over into budgetary constraints and fiscal crisis.[167]

The causes of urban shrinkage can be traced to a multitude of economic and demographic processes, and the separate impacts of these processes are often compounded when they overlap in specific combinations in local settings.[168] In

Eastern Europe and Russia, the dissolution of state socialist systems (and state-planned economies) and the transition to market regimes coincided with declines in fertility and considerable out-migration from cities.[169] This out-migration was particularly acute in geographically remote and peripheral cities, where the collapse of state-sponsored manufacturing and industrial production resulted in plummeting prospects for continued employment.[170] In cities in northern and Western Europe (and Japan), low fertility rates (sometimes described as a demographic shock) and aging populations (particularly in Germany) have magnified issues related to shrinkage, with demographic decline and deindustrialization reinforcing each other.[171]

In contrast to Europe, urban shrinkage in the former industrial heartlands of the U.S. Rust Belt, the mid-Atlantic, and the eastern seaboard has conformed to a longer historical arc of gradual decline that first became readily apparent in the decades after World War II.[172] Unlike the patterns of urban shrinkage in Eastern and Central Europe, linked primarily to the collapse of state socialism, the trajectory of decline in U.S. cities was not linked to historically specific conjunctures and dramatic political-economic upheavals. In the old industrial cities of the U.S. Rust Belt, the global restructuring of industrial production that began slowly in the second half of the twentieth century but quickly accelerated in the last couple of decades resulted in massive deindustrialization and job losses, leaving abandoned central cities and neglected inner-ring suburbs in their wake. Shutdowns, bankruptcies, and layoffs—exacerbated by pressures of municipal fiscal austerity—became the norm from the 1960s onward as new technologies, offshoring, and competition from overseas corporations led to declining local jobs in manufacturing and related activities. Many large and medium-sized cities that were once heavily dependent on industrial production (particularly those engaged in the production cycle of fossil fuels, steel and other metals, and the automobile industry)—places like Youngstown and Cleveland (Ohio), Erie and Pittsburgh (Pennsylvania), Flint and Detroit (Michigan), Gary (Indiana), Bridgeport (Connecticut), Camden (New Jersey), and Buffalo and Syracuse (New York)—suffered the devastating consequences of property abandonment, neglected infrastructure, and fiscal austerity.[173] More recently, shrinking population in the Sunbelt cities of California and Florida in connection with the subprime mortgage debacle and the great recession of 2006–2008 has led some urban scholars to suggest new geographic fault lines for urban shrinkage.[174]

In the U.S. Rust Belt, urban shrinkage in metropolitan areas is associated with a historically specific set of characteristics, particularly the impact of (1) deindustrialization coupled with declining opportunities for stable employment, (2) "the pernicious combination" of unchecked suburban sprawl and jurisdictional fragmentation, and (3) racial segregation caused in part by white flight from inner cities to the outer suburbs (in St. Louis, Camden, Detroit, and elsewhere).[175] The outer-ring suburbs (so-called bedroom communities) that surrounded the manufacturing cities of the industrial heartlands of the U.S. Midwest were able to incorporate themselves as autonomous municipal jurisdictions, thereby insulating themselves from having to share tax revenues and contribute to public expenditures such as social services and school systems in the inner metropolitan core.[176]

The strong lure of suburbanization brought a centrifugal migration (of people, business office complexes, and commercial facilities) away from the inner-city core. By providing not only new places of residence but also expanding job opportunities, suburbanization played a significant role in the decline of downtown urban cores. Suburban sprawl has triggered a massive territorial rescaling, with outer-ring suburbs expanding in population while inner-city cores have generally declined. Job growth has pivoted away from inner-city cores to residential suburbs and edge cities, triggered in large part by widespread automobile ownership and relatively easy access to home mortgages to finance new suburban homes.[177] In addition, the introduction of a whole range of public policies favoring suburbanization and punishing inner cities tilted financial support toward the get-out-of-the-city juggernaut.[178] This combination of growing middle-class suburban peripheries, declining investment in urban centers, and rising rates of poverty in inner-city neighborhoods strangled downtown growth and led inevitably to urban decline.[179]

When all is said and done, long-term structural transformations, in combination with more immediate circumstances of local origin, have typically accumulated to produce similar results: population shrinkage, declining opportunities for stable employment, neglected infrastructure, and neglected properties, leaving in their wake cities with vacant wastelands and abandoned landscapes.[180] The steady accretion of problems has drained essential resources from many metropolitan areas, leaving cities with diminishing fiscal resources and austerity budgets.[181]

THE SHAPE OF SHRINKAGE

Viewed at the level of the metropolitan region, patterns of urban shrinkage vary considerably.[182] Some metropolitan regions in the U.S. Rust Belt, such as Buffalo/Niagara Falls and perhaps Pittsburgh, were subjected to a slow process of population loss in both older downtown cores and inner-city suburbs, and even outer-ring suburbs, over many decades.[183] This scenario can be contrasted with a distinct spatial patterning known as the "doughnut effect," in which the historic inner-city core begins to contract or to become economically disassociated from the satellite edge cities developing around it. This kind of hollowing out has occurred in metropolitan areas in the United States (such as Cleveland, Detroit, and Houston), in Australia (where about two-thirds of all population growth between now and the year 2030 is expected to occur on the fringes of the five largest cities), and in Europe (for example, Glasgow), where robust growth in edge cities has lured business office complexes and industries away, along with middle- and upper-class residents, leaving behind those unable to afford to leave the older core city.[184] The recent back-to-the-city surge—often associated with gentrification—has offset this doughnut pattern in some cities.[185]

Like many postindustrial cities of the Global North, central cities in many Latin American metropolitan regions—such as São Paulo, Santiago, Mexico City, and Guadalajara—that developed strong industrial production sites during the period of import substitution industrialization (ISI) have undergone accelerated restructuring processes characterized by shrinking metropolitan cores and inner-city suburbs combined with rapidly expanding peripheries after the adoption of neoliberal macroeconomic reforms in the 1980s.[186] This pattern of shrinkage (what has been called the *vaciamiento urbano*), where decentralizing pressures have resulted in declining downtown cores and a corresponding expansion of industry, commerce, and residential accommodation toward the peripheries, conforms to the hollowing-out or doughnut effect.[187]

Nevertheless, in the United States and Europe, there are many examples where the doughnut effect is not evident. In Paris, for example, a stable core has remained, surrounded by shrinkage in the outer suburban rings, which were the traditional location of large-scale industry.[188] In eastern Germany, cities have declined unevenly, producing a more finely grained perforated pattern, with areas of growth and decline in close proximity to each other throughout the city.[189] The emergence of voids, empty spaces, or what Timothy Moss has called "cold spots," has produced an urban mosaic that defies simple classification.[190] Viewed at ground level, this checkerboard pattern is also seen in some cities like Detroit, where in some neighborhoods residents have commandeered adjacent vacant lots, creating areas of suburban-like, low-density development.[191] In St. Louis, pockets of prosperity and gentrification have persisted within the central city while the overall pattern points to continued population loss and neighborhood decline.[192]

In contrast to many Rust Belt cities, where suburban sprawl was integrally connected with urban shrinkage, shrinking cities in Central and Eastern Europe experienced different patterns of growth. Cities and towns in Central and Eastern Europe underwent outward expansion on their fringes, but this horizontal movement was primarily due to the pressure of urban growth rather than to centrifugal migration (either of people or business-commercial facilities) or "consistent deconcentration."[193] In eastern Germany after the collapse of the socialist command economies in 1990, suburban sprawl (in the more conventional use of the term) occurred in conjunction with demographic shrinkage and economic restructuring "on a scale unprecedented in terms of speed, depth and breadth." For example, Leipzig lost almost a fifth of its inhabitants within a period of less than a decade after 1990. Approximately half of this population loss was due to out-migration to the more economically prosperous cities of western Germany, but the other half was a product of rapidly expanding residential suburbanization. When the Leipzig municipality incorporated several surrounding towns and villages after 1998, it reclaimed its earlier population losses and increased its size to close to half a million residents. At the start of the twenty-first century, the flow of residents to the suburban peripheries gradually came to a virtual standstill.[194] In Central and Eastern European cities like Poznań (Poland), the metropolitan core areas have declined in population while the suburban peripheries have expanded.[195]

Similarly, in heterogeneous, polycentric metropolitan regions, growth and shrinkage have taken place side by side and at the same time. For example, in the Berlin/Brandenburg metropolitan region, centralization and decentralization have occurred simultaneously, producing a heterogeneous patchwork pattern that appears on every scale. While the suburbs of Berlin can be considered poles of stability, they exist in a regional landscape of shrinkage.[196] In cities like Cleveland, the fastest growing and fastest declining parts of the city are right next to each other, often intertwined. In this sense, urban shrinkage is not an abnormal condition but an integral, indeed inevitable, part of the growth–decline–renewal cycle of every city.[197]

SHRINKING CITIES: USEFUL FRAMEWORK OR FUZZY CONCEPT?

Broadly speaking, the term "urban shrinkage" refers to urbanized territories experiencing socioeconomic distress and demographical decline. While questions related to spatial polarization, increased differentiation between areas, and the balance between growth and decline are not new, the idea of shrinking cities has reentered scholarly discourse as a topic of special interest for researchers, planners, and urban designers. In its broadest definition, "shrinking cities" refers to metropolitan areas that have undergone the combined force of considerable population loss and socioeconomic decline over a prolonged period.[198] Urban shrinkage is a global phenomenon, interwoven with national, regional, and local circumstances, but the discussion has taken place in a range of geographic contexts, using several languages (English, German, French) and flexible terminology. As a result, national debates have occurred independently of one another, thereby inhibiting the potential for eliciting cross-national knowledge. Specific place-based shrinking processes may look similar, but the underlying causes may be considerably different.[199]

Terms like "shrinking cities," "urban shrinkage," "urban decline," "cities after abandonment," "cities in distress," and "postindustrial cities in decline" are often used interchangeably. While these terms have a great deal in common, each originated out of a different sociohistorical context. They each carry unstated assumptions and convey different meanings. In the United States, "urban decline" is the term typically used to describe neglect and abandonment in inner-city neighborhoods in the post–World War II period. In Germany, the term "urban shrinkage" originated at a later date and has featured prominently in debates about revitalization and postindustrial development. In its most common usage, "urban decline" refers to the spatial concentration in large cities of social, economic, and environmental problems, such as high levels of unemployment and poverty, social exclusion (particularly patterns of racial segregation), housing deterioration, and broken-down urban infrastructure. While urban problems may cohere with declining population and job loss, there is no necessary causal connection.[200]

Similarly, the term "urban shrinkage" also lacks precise meaning. While the term "shrinking cities" is associated with conditions ranging from low fertility rates to out-migration, population loss, suburbanization, and socioeconomic decline, no definitional clarity or consistency binds these elements together in a coherent package. As a consequence, the term suffers from loose usage and sometimes conveys contradictory meanings. Definitions that focus on population loss as the major dynamic characterizing shrinking cities "build a dividing line along a single dimension (population losses) which sorts cities according to a bimodal continuum of 'growth' and 'shrinkage.' "[201] Joseph Schilling and John Logan, for example, define shrinking cities as those older industrial areas that have lost more than 25 percent of their population over the past forty years and are characterized physically by abandoned properties, vacancies, and blight.[202] In contrast, Philipp Oswalt and Tim Rieniets regard a population loss of 10 percent as a sufficient threshold to qualify for "shrinking city" status.[203] Still other scholars have used much looser definitions, such as sustained population loss over an indeterminate time frame.[204] Yet, as Mattias Bernt has persuasively argued, the lack of consensus on threshold values—how much population loss constitutes urban shrinkage—leads to insurmountable methodological difficulties, especially regarding the possibility of comparative analysis, because of their arbitrary criteria.[205]

The term "shrinking cities" has entered into both popular and scholarly discourse as a kind of sweeping polysemic category that describes widely diverse cities with little in common. Current usage of the term is broad and fairly inclusive. In the United States, it has been stretched to encompass all manner of downward demographic shifts that have taken place beyond the typical Rust Belt localities, afflicting metropolitan regions and both large and medium-sized cities of a much more heterogeneous sort, including such cities as Fresno, Stockton, San Francisco, and San Jose.[206] The term "urban shrinkage" has also been applied to natural or human-induced disasters such as the devastating effects of Hurricane Katrina on New Orleans. While the broad usage of the term clearly "rounds up the usual suspects for urban shrinkage," it also incorporates a much wider range of cities and circumstances, raising the important question of whether it may well be too inclusive to be useful.[207]

In its typical usage, the term "shrinking cities" refers to a combination of conditions (for example, deindustrialization, contraction of employment, out-migration, population loss, housing vacancies, and fiscal stress) which, taken together, produce stagnation and decline leading to a socioeconomic crisis. Yet this list of problems is used inconsistently and is very much dependent on the local circumstances and historical context. Besides, there is no agreement about how little or how much of these conditions is required for a city to be included under the broad umbrella of shrinkage. For example, scholarly studies of cities in eastern Germany tend to focus largely on housing vacancies and property abandonment. In contrast, scholars exploring struggling cities in decline in the United States have stressed problems associated with deindustrialization, job loss, poverty, racial segregation, and crime.[208] As a result, the place of shrinkage-related problems in the definition of shrinking cities is arbitrary and inconsistent. On the one hand, there appears to be considerable agreement in the scholarly literature that a relationship exists between essential features of urban shrinkage like population losses and resulting phenomena like housing vacancies, property abandonment, and poverty. On the other hand, the "shrinking cities" literatures does not seem to indicate necessary connections between what appear to be very different features.[209]

While this litany of characteristics may help us to understand the occurrence of urban shrinkage and decline, they do not tell us about the degree or extent of these, or at what speed these processes have taken place. To give an example, mining regions in Europe, such as the Saarland region in Germany and the Black Country in the Midlands in the United Kingdom, have suffered from the closure of the mines.[210] Employment has decreased significantly, and people have consequently migrated to other regions, thereby causing populations to shrink. It is clear that urban and regional decline has occurred. Nevertheless, the degree of decline—something that really matters to local communities—differs strongly between the regions and also between cities within these regions. The closure of the mines alone cannot explain this variation. Concentrating on general causes makes it easy to overlook the influence of the particularities between regions and within regions—which may be crucial for understanding the variations among growing and declining cities.[211]

For both urban decline and urban shrinkage, the causes (deindustrialization, low fertility rates and aging populations, the centrifugal force of suburbanization) are as numerous and complex as the consequences (for instance, property vacancies, social segregation, crumbling infrastructures, and declining tax revenues).[212] But the frequent conflation of causes and consequences has produced even more ambiguity in the common, everyday use of these terms. This mixing of causes and consequences has led some scholars to suggest that the shrinkage literature has engaged in a kind of circular reasoning.[213] While the elasticity and flexibility of the term "shrinking cities" is useful for the productive employment of the concept, it is nevertheless necessary to tighten its definition and restrict its application. Otherwise, we risk diluting and overextending the concept of urban shrinkage to the point that it loses its heuristic usefulness in describing vastly different trajectories of differing urban experiences.[214]

In a similar vein, the use of the term "urban decline" suffers from imprecision. The result of a blunt approach to defining this term precisely and narrowly would be to yield a single, clear definition, but, as Robert Beauregard has argued, "one that strips away the multitude of elusive meanings that are essential to our understanding of urban decline."[215] Perhaps urban decline is best understood as "a dynamic matrix of meanings, an elusive rhetorical structure that entraps all sorts of notions related to the fate of cities."[216]

There are two main lacunae in the scholarly understanding of urban shrinkage. The first relates to the temporal dimension—that is, the need to disentangle cause and effect in the shrinkage process. Scholarly research and writing have done an admirable job of uncovering the causes of urban shrinkage but have lagged behind in unpacking and unraveling the sequence of urban shrinkage, the cumulative processes of overlapping and intersecting relationships that unsettle cause and consequence, and the pace at which shrinking has unfolded.[217] Triggering mechanisms influencing shrinkage include declining fertility rates and aging populations, economic restructuring (such as deindustrialization), revenue shortfalls resulting in austerity, and reordering of political systems (such as the collapse of state-centered planning under socialism).[218] The order in which these processes occur and influence one another, however, has remained largely under-investigated. The interdependencies between causes and effects of shrinkage and the feedback mechanisms have produced complex patterns of decline. The non-linear dynamics of urban shrinkage have ensured that the speed and duration of decline vary greatly from city to city.[219]

The second lacuna concerns the spatial dimension of the shrinkage process. The scholarly literature has identified differences in causes of shrinkage in different places. However, those macro trends cannot explain why cities within one region have experienced differing rates of shrinkage or growth. Research and writing have largely overlooked patterns of regional differentiation in both the causes and the degree, or extent, of shrinkage.[220] Something that further complicates efforts to settle on single explanatory frameworks and universal models is that processes of growth and shrinkage often occur simultaneously in tandem.[221]

Equally important, the scale on which population losses are observed is also inconsistent in scholarly accounts. What constitutes the administrative unit that is classified as a city is arbitrary. The administrative and functional borders of cities are subject to alteration by political fiat. In fact, as idealized constructs, cities are fuzzy notions. For example, political observers often point out that the City of Detroit declined in population from a highpoint of 1.9 million in 1950 to less than 700,000 in 2017. Yet when the metropolitan region becomes the unit of observation, population decreases amounted to only around 4 percent over the same time frame.[222] Thus, in a real sense, scale matters. The use of quantitative approaches, including consistent measurement techniques and systematic methods—or what can be called "spatial metrics"—has lagged far behind cavalier pronouncements about the scale and degree of urban shrinkage. Spatial metrics used in studies of urban shrinkage are underdeveloped and hence not sufficiently powerful to provide a comprehensive assessment of the spatial patterns of shrinking.[223] As Mattias Bernt has suggested, "it becomes unclear whether a statistically observable difference [between administrative units called cities] is a consequence of qualitatively varied developments, or of difference in the territorial definition of the objects of research." The greater the differences in administrative structures and settlement patterns between cities, the less meaningful become comparisons based on threshold criteria and numerical data. The varying availability of empirical data and diverse standards of measurement, methods of research, and key variables among different cities in different countries "tend to be the rule rather than the exception."[224]

Despite the robust discussion and debate focusing on urban growth and its different patterns, it is not possible to identify a coherent theory of shrinkage.[225] The idea of shrinkage has not been able to overcome its current use as a collection of negative symptoms largely associated with stagnation and decline.[226] While the idea of shrinking cities has heuristic value as a descriptive category, its theoretical limitations as an analytic tool greatly restrict its usefulness as mode of explanation.[227] The term "shrinkage" is empirically imprecise, possibly because it may be a statistical artifact of measurement techniques. Further, it is theoretically incoherent (a chaotic conception) because it lacks definitional clarity and internal consistency.[228] In sum, despite its continued usage, the term is conceptually hollow, analytically vacuous, and lacking in explanatory utility.[229] Alternative terms like "struggling cities in distress" and "cities after abandonment" also lack analytic clarity but offer perhaps a more polycentric approach than simply focusing on shrinkage as population contraction and job loss.[230]

Matthias Bernt has argued that "the state-of-the art research on 'shrinking cities' suffers from a fundamentally misleading conceptualization of shrinkage which forces essentially different urban constellations onto the Procrustean bed of a universal model of 'shrinkage.' " In short, conceptualizing shrinkage as a universal phenomenon with local specificities and variations erases fundamental differences that lie beneath the common characteristic of population loss.[231] Research and writing on shrinking cities seems to have coalesced around "a tacit, yet unspecified, common understanding" of the problems related to urban shrinkage. Yet unfortunately, "this understanding is neither grounded in consistent theoretical reasoning, nor on much comparative empirical research." The result of this approach is "the dominance of single-case studies" and an arbitrary mix of theoretical arguments combined with empirical evidence, all placed under an expansive umbrella term.[232] For the term to remain relevant, the most fruitful approach to understanding urban shrinkage is to move away from an outcome-orientated approach toward a process-orientated research agenda. Unpacking the pluralist world of urban shrinkage requires us to bring together causes for urban shrinkage and decline, with their multiple dynamics and uneven impacts, and situate these in the context of locally based urban trajectories.[233]

Local strategies seeking to combat urban shrinkage have tended to follow a similar path: the urge to restore the fortunes of declining cities often leads local "growth machines" to favor market-led solutions, particularly gentrification or selective rebuilding of singular projects. These strategies are not always successful, but when they are, they often result in increased segregation on a micro scale or in the eviction of the most vulnerable, fragile, and marginal populations.[234] Some postindustrial cities that have experienced decline have undergone modest regeneration of sorts. These resurgent cities are often characterized by deep inequalities that have accompanied the asymmetric patterns of restored prosperity.[235]

ASYMMETRICAL URBANISM: THE STRANGE CHOREOGRAPHY OF GROWTH AND DECLINE

Cities always contain elements of instability, which by their very nature are liable to produce rapid growth under favorable circumstances and to result in stagnation or contraction when conditions change. The combined effects of structural economic weakness, a lack of real employment options with opportunities for upward mobility, the departure of the young and the skilled (leaving the unskilled, the old, and the jobless), empty and decaying housing, entrenched poverty, dwindling tax revenues to pay for requisite social services, and deteriorating infrastructure combine to reinforce decline, producing an ever-widening downward spiral of neglect and abandonment.[236]

As a general rule, scholarly research has identified the general forces driving postindustrial decline, but the recognition of structural determinants alone cannot explain why metropolitan regions and cities that have been exposed to similar pressures still differ in terms of outcomes. The extent to which such social forces as deindustrialization, declining fertility rates, and job loss produce urban shrinkage depends on how these pressures are filtered through, first, the institutional arrangements across cities and regions, and, second, the spatial context within which the metropolitan regions or cities are located.[237] Equally important, the temporal dimension of decline—speed, tempo, and rhythm—can significantly affect the patterns of abandonment.[238]

Always and everywhere, growth and decline are interwoven processes that take place in tandem. Yet once a tipping point has been reached (Manuel Wolff and Thorsten Wiechmann have likened it to a self-reinforcing avalanche), the downward slide toward a condition of permanent stagnation seems to take on a life of its own.[239] Neglect and abandonment are at once spatial categories and temporal ones. The spatiotemporal dynamics that characterize postindustrial cities in decline revolve around the decelerating pace of turnover time in the marketplace for landed property and the degradation (sometimes abrupt and sometimes gradual) of the built environment. The imbrication of temporal slowdown and spatial deterioration does not follow a singular logic or general pattern but instead unfolds unevenly in accordance with differing circumstances and historically specific conditions. The pace and scope of decline are critical elements in understanding what is happening and why.[240]

Shrinkage and decline are both expressions of urban transformation, but different in their underlying causes and consequences. Generally speaking, urban decline refers to episodic, cyclical, and locally specific circumstances that arise from cyclical patterns of urban development under the rule of real estate and competitive property markets. In contrast, the shrinking cities syndrome consists of a permanent and seemingly irreversible downturn in socioeconomic fortunes with seemingly no prospect for restoration of the status quo ante.[241] Cast in bold strokes, shrinkage refers in the first instance to two concurrent and intertwined processes—loss of population and socioeconomic malaise and deterioration—that have wide-ranging and far-reaching effects on the functioning and governance of local communities and even whole metropolitan regions.[242]

CHAPTER 5

SPRAWLING MEGACITIES OF HYPERGROWTH

The Unplanned Urbanism of the Twenty-First Century

The unprecedented scale, scope, and pace of urbanization over the last three or four decades fundamentally reshaped urban life at the start of the twenty-first century and, as such, made it necessary to rethink our conventional understanding of the modern metropolis. At a time when the unprecedented scale of worldwide urbanization has pushed humankind over the symbolic threshold where now a majority of the world's population lives in urban areas, what it means to talk of sustainable urban development and to provide for human security in fragile urban environments has assumed enormous significance. Scholarly forecasts have projected that in the foreseeable future an estimated 93 percent of population growth in cities will occur in the so-called developing nations, with 80 percent of that taking place in Asia and Africa. It has been estimated that by 2030 between a third and half of the world's population will be subjected to a precarious, and often abject, everyday existence in the neglected interstices of these rapidly growing metropolitan regions.[1] At present, there are more than four hundred cities with a population of more than one million and nineteen cities over ten million. In 1975, just three cities could be classified as megacities—those with populations of ten million or more—and only one was in a (so-called) less developed country. In 2000, megacities numbered sixteen. By 2025, it is estimated that twenty-seven megacities will exist, twenty-one of them in less developed countries.[2]

This unprecedented shift in the demographic patterns of global urbanism has prompted some urbanist observers to rather cavalierly dismiss the great metropolitan centers that once represented for much of the world the imagined future of late modernity—London and Paris in the nineteenth century, New York, Tokyo, and Los Angeles in the twentieth—as relics of a vanishing past.

These once magnificent model cities of the modern epoch appear as somewhat anachronistic artifacts of a bygone era, while fast-growing cities like São Paulo, Lagos, Mumbai, Karachi, Mexico City, Kinshasa, Cairo, and Shanghai seem to presage what is to come.[3] Most significantly, the world's fast-growing cities—the so-called megacities of hypergrowth located primarily in the peripheral zones of the world economy—have absorbed the lion's share of these urban population increases and are expected in the foreseeable future to take in nearly all additional world population growth. In the twenty-first century, it is hard to avoid the startling conclusion that the expanding megacities of eight to ten million inhabitants or more, and the astounding hyper-cities of twenty million or more, have already become the new cities of the future.[4]

UNREGULATED URBAN FUTURES

Branded as exemplary expressions of failed urbanism in need of remediation, these rapidly expanding cities have buckled under the strain of crumbling infrastructure, inadequate regulatory regimes, and overtaxed service delivery. Yet they represent the main driving force for the future of urbanism on a global scale.[5] Sprawling megacities of hypergrowth consist of asymmetric and highly unequal urban landscapes marked by severe socioeconomic disparities. Evolving dynamics of global urbanism have produced uneven patterns of urban transformation in which newly created enclaves boasting of towering skyscrapers, upscale malls, luxurious residential dwellings, and corporate office blocs jostle incongruously with derelict zones of abandonment, sprawling informal settlements, and makeshift shack encampments.[6] The great majority of new urban residents who inhabit these fast-growing megacities are poor and reside in impoverished conditions. Without the social safety net available to marginalized urban residents in relatively vibrant and heathy cities, ordinary people living in struggling cities of hypergrowth face precarious circumstances that are not merely a socioeconomic condition but an existential threat to well-being that prevents their incorporation into the mainstream of urban life.[7]

These cities have become embedded, in varying degrees and at different scales, in globalized value chains while at the same time unskilled wage earners have become increasingly redundant and superfluous to the efficient functioning of formal economies, leading to diminished opportunities for regular work and a widening and deepening of unregulated and unauthorized income-generating practices. The sheer scale and intensity of informal practices in securing housing and tapping into alternatives to wage-paying work suggest that the routine focus of conventional urban studies on formal economic transactions, built infrastructures, and officially sanctioned regulatory regimes through the narrow prism of the planned city fail to capture the complexity and dynamism of everyday life in struggling cities under extreme stress.[8]

These megacities of hypergrowth are exemplary expressions of runaway urbanism. A key component of runaway urbanism is what is left behind: the ruins of failed building, contaminated zones filled with toxic waste, and abandoned people (or, as Zygmunt Bauman has argued, "the outcasts of modernity").[9] The disruption, distress, and agony of rapid urbanization leave visible reminders embedded in the urban landscape. "Marked by footprints of ecological exhaustion, poverty and violence," these abandoned "leftover spaces" constitute the sites of habitation for the most vulnerable and marginalized urban dwellers.[10]

As a general rule, newcomers to cities experiencing rapid population growth face housing shortages, dwindling public space, poor transportation networks, inadequate social services, crumbling infrastructure, and countless other social, aesthetic, and environmental challenges. With few opportunities for formal wage-paid employment and with limited access to decent housing, recent migrants and expectant work seekers are more or less condemned to a permanent existence outside the mainstream of urban life. They face the daunting challenge of finding ways to generate income in the shadows of the formal economy and finding shelter in informal settlements, squatter camps, and slums. According to authoritative estimates, about three in ten urban residents in these overcrowded megacities are squatters who reside in unauthorized and illegal housing arrangements.[11]

References to the megacities of hypergrowth, whether scholarly or journalistic, tend to be dominated by two contrasting image categories.[12] The first is an apocalyptic vision of an out-of-control "planet of slums" where increasing numbers of urban dwellers survive at the margins of human decency, imprisoned in an exceptional state of deprivation and trapped in a downward spiral of degradation and abjection. These sprawling megacities of hypergrowth have become symbols of seemingly permanent underdevelopment and a shorthand for dysfunctional and failed urbanism. They are the recognizable frame through which to locate and map deviations from the allegedly normal and expected pathways of development and progress.[13]

The second is a populist imaginary of entrepreneurial "shadow cities" infused with the alchemic energy of enterprising individuals struggling to unburden themselves of stifling regulatory regimes that serve no purpose but to ensure that slum dwellers remain in a permanent state of impoverishment. As a number of scholars have argued, the increasingly popular genres of slum tours, artistic exhibitions, and filmed documentaries have produced particular uplifting narratives and roseate images of the slum that express a kind of "admiration close to amazement at people's resilience, enterprise and creativity."[14] By offering a seemingly authentic experience of poverty, slum itineraries have become an avenue for contesting the dominant stigmas of indolence, moral turpitude, and predatory behavior by presenting an alternative view that stresses dignified, energetic, and industrious people imbued with a genuine sense of community and of belonging.[15]

As rhetorical devices, these two strains of thought largely tap into preordained beliefs about globalization, the world economy, and global capitalism. The essentialism embedded in these alternative perspectives leaves little room for nuance and subtlety. While each contains elements of truth, they alone cannot provide a complete portrait of the complexities of these sprawling cities seemingly without geographic limits.[16]

In so many ways, the sprawling megacities of hypergrowth are among the most innovative and vibrant places on the planet. Much larger than the celebrated modern metropolises of North America and Europe, with wholly different textures of density than the seemingly limitless ("100-mile") post-metropolises of the American Sunbelt, megacities located in the peripheral zones of the world economy have opened up entirely new frontiers of urban life, not only extending the limits of deprivation and extreme impoverishment but also inaugurating new methods of accumulating capital and triggering innovative experiments in nonformal economies and modes of everyday living outside of the disciplinary logic of market rationalities.[17] The inventive use of pirated technologies,[18] experimentation with cutting-edge building design practices,[19] and creative adaptation of sustainable systems of low-cost infrastructure have enabled some ordinary cities of the Global South to leapfrog over the historic cities of the Global North and to carve out niches for themselves at the forefront of global innovation.[20]

Yet there is no shortage of caricature, hyperbole, and unreflective sensationalism about what makes these large, sprawling megalopolises in the developing world such quintessential exemplars of stalled, failed, interrupted, or distorted urbanism.[21] Even a cursory review of popular portrayals of everyday urbanism in the sprawling megacities of hypergrowth—whether in journalistic accounts, documentary and commercial films, fiction writing, or photographic exhibitions—reveals a clear preoccupation with the chaotic, disorderly, and dangerous features of city life.[22] Seen within the context of contemporary globalization, these struggling cities certainly face a host of seemingly insurmountable challenges, including stalled economic growth and restricted opportunities for wage-paid employment, collapsing infrastructure, massive rural-to-urban migration, environmental degradation, lingering social conflict over scarce resources, and state violence directed at controlling social movements that have asserted their "right to the city."[23] Framed through the restrictive lens of perpetual and permanent crisis, these struggling cities under stress are characterized by what they lack: good governance, vibrant place-space, efficient transportation systems, adequate infrastructure, functioning services, and aesthetically pleasing environments. This steady accumulation of dysfunctionalities has prompted many urban theorists to regard these Third World cities of the developing world as globally insignificant and structurally irrelevant. As the physical footprints of sprawling megacities have expanded with "unprecedented ferocity," scholarly attention has largely focused on the ensuing conflicts over available land, access to housing, ecological degradation, and inadequate physical infrastructures that seem always to accompany "runaway" growth.[24] This stress on their alleged failures and deficiencies has meant that these untamed megacities of hypergrowth have been viewed largely through the normative lens of developmentalism, a prescriptive approach that broadly understands these places to be sorely lacking in the normal (and expected) qualities of urban vibrancy and that favors remedial policy recommendations directed at improving capacities for good governance, service delivery, and economic efficiency.[25]

THE COMPLEXITY OF GLOBAL URBANISM AT THE START OF THE TWENTY-FIRST CENTURY

The fastest growing cities in the world are located outside the core areas of the capitalist world economy in what is commonly referred to as the Global South. Yet mainstream scholarship in urban studies has stubbornly refused to move too far away from the long-standing separation between the theoretical work that derives its insights from focusing on the Euro-American experience of urbanism and policy-oriented research concerned with urban development in cities struggling with not enough wage-paid employment opportunities to absorb available work seekers, with shortfalls of decent housing, and with regulatory regimes stretched to the breaking point.[26] This assumed dichotomy between innovative globalizing cities with world-class aspirations in the core areas of the world economy and imitative Third World cities suffering from crippling underdevelopment has hampered a robust theoretical understanding of contemporary urbanism. As Colin McFarlane has argued, "a vast terrain of urban life has been for some time" cloaked in "the shadows," pushed to "the edge of urban theory."[27] As telling exemplars of stalled or failed urbanism, the megacities of hypergrowth do not conform to the expectations of conventional urban theory. As objects of inquiry, they typically function as negative examples of what can go wrong when urban growth is not planned or managed.[28] The challenge for urban studies and planning theory and practice is to move beyond the binary opposition that First World cities offer models to be emulated and Third World cities represent problems to be fixed.[29]

Despite being overlooked as sources of theoretical insight, the burgeoning metropolises of the Global South—what Doreen Massey has called the "big but not powerful" megacities of the Third World—have become objects of a great deal of curious fascination.[30] The idea that megacities located in the peripheral zones of the world economy offer a glimpse of the urban future is a compelling trope that has succeeded in capturing the imagination of a wide range of urban commentators in Europe and North America.[31] In *Planet of Slums*, Mike Davis provocatively suggests that "the cities of the future, rather than being made out of glass and steel as envisioned by earlier generations of urbanists, are instead largely constructed out of crude brick, straw, recycled plastic, cement blocks, and scrap wood."[32] In a similar fashion, Robert Neuwirth—after having lived in squatter settlements in megacities on four continents—reached the conclusion these places provided a proleptic vision of the "new urban world" to come.[33] However, where Neuwirth finds reason for optimism in a future of industrious squatters constructing lasting and viable communities under the most adverse and precarious conditions of existence, Davis projects a disturbing image of these megacities-yet-to-come. "Instead of cities of light soaring toward heaven," he forecast, "much of the twenty-first century urban world squats in squalor, surrounded by pollution, excrement, and decay."[34]

As a general rule, the analysis and interpretation of sprawling megacities of hypergrowth have careened wildly between these two polar extremes. On the one hand, there are hopeful visions of a new, vibrant urbanism found in the inventiveness of informal markets, the ingenuity of self-built housing, and the unquenchable spirit of bootstraps entrepreneurialism; on the other hand, there are eschatological evocations of urban apocalypse grounded in the permanent misery of the slums.[35] Optimists envision new networks of powerful, stable, and prosperous city-states, each bigger than many small countries, where the benefits of urban living, the opportunities for socioeconomic advancement, and the relative ease of delivering basic services compared to underresourced rural zones combine to raise living standards for new arrivals.[36] In contrast, pessimists make alarmist predictions of a dystopic future in which large numbers of people fight over scarce resources in sprawling, divided, anarchic "noncommunities" ravaged by disease, environmental degradation, and violence.[37]

Yet while megacities of hypergrowth are occasionally portrayed in a positive and sometimes even romantic light, more often than not they treated as a cause for alarm. These unplanned, fast-growing conurbations are routinely indexed as expansive sites of explosive population growth and massive concentrations of poverty, miserable conditions that are only exacerbated by the constant threat of environmental deterioration and natural disaster.[38] Despite different angles of vision, scholars, journalists, and documentary filmmakers have tended to converge around a shared view that sees megacities of the Global South as dismal failures beyond remediation or, as Matthew Gandy has put it, "a realm of irrationality beyond the reach of human agency or any realistic prospects of improvement."[39] Rather than framed in the language that characterizes the utopian ideal of the modern metropolis, sprawling, unplanned "slum cities" appear as indeterminate liminal spaces where law, extralegality, and illegality are braided together to produce the other side of universality—a legally ambiguous space outside the domain of moral legitimacy, "where universality finds its spatial limits." These slum cities constitute urban landscapes that signal a radical departure from the original and conventional meaning of "the city"—expectant places open for the flowering of public culture, the embrace of civic life, and the nurturing of citizenship.[40]

Seen through the dystopian lens of chaos and near total collapse, the image of these "feral cities"—"where the rule of law has long been replaced by near-anarchy"—contains all the *noir* elements of disaster, ranging from environmental degradation to murderous gangs competing over scarce resources, disease, and death.[41] For example, writing in the *New Yorker*, George Packer has argued that "Lagos has become the archetype of the megacity, perhaps because its growth has been so explosive, perhaps because its cityscape has become so apocalyptic."[42] In warning of "the coming anarchy," Robert Kaplan concludes that the megacities of West Africa have become "the symbol of worldwide demographic, environmental, and societal stress."[43] Grounded in a kind of naturalist determinism, arguments of this sort ignore the impact of networks and circuits of capital investment and unequal trade that reproduce wealth at one pole and impoverishment at the other.[44] Whether framed as a kind of "return of the repressed" (Mike Davis) or unwelcome "barbarians at the gate" (Robert Kaplan), urban commentators from widely divergent political persuasions foresee disaster eventually coming "home"—arriving unceremoniously and without warning at the doorstep of Western civilization.[45]

This compounded anxiety—a dystopia that is not just discovered "over there" but that will eventually migrate "over here"—has become almost commonplace in commentaries about sprawling megacities of hypergrowth. When focusing on these giant, unplanned conurbations growing at a pace that seemed unimaginable not long ago, journalists and scholars alike tend to imagine a worldwide urban future of crisis, chaos, and collapse.[46] More than a quarter of a century ago, Eugene Linden issued a dire warning that "the developed world ignores at its peril the problems of Third World cities."[47] In a similar vein, Matthew Power has suggested that "in its poverty, enormity, utter squalor, and lack of services, [metropolitan Manila] perfectly represents the catastrophic twenty-first-century vision of the megacity."[48] Equally telling, Robert Kaplan has suggested that megacities are "an appropriate introduction to the issues, often extremely unpleasant to discuss, that will soon confront our civilization."[49]

Whether these forecasts have utopian or dystopian inflections, what these visions have in common is their shared sense of the future pathways of global transformation, in which the "prevalence of slums—barrios, favelas, shantytowns, squatter settlements, *urbanizaciones piratas*, *barriadas*, *bidonvilles*, or whatever name one chooses to use—have already become the paradigmatic settlement spaces of the twenty-first century."[50] As world-renowned architect Rem Koolhaas has suggested of the largest megacity in Africa: "Lagos is not catching up with us. Rather, we may be catching up with Lagos."[51] Given how often this point of view is repeated, and hence how embedded it has become in popular writing, one needs to inquire, "according to what relationships between history and geography, between time and space," do megacities and their slums acquire such proleptic roles? "What is at stake in claiming that familiar developmental narratives have collapsed," as Aaron Zeiderman has suggested, "such that places once thought to be advancing towards the so-called great modern cities of Europe and the United States have now come to represent to the North its own future?"[52]

Seeing sprawling megacities of the Global South as "cities of the future" expands the conceptual boundaries of urban modernity to include places that conventional urban theories have typically treated as either outside of historical time or lagging behind it.[53] At first glance, this reframing seems to suggest a radical reordering of the presumed relationship between history and geography so central to European modernity. In following the logic of what James Ferguson has called a "non-progressive temporalization," in which history seems to move "backward" in the direction of Karachi, Manila, and Lagos rather than "forward" toward Paris, London, and New York, the cities-of-the-future narrative paradoxically retains the singular logic of historical time unfolding from one stage to the

FIGURE 5.1 Megacities of Hypergrowth: Kibera, Nairobi, Kenya. Aerial view of Kibera, the sprawling "slum city" located southwest of downtown Nairobi. Martin Murray Photograph

next, with urban transformation the "inevitable culmination of a singular and universally unfolding history."[54]

Yet this discourse, as Austin Zeiderman has argued, "re-inscribes and reinforces, even as it seems to rearrange and challenge, the historical logic and geographical order" that associates modernity with development.[55] In an ironic twist, the "progress" so central to the conventional understanding of modernity is turned on its head.[56] In a strange kind of introverted logic, this discourse renders the developmental promises and predictions of (eventually) "catching up" with the West a totally hopeless and obsolete pipedream. Equally important, it inverts "the order in which cities of rich and poor countries are expected to converge," and it replaces the ambition, aspiration, and expectancy of development in the Global South with fears of degeneration and decline in the Global North.[57]

Imagining these unplanned megacities of hypergrowth to represent the imminent dystopic future of First World cities, rather than viewing them as struggling to keep pace with the West, is merely a variation on the regulative fiction at the core of developmentalist thought and modernization theory that cities occupy different stages in the linear unfolding of singular historical time.[58] In challenging this introverted and dubious logic, which reverses the modernist conception of progress, Matthew Gandy, writing about Lagos, argues against the provocative claim that "to write about the African city is to write about the terminal condition of Chicago, London or Los Angeles."[59] Yet to adopt this position is to ignore "the fact that every extremity of Lagos's deterioration over the past quarter century has been linked, in inverse proportion, to the capital accumulated in Chicago, London or Los Angeles."[60] While megacities of hypergrowth may not occupy different temporal stages along a singular pathway toward development, their place in the world of cities is asymmetrical with respect to how they are differentially integrated into the globalized circuits of capital and commodity flows that create the spatial unevenness of the world economy.[61]

RETHINKING CONVENTIONAL URBAN THEORY: CONCEPTUAL BLIND SPOTS

Rather than objects of knowledge, the megacities of the Global South are typically framed as sites of intervention—dystopian places in need of remedial action to address failed municipal management, broken-down infrastructure, overcrowded streetscapes, and derelict housing.[62] The developmentalist framework is unable to grasp and accommodate the complexity and diversity that characterize sprawling megacities of hypergrowth. Looking at these cities in their own right—rather than as pale imitations of "real urbanism"—offers an opportunity to revise our understanding of urban development not as a staged trajectory moving inexorably toward a common end point, but as a complex process with multiple pathways toward indeterminate outcomes, and to bring into view new practices and experiences suppressed by the developmentalist discourse of remedial intervention. Viewing these megacities horizontally as part of a world of cities, rather than vertically in a ranked hierarchy of importance, provides the occasion to rethink "the contours of modernity in a global age."[63]

The bulk of research and writing on slum cities focuses on the multiple (and intersecting) dimensions of poverty and deprivation, socioeconomic marginalization, institutional breakdown, and administrative dysfunctionality.[64] Understanding how and why these cities have failed to live up to the expected standards of the modern metropolis is typically framed through the prescriptive lens of developmentalism: what do these cities have to accomplish in order to "catch up" with the West, and what backward features do they need to jettison in order to become modern and cosmopolitan? This kind of applied policy-oriented research typically takes its cue from conventional development studies and, as a consequence, focuses on the urban dimensions of economic growth and development.

Themes that typically arises from such an explicit policy orientation stress the enabling role of smoothly functioning land markets that, in conjunction with connected infrastructures, are capable of facilitating and catalyzing increased investment, productivity, and access to distant markets, along with prescriptions for the institutional embeddedness of private property and the rule of law.[65]

Once one abandons the presumption that cities are best understood when arranged in ranked hierarchies in accordance with their structural importance in the world economy, it becomes possible to dispense with such regulative fictions as First World and Third World cities permanently ensconced in the developed and the developing world.[66] The conjoined ideologies of modernization and development rest on a conception of differential time with their classification schemes of leads and lags, forward and backward, movement and stasis. Yet adopting an alternative synchronic (or historical-structural) framework challenges these presumptions of differential time embedded in the discourses of modernization and developmentalism. This paradigm shift is necessary to fully comprehend the asymmetric relations that unevenly divide regions, places, and locations on a world scale. Correlatively, it enables us to understand how all sorts of different cities—from the so-called global cities to the sprawling megacities of hypergrowth, from vibrant "competitive cities" to unhealthy "uncompetitive cities," and from postindustrial cities aspiring to world-class status to deindustrialized "loser cities" suffering from neglect and abandonment—coexist simultaneously, embedded within the same temporal frame in the same historical time.[67] Properly understood, the current condition of megacities of hypergrowth represents "*neither* an earlier stage *nor* the ultimate end point on a supposedly global time line of urbanization."[68]

The growing recognition of the need to theorize the diversity of urban trajectories cannot be accomplished without unsettling and disrupting conventional approaches to understanding global urbanism. It is necessary to resist the regulative principle, guiding much of conventional urban studies, that cities of the Global South can be studied on the basis of established knowledge about universalizing processes of urban growth and development taking place in linear time along predetermined pathways laid down by the great metropolises that came before. When historical time is "re-temporalized" in ways that dispense with necessary linear stages, cities that were once conceived as advancing (slowly or rapidly) toward a kind of universalizing modernity along routes laid down by mature cities that came before suddenly find themselves in a non-hierarchically arranged world of cities.[69]

"Once modernity ceases to be understood as a telos," James Ferguson has argued, "the question of rank is de-developmentalized, and the stark differentiations of the global social system sit raw and naked, no longer softened by the promises of the 'not yet.' "[70] Accordingly, the inevitable passage of time "is no longer expected to transform poor cities into rich ones, and thus waiting patiently in anticipation of progress makes little sense."[71] As Ferguson and others have argued, the developmentalist narrative that defined the era of high modernism has "in some measure lost credibility in recent years." The failure of the developmentalist models to deliver on their promises of convergence is visible in "not only in the domain of academic theory but also in practical economic terms." To paraphrase Ferguson, no one talks about sprawling megacities of the Third World following in the footsteps of the developed cities and eventually converging with the mature modern metropolis anymore.[72]

If the ontological status of megacities of the Third World is fixed outside narratives of development, and the globalizing cities of the First World project from them their own possible unstable future, then hopes of development in the struggling cities of Global South are transformed into fears of degeneration in the mature cities of the Global North.[73] The conjoined anxieties brought about by the unwanted influx of poor immigrants to North American and European cities coupled with growing urban inequalities have triggered a visceral response, that has taken the form of calls for strengthened borders as the first line of defense against Unwanted Others. The heightened apprehension associated with preventing the well-functioning and planned cities of the North from becoming just like the disorderly, unplanned, and malfunctioning megacities of the Global South has stimulated "the proliferation of numerous technologies of fortressing and exclusion."[74]

Originating in the collective imagination of conventional urban studies, the twin foundational tropes of "slum" and "informality" are, in one way or another, the preeminent socio-spatial constructs that purport to explain the backwardness, dysfunctionality, and stasis of unplanned cities of hypergrowth.[75] These dual optics form the empirical and analytical point of departure for much of the research and writing on these unplanned cites in distress as their objects of inquiry.[76] Perhaps above all else, these ubiquitous terms operate as a convenient shorthand for the conditions of bare life and sheer survivalism that characterize struggling cities at the margins of the world economy. Framed as an essential part of the normative discourse of development, slum and informality function as equally fungible terms for underdevelopment and the inherent backwardness of Third World urbanism.[77] As metonyms, they conjure up an image of an abject but uplifting human condition, where "the wretched of the earth" conduct their everyday lives in deplorable circumstances, yet still find the wherewithal to survive and sometimes even to thrive.[78]

More often than not, scholars and policy makers have flattened and homogenized the terms "informality" and "slums," overlooking internal variation and difference. Only a nuanced understanding can provide us with the tools necessary for careful analysis. As Robert Fishman has suggested in another context, "As always, language lagged behind reality."[79] Names and label do matter. To understand the meaning of terms, "we need to determine who has used them, and how and why they have done so."[80] Rather than trying to simplify what we mean by slums and informality, it is more fruitful to welcome complexity.[81]

SLUM CITIES

As a general rule, mainstream urban studies literature locates the so-called slum cities of the Third World on a "continuum from nascent/informal to developed/post-industrial," where the latter represents the apex of achievement in urban development and city building.[82] Implicit in this linear trajectory of urban development is a normative prescription that calls for progressively replacing "the chaotic, malfunctioning city of informality" with an orderly, coherent, optimally functioning modern metropolis with formal institutions and formal employment. This narrative—originating out of 1950s–1960s modernization theory and developmental studies—tends to function, as Edgar Pieterse has put it, on the assumption that modern, gleaming, skyscraper-filled cities, with adequate networked infrastructures to support them, are the "only and ineluctable way into the urban future."[83] Equating the formal with modern, developed, and mature and, conversely, the informal with not-yet-modern, underdeveloped, and backward forecloses, a priori, a more nuanced understanding of how informality and formality are inextricably interlinked in a symbiotic relationship of dominance and dependence.[84]

As designated locations, slums are such a recognized object of inquiry that they have acquired their own specialized vocabulary: *favelas* (Brazil), *villas miseries* (Argentina), *kampungs* (Indonesia), *barrios* (Colombia and the Dominican Republic), and *banlieues* (France). It is this deplorable space—whether referred to as shantytown, squatter camp, informal settlement, outcast ghetto, or whatever name seems to fit—that constitutes "the iconic geography of this urban and human condition."[85] In conventional urban studies, the slum functions as the foundational master trope of the Third World city. As a demographic, legal, and territorial construct, the slum is the concrete manifestation of extreme overcrowding, dysfunctionality, and urban pathology.[86] The Third World slum is the recognizable

FIGURE 5.2 Megacities of Hypergrowth: River Crossing, Kibera, Nairobi. One of the few cement bridges crossing the Nairobi River that runs through Kibera on the outskirts of Nairobi. Martin Murray Photograph

normative frame through which the sprawling megacities of the Global South are perceived and understood, and their differences mapped and located.[87]

Slums are not merely material locations or topographical containers of social processes and spatial relations. They are also imaginary objects that are socially constructed around their associations with "risk and volatility rather than order and certainty."[88] To designate slums as vulnerable spaces "at risk" opens the door to top-down strategic interventions designed to mitigate harm and minimize misfortune. In this sense, to be at risk becomes an opportunity for large-scale real estate developers to propose the construction of megaprojects like airports, shopping malls, and technology parks as viable solutions to vulnerability. The subsequent displacement and resettlement of ordinary people from precarious neighborhoods and sites that municipal authorities judge to be hazardous and dysfunctional effectively unlocks landed property to fresh capital investment. Large-scale property developments create "engines of value production" where the formal calculation of capitalist rationality replaces informal use values.[89]

No matter how we define and describe them, slums have become the most visible expression of urban marginality and exclusion in the struggling megacities of hypergrowth. In its typical usage, the pejorative label "slum" has come to stand for the antithesis of the lofty ideals and aspirational goals of politicians, urban planners, and policy makers alike.[90] Regardless of the name we use to refer to them, slums reveal the spatial dimensions of uneven urban development associated with growing inequalities across multiple fronts. Social disparities have created conditions that are extreme by any standard: acute deprivation juxtaposed against excessive wealth, affluent residential estates contrasted with squatter settlements and homeless encampments, luxurious shopping meccas balanced against informal street trading, and Disney-like entertainment extravaganzas as opposed to rock-strewn playing fields. The evidence of hardship and enduring inequality is everywhere, sometimes painfully obvious in plain view, at other times conveniently hidden from sight, pushed to the edges and margins of vibrant city centers.[91]

According to developmentalist thinking (with its roots in post–World War II theories of modernization), slums are a temporary spatial typology "characteristic of fast-growing [urban-based] economies, and they progressively give way to formal housing as economic growth [eventually] trickles down" to the poorest of the poor. Even if slums appear as persistent features of the urban landscape, the modernization argument suggests, slum living only represents a transitory phase in the life cycle of newcomers to the city, as successive generations gradually assimilate into the mainstream of urban life. Even in the face of considerable evidence to the contrary, this modernizing myth has proven remarkably resilient in both scholarly and policy-making circles.[92]

Despite the scholarly attention devoted to describing and understanding irregular and makeshift housing arrangements in cities, there is no firm agreement on how to represent slums in the wider discourse on global urbanism.[93] Alongside the familiar list of pathologies, the slum has appeared once again in scholarly debate and discussion as a term used to refer to deprived locations with certain intrinsically debilitating features that set it apart from generic descriptions of standard low-income housing. Slums are invested with all sorts of qualities that careen back and forth between characterizations as vibrant places of innovation to abject sites of depraved indifference.[94] Representations of the slum—that is, how we recognize its place and meaning in critical urban studies—is as much about mapping its objective features as it is about imagining the invisible features of everyday life.[95]

THEORETICAL CONSTRUCTION OF "SLUM"

Both in the mainstream urban studies literature and in popular accounts, the term "slum" functions as the dominant spatial paradigm for the most resource-deprived zones of the sprawling megacities of hypergrowth. Like similarly value-laden terms such as "underclass," "culture of poverty," and "ghetto," the word "slum" is almost always used pejoratively, a negative stereotype that evokes disapproval and even condemnation in ways that only serve to re-create and reinforce unfounded myths about poor people living in indecent conditions. Yet "despite all of its inglorious associations over a long and disreputable history," the term slum has continued to be embraced by scholars and policy makers alike, with few apologies or reservations.[96] In contrast, critics charge that the uncritical and careless use of the term "slum" is epistemologically inadequate as an analytic concept. Scholars and policy makers often deploy the term as a homogenizing label denoting a uniform and undifferentiated condition of urban impoverishment, thereby ignoring and obscuring the underlying complex configurations of socio-structural inequality and asymmetrical power relations that often accompany living under deplorable circumstances.[97] The misrepresentation of the complexity of everyday life in slums leads invariably to exaggerated caricature and to distortions in crucial policy-making decisions.[98]

Critics who object to the appropriateness and even suitability of slum terminology claim that the term is all too often used in simplistic, potentially unhelpful, and even "dangerous" ways.[99] Allen Gilbert, for instance, has warned against bringing the problematic language of the slum back into respectful scholarly conversation. He argues that the return of slum terminology signifies a kind of undiscerning and casual sloganeering that paints endemic urban poverty in rather broad brushstrokes, thereby reducing the lives of all poor people, despite their differing circumstances, "to the lowest common denominator."[100] It is not merely a matter of language that is troubling "about this newfound charm for a rather old-fashioned way of classifying the urban poor," but that its indiscriminate, incautious use has become "an epistemological shorthand for portraying the problems" associated with urban poverty.[101] Thus, for Gilbert, the reintroduction of the language of slums not only creates and reinforces misleading stereotypes about poor and marginalized urban residents but also glosses over highly differentiated circumstances that accompany urban poverty.[102] What makes the term "slum" particularly dangerous is the steady stream of "negative associations that the term conjures up, the false hopes that a campaign against slums raises, and the mischief that unscrupulous politicians, developers and planners may do with the term."[103] Its perfunctory use as a metaphorical surrogate for a complex combination of makeshift (often self-built) housing, resource-scarce environments, nonexistent physical and social infrastructure, and lack of opportunity often goes hand in hand with top-down, short-term, and rather simplistic responses from municipal authorities, ranging from assumptions about the sociocultural homogeneity of all residents to generic, one-size-fits-all solutions that are supposed to work well everywhere, and from assumptions of shared circumstances and common plight to the implementation of "harsh and often violent slum eradication strategies."[104] Whether they are called slums, informal settlements, shantytowns, or squatter encampments, those impoverished areas where the urban poor reside consist of a withering variety of building typologies, dissimilar life chances and opportunities for advancement, clouded property relationships, and highly unequal access to infrastructure and services. Generalizations are difficult if not impossible to make.[105]

David Simon insists that the language of slums is also essentialist, since the diversity of slum ecologies and situations worldwide is so great as to make meaningful definition almost impossible, especially in singular and universal terms. He suggests that it is possible to identify at least four different categories and meanings of term "slum": first, a literal definition, one that identifies slums in terms of the lack of key services or resources; second, an aesthetic definition, one in which they are distinguished in terms of appearance (the use of particular construction materials, the lack of physical infrastructure, and overcrowded living conditions); third, a legalistic definition, one in which they are classified in terms of land that is occupied outside of formal statutes and without secure legal tenure; and, fourth, an emotive or pejorative definition, one in which they are cataloged in terms of ascribed attributes that may be highly inaccurate.[106]

Despite the cautionary warnings that the return of the slum is an invidious trend that needs to be exposed and challenged, scholars such as Vyjayanthi Rao make the opposing claim that there is a broader theoretical value in applying the term "slum" in a normative sense, because it offers a new analytic framework for understanding at least one major dimension of the rapidly growing cities of the Global South.[107] In similar fashion, Ananya Roy introduces the idea of "subaltern urbanism" as an alternative framing device that stands in opposition to the conventional understanding of "slums" and "slum ecologies." Writing against such apocalyptic and dystopian narratives of the megacity and its ubiquitous and homogeneous slums, she suggests instead that these subaltern spaces function not only as a genuine "terrain of habitation, livelihood, and politics" but also as vibrant expressions of improvisational activities and the entrepreneurial spirit. She sees this alternative use of the term "slum" as a "vital and even radical challenge to dominant narratives of the megacity."[108] Subaltern urbanism is thus a contrasting paradigm that seeks to "confer recognition on spaces of poverty and forms of popular agency that often remain invisible and neglected in the archives and annals of [conventional] urban theory." The central question is this: how can we understand the heterogeneity of subaltern urbanism outside the familiar metonymic categories (and language) of megacity or "slum"? How can we create new frames of reference, and make use of reconfigured concepts, to chart new itineraries of research and analysis?[109]

THE MEGACITY AS THEORETICAL OBJECT: LOCATING "SLUM CITIES"

As urbanization has continued to spread inexorably across the globe, the distribution of spatial and socioeconomic resources has remained highly unequal.[110] Marooned on the outskirts of the law, more than one billion people worldwide live in urban slums and squatter settlements, mostly in the megacities located in the peripheral zones of the world economy.[111] These dismal places have become huge containment zones for disposing of "a super-abundance of [surplus] labor."[112] Slums constitute an extralegal liminal zone that exists in a regulatory vacuum, where predatory entrepreneurs, rapacious landlords, and hustlers of all kinds operate with virtual impunity, "unfettered by law or public scrutiny," and where, at one and the same time, ordinary people improvise and inventively make use of available resources in order to "make do" and "get by." The imbrication of these dual dynamics makes it difficult to locate singular logics governing social life in the slumlands in struggling cities.[113]

The rule of law, extralegality, and illegality "commingle haphazardly in urban slums to produce spaces and subjects" outside the mainstream of urban life.[114] Located at the margins of legality and legitimacy, slum dwellers benefit from few legal rights and survive outside the opportunities of formal economies. Subjected to the force and violence of the law yet not afforded any of its protections, those who live in informal settlements and slums become trapped in a liminal existence—neither here nor there in the grand scheme of the regulated order and stability of place that define the modern metropolis. In analogous terms, there is a striking similarity between slums and other sites of extralegal exception, such as internment camps, refugee settlements, and embattled war zones.[115] But to unduly stress the formal—or paradigmatic—equivalence can easily overlook fundamental differences between these different kinds of locations.[116]

Three enduring and interrelated features of global capitalism—extralegal force and violence, vast reserve armies of labor outside the formal waged economy, and informality—produce and sustain urban slums on a grand scale. The genesis and persistence of slums and slum dwellers provide ample testimony to the hidden hand of market competition in the service of the ongoing accumulation of capital

FIGURE 5.3 Megacities of Hypergrowth: Nairobi River, Kibera. Shack dwellings along the Nairobi River, a polluted waterway filled with contaminated materials and garbage that bisects Kibera. Martin Murray Photograph

working in concert with the iron fist of state repressive apparatuses. While business enterprises have found ways to tap into inexpensive reserves of disposable labor housed in informal settlements, state policies have stressed containment and quarantine over upgrading and service delivery. Over the past thirty years, neoliberal restructuring of the world economy and the reordering of the responsibilities of public administration through structural adjustment programs have accentuated the process of slum expansion. As a result, slums in the peripheral zones of the world economy have grown at an exponential rate without any indication of slowing down or even reaching a saturation point.[117]

In its current neoliberal incarnation, the predatory forces unleashed by global capitalism have pursued a logic of extraction in which natural resources, available land, and surplus labor are subjected to systematic exploitation. Taken together, this voracious appetite for ceaseless accumulation of capital has resulted in a sharpening of structural inequalities on a world scale.[118] All around the world, these sprawling sites of deprivation have become vast dumping grounds for warehousing "surplus humanity"—disposable people excluded from the productive circuits of formal work and legal protections.[119]

SITUATING SLUM ECOLOGIES

The idea of slums as exemplary expressions of precarious settlements of habitation has long functioned as an important foundational trope in conventional urban studies. Framed within the parameters of modernization theory and development practice, slums represent the accumulation of spatial and behavioral pathologies that stand in the way of progress. From Friedrich Engels in the nineteenth century and Jacob Riis in the early twentieth century, and to the 2003 UN-Habitat *Report on Slums*, the notion of slums appears time and again as a distinctive spatial ecology denoting highly circumscribed modes of habitation characterized by deprivation and want.[120] The vast slumlands that have proliferated on the fringes of megacities of hypergrowth consist primarily of unauthorized agglomerations of self-built shacks located on tiny, unserviced plots lacking the basic social amenities required to secure a decent life.[121] As the paradigmatic expression of chaotic, disorganized, and amorphous urbanization, they signify what can happen when urban landscapes are not planned and growth is not regulated.[122]

Mainstream scholarly research and writing typically look upon megacities of hypergrowth as synonymous with the unfettered horizontal expansion of unplanned and unregulated patterns of human settlement.[123] For scholars imagining the global urban future at the start of the twenty-first century, these "slum cities" have become metaphorical shorthand for the deplorable human condition of the Global South.[124] Seen through the normative lens of developmentalism, these overcrowded, unregulated metropolitan agglomerations, expanding horizontally at an alarming rate, are characterized primarily by herculean problems of underdevelopment—crumbling or nonexistent infrastructure, environmental degradation, poverty, and disease—that call out for remedial action.[125] Megacities can thus be understood as the "constitutive outside" of conventional urban studies, existing at the lower rungs of contemporary global urbanism while the much-celebrated global cities occupy the upper echelons of world-class success.[126]

Slums largely consist of self-built housing constructed of impermanent building materials such as plastic, corrugated iron, plywood sheets, and mud bricks.[127] As the mirror opposite of middle-class suburban sprawl, horizontal expansion of vast slumlands, shack settlements, and squatter encampments on the peri-urban fringe has become the predominant building typology in struggling megacities of hypergrowth. In response to an "illimitable appetite for space," slums expand outward, following pathways of least resistance, invading unoccupied land, and colonizing available niches. More often than not, squatter settlements, in the words of Mike Davis, come to occupy "low-value urban land, usually in hazardous or extremely marginal locations such as floodplains, hillsides, swamps, or contaminated brownfields."[128] As Eileen Stillwaggon has observed, "Essentially squatters occupy no-rent land, land that has so little worth that no one bothers to have or enforce property rights to it."[129]

Yet this situation is not always the case. Squatters sometimes invade underutilized or temporarily vacant land, only to discover that large-scale real estate developers have set their sights on these prime locations for redevelopment projects. Likewise, the desperately poor sometimes built makeshift shelters on top of unstable land susceptible to mudslides or sinkholes, astride rivers prone to flooding, or alongside polluted water courses.[130] Under these circumstances, squatters are often subjected to the unforeseen vagaries of nature, and even to deliberate forcible removal and eviction.[131]

In order to understand slum typologies, it is important to distinguish between the *process* through which informal settlements come into being and the eventual *outcomes* (e.g., spatial form, vernacular design, building materials and characteristics, and grid patterns). Spontaneous settlements do not just happen without any forethought or purpose. Whether the result of collective engagement or individual initiative, they take shape in accordance with a long line of path-dependent "choices among available alternatives."[132] Land invasions and self-built shelters range from those carefully prepared in advance to those that are the consequence of incremental accretion over time.[133] Informal settlements exhibit a "highly adaptive, deliberative, and context-sensitive mode of reasoning put into practice"—what Raul Lejano and Corinna Del Bianco have called an improvisational "logic of enactment."[134]

All too often, scholars, policy makers, planners, and journalists have portrayed the continued existence of slums as the dystopian face of distorted urbanization, where socioeconomic insecurity, competition over scarce resources, and violence (both structural and individual) overshadow whatever associational life, communitarianism, and mutual assistance manage to survive. The typical portrayal of slums reveals layers of multiple deprivations, including chronic underemployment and unemployment, income insecurity, fractured familial relations, social exclusion, unhealthy environments, endemic violence, and predatory crime. In this disturbing imaginary of absolute abjection, slum dwellers operate as the "hunters and gatherers in the urban jungle."[135] These degrading and deplorable conditions radiate outward to affect surrounding neighborhoods and penetrate downward into interpersonal relations.[136] These resource-deprived areas are identified through a great variety of "equivalent words in different languages" and across very different geographic regions.[137]

Put in simplest terms, the essential features of slums consist of low-income and impoverished neighborhoods characterized by overcrowding and high population densities that exceed the carrying capacity of the built environment, where residents lack legal rights to place of residence or security of tenure and suffer from inadequate social services and rudimentary infrastructure. There is little agreement about what characteristics constitute the key determinants of slums because of their complex spatial layouts and social complexion, the scale and variety of local differences, and the pace of expansion and transformation. What is not in dispute is that slums are multidimensional in nature, with different quantitative and qualitative characteristics that are often difficult to classify and define.[138] As a general rule, the residents of slums live in makeshift and poor-quality housing that is often self-built, are never far removed from the threat of forced removal, lack reliable access to clean water and adequate sanitation facilities, and cannot escape from unhealthy living conditions and hazardous locations that render them extremely vulnerable to chronic illness and debilitating disease.[139] As they evolve and mutate over space and time, slums set in motion the conjoined processes of continuous displacement, ceaseless movement, and constant vigilance. They reflect the uprootedness of precarious urban lives.[140]

The formulation of this urban catastrophe genre—with the slum as the centerpiece of an apocalyptic future—unnecessarily (and inaccurately) flattens out differences in the historical evolution of distinct social-spatial forms, homogenizes and stigmatizes the experience of slum dwellers, and conjures up *noir* images of "huge undifferentiated neighborhoods filled with hopeless underemployed masses."[141] These alarmist forecasts of an impending urban cataclysm "reproduces an over-determination of urban poverty" that conceives of slum dwellers solely as more or less passive victims of circumstances instead of active agents in survivalist economics.[142] While much has been written about broken-down infrastructure, overcrowded streetscapes, and lack of municipal oversight in slum cities located at the margins of modernity, much less is known about the everyday lives of ordinary urban residents. In focusing almost exclusively on the common lot of misery and deprivation, scholars often pay scant attention to the hierarchies and divisions that separate slum dwellers into distinct social categories of opportunity. Most important, these scholars tend to overlook the dynamics of cooperation and mutual assistance that characterize the daily existence of the urban poor.[143] What is easily ignored are the endurance and persistence of these informal modes of habitation, the complexity of livelihood strategies of slum dwellers, and the forms of popular contestation through which marginalized urban residents establish a presence for themselves in struggling cities around the world.[144]

Terms like "squatting," "informal settlements," and "irregular housing" are perhaps too bland and all-encompassing to account for the almost limitless variety of possibilities included under the framework of unauthorized self-built housing. What is often overlooked in the narrative of informal housing is the range of strategies that the shelterless poor adopt in order to secure access to places to live.[145] At one end of the spectrum, new arrivals to the city squeeze into already overcrowded housing that is ostensibly covered under some sort of formal regulatory framework, even if official enforcement of building regulations is a rare occurrence. This type of shelter includes such officially recognized options as hand-me-down housing, hostels for temporary workers, and purpose-built tenements. At the other end of the spectrum, such out-of-the-way places of last resort as bedding down under bridges, sleeping under cardboard in doorways, and breaking into abandoned buildings represent perhaps the most abject form of urban precariousness. Reduced to bare life, those legions of homeless wanderers (whether individuals or small groups) eke out a meager and largely invisible existence outside the law and without its protection in zones of extreme biopolitical abandonment.[146]

Unauthorized self-built sites of human habitation cannot be easily classified. Other modes of securing irregular accommodation appear under various guises on the fringes of the mainstream of urban life. These include pirated subdivisions of existing informal settlements and makeshift additions to existing structures. Correlatively, the shelterless poor often engage in extralegal land seizures, invading underutilized land and squatting on public or private property without authorization.[147] Terms like squatting, informal housing accommodation, and shack settlements are often treated as if they suggest similar or homogeneous modes of precarious living under conditions of extreme deprivation. Yet ontologically speaking, modes of irregular housing accommodation forged outside official sanction depend on a broad constellation of shifting relationships, including various kinds of security of tenure, private property ownership (and the power to extract rents), and the right to claim rights (official recognition under the rule of law). Land invasions and unauthorized occupation of contested territories often take place under circumstances in which ambiguous and incongruous systems of land tenure and formal rights of private property ownership create regulatory voids that are often subject to inconsistent interpretations, resulting in continual land disputes and amounting to a situation of widespread tenure insecurity. The capricious enforcement of ill-defined regulations and decrees provides little or no incentive for unauthorized squatters to participate and comply with official procedures. Under these circumstances, formal land titling appears as much a pipedream as a realistically obtainable goal.[148]

Rather than a finished product or static condition, makeshift housing is an experimental practice, involving incremental processes of gradual improvement, cumulative adjustment, and incremental adaptation.[149] The "sheer plurality of practices and trajectories" of unauthorized dwelling and makeshift living arrangements outside official sanction makes it nearly impossible to develop a "strict homogeneous theory" or uniform understanding of squatting, informal settlements, and slums.[150] While these precarious sites of habitation may look alike, their logics of incorporation, rules of engagement/governance, and modes of occupation may be strikingly dissimilar. AdouMaliq Simone, for example, has cautioned against efforts to impose strict conceptual categories—terms like "irregular," "provisional," and "informal"—on the fluid textures and malleable surfaces of adjustment, experimentation, and innovation that sustain everyday practices of securing a place in the city.[151] For Simone, these efforts have often tended to "oversimplify, normalise, or occlude methods of composing everyday life that entail much less stability or calculation than those terms would seem to connote."[152]

In actuality, informal settlements and squatter encampments often take the physical form of strangely configured patchwork patterns of segmented localities in which spaces of income generation and irregular work overlap to varying degrees with places of dwelling.[153] By the same token, while squatters are aware that their illegal occupation of land (whether public or private) amounts to a collective contravention (or violation) of property laws and formal civic regulations, they nevertheless claim an entitlement to residence and livelihood as a matter of civic right.[154] A variety of informal associations and organizations that represent the collective interests of landless squatters, illegal traders, and informal entrepreneurs frequently engage in "the untidy contradiction of such contested space," since their claims can "only be made on a *political terrain*" of compromise and negotiation, "where rules may be bent or stretched, and not on the terrain of established law or administrative procedure."[155] This kind of urban political contestation demonstrates how state statutes and regulatory policies at times operate with impunity to impose draconian measures from above, while at other times they are extended, stretched, or contravened under circumstances in which the anticipated costs of strict enforcement of existing rules and regulations might presumably exceed the possible beneficial outcomes.[156]

These informal settlements of precarious habitation function as platforms for unfolding modes of improvisational urban practice, including the crafting of alternative livelihood strategies, the forging of new social networks, and the fostering of new "provisional and informal modalities" of associational life.[157] Slums, squatter encampments, and other informal modes of habitation provide

FIGURE 5.4 Megacities of Hypergrowth: São Paulo, Brazil. A new favela, or land occupation and informal settlement, in its early stages of growth at the edge of São Paulo. Maria Arquero de Alcarõn Photograph

relatively inexpensive housing for the shelterless poor and facilitate social support networks for newcomers to the city. As vast catchment areas for reserves of low-cost labor, these precarious settlements provide opportunities for petty entrepreneurs to undertake small-scale manufacturing and handicrafts production and for informal traders to sell their wares.[158]

The people living in informal settlements work together to redefine the conditions of urban life, often in remarkably creative and inventive ways. For Asef Bayat, slums are abject spaces of deprivation, yet they seem to organically engender a particular type of social agency: those spontaneous and improvisational activities that he has termed "informal life." Defined by flexibility, pragmatism, and negotiation, informal life signals a "constant struggle for survival and self-development."[159] For Bayat, Simone, and others, the undue stress on chronicling various survivalist strategies through which slum dwellers are able to eke out a bare existence frames the urban poor too much within a compressed narrative of victimhood.[160] Only able to react to circumstances, the urban poor—in the analytic framework of victimology—lack active agency and rational calculation in selecting from a narrow range of options to endure and survive. Yet there is an alternative view. Viewed through a wide-angle lens of engaged contestation, "marginalized subalterns" craft what Bayat has called "street politics," a mode of deliberative informal action that represent a field of discontent or a politics of redress, but not overt protest. These informal street politics represent a mode of "non-collective but prolonged direct action" aimed at achieving gains, interspersed with episodic moments of collective action and open protest in defense of past achievements.[161] These everyday practices—which Bayat has termed "the habitus of the dispossessed"—are a key feature of what might be called makeshift urbanism, subaltern urbanism, the improvisational city, or spaces of insurgent citizenship.[162]

In seeking to capture the stealthy, opaque tactics through which marginalized urban residents carve out usable spaces for themselves in uninviting urban landscapes, Bayat has coined the phrase the "quiet encroachment of the ordinary."[163] When these silent encroachments spill over the tacit boundaries of unofficial tolerance, the regulatory apparatuses of the state administration "kick into gear and retaliation can be swift and violent."[164] As a productive force, the spatial practices of informality thus shape urban landscape as a patchwork of segregated and divided localities.[165] Taken together, the everyday politics of redress, along with sporadic episodes of collective violence, often lead to spatial sorting along the lines of race, ethnicity, gender, age, and political loyalty.[166]

The evolving patterns of urbanization always involve multilayered and superimposed processes which produce complex, asymmetrical urban landscapes. As James Holston and Teresa Caldeira have suggested, innovative practices often come to life at sites of degradation. Uncovering these types of improvisational inventiveness require on-the-ground ethnographic inspection. The top-down gaze from afar (the bird's-eye view) cannot reveal slum spaces as arenas "in which residents use their ingenuity" to forge everyday worlds of adaptations, connections, and strategies within which to inhabit urban landscapes on terms not foreseen by the powerful forces of private property and municipal regulatory machineries.[167]

To focus on this inventiveness is not to ignore the powerful forces of global capitalism, the absence of stable wage-paid employment, and the erosion of the capacity of poor people to participate in social networks of support.[168] Nor is it to deny the deleterious effects of micro-exploitation in shaping life chances and the burning cauldron of prejudicial loyalties that sow discord and dissension in local settings. But it is to emphasize the capacity of slum dwellers to produce something distinct that "cannot readily be assimilated into established conceptual frameworks." Recognizing the creativity and invention of slum dwellers draws attention to active agency and rational calculation in a story all too often dominated by tales of passive victimhood. To bring to light the "creativity-of-practice" is also to make visible alternative possibilities. In this way, developing a paradigm of analysis of informal (and makeshift) urbanism that reveals the active agencies of slum dwellers suggests a conceptual approach that is not at once reductionist, totalizing, and deductive.[169]

THEORIZING THE FORMALITY/INFORMALITY DIVIDE

The distinction between informal and formal is one of the most enduring dividing lines, or boundary markers, in urban studies and development theory.[170] Debates, often contentious and heated, have largely centered not only on the descriptive capacity of this distinction to accurately identify actual conditions on the ground but also on its heuristic value as a useful dichotomy in planning theory, policy studies, and the practice of urban governance.[171] In his 1973 study of Accra (Ghana), Keith Hart first coined the term "informal economy"

FIGURE 5.5 Megacities of Hypergrowth: Mumbai, India. The Dhobhi Ghat, an open-air laundromat in the middle of Mumbai, where workers handwash garments for nearby facilities. More than two hundred families live in this area in illegal shack dwellings, surrounded by the city's skyscrapers. Mystical Mumbai

to distinguish self-employed individuals engaged in unenumerated, small-scale income-generating activities from the formal sector consisting of enumerated, large-scale, capital-intensive firms operating in regulated markets. Hart explored what largely rural-to-urban work seekers did in the interim before formal wage-paid opportunities became available.[172] In its conventional usage, the informal economy generally refers to small-scale or family ownership of enterprises operating outside of administrative regulation and public oversight, characterized by ease of entry, high turnover, reliance on local resources, small-scale operations, labor intensiveness, and adapted technologies. Yet perhaps because of their overuse, terms like informal economy, the informal sector, and informality have become highly elastic, with vague and imprecise meanings.[173]

Ostensibly, this distinction between formality and informality is simply a descriptive marker, a way of categorizing and classifying different types of activities, a shorthand device for distinguishing different modes of income generation or access to housing, or a means of shedding light on unregulated and even authorized practices often hidden from view. Yet even a cursory review of the scholarly literature reveals considerable disagreement over the usefulness of the concept of informality and the extent to which it accurately identifies a discrete type of socioeconomic activity.[174] The early use of the term "informal sector" as a separate and autonomous sphere of economic activity has turned out to be unproductive and, as a consequence, to have questionable utility with regard to its explanatory power.[175]

The basic on-the-ground realities, however, that stimulated interest in the formal/informal distinction in the first place have remained a serious challenge for urban theorists, planners, and policy makers alike—namely, the persistence of a wide range of activities that do not easily fit within the definitions and analysis of conventional economics. In seeking to account for underdevelopment, mainstream economists have associated informality with extremely low productivity and inefficiency (in comparison with formal economies), stagnant growth, and oversaturated markets, with microenterprises and shoestring businesses producing "low-quality products for low-income customers using little capital and adding little value."[176] Similarly, in conventional urban studies literature, the term "informality" typically functions as a metonym for unauthorized and unregulated activities that take place on the margins of official sanction and outside of formal regulatory regimes and existing institutional rules.[177]

When it entered the scholarly vocabulary of urban studies in the 1970s, the formality/informality divide served as the main conceptual framing device for understanding unregulated work and unauthorized settlement patterns. In spite of its steady expansion as a mode of livelihood for the urban poor in struggling cities, the informal economy before the onset of liberalization policies in the 1980s was typically seen in terms of the structural weaknesses of the formal sector and its inability or failure to absorb available surplus labor and to adequately address the need for affordable shelter. Labor relations in the informal sector were considered restrictive and exploitative, causing economic growth in general to slow down, thereby impeding improvements in income, purchasing power, and upward mobility.[178]

Despite its widespread and continued popularity in scholarly literature and policy formulations, the term "informality" does not appear to be matched by any widespread agreement concerning its precise meaning or points of reference. The proliferation of all sorts of competing alternative definitions has only added to the conceptual confusion. The usefulness of the idea of informality must be found not in its analytical clarity and conceptual precision, but rather in its continuing capacity to evoke vague and generalized images that highlight the everyday realities on the ground for urban dwellers without regular wage-paying work and without legally enforceable claims to residential accommodation. These images have proved so powerful that the term has continued to be used with regularity. While there is little consensus on a common definition, critics and advocates alike tend to assume the basic utility of the idea of informality as indicating certain types of activities outside official sanction and formal regulation. What has remained problematic—and a matter of debate—is how best to conceptualize the distinction between formality and informality.[179] While some scholars are content to work with a distinction (however vague and imprecise) between formality and informality, others have suggested that "contemporary urbanism in both the global north and south can be better understood by unraveling the association between the formal and informal."[180]

RETHINKING/REFRAMING INFORMALITY

In scholarly research and writing, informality is more often than not associated with irregular and casual work, persistent underemployment, lack of wage-paying jobs, and poverty, on one side, and social exclusion, marginality, and socioeconomic inequality, on the other.[181] Seen through the lens of makeshift urbanism, informality is a central organizing principle for income-generating strategies such as microenterprise, self-exploitation, small-scale trading, and the do-it-yourself provision of shelter and services. The turn toward informal modes of self-help (whether through hustling or petty entrepreneurialism) reflects the growing formlessness and unpredictability of "getting by" outside the mainstream of urban life.[182] Rather than conceiving of informality as a residual or transitional category (a key feature in conventional dual-economy models), it is more fruitful to treat informal practices as a constituent component of the dynamics of city building on a global scale. Nonstatutory management of land use, clouded and ambiguous property rights, unauthorized supply of infrastructure and services, and flexible, negotiated, and deregulated access to sites, facilities, and vacant spaces have become key elements in everyday, makeshift urbanism.[183] As a "generalized mode of urbanization," informality modifies and embeds itself within the transactional exchanges that circumscribe livelihood strategies, access to material and social infrastructures, claims to land tenure, the legitimacy of property rights, and housing provision.[184]

The practices that are often subsumed under the blanket category of 'informality" vary considerably in terms of structural positionality, life chances, and social status. First and foremost, it is important to distinguish between the informal workers (the so-called informal proletariat) who are employed in informal jobs outside official sanction and legal protection and those wageless toilers who are self-employed in unregulated or unregistered enterprises.[185] In this age of rapid urbanization, when the steady influx of work seekers to cities cannot be absorbed into the world of regular wage-paid employment, informality has become the dominant mode of urban life.[186] The conjoined forces of globalization, the triumph of market logic, and the withdrawal of public authority from municipal governance have left in their wake countless overburdened cities that are incapable of managing urban space and providing public resources for poor urban residents.[187] Distressed cities on the periphery of the world economy have experienced increasing socioeconomic inequalities, new kinds of socio-spatial segregation, simmering conflict and violence, broken-down infrastructure, dwindling social services, and the collapse of conventional mechanisms of governance. Under these circumstances, informal practices have come to dominate the everyday lives of the urban poor. It is both an everyday mode of income generation, where opportunities for formal wage-paid employment are scarce and precarious self-employment is the norm, and a form of shelter and service provision, where decent affordable housing is beyond the reach of the urban poor.[188]

As a general rule, scholarly research and writing on informality tend to fall into two separate fields of inquiry: the operations of unregulated labor markets, informal work, and irregular employment, on the one side; and the auto-construction of housing and the self-provisioning of services, on the other. One unfortunate consequence of this artificial bifurcation is that the politics of informality—that is, the social practices through which urban residents assert their rightful place in the city—are typically classified in accordance with the separate spheres of work and income, on the one side, and housing and dwelling space, on the other. Despite the fact that "the everyday lived experience of informality"—or "the spatial practices of informality"—systematically undermines any alleged analytic separation between irregular work and self-provisioned housing, planning and policy recommendations typically focus on one sphere to the exclusion of the other.[189]

A strong current in the scholarly literature, spanning the political spectrum from the neoliberal populism of Hernando de Soto to various Gramscian and neo-Marxist formulations, views informality as more or less a separate and bounded sector of unregulated work, self-organized enterprise, and irregular residential settlement. This framework treats informality as an extralegal domain outside of state regulation and apart from formal market economies. The policy interventions that follow from this formulation call for the integration of multiple informal practices into "the legal, formal, and planned sectors of political economy."[190]

Even a cursory review of recent scholarly writing in planning theory and public policy reveals a persistent attachment to the formal and informal as separate and detached spheres of activity. Critics of this conventional dualist thinking, such as Timothy Mitchell, have explored the relationship between formal and informal activities through an analysis of the boundary between capitalist market economies, on the one hand, and noncapitalist activities beyond the operative logic of market exchange (that is, activities outside the formal limits of market-driven and profit-oriented capitalist economies), on the other. While he rejects this approach, Mitchell argues that the idea of a clear line of demarcation between the capitalist and the noncapitalist spheres provides a common way in mainstream policy analyses not just to think about formality and informality, "but to diagnose their problems and design remedies."[191]

Conventional planning theory and policy analysis typically operate with the presumption that the capitalist economy is a self-contained, tightly circumscribed, and autonomous sphere of market-based activities, outside of which exists the noncapitalist and nonmarket world of barter and reciprocity.[192] The task of conventional development economics is to find ways of extending the institutional rules of private proprietorship, the market exchange of commodities, and capitalist enterprise into these nonmarket spheres. Policy prescriptions, ranging from deregulation to liberalization to privatization of collectively held assets, foster market solutions to urban poverty by both promoting economic competition and unleashing the enterprising energies of microentrepreneurs.[193]

While the notion of informality seems useful as a way of highlighting the absence or ineffectiveness of formal regulation by legal and bureaucratic institutions in governing such matters as labor contracts and setting minimum standards for working conditions, to characterize the informal as simply the mirror image of its formal counterpart is unjustified and hence unhelpful. Such a characterization not only overlooks the heterogeneity of (unregulated) informal activities and their intimate links with (regulated) formal activities but also implies—as imagined by neoclassical development economists like de Soto—that the unleashing of the capitalist logic of market exchange enables the "dead capital" trapped outside the formal economy to enter the capitalist marketplace as a source of new wealth. Among other things, what this assessment misses is that so-called free-market competition is never completely unrestrained but is always structured by informal institutional constraints. An exclusive focus on formal institutions and regulatory rules offers only a shortsighted approach to understanding how labor markets actually function. As Ragui Assaad has persuasively argued, "informality is a quality that permeates all spheres of competitive market economies to a greater or lesser extent."[194]

INHERENT AMBIGUITY

Informality is a notoriously slippery term that defies simple classification. It is difficult to define because of the multiple meanings it assumes in everyday usage and the selectivity and pragmatism of its application in various disciplinary fields of inquiry.[195] The importance of informal practices as key elements in urbanization is almost universally acknowledged across a wide range of academic fields, including development studies, planning theory, and urban policy research, but there is considerable disagreement on how informality and informal practices should be defined and approached as objects of study in their own right.[196]

When it is conceptualized in the field of economics, informality typically refers to survival strategies of the wageless poor or to unauthorized enterprise operating in the extralegal shadowlands outside of discriminatory statutory regulations and legal barriers.[197] Similarly, in conventional urban studies literature, informality typically refers to strategies that enable excluded and marginalized urban dwellers to make claims to land, housing, infrastructure, and services that are beyond their reach in formal market exchanges. In its much narrower sociological usage, informality takes on a somewhat selective, subjective, dynamic, and multidimensional meaning. Some scholars equate the idea of informality with more spontaneous, flexible, personalized, and context-dependent social interactions that enable innovation and improvisation, thus opening a multiplicity of possible ways of creatively operating outside of official sanction and state regulatory authority.[198]

Challenging the dichotomous distinction between formality and informality does not mean that concepts like regulated and unregulated, legal and illegal, and sanctioned and unsanctioned should be discarded as useless baggage. Yet the formal and the informal should not be understood as opposite and separate domains.[199] As Kim Dovey has argued, the informal/formal distinction is both fundamental and nondichotomous: "it is a single twofold concept rather than two concepts in opposition."[200] Likewise, the informal, the nonlegal, the extralegal, and the illegal coexist and intermingle with the formal and the legal.[201]

The legacy of dualistic models—with their sharp distinctions between traditional and modern economies and their clear divisions between regulated and unregulated activities—has lingered for far too long in mainstream research and writing on informality. This binary thinking that at first dominated definitions of informality has begun to fade, as it has become abundantly clear through concrete empirical research that actual conditions on the ground rarely conform to any fixed or rigid boundaries separating legality from illegality and officially sanctioned from unsanctioned practices.[202] Scholarly research and writing have moved beyond a narrow understanding of informality simply as an alternative way of coping with deprivation in the absence or weakness of functional formal systems of regulation.[203] A growing number of scholars have offered a more nuanced understanding of informality, suggesting that it is more appropriately conceived as a "highly differentiated process embodying varying degrees of power and exclusion."[204] In challenging the idea of the informal as a distinct sector or sphere of activity, Anaya Roy has called for the reconceptualization of informality as "an organizing logic, a system of norms that governs the process of urban transformation itself."[205] In this new way of thinking, informality offers urban residents a range of opportunities to tap into formal institutions, bending and molding these to fit with informal practices in order to take advantage of various income-generating streams that offer more than mere survivalist economics. As an idiom of everyday urbanism, informality signifies a logic through which the production and management of space (as a site of livelihood and place of habitation) take place.[206]

Seen in this light, informality is not restricted to the bounded (and artificially circumscribed) space of the slum or to the life-worlds of unregulated work and income generation outside of wage-paid employment. Instead, it is a mode of social organization that connects people and places through webs of interaction, market transactions, and mobilization of resources. Understood in this way, informality signifies a complex bundle of transactions that connect different socioeconomic activities and spaces to one another. The forms of urban informality run the gamut from illegal squatter land invasions to the extralegal acquisition of land for luxury estates, shopping malls, and corporate business parks. In the megacities of the Global South, land use typically violates some planning or building code, so that the construction of everything from unauthorized self-built shacks to luxury apartment complexes takes place outside formal regulatory frameworks.[207] Planning by exemption has replaced single regulatory frameworks, both in the construction of luxury megaprojects and in the authorization to carry out slum clearances.[208] As Roy has argued, "the valorization of elite informalities and the criminalization of subaltern informalities produce an uneven spatial geography of spatial value."[209]

In this sense, the concept of informality cannot be understood in strictly ontological or topographical terms, as either a state of being that defines particular types of activities or distinct practices that take place in specified locations. Instead, it is a heuristic device that works to uncover "the ever-shifting urban relationship between legal and illegal, legitimate and illegitimate, authorized and unauthorized."[210] Where these relationships began and end is malleable and fluid. In everyday discourse, what is legal and illegal, legitimate and illegitimate, authorized and unauthorized appears to be more a matter of naming and classification than designating a real existence. Despite the differences typically implied by these dichotomies, these relationships are never stable and fixed but instead are always arbitrary and uncertain, with negotiation between them often accompanied by displays of power and violence.[211]

Rather than falling unambiguously under the designation of illegal, informality is more fruitfully understood as a kind of unregulated activity under circumstances "where similar activities are regulated."[212] As Nezar AlSayyad and Ananya Roy have persuasively argued, this type of unregulation is in itself a distinct form of regulation, a set of practices that establish informality (or exemption or exception) as a mode of governance. As such, informality "is a process of structuration that constitutes the rules of the game," setting limits on what is possible and, simultaneously, opening up fresh opportunities. If formality operates through the "fixing of value" (by establishing a legible set of institutional rules), then informality acts through the "constant negotiability of value"—the breaking or ignoring of established rules and the introduction of ad hoc exemptions to conventional normative expectations.[213] In assuming the exclusive power to determine the contours of the playing field by distinguishing between the norm and the exception, formal regulatory regimes determine how and when to enact and, conversely, to suspend the rule of law.[214] In effect, this power to "determine what is informal and what is not" amounts to deciding which kinds of informality are allowed to remain, and perhaps thrive, and which should be condemned as illegal, and hence subject to state-sponsored law enforcement. This institutional capacity "to construct and reconstruct categories of legitimacy and illegitimacy" epitomizes the power to establish what is formal and what is not. The power to establish state-sanctioned formal regulation is thus the key to deciding what can be classified under the rubric of informality and what is not.[215]

URBAN INFORMALITY AS A MODE OF REGULATION

Informality is certainly not a new phenomenon, nor does it stand outside (and alongside) formality as a distinct and neatly packaged category as it was originally conceived.[216] The origins of informality cannot be attributed to underdevelopment or "lagging behind." Similarly, to portray informality as a sign of backwardness is to misconstrue its operations as somehow marginal to the dynamics of the so-called development process. The formal/informal divide does not indicate opposing dynamics or competing logics. On the contrary, informality and formality are inextricably intertwined processes that work together as integrally connected modes of organizing urban life. Informality is an ever-present condition, sometimes largely invisible and somewhat inoperative and other times very visible and obvious.[217] Given the multiple ways in which formal and informal overlap and intersect, it is virtually impossible to disentangle the two processes in the actual practices of everyday life. In writing about Rio de Janeiro, Janice Perlman has concluded that "all the conventional distinctions between the 'formal' and 'informal' city have begun to blur," and the separation of urban space into formal and informal "is no longer applicable, if it ever was."[218]

The dualistic thinking behind such hard-and-fast distinctions as formality versus informality turns "the merely different into absolutely other."[219] As Carmen Gonzales has persuasively argued, the sometimes-intense debate over formality versus informality in many ways represents a tempest in a teapot. Far from signifying a wholesale failure of the formal rule of law, "informality constitutes a parallel and intersecting system of law" that the urban poor resort to in the face of daunting socioeconomic hardship and deprivation. For those without regular work and without decent shelter, informality is a rational and calculated response to marginalization and exclusion from the mainstream of urban life.[220]

Properly understood, informality cannot simply be conceptualized as a practice that occurs outside and independently of the regulatory reach of the state administration, but rather must be seen as a process produced by state action or inaction.[221] The laxity of municipal authorities to regulate urban informality should not necessarily be interpreted as a "colossal public policy failure," a "random occurrence" or an accidental oversight, or even the "unwelcome result of scarce resources."[222] Rather, informality may at times function as a deliberately flexible strategy aimed at fostering and maintaining a fluid condition of uncertainty and legal ambiguity. This environment of ambiguity and illegibility enables municipal officials charged with the enforcement of formal rules and regulations to operate with a significant degree of flexibility and discretion, and hence impunity.[223] In her work on Calcutta, for example, Ananya Roy has shown how the tactical mobilization of extralegality in the informal recognition of land occupation provided political authorities with a "technique of discipline and power" to manage landless squatters.[224]

Formality/informality, legality/illegality, and legitimacy/illegitimacy are not discrete categories with a fixed ontological status but fluid conditions that reflect differential access to power, property, and resources. Formal regulations codified in legal statutes do not function simply "like a blueprint structuring social action and spatial form." Instead, they exist mostly as a point of departure that opens possibilities for spatial negotiations and maneuvering.[225] The fluid and contingent enforcement of legal rules and regulations produces "a novel, extra-legal technique of spatial management."[226]

As a general rule, conventional definitions locate informality in terms of socioeconomic activities that lie outside formal systems of recording, remuneration, legal recognition, and state administrative control.[227] Seen from this vantage point, a wide variety of activities and enterprises—ranging from street vendors and itinerant hawkers to small-scale artisans producing handicrafts and even some illicit activities, such as the drug trade, dealing in stolen goods, and prostitution—are collectively cast in stark opposition to formal employment and regular wage-paying work. Conventional urban studies scholarship and planning theory operate with the assumption that formal regulatory regimes make use of various "technologies of visibility, counting, mapping, and enumerating" to bring stability, legibility, and predictability to urban landscapes. In this way of thinking, formal planning represents order, while informality suggests disorder and the lack of regulation.[228]

However, there is also a well-established body of scholarly writing that seeks to move beyond the simplified dualism of formal versus informal, focusing instead on the interpenetration and overlapping of formal and informal activities that result in the coproduction of urban space.[229] Without a clearly demarcated dividing line between formal and informal, a number of scholars have suggested that it is more fruitful to think of activities as forming part of a continuum, rather than viewing these in dichotomous terms.[230] Equally important, recent scholarship has challenged the uncritical mapping of informality onto poverty, illegality, and disorganization, viewing informality instead as an idiom, a governing tool, and a negotiable value of contemporary urbanization.[231]

Recognizing the transactional and relational nature of formal and informal practices enables us to see that informality does not exist as a separate sphere outside of officially sanctioned regulatory mechanisms but is instead thoroughly imbricated within them.[232] Formal regulatory regimes and their accompanying legal norms are "in and of themselves permeated by the logic of informality."[233] In challenging the conventional view that the presence of informal practices, such as unsanctioned housing or unauthorized income-generating activities, is a straightforward indication of the absence of formal regulation, Ananya Roy has argued that informal practices are best understood as an outcome of deregulation, or the calculated withdrawal or relaxation of officially sanctioned regulatory powers. As the embodiment of "a distinctive form of rationality," deregulation "indicates a calculated informality," one that involves purposive action and inaction, and "one where the seeming withdrawal" of formal regulatory procedures "creates an [alternative] logic of [asymmetrical] resource allocation," different modes of authority, and differential possibilities for accumulation.[234]

Modes of deregulation enable us to understand how modes of urban governance are thoroughly imbricated in the "unstable relationship between the legal and the illegal," or what James Holston has called the "misrule of law."[235] Deregulation ensures that property relations, ownership of resources, and rights of occupation "cannot be fixed and mapped according to any prescribed set of regulations or the law."[236] In this sense, formal rules and legality are flexible and malleable, or "perhaps better understood as fictions, as moments of fixture in otherwise volatile, ambiguous, and uncertain systems of planning."[237]

For Roy, informality does not arise "from the failure of [formal] planning" or the absence of officially sanctioned regulatory authority. Instead, in her view, informality originates out of deregulation and hence "can be thought of as a [largely invisible] mode of regulation." Deregulation thus shifts the burden of survival to the urban poor, who are compelled to get by on their own, excluded from the institutional protections provided under formal regulatory regimes. Seen from this angle of vision, the ambiguous and inconsistent application of formal regulations provides a mechanism for disciplining the urban poor.[238] For example, as Jonathan Shapiro Anjaria has shown for street vendors in Mumbai, city officials gain from not legitimizing the rights of street vendors to operate openly, instead keeping them in "a constant state of flux" in order to use the threat of eviction as leverage for extortion.[239]

In a similar vein, Oren Yiftachel locates urban informality in the "gray spaces" between the "whiteness" of legality, approval, and safety, and the "blackness" of eviction, destruction, and death. These in-between spaces are neither fully integrated into the mainstream of urban life nor eliminated altogether. They form the semipermanent but malleable margins of cities struggling with unemployment and insalubrious housing. These gray spaces include the practices of self-built housing, squatter encampments, microenterprises, and informal economies, residing uncomfortably (and literally) "in the shadow" of the formally planned city, with its legally sanctioned regulatory regimes and formal economies. Yiftachel suggests that these intermediate gray spaces make possible the imposition of various modes of domination, mechanisms of differentiation that are managed by urban regulatory regimes and facilitated by the tools and technologies of formal planning.[240]

For Yiftachel, municipal authorities not only tolerate and endorse informal activities of rich and powerful urban residents but also support—in effect, "whiten" them and grant them legitimacy—through the extension of physical infrastructure and the provision of services. At the same time, these same municipal authorities actively criminalize—or "blacken"—informal activities of urban dwellers in order to delegitimate them.[241] Seen in this way, informality is a mode of urban governance and an expression of variegated constellations of power.[242]

As active agents overseeing state regulatory regimes, municipal authorities and city officials play oversize roles in framing the structural conditions and the organizing logic inherent in urban informality.[243] In her critical evaluation of informal economic activities in sub-Saharan Africa, Kate Meagher has shown how the expansion of informality is not a process occurring outside the regulatory authority of the state administration, but instead takes place in an environment involving "state complicity." She thus underscores the usefulness of moving beyond the understanding of informality as a separate and autonomous sphere of marginality toward a structuralist approach that conceives of "informalization" as an essential component of a wider defensive response to persistent socioeconomic crisis.[244]

THEORIZING INFORMAL URBANISM

In thinking about the dynamics of global urbanism, Rahul Mehrotra distinguishes between the "static city" of permanence and stability and the "kinetic city" of ephemerality and constant motion. These two modalities of urbanism coexist, occupying the same space and collapsing together into a "simultaneous—often kaleidoscopic" assemblage of overlapping and competing logics and visions. While those powerful city builders who manage the static city aspire to erase all manifestations of instability and irregularity and to reconfigure these to conform to formal order and rational use of space, the kinetic city operates on the terrain of mobility, transience, and impermanence. These irresolvable tensions give rise to all sorts of contradictions, conflicts, and power struggles.[245]

The distinction between the static city and the kinetic city corresponds to the intersecting dynamics of formal urbanism and informal urbanism. The volatility and capriciousness of the kinetic city enables one to more fully comprehend "the blurred lines of contemporary urbanism." The messy, unregulated informal practices that violate the abstract spaces of the planned city—with its land-use provisions, zoning regulations, and nuisance policies—illustrate "the collapsed and intertwined existence" of the static city and the kinetic city.[246]

In its original conception, informality "stems from the essential conditions of correcting or compensating for the unequal distribution of resources" (regular work, decent housing, and access to urban services). The concept of informal urbanism extends the idea of informality beyond the conventional fields of unregulated housing and irregular work to include the plethora of experimental practices and improvisational tactics through which urban residents across the class spectrum establish operational networks, recycle disused materials, and rebuild the city in the interstices (and blind spots) of existing regulatory frameworks.[247] Conventional city-building practices—focused almost exclusively on formal regulation of land use and top-down planning—have proved "poorly equipped to cope with the dynamism, complexity, and resilience of informal urbanism."[248] The kinetic quality of informal urbanism does not allow formal regulatory mechanisms to keep pace with or respond in a meaningful way to the spontaneity and versatility inherent in inventive extralegal tactics. In the sprawling megacities of hypergrowth, reaching the ultimate goal of formal regulation of the use of space amounts to an unobtainable "precarious achievement" that is both provisional and exceptional.[249] In this sense, as Andy Pratt has suggested, formality is the exception and informality is the norm.[250]

Informal urbanism is a crucible of innovation that cannot be classified under the rubric of mere survivalism.[251] The informal city consists of a variegated network of everyday activities, mobilization of material resources, and recycling of infrastructures. These flexible, temporary tactics take place alongside the formal economies of the capitalist marketplace, outside the disciplinary logic of regulatory regimes, and beyond the scope of legal recognition.[252] Informality is a mode of social organization and a mechanism for furthering collaboration. From the construction of unregulated settlements and unauthorized land invasions to self-organized economies and curbside trading, informal urbanism is an integral feature of the sprawling megacities of the Global South.[253] The "lived spaces of informality"—or the "space of localities"—and not just vectors of employment and housing are key elements in understanding the dynamics of informal urbanism. These terms better capture the entangled sites and interlocking networks of home-based work, small-scale manufacturing, and shelter that are the lifeblood of wageless laborers excluded from the mainstream of urban life.[254]

Property understood, space is not just a passive stage set where urban activities take place but an active agent that not only defines limits of social action but also reinforces inequities in power and the allocation of scarce resources. The production of space and the politics of informality work simultaneously to open opportunities and to restrict movement and mobility. In this sense, the space of localities plays a crucial role in shaping the possibilities for gaining access to power, contravening restrictive rules, or engaging in resistance. The everyday politics of stealth, survival, and encroachment both produce and reconfigure urban space as an uneven patchwork of deeply segregated but malleable localities and reclaimed territories.[255] The informal urbanism that has emerged in struggling cities is one of "productive disorder" or the "unhinged proliferation" of home-based workshops, impromptu markets, itinerant street hawkers, labor-intensive factory sites, small and large settlements housing the working poor "spread all over the planned metropolis," a "pirate culture" of recycling, retrofitting, and adaptation.[256]

CHAPTER 6

BUILDING CITIES ON A GRAND SCALE

The Instant Urbanism of the Twenty-First Century

The modernism of underdevelopment is forced to build on fantasies and dreams of modernity, to nourish itself on an intimacy and a struggle with mirages and ghosts. In order to be true to the life from which it springs, it is forced to be shrill, uncouth and inchoate. It turns in on itself and tortures itself into extravagant attempts to take on itself the whole burden of history. It whips itself into frenzies of self-loathing and preserves itself only through vast reserves of self-irony. But the bizarre reality from which this modernism grows, and the unbearable pressures under which it moves and lies—social and political pressures as well as spiritual ones—infuse it with a desperate incandescence that Western modernism, so much more at home in the world, can rarely hope to match.[1]

In the nineteenth and twentieth centuries, the powerful social forces associated with industrialization and modernization not only greatly accelerated the pace and scale of global urbanization but also significantly redefined the nature of city life. City builders around the world sought to emulate what they considered the most striking and progressive features of the modern metropolis: towering skyscrapers, splendid iconic buildings, legible street grids, distinct districts characterized by functional specialization, efficient traffic corridors, and vast public sites of grand monumentality. For all intents and purposes, mid-twentieth-century Manhattan (and its surrounding boroughs) exemplified this exhilarating vision of metropolitan modernity.[2]

At the dawn of the twenty-first century, the centripetal pressures that drew industry, finance, and technical expertise—along with money, wealth, and power—to key urban-industrial city-regions at the core of the capitalist world economy have come unglued. Contemporary concepts like competitive cities, postindustrial cities, and tourist-entertainment machines have become key watchwords signifying a profound shift in both the form and function of global urbanism. At the macro level, this shape-shifting has unsettled not only our conventional understanding of the place of particular cities within a world of cities but also what it means to talk about globalizing cities with world-class aspirations.[3]

With the exponential growth of urban populations, it is expected that countless numbers of new towns and city extensions will come into existence in the coming decades. While public involvement in closely monitored urban redevelopment projects has declined, large-scale private real estate developers have stepped in to fill the gap.[4] Private real estate development is nothing new, but the scale and ambition have expanded considerably. Large-scale private corporations have entered the city-building business, vying to construct entirely new cities out of whole cloth, packaging their services (from design and financing to actual construction and management) in order to replicate their generic spatial products elsewhere. To address the complexity of these city-building efforts, these private companies have adopted new organizational models and buy-in schemes and have invented new financial tools to pay the costs of the initial rollout of basic infrastructure. The dynamics among profit-seeking investors, real estate developers, design and engineering firms, architects and clients, and end users are constantly shifting in relation to changing circumstances.[5]

INSTANT CITIES AND FAST-TRACK URBANISM OF THE HYPERMODERN AGE

Over the past several decades, the steady increase in the number of new master-planned, holistically designed satellite cities have not only significantly reshaped existing urban landscapes but also reinforced the spatial boundaries separating affluent zones from impoverished wastelands. What makes these recent city-building efforts different from earlier attempts is that they involve constructing entirely new cities out of whole cloth rather than rehabilitating the existing built environment. Unwilling to take up the challenge of refurbishing existing large metropolises, private real estate developers have begun to construct new cities built entirely from scratch. Fashioning new urban landscapes de novo enables city builders to bypass the messy problems associated with the current state of urbanism in the (misnamed) developing world: crowded streetscapes, lack of land-use zoning and code regulation, traffic gridlock, poor service delivery, and street crime. Located on the edge of existing metropolises, these new satellite cities promise to deliver up-to-date services, high-quality infrastructure, and safety and security in cocooned environments that gesture more in the direction of so-called world-class cities than toward their immediate surroundings.[6]

The unprecedented scale and scope of fast-track urbanism in the Asia Pacific Rim, the Persian Gulf, and elsewhere at the start of the twenty-first century pose serious challenges to the customary image of the modern metropolis—and even the very idea of the city.[7] These instant cities defy the conventional understanding of urban growth and development as a slow process of layering and retrofitting, in which creative destruction of the built environment, incremental rebuilding, and constant erasure and reinscription produce an ever-evolving urban form that combines vastly different building typologies and architectural styles and where the old and the new are cobbled together in uneven and hybrid configurations.[8] As consciously designed spatial products that deliberately seek to avoid "the messiness of urbanity" by aspiring to be "earthly utopias," these experiments with fast-track urbanism force us "to rethink what a good city is and what it should be."[9] The rapid transfers of investment capital, cutting-edge information technologies, and expert advice (often referred to as traveling ideas) have created possibilities for compressed or telescoped urban development, enabling city builders to begin afresh by bypassing existing derelict environments and leapfrogging over the unwanted detritus of past experiments with (failed) modernity.[10]

From the outset, it should be clear that these instant cities have not followed established routes or pathways for expected urban transformation laid down elsewhere and at earlier times.[11] In short, the experimental city-building processes that have produced instant cities and fast-track urbanism are not at all reducible to antecedents in North America and Europe—or anywhere else for that matter.[12] What remains an open question is the extent to which construction of completely new cities de novo foreshadows city building to come over the foreseeable future. What defines this instant urbanism—or what some have labeled "test-bed urbanism" or ready-made "cities-in-a-box"—is not linear temporal evolution or incremental gradualism but spatial transformation that conforms to a tabula rasa prototype.[13] Unlike the long processes of incremental evolution that shaped urban transformation in earlier centuries, instant cities are the manufactured outcome of a super-fast urbanism, lurching forward at breakneck speed.[14]

Conventional urban studies scholars and planning theorists have no agreed-upon, ready-made models or standardized formulas appropriate for addressing this new type of fast-track urbanism. In seeking to locate themselves at the cutting edge of new approaches to city building, these instant cities both embrace and reject modernist principles that informed architectural stylistic innovations and planning protocols over the last century. By fostering the separation of urban landscapes into precincts for work, recreation and leisure, dwelling, and transportation, the design specialists who build instant cities have implicitly endorsed the rigid modernist principles of land-use zoning and functional specialization. Yet with no clear hierarchy between center and margin, these instant cities confound the modernist expectations of abstract rationality, spatial differentiation, and heterogeneous urban form.[15] These new instant cities are produced by industrial machine-age processes so cherished by modernist city builders, but they are too rapidly built and too dynamic to be structured or controlled by the rigid modernist principles of rational order, functional specialization, and spatial coherence.[16]

Yet if modernist city building is clearly not the appropriate model for making sense of these new urban conditions, then the principles of postmodern urbanism have even less to contribute. Instant cities appropriate the semiotic strategies of postmodernist thinking, but they are unable to produce the kinds of neo-traditionalist attachment to place, the sense of belonging, so favored by postmodern urbanism.[17] If postmodern urbanism dispenses with seeking ideal solutions on a grand scale and is associated with the characteristics of human-scale miniaturism, contextualism, pluralism, eclecticism, bricolage, "location-persuasion," and playful historicism, then instant cities replicate the modernist enchantment with verticality and monumentality, using sheer size, scale, and spectacle as mechanisms to mimic the modernist quest for an overarching universal utopia, and, in so doing, disconnect the urban realm from history and spatial context.[18]

Large-scale, master-planned, and holistically designed satellite cities have become an increasingly common urban development concept; originating in the Persian Gulf and the Asia Pacific Rim, it quickly spread to India, Africa, the Middle East, Latin America, and elsewhere.[19] For the most part, real estate developers—often operating in rough synchronization with municipal authorities but sometimes working almost completely on their own accord—have followed a similar spatial pattern of building entirely from scratch instead of trying to retrofit the existing built environment or insert their megaprojects into existing urban landscapes.[20] What distinguishes these new master-planned satellite cities—or what Gavin Shatkin has called "bypass-implant urbanism"—from the conventional understanding of city building and the modern metropolis is that they are constructed entirely free from the spatial, institutional, and logistical constraints associated with the existing urban form.[21] Constructing these large-scale urban development projects does not conform with the modernist principles of erasure and reinscription, in which city builders incrementally rebuild the urban social fabric by tearing down outdated parts of the existing built environment to make way for something new; instead, they are master-planned and holistically designed enclaves—what can be called "urban integrated megaprojects"—with a deliberately conceived assemblage of distinct parts fitting into a functionally integrated whole.[22] Building at such a large scale and at such a quick pace has produced a kind of fast-track urbanism in which instant cities seem to emerge out of nowhere like alien spaceships unexpectedly coming to ground without much forethought or deliberate preparation. As a general rule, these new satellite cities are strategically located outside the historic central core of existing metropolises, often with easy access to sparkling new airports and tourist-entertainment centers designed to attract corporate business travelers and the cosmopolitan consumerati. In a departure from past efforts at state-driven master planning and (paternalistically inclined) new town developments, these large-scale, self-contained cities are constructed as profit-making undertakings tout court, often the brainchild of a single prophetic real estate developer or a consortium of property investors, sometimes in partnership or alliance with municipal or state authorities.[23] These large-scale self-contained, mixed-use urban enclaves "represent a vision for the transformation of the urban experience through the wholesale commodification of the urban fabric."[24]

The new instant cities are high-modernist experiments in which the assembled pieces are imagined as something akin to machine parts for creating a technological and ecological utopia. These cities are brought to life not by the needs and desires of ordinary residents but by the material interests of global real estate developers and state policy makers. Unburdened by tradition and historicism, these cities are branded as global destinations, whether as business hubs, high-tech incubators, tourist getaways, or gateways to somewhere else.[25]

The discourse surrounding fast-track urbanism is devoid of the kinds of social consciousness and engagement that historically gave substance to the notions of the common good and the public realm. Instead, the corporate interests behind these new city-building projects are mesmerized by the technological fix, itself an abstract and utopian view of rational efficiency in which stylized images of infrastructure as "smart" and manicured "nature" as playground have become permanent features of branding exercises. These instant cities have reinvented the idea of up-to-date urbanism in conformity with the market logic of commodity exchange, benefiting only the privileged few who can afford the high costs of living in such streamlined technological bubbles.[26]

The main objective behind fast-track urban experiments is not to rationalize urban space by bringing it into conformity with up-to-date international best practices but to create opportunities for maximizing profitability. Instant cities coincide with a model of urban development that entails the synchronized assemblage of spatial components: hard and soft infrastructure, the provision of requisite services, and prepackaged building typologies are rolled out simultaneously as a fitted ensemble of integrated parts that combine to produce a preconceived whole package.[27] What propels this kind of fast-track urbanism is the wish image of contemporary global capitalism: the wholesale transformation of urban space into a profit-making machine.[28] City builders have become fixated on using the sights and sounds of faux urban life—the spectacle, enchantment, and excitement, or what Richard Lloyd and Terry Nichols Clark have called the "city as entertainment machine"—as a source of capital accumulation in and of itself.[29]

EARLY PROTOTYPES: HISTORICAL ANTECEDENTS

Holistically designed experiments in planned urban living are certainly not new in the history of city building. Utopian visions of the master-planned "good city" have a history going back centuries in many parts of the world.[30] One particularly important early prototype for master-planned urban enclaves can be found in the construction of philanthropic "company towns" in Europe and North America.[31] As a general rule, settings for company towns were in out-of-the-way locations where extractive industries (such as coal fields, mines, and lumber) had established a virtual monopoly franchise. Margaret Crawford has explored the historical transformation of American company towns as a distinct urban form spanning a 150-year time frame.[32] The conscious planning of company towns evolved from a vernacular building activity to a professional design task, undertaken by architects, design specialists, and town planners. After all, company towns were the spatial manifestation of a social ideology of paternalism embedded in the crass economic rationale of profit maximization.[33] Perhaps not so surprisingly, this idea of planned settlements as self-contained enclaves has not disappeared. At the start of the twenty-first century, company towns attached to corporate enterprise and extractive industries such as oil, minerals, and agricultural commodities are spread haphazardly around the globe.[34]

In similar fashion, the Garden City movement that blossomed in the early twentieth century sought to offer an idyllic, community-based, almost utopian alternative to the overcrowded, polluted, chaotic, and unhygienic conditions of the disorderly modern metropolises produced under the unbridled excesses of the Industrial Revolution.[35] Inspired by Sir Ebenezer Howard, the Garden City movement conjoined two different types of late-nineteenth-century experimental communities, creating a tension that was never fully resolved.[36] The first type—utopian, idealistic, and radical in nature—challenged conventional laissez-faire values of individualism by endorsing forms of small-scale, communitarian social life that rested on cooperation and mutual aid, fraternity, and shared economic progress. The second type found expression in the model industrial towns of self-appointed enlightened capitalists. This approach reinforced a rule-oriented, hierarchical social order epitomized by the values of top-down paternalism.[37]

As city builders exported their models and blueprints for master-planned cities to the far reaches of European colonial empires, they produced hybrid concoctions that blended idyllic garden city motifs with both racial segregation of residential accommodation and rigid controls over social life and social movement.[38] In particular, garden city experiments in overseas colonial settings played a crucial role in the creation of racially polarized urban environments that scholars have referred to under the broad brush of fractured "dual cities."[39] By contributing to racial segregation in its many guises, the imposition of garden city designs in colonial urban space represented, in contrast to the original vision espoused by Ebenezer Howard, essentially a racially inscribed variant of exclusivist city building.[40]

In the Great Depression of the 1930s, Frank Lloyd Wright introduced the idea of Broadacre City, modeled on a romanticized notion of "small is beautiful." Along with comparable modernist visions, Broadacre City shared a set of beliefs in rational solutions to building planned communities, including a stress on central administrative authority, an emphasis on transportation networks, the use of the machine as a metaphor for progress through industrial technology, and the separation of urban space into discrete zones for leisure and work activities.[41] The utopian goal of reducing—or even eliminating—disorder "reached an unprecedented level at the high-noon of modernism in the [early] twentieth century." In particular, Le Corbusier's all-encompassing plans to reshape cities in ways that conformed to the principles of efficiency and rational use of space appealed to municipal authorities "eager to display their modernist credentials."[42] To this end, new cities built from scratch, such as Brasília and Chandigarh, were overburdened with an excess of symbolic importance meant to indicate the kind of forward-looking progress associated with the modernist age.[43] The New Town initiative (like Milton Keynes outside London) that flourished in postwar Great Britain from the 1940s to the 1960s was one of the largest public housing programs of its kind, representing yet another iteration of large-scale planned communities that offered an alternative to disorderly industrial cities.[44] In the second half of the twentieth century, urban planners, infused with the spirit of high modernism, sought to decongest historic city centers by uprooting the working poor and relocating industry to the urban periphery.[45]

With all its pretensions of expert knowledge and rational order, the utopianism of the late nineteenth and early twentieth centuries spoke of a common good, popular governance, and broader intentions to alleviate social problems.[46] By the mid-twentieth century, the realities of capitalist industrialism had transformed garden cities into what Robert Fishman termed "bourgeois utopias" where the antiurban impulses captured in suburban enclaves represented "a collective assertion of class wealth and privilege as impressive as any medieval castle."[47] For many progressive and radical thinkers, the utopian dreamscapes of the nineteenth century had fallen into disfavor, reborn as dystopian nightmares in the second half of the twentieth.[48]

These early utopian city-building efforts and the new master-planned satellite cities that blossomed at the start of the twenty-first century have much in common: both are large-scale, holistically designed, functionally integrated, and purpose-built enclaves attached to the edge of existing metropolises as stand-alone (self-contained) entities. Both are very much the product of their time, each in its own way offering a forward-looking antidote to the perceived distressed urbanism that surrounded them. But ultimately both the new master-planned satellite cities and earlier efforts at building autonomous "off-grid" communities amounted to open-ended, quasi-utopian experiments. The new corporate-led urban enclaves that originated in the Persian Gulf and the Asia Pacific Rim and quickly spread everywhere have emerged as aspirational epicenters for privileged stakeholders striving for commercial success in the highly competitive circumstances of global capitalism. But with so many of these new master-planned megaprojects in the pipeline, it is difficult to predict the outcome of so much feverish city building at a time of intense competition for a place in the rank order of world-class cities. A vast global urban experiment is currently underway, with not nearly enough forethought or analysis as to how these satellite cities will eventually affect the economies, physical environments, and everyday lives of people who will live both in and outside of them.[49]

Examining this new prototype for city building at the start of the twenty-first century requires a counterintuitive excursion into a retro-futuristic global urbanism that brings to mind surrealist painting or science fiction novels. These new instant cities are eerily reminiscent of such widely divergent urban experiments as walled medieval cities and autonomous city-states; the confederation of merchant towns assembled under the banner of Hanseatic League that dominated the Baltic maritime trade from the fourteenth to the seventeenth centuries; the colonial "free ports" of the eighteenth and nineteenth centuries; the master-planned garden cities popularized by Ebenezer Howard at the start of the twentieth century; the high-modernist cities built out of whole cloth as symbols of national destiny (Brasília, Abuja, and Chandigarh); the fast-track urbanism of the Persian Gulf (and its spinoffs loosely termed Dubaization); and the towering "skyscraper cities" built virtually overnight in the Pearl River delta of southern China. Yet what makes these instant cities strikingly different from earlier experiments with master-planned, holistically designed city-building efforts is that they consist of a seemingly novel mixture of enclave capitalism, speculative projection, unregulated neoliberalism, and generic model-city designs.[50] These new cities built from scratch bring to mind an aphorism coined by Robert Hughes to describe the fabrication of Brasília: "Nothing dates faster than people's fantasies about the future."[51] Without continuously reinventing promises of the future-yet-to-come, these instant cities—as exemplary expressions of global urbanism in the neoliberal age—become passé and outdated. These efforts at large-scale reurbanism gesture toward genuine city-ness, but the satellite cities seem more like self-contained platforms for projecting a roseate image of urbanity that can only exist in microcosm and in relative isolation.[52]

READING FAST-TRACK URBANISM

Reading these fast-track satellite cities as narratives—or, more precisely, as spatial stories, embedded as they are in utopian visions about the future—enables us to decipher the deeply rooted reservoir of cultural conventions associated with concepts of novelty and progress. As Walter Benjamin suggested in his unfinished masterwork *The Arcades Project*, "every epoch appears to itself [as] inescapably modern."[53] Seen from this angleof vision, these instant cities contain hidden sociocultural messages that seek simultaneously to distance themselves from the allegedly outdated past and to foretell the radiant future. These self-contained, cocooned enclaves envision an integrated lifestyle approach to city living, in which material scarcity is nonexistent and strict regulatory regimes ensure predictability, stability, and rational use of space. It is perhaps not surprising that past "utopias of social space" (City Beautiful, the Garden City, the Radiant City, the City of Tomorrow) have shown themselves to be remarkably resilient as models of city building that, without the slightest gesture toward critical self-reflection, can be bent, folded, and repackaged to serve a variety of contemporary purposes. The acknowledgment that holistic master planning lies at the heart of these new instant cities suggests that the modernist-inspired impulse toward "starting over"—the relentless production of the new by dispensing with what came before—has remained a powerful force in imagining the future city. Yet the history of city building is cluttered with the abandoned ruins of once-vibrant dreamscapes, discarded relics of bygone eras that no longer resonate and that have lost their raison d'etre.[54]

What is striking about the spatial development models, architectural designs, and artistically imagined renditions that appear in glossy promotional materials and visual presentations is the singular message that these new cities built from scratch promise a future free from the broken-down infrastructure, overcrowded streetscapes, and unplanned sprawl that mark the present situation in struggling cities at the margins of modernity. A once-enduring faith in the modernist teleology of the gradual yet inevitable march of progress has evaporated in the face of the failure of modernism to fulfill its promise of trickle-down upliftment. Nostalgia for the future—to borrow out of context an apt phrase from Charles Piot—reflects deep-seated anxieties about the troubled urban present. With the visible ruins of earlier failures standing in the way of forward-looking visions, the language of building from scratch and starting afresh provides an available, uplifting discourse for imagining a future unencumbered by the past.[55] These calls to wipe the slate clean, to leap over the historical past, and to achieve in a single bold move the coveted status of world-class city reflect a longing for a frictionless utopia, a worldly paradise where cities buzz and hum with excitement, offering ceaseless opportunities for prosperity and advancement.[56]

EXPERIMENTAL URBANISM: BUILDING CITIES FROM SCRATCH

While modernist approaches to a comprehensively planned urbanism as a strategy and method of state-led intervention have come under severe attack for their rigid adherence to strict design protocols, the modernizing impulses that gesture toward the imagined ideal of a future urban utopia haves not disappeared.[57] At a time when state-sponsored modernist visions of city building

FIGURE 6.1 Instant Cities: Doha, Qatar. Aerial photograph of the rows of towering skyscrapers along the shoreline of Doha. Ramaz Bluashvili

have lost traction around the world, large-scale corporate enterprise has stepped into the breach, promising to start afresh and build new satellite cities that are efficient, sustainable, and well-managed. At the start of the twenty-first century, large-scale corporate enterprises have unveiled a plethora of audacious schemes that include ambitious visions for master-planned, holistically designed, private cities cocooned under the umbrella of autonomous zones, privately sponsored infrastructure projects on a grand scale, and the unprecedented empowerment of corporate actors in urban governance.[58]

Constructing entirely new cities from scratch represents a powerful new trend in city building at the start of the twenty-first century. Instant cities appear as if out of nowhere. While tinkering with the incremental rebuilding of existing urban landscapes remains the dominant mode of urban transformation in cities around the world, the turn toward building master-planned, holistically designed satellite cities has significantly reshaped thinking about urbanism and urban life. Despite the celebratory praise showered on such acclaimed instant cities as Dubai, Doha, Abu Dhabi, Masdar, Astana, and Shenzhen, this experimental approach to city building remains largely in its infancy.[59] Stories of remarkable success are paralleled by sobering tales of failure, where overconfident city builders have been unable to make good on initial promises.[60] Speculative urbanism—what Linda Samuels refers to as a "particular brand of intentional, overzealous development as a strategy of economic growth"—invariably produces periods of intense overdevelopment and underutilized real estate.[61] What remains an open question is the extent to which construction of completely new cities de novo foreshadows city building to come over the foreseeable future.[62]

THE UTOPIANISM OF TABULA RASA CITY BUILDING

Put in historical terms, the practice of city building has never really been far removed from grand, utopian ideas for how best to use the tools of architecture, physical design, engineering, and urban planning to bring about the improvement of urban living. In the simplest terms, utopian urbanism refers to efforts to build an idyllic urban milieu through the deliberate design of the built environment. Leading figures in the urban planning canon, including such visionaries as Ebenezer Howard, Frank Lloyd Wright, and Le Corbusier, soundly rejected the gradualist idea of incremental improvements to the built environment of existing cities and instead embraced a tabula rasa approach that called for building new cities from scratch in accordance with holistic design specifications.[63] As a general rule, utopian planning ideas tend to reflect broader socioeconomic trends and public concerns of their time.[64] While the inspiration for cities built from scratch has its roots in both the Garden City (built to reflect harmony with nature) and the purist Machine-Age City (the efficient engineering marvel associated with Le Corbusier), this new prototype for preplanned city building represents a kind of futuristic escapism only possible in the information age of hypermodernity and neoliberal globalization.[65]

By definition, building new cities from scratch involves the invention of completely new spaces where nothing came before. This "starting afresh" approach has been a persistent theme in large-scale city-building projects from Canberra to Chandigarh, from Astana to Abuja, and from Islamabad to Brasília. In seeking to invent new spaces that can give rise to an ideal community, these modernist predecessors to contemporary instant cities all sought to create an orderly urban life in a deliberately planned environment that bound the lives of residents within an idealized physical setting.[66] Predictability, efficiency of movement, and rational use of space were the key watchwords and the underlying driving force behind this modernist planning ethos. While they may strive to distance themselves from the excessive rigidity and brutalist rationality of these modernist doctrines, the designers of twenty-first century fast-track urbanism have borrowed a great deal from these earlier modernist-utopian traditions. In their overall structures and core functions, and their aesthetic appeal and stylized design, instant cities resemble a kind of hybridized Euro-American (Westernized) urbanism, in which a variety of activities come into play across the socially congregating spaces of the cityscape and the normative axiom of "planned heterogeneity" is embedded as a guiding principle. By attaching themselves to every "good" and "successful" element found in a historicized idealization of Western urbanism, the city builders behind the new satellite cities that have blossomed on the outskirts of existing metropolitan landscapes in southeast and northeast Asia, Latin America, and Africa have sought to blend such aesthetically pleasing features derived from their interpretation of the Collage City, Manhattan, and Garden City models with a forward-looking stress on green urbanism, eco-friendliness, and sustainability.[67]

Sponsored by large-scale real estate interests with either the active guidance or passive acquiescence of state authorities, the city-building efforts that have gone into the creation of these new satellite cities amount to profit-making endeavors couched in the popular discourses of sustainable urbanism, the smart city, and ubiquity of techno-environments. Wrapped in the glossy patina of a green, globally connected, functionally defined, and highly controlled spatial imagination, this futuristic urbanism offers a sanitized blueprint of urban space that conforms to the idealized vision of the perfect neoliberal city where entrepreneurially minded residents pay for what they get and get what they pay for. Planning for everyday life in these new instant cities cavalierly overlooks remedies for how residents who cannot afford the price of living in such an urban Valhalla "can create their own public spaces and enjoy them in the city."[68]

INSTANT URBANISM: THE RAPID PACE OF URBAN TRANSFORMATION ON A GLOBAL SCALE

Not so long ago, the renowned architect Rem Koolhaas drew attention to what he termed the generic Gulf City (of which Dubai is the apotheosis). In his words, "The emerging model of the city is being multiplied in a vast zone from Morocco in the West, then via Turkey and Azerbaijan to China in the East. The Gulf is not reconfiguring itself, it's reconfiguring the world. This may be the final opportunity for a new blueprint of urbanism." Despite the rhetorical flourishes and characteristic hyperbole, Koolhaas clearly identified an emergent trend in city building that suggests a possible future for urbanism.[69]

As if by the wave of a magic wand, Dubai suddenly burst onto the global urban scene as an instant iconic symbol of size and spectacle.[70] With the discovery of oil in 1968, Dubai was transformed virtually overnight from an obscure fishing village and isolated smuggling port located on what was historically referred to as the Pirate Coast of the Arabian Gulf into a gigantic hypermodern metropolis, growing at breathtaking speed and attracting a huge influx of corporate executives, financiers, consumer-driven tourists, and guest workers from all corners of the world. Almost immediately, it became a global destination of monumental scale, a desired brand, and a paradigmatic model that city builders around the world were quick to emulate.[71] "More than any other place on earth," Christopher Dickey has suggested, "this city-state in the United Arab Emirates is the creation of worldwide commerce, a specialty-built magnet for the kind of hot money that seeks the quickest, highest profits and then moves on when they disappear."[72]

The rapid-fire urbanism that characterized the building of Dubai exemplifies this emergent pattern of building completely new cities entirely from scratch. Dubai is the reigning exemplar of an aspiring world-class city that represents the truly generic condition of contemporary fast-track urbanism. It has often been asserted and assumed that if Manhattan represents the apotheosis of twentieth-century congested urbanism, then Dubai has become the emergent

FIGURE 6.2 Instant Cities: Dubai, United Arab Emirates (UAE). Aerial photograph of Dubai, highlighting the Burj Khalifa, the tallest building in the world. Mo Ismail

metropolitan prototype for the twenty-first century—a futuristic oasis floating unanchored above and beyond the outdated urbanism of yesteryear.[73]

The initial success of Dubai's development model for establishing a regional business hub by combining market liberalization and deregulation with explosive urban growth triggered a kind of copycat urbanism and building frenzy in which cities like Doha (Qatar) and Abu Dhabi sought to carve out specialized niches in the global marketplace. One distinctive feature of contemporary urbanism in the Gulf is the use of large megaprojects and the installation of state-of-the-art infrastructure as catalysts for accelerating the pace of urban growth and branding these newly minted cities with a global image as vibrant business centers.[74]

Skeptics, however, have dismissed Dubai as an unsustainable experiment in global excess, an enclosed gated community for the super-rich, rooted in mindless consumption and economic injustice and preserved by underpaid indentured laborers who are subjected to arduous work with long hours and physically segregated on the outskirts of the city in featureless, barracks-like accommodation with few amenities.[75] Juxtaposing the experiences of wealthy business elites and mobile travelers who enjoy the visible city of exclusive enclaves, skyscrapers, and artificially created offshore islands with the everyday lives of impoverished immigrant workers who inhabit the invisible city of miserable labor camps illustrates the indisputable condition that Dubai is actually two separate cities occupying the same space.[76] It is this seamy side of such city-building efforts—the "grit beneath the glitter," as Hal Rothman and Mike Davis suggest for Las Vegas—that enables us to counter the overly romanticized rhetoric of successful urbanism that often accompanies one-sided, celebratory stories about Dubai and other instant cities of the Gulf states.[77] A fundamental question that arises in urban studies and planning theory concerns not only the deleterious environmental impact of cities like Dubai that consume more than they produce, but also the extent to which city building of this nature could work as a model of sustainable urbanism on a global scale.[78]

Yet the emergence of these instant cities undermines long-standing ideas embedded in the conventional urban studies literature that suggest that "underdeveloped cities" in "developing areas" proceed through distinct stages of growth along a single pathway in linear time. These new cities are cobbled together in sometimes strange configurations from generic models derived from elsewhere.[79] The secret key to the success of these city-building efforts is serial repetition. While variation certainly occurs, city builders tend to make use of complementary architectural stylistics, branding ideas, functional components, environmental programs, building typologies, legal envelopes, aesthetic formats, financial instruments, and standardized construction technologies to piece together assemblages of strikingly similar parts. Working hand in glove with international architects and urban design specialists, real estate developers have borrowed, plagiarized, and ransacked planning motifs from generic prototypes. In transferring design specifications from the global to the local, real estate developers cherry-pick the signature features of universally acclaimed cities and recycle them as the basic building blocks of these new instant cities.[80]

These city-building projects conform to the new urban design politics of bringing together aesthetically pleasing architecture and entertainment spectacles.[81] The serial reproduction of business conference centers, up-to-date transportation infrastructure, glitzy airports with global reach, corporate office complexes, luxury shopping and entertainment districts, and towering skyscrapers offers amenities comparable to those of First World cities. It is not unusual for these blatant copies to surpass the original in terms of grandeur, aesthetics, and style.[82]

The current frenzy for constructing spectacular architecture goes well beyond functional use to the creation of an assemblage of iconic "trophy buildings" with visual appeal that far exceeds any operational purpose.[83] For aspiring world-class cities, no building typology embodies the heightened importance of symbolic meaning in architecture quite like the high-rise skyscraper.[84] It is here, in gesturing toward sheer verticality, that architecture breaks from the austere modernism of the mid-twentieth century.[85] This almost obsessive fixation on the vertical dimension of city building goes hand in hand with high-speed hypermodernity.[86] A high-rise office tower is more than a marvel of engineering prowess; it is also a symbol of urban, national, and corporate identity—a claim to up-to-date, cosmopolitan modernity.[87] City boosters use spectacular architecture as a way to announce their hoped-for emergence on the world stage.[88] As exemplars of the principle that height is a symbol of power, unusually tall buildings like the Burj Al Arab Hotel in Dubai, the Pentronas Towers in Kuala Lumpur, and Taipei 101 in Taiwan, along with other equally spectacular architectural monuments, have greatly enhanced the cachet of aspiring world-class cities seeking to secure their place among the leading metropolitan centers.[89]

Yet in the hope of "keeping up with the West," spectacular high-rise buildings are only one route along the march toward modernity.[90] The city-state of Singapore, for example, has carved out a distinct niche for itself as a laboratory for a distinct Asian brand of green urbanism—the self-styled "City-in-a-Garden."[91] Convinced that planning guidelines and innovative practices in Singapore have produced successful urbanism, state-owned and private architectural and planning consultancies have packaged and aggressively marketed their environment expertise and infrastructure planning for export to cities throughout Asia.[92]

Skeptics, however, have suggested that while the Singapore brand has become a recognized global technology, it is a composite of proposed remedies that can be invoked and emulated but cannot really be reproduced.[93]

THE WONDER WORLD OF FAST-TRACK URBANISM: DUBAI AS EXEMPLARY EXPRESSION AND FORERUNNER

The glitzy experimental city of Dubai provides an archetype for this kind of fast-track, overnight city building.[94] Large-scale real estate developers, financiers, and private design firms have come to play an increasingly vital role in the core functions of urban planning, particularly in the comprehensive implementation of regulatory regimes governing land use, overseeing building typologies, and enforcing legally sanctioned codes. This privatization of planning has undermined the public administration of urban space and replaced municipal authority with private governance.[95] Equally important, these new megaprojects represent new ways of thinking about city planning and the management of urban space not as the almost exclusive terrain of public authority but as the private prerogative of real estate developers.[96]

The proliferation of these master-planned, profit-oriented, and privately managed satellite cities represents a new type of urban real estate development in which large-scale corporate enterprise has become directly involved in creating urban enclaves that seek to mimic business-led growth strategies in aspiring world-class cities around the globe. These enclaves rest on the distinctive feature of being globally connected and locally disconnected, both physically and socially. With their free-enterprise zones, relaxed financial regulations, high-technology and export-oriented production facilities, corporate office complexes, luxury residential accommodation, and upscale leisure and entertainment opportunities, these island-like enclosures are detached from their local surroundings and are instead oriented to the key global command-and-control centers of the world economy.[97]

What distinguishes the uninhibited freedom of blank-slate city building from conventional practice of incremental layering are the opportunities for architects, design specialists, and urban planners to make decisions with little mediation or compromise.[98] Empowered to act with virtual impunity, they can build what they want with little or no oversight from public authorities. While collaborative and participatory approaches to planning practice regard inclusionary decision-making processes and public participation as an unequivocal benefit, architects and design specialists often see this process of negotiation as interference with their creative vision, and real estate developers rankle at compromise because it threatens to undermine the bottom line of profitability.[99] At a time when competitive advantage has become the new mantra of urban governance, large-scale megaprojects not only offer a built environment geared to the highest global standards of up-to-date infrastructure and sophisticated technology but also provide "a specific kind of global image that can be marketed in the global market place."[100]

In writing about the "art of being global," Aihwa Ong has identified the historically specific characteristics of contemporary city-building practices in rapidly urbanizing Asia.[101] She has defined two distinct logics governing hyper-building: first, frenzied, speculative overbuilding in anticipation of rising real estate values and a profitable future; and second, "inter-referencing" of spectacular architecture copied from elsewhere. From Dubai to Shanghai, from Singapore to Hong Kong, and from Doha to New Songdo City (Seoul), architects and designers have mimicked and even plagiarized design motifs and aesthetic styles from each other. As Ong has argued, city builders regard visually stunning projects as what amount to "leveraging practices that anticipate a high return not only in real estate but also in the global recognition of the city."[102] Landmark buildings, glittering airports, and globally connected business hubs in some aspiring world-class cities have spawned overnight copycat projects elsewhere. In this sense, hyper-building becomes an end in itself.[103]

It is certainly the case that for the rapidly globalizing cities of the Asia Pacific Rim and the Persian Gulf, this kind of fast-track urbanism is a state-driven technique of nation building directed at building a boosterist consensus for "catching up" with the West. New satellite cities that have sprouted virtually overnight near Saigon (Phu My Hung), Kuala Lumpur (Petaling Jaya and Subang Jaya), Jakarta (Lippo Karawaci), Phnom Penh (Camko City, Grand Phnom Penh International City, and Koh Pich [Diamond Island City]), Seoul (New Songdo City), Macao (Macau Cotai Strip), and elsewhere play a crucial functional role in the restructuring of global hierarchies and the global marketplace, similar to the autochthonous role of the Italian city-states of Venice and Genoa in the historical origins of early capitalism.[104]

TIME, CHRONOLOGY, AND HISTORY: THE TEMPORAL DIMENSIONS OF FAST-TRACK URBANISM

The idea of progress was once regarded as the most fulfilling manifestation of radical optimism and a promise of universally shared and lasting improvement.

At the start of the twenty-first century, reference to progress conjures up the opposite sentiments. It has become a metaphor for a kind of dystopian and fatalistic resignation in an age where there is no alternative. As Zygmunt Bauman has argued, progress "now stands for the threat of a relentless and inescapable change that augurs no peace and respite but continuous crisis and strain—and forbids a moment of rest; a sort of musical-chairs game in which the moment of inattention results in irreversible defeat and in the no-appeal-allowed exclusion." Instead of great expectations and confident predictions of betterment, "progress" evokes nightmares of failing to keep pace, of falling behind in the competitive race to stay on top.[105] The rhetoric of fast-track urbanism associates instant cities with progress via instant gratification. Yet as powerful and compelling as it happens to be, the discourse of instant urbanism should also be approached with care, lest it turns into an abstract metanarrative that obscures more than it reveals. It is true that city building on a global scale has taken place at an accelerated pace, but the overall costs of this time-space compression are as yet unknown.[106]

However it is measured, speed is the key metric of circulation. Broadly speaking, circulation has two dimensions: first, the velocity of moving from place to place—that is, speed across spatial distance; and, second, the pace of processing and consumption—that is, speed intensifying time. The first relates to the extension of spatial horizons; the second relates to the foreshadowing of temporal ones. Embedded as they are in the contemporary age of globalization, instant cities combine these two dimensions in novel ways not shared by alternative modes of urban transformation.[107] Fast-track urbanism relies on the kind of time-space compression that brings about the reordering of distance, the overcoming of spatial barriers, and the shortening of time horizons.[108] For instant cities, the stress on immediacy replaces speed with proximity and instantaneity.[109]

What might be called instantaneous urbanism conflates history and chronology. Unlike the fits and starts, the stalled moments of inactivity, and the reversals that characterize the temporality of history, chronological time conveys the idea of sequential movement along a linear pathway toward a distant yet approaching end point. As Siegfried Kracauer suggested decades ago, chronological time is a "homogeneous medium" that marks "a flow in an irreversible direction."[110] While it may appear as natural and value-neutral, linear time usually conveys an evaluative judgment: "new is better than the old; recent is superior to the past; and the future will be better."[111] Framed in terms of what Reinhart Koselleck called the "horizons of expectations" (where the future as it is revealed is made present), instant urbanism conceives of city building as a continuous process unfolding in linear or chronological time along a preordained route from inception to completion.[112]

Instant cities and fast-track urbanism project images of sudden, almost instantaneous transformation in which something is created (as if my magic or divination) out of nothing. These depictions of building on a blank slate provide grounds for dismissal of instant cities as exemplary expressions of a false modernity. In this reading, instant cities are not real but are instead artificial or synthetic creations, an otherworldly kind of hallucinogenic phantasmagoria. As Natalie Koch has suggested, the "Disney stigma [effectively] casts the city as a theme park," an imagined but not an actual place. To be sure, instant cities sorely lack the layered accretions of the historical past and collective memory that shape and characterize conventional urban settlement patterns.[113] The fictive, artificial tableaux that characterize instant cities endow them with "real-not-real" qualities without historical context.[114] But to frame instant cities exclusively through the stigmatizing discourses of phantasmagorical, counterfeit modernity and Disney-like imaginaries obscures the "complex geographies of power" and the "more complicated political-economic relations that accompany and give rise to these spectacular urban development projects."[115]

In certain respects, the master-planned, holistically designed instant cities that have blossomed at the start of the twenty-first century bear a striking resemblance to their modernist and high-modernist forbearers.[116] The dreamscape of building brand new cities is strongly linked to the persistent myth that the cure for the paralyzing illness of urban disorder can be found in simply abandoning existing built environments, building protective walls and barriers to exclude the unwanted, and starting afresh. Following Zack Lee, it can be argued that instant cities can be fruitfully understood as diverse assemblages of worlding practices, in which city builders cobble together diverse ideas, features, formats, and programs to fashion their peculiar version of world-class urbanity linked to the global economy.[117] At a material level, city builders often use architecture "as a surface for projections to another real space, more perfect and better arranged, that should become actualized in reality."[118] In thinking about up-to-date design, they often tap into popular traveling ideas and international best practices as sources of inspiration.[119]

THE LIBERTARIAN OPTION: CHARTER CITIES AND THE DREAMSCAPE OF BLANK-SLATE URBANISM

The thinking behind instant cities cannot trace its origins to a single fountainhead of ideas, but instead has borrowed promiscuously from a wide variety of different sources. Extreme libertarian versions of pure public-choice theory periodically appear in idealistic proposals to create new start-up cities that are not—in theory at least—encumbered by the clumsiness and inefficiencies of interest-group politics that function through bargaining and compromise.[120] These libertarian thinkers wish "to wipe the messy slate clean."[121] In the recent past, a range of emerging utopian discourses have called for the creation of autonomous libertarian enclaves where the principles of *laissez-faire* economics provide the platform for new kinds of urban governance. The free-market thinkers behind these imagined utopias advocate a rethinking of conventional public administration as a "private government service provider" and a rethinking of citizens as mobile consumers of these private government services. Seen in this light, citizens are encouraged to "vote with their feet" by "opting-in to the jurisdiction that best fits their needs and beliefs."[122] This enduring fantasy of starting over with autonomous, self-governing enclaves has its roots in utopian thinking that curiously has animated both free-market libertarians and communitarian anarchists.[123]

A prominent U.S. economist turned policy entrepreneur, Paul Romer, has proposed the concept of Charter Cities as a radical solution to the problems associated with rapid and unregulated urbanization on a global scale. As a framework for addressing dysfunctional urbanism, the Charter Cities initiative focuses on the potential for so-called start-up cities to fast-track the kinds of institutional reforms conducive to individual ownership, private enterprise, and entrepreneurship. Charter Cities function in strict adherence to institutional frameworks that establish new rules of the game governing market-based transactions and structuring all modes of social interaction. The ultimate goal of the Charter Cities movement is to establish privately owned and managed urban enclaves that are geographically, administratively, socially, fiscally, and, ultimately, politically separate from the host nations within which they are located.[124]

The key to the success of the Charter Cities model is to establish clear institutional rules beforehand that apply uniformly to all who decide to participate and to effectively enforce these regulations so as to reinforce long-term commitments.[125] The effective enforcement of these rules—primarily geared toward fostering competition and free choice—creates the solid bedrock of trust that private investors and property-holding individuals require in order to make long-term commitments to engage in marketplace exchange. In the model developed by Romer (and grounded in post-neoclassical endogenous growth theory), Charter Cities bear a striking resemblance to special economic zones—cocooned enclaves that adopt strict rules for managing specific economic goals in ways that differ from those governing their host countries. Yet unlike special economic zones, Charter Cities move beyond a narrow focus on the management of specific economic goals by expanding the scope of institutional rules that allow for the support a broader range of market transactions and exchanges in a modern capitalist economy and to structure social interactions appropriate for a well-run city.[126] In short, these Charter Cities function as special administrative zones (or special reform zones) with sound rules and legal institutions, pro-growth and investment-friendly policies, and environmentally sustainable practices. Outfitted with such ring-fencing operations, privately managed Charter Cities belong to that distinct subcategory of self-governing enclaves that exist in the extraterritorial realm of pure administrative autonomy.[127]

This attachment to *laissez-faire* ideas underpinned an ambitious scheme (initiated in 2011–2012 but eventually abandoned amidst a great deal of public outcry and protest) to launch several model "free market cities" in Honduras.[128] Sometimes referred to as Free Cities, Free Towns, Charter Cities, Future Cities, Enterprise Zones, and Special Development Regions, these utopian city-building projects fashioned on *laissez-faire* principles were envisioned as "special development regions" located on unoccupied land roughly the size of Hong Kong and Singapore and administered "by an extra-governmental committee that would not be subject to any existing [regulatory and] institutional constraints."[129]

While the construction of "free cities" may have stalled in Honduras, real estate developers have already begun construction of a private city on the outskirts of Guatemala City. Hailed as a safe haven from the crime-ridden capital, this new self-enclosed city, called *Paseo Cayala*, provides a full range of amenities, including apartments, boutiques retail shops, nightclubs, restaurants, and storefronts designed in a Spanish colonial style and contained inside a fourteen-hectare compound encircled by high white stucco walls. The cheapest home prices are seventy times the average annual wage in Guatemala. Developers have already unveiled plans to expand the size of the privately owned and privately managed city into a fully operational 352-hectare alternative to the dangerous and congested streets of Guatemala City. The single gated access point leads to an underground parking garage, from which residents and visitors proceed through gazebo-covered escalators decorated like art nouveau Parisian metro stations onto uncluttered streets patrolled by armed private security guards.[130]

The appeal of various proposals to create charter cities—model cities "featuring economic and social institutions conducive to growth and development" that are available to all willing to voluntarily abide by preestablished institutional rules—is rooted in a utopian vision that promises such "substantive and procedural safeguards" as the equal protection of basic rights and the impartial administration of legal remedies for the redress of grievances.[131] Proponents have lauded the Charter Cities model as "the world's quickest shortcut to economic development."[132] But skeptics have warned that Charter Cities amount to neoliberal experiments with new kinds of paternalistic overrule.[133]

VARIETIES OF MASTER-PLANNED CITIES BUILT FROM SCRATCH

No single formula or shared blueprint captures the variety of possible iterations for this tabula rasa approach to city building. In some instances, these instant cities are genuine expressions of simply starting over, building an entirely new metropolis out of whole cloth. The feverish construction of Astana, the capital city of Kazakhstan, reflects this pattern.[134] In other cases, these newly minted cities take shape as privately administered, semiautonomous mini-states, offering the promise of a well-managed alternative to the messy realities of everyday urbanism that surrounds them. In still others, they represent the creation of well-functioning economic hubs, technologically advanced full-service cities that can successfully compete for private financing, knowledge production, and talent and expertise in the global marketplace. These fast-track satellite cities acquire multiple personalities, assuming, for example, the creative-industry patina of antiquated industrial buildings and reused warehouses, the postmodern new urbanist patina of "Italianate lake-town romanticism," the high-modernist expressions of global finance, and hi-tech smart city "innovation districts" built around information communications technologies (ICT).[135] All in all, these new city-building projects appear in many guises, sizes, and shapes, but the common denominator is that private companies instead of municipalities and their planning departments have actively started to construct and manage cities, including initial conceptualization and spatial design, the mobilization of investment and financing, and actual construction of the built environment and installation of services—in short, "the full package or 'city in a box.' "[136]

Master-planned, holistically designed smart cities (eco-cities, intelligent cities, sustainable cities, ubiquitous cities, or whatever other trendy name seems to fit) are projected to be the new urban utopias of the twenty-first century.[137] By integrating comprehensive spatial planning with up-to-date digital technologies and information systems, smart cities (and their similarly conceived brethren) have become the new urban laboratories for technical innovation, surging to the forefront of worldwide marketing efforts as solutions to the challenges of rapid urbanization and sustainable development.[138] In China, Southeast Asia, and

India (with more than one hundred smart cities under construction or in the advanced stage of planning), city builders have turned to some variant of smart cities as launching pads for spurring economic growth.[139]

Smart cities, eco-cities, and other variants of the current fixation with sustainable urbanism are part of a longer genealogy of utopian urban planning ideas that have emerged and taken root as a response to the difficult challenges of development and modernity in the contemporary era. Built in the shadow of Dubai and other successful instant cities that came of age during the era of neoliberal globalization, these master-planned, holistically designed smart cities reflect a sharp turn toward "entrepreneurial urbanism"—a wholesale shift in urban governance that signals the triumph of a new urban imaginary of cities as business enterprises rather than experimental models for promoting social justice and social inclusion.[140]

THE CORPORATE CITIES PARADIGM

The steady growth of privately managed "corporate cities" is a disturbing harbinger of twenty-first-century urbanism. As part of a global phenomenon, real estate developers and corporate property owners assume an expanded role in determining what form the built environment takes and in creating regulatory (management) regimes that serve their narrow preoccupations above any consideration of the public interest.[141] In the typical case, their spatial layout consists of assemblages of districts, zones, or thematic enclaves, with names that evoke familiar images of rewarding work, domiciliary tranquility, and leisurely playfulness. Real estate developers market these self-contained corporate enclaves as mythologized "fairy-tale urbanism," where daily life is carefree and where residents and visitors alike are spoiled by a plethora of equally enviable choices.[142]

Corporate cities, entrepôts, bubble urbanism, sequestered enclaves—the names are interchangeable, but the spatial and organizational formats are strikingly similar. These corporate cities occupy a discrepant territory between two different species of urbanity: open and cosmopolitan, on the one side, and secretive and opaque, on the other.[143] These corporate precincts have succeeded in combining the "connected expansiveness" of the networked global marketplace and transnational business with the sequestered-enclave format that enables those inside to maintain both physical and psychological distance from what happens outside their perimeter. These corporate enclaves operate within the realm of extraterritorial sovereignty in the historical tradition of free trading ports, Maroon communities, offshore banking, pirate hideouts, and frontier colonies.[144] As an evolving urban paradigm, these large-scale urban enclaves function as preprogrammed, self-regulating, *faux* city-states with inexhaustible possibilities for business and trade, precisely because of the lack of public authority and municipal oversight. Such smart eco-cities as Songdo IBD, Dholera Smart City (Gujarat), and Masdar are marketed as problem-free enclaves.[145] Corporate cities have manufactured their own regulatory powers outside conventional norms of public administration that enable them to bypass procedural rules of democratic decision making.[146]

Across the emerging urban landscapes of the Global South, city builders, including municipal officials, design specialists, and key political figures, have begun to construct and package, promote and market, brand-new cities at a frenzied pace. This emergent economy of fast-tracked, assembly-line cities has become the new mantra for large-scale corporate enterprise. Such well-known multinational corporations as Cisco Systems, IBM, Siemens AG, Gale International, Landor, KPF, and Living PlanIT have entered into the market for building cities de novo.[147]

Naming and labeling are part and parcel of elaborate branding exercises designed to draw attention to the involvement of large-scale business enterprise. The corporate giant Cisco Systems coined the term "city-in-a-box" as a metaphor to describe their approach to constructing from scratch the new smart city project called New Songdo City outside Seoul.[148] Building cities as if they were assemblages of standardized parts cobbled together at breakneck speed has become a multibillion-dollar business. Countless new satellite cities, built from scratch at a frenzied pace, have appeared across Asia, Africa, and Latin America. In justifying this fast-track approach to assembly-line city building, political figures and public officials contend that cities have to be built very quickly and with smart design features in order to keep up with the accelerating demands of a rapidly urbanizing global population.[149]

Many of these imagined future cities have not advanced beyond the initial planning stages.[150] Early city-building efforts include Clark Green City (Philippines), Colombo Port City (Sri Lanka), Gujarat International Finance-Tec City (India), Iskandar Malaysia (Malaysia), Jazan Economic City (Saudi Arabia), Kabul New City (Afghanistan), King Abdullah Economic City (Saudi Arabia), Lavasa (India), Rawabi (Palestine), and the Sino-Singapore Tianjin Eco-City (China). Nevertheless, a number of exemplary prototypes for this kind of city building—including Masdar City (Abu Dhabi), Songdo IBD (outside Seoul), Astana (Kazakhstan), Dongtan Eco-City (outside Shanghai), Lavasa (outside

Mumbai), Lusail City (Qatar), Dholera Smart City (Gujarat), Waterfall City (Johannesburg), and Steyn City (Johannesburg)—have come into existence.[151]

Corporate cities cannot operate without duplicity, secrecy, and subterfuge. Theories of globalization—with their fanciful fables about the waning influence of the nation-state and the triumph of a borderless world—are the perfect camouflage for a globally ascendant corporate business culture "that clearly prefers to manipulate *both* state and non-state sovereignty, alternatively releasing and laundering their power and identity to create the most advantageous political or economic climate."[152] Corporate cities operate between state and nonstate jurisdictions, seeking the extra-jurisdictional spaces (like export-processing zones, free-trade zones, or special economic zones) that provide a relaxed atmosphere of legal immunity.[153] They aspire to lawlessness, not in the sense of anarchy or the nonrecognition of authority, but in the legal tradition of a state of exception—a circumstance where regulatory regimes operate with rules that apply inside but not outside their own borders.[154]

The expansion of privately managed corporate cities depends upon the adoption of new modes of urban governance premised on the delegation to private business interests of decision-making powers once reserved for public authorities. The implementation of what Ananya Roy, Gavin Shatkin, and others have described as "new planning regimes" (planning by exemption or exception) have transferred the provision of large-scale infrastructure projects and urban management services into the hands of private real estate developers.[155] This shift of urban planning functions into corporate hands creates a new geometry of spatial production that privileges privatized spaces for elite and corporate consumption, large-scale megaprojects intended to attract corporate investments, and deregulated localities such as free-trade zones, enterprise zones, and business improvement districts. Whatever the differences among these places, the suspension of conventional regulatory powers enables these zones of exception to provide special benefits for private corporations.[156]

GENERIC CITIES

In the opening line to his often-quoted essay "The Generic City," the architect Rem Koolhaas inquires in a peculiarly provocative way: "Is the contemporary city like the contemporary airport—'all the same'?"[157] Koolhaas postulates a convergence in the contemporary practice of city building, in which relentless urban expansion on a global scale subsumes the historical city into repeatable spatial products and results in the production of generic "cities of sameness." In contrast to Le Corbusier's universal global ideal of standardized reproduction, this new global sameness of city building is not about the serial repetition of identical "cloned cities." The generic city recognizes and even celebrates the unruliness and chaos of contemporary urbanity. Sameness represents the lack of recognizable difference among cities around the world as the mobility of traveling ideas, international best practices, global branding, cultural connectivity, and material overabundance have created generic cities indifferent to the past and to historical context. They thrive on the spectacle of bigness, novelty, and speed. In short, the more cities change, the more they seem to look alike and the less they are able to retain a distinctive sense of place. By replicating a look-alike "state of indifferentiation," generic cities resemble a kind of "traveler's space," or what Marc Augé called "non-places."[158]

In the contemporary era of globalization, instant cities typically combine the rigidity and inflexibility of modernist planning principles with the postmodernist stress on playfulness, visual spectacle, and lifestyle choices. What distinguishes this novel kind of fast-track urbanism from prototypical modern metropolis is its dependence on a pirated or recycled modernity, unconcerned with modernity's classic search for genuine originality. This idea of "recycled urbanism" may conjure images of a borrowed, unoriginal modernism. Originality—the constant search for newness—was the core idea behind city building in the nineteenth and twentieth centuries and the wellspring of its claim to dynamism. But the claim to modernity that undergirds the planning of instant cities relies less on an architecture of novelty than on an engagement with speed.[159] The recycled urbanism that underlies the construction of instant cities is not a simple process of achieving more of the same by simple replication or mimicry; rather, it works as a complex "difference engine" (Charles Babbage) in which each copy is different from its predecessor because of variation and recombination of assembled parts.[160]

BUILDING INSTANT CITIES FROM SCRATCH: TRENDS AND POSSIBILITIES

Instant cities constitute an emergent prototype for market-led city building in the twenty-first century, creating "prosthetic and nomadic oases" that assume the form of vast urban enclaves, either standing alone in defiant isolation or inserted incongruously into existing urban landscapes.[161] As experimental sites for the material realization of neoliberal utopia, these cities represent innovative platforms for planning practice in building future cities under the hegemony of market-led urban growth.[162] Lacking any historical identity and with few gestures to the past, these hyperactive cities thrive on newness, ephemerality, bigness, and unparalleled optimism in a roseate future; spectacle and enchantment replace the everyday urbanism of compact neighborhoods with their tight integration of mixed land uses, slow pace, and quaint streetscapes.[163] Built to resemble large-scale theme parks or gigantic stage sets, these instant cities consist of clusters of "destination precincts" held together by virtue of proximity and defined by their functional specialties: work and business, home and privacy, recreation and distraction, leisure and entertainment. With fast-tracked urbanism as the operative principle, the serial reproduction of standardized modules cobbled together to resemble a functional city has increasingly become the dominant mode of city building.[164]

City builders use the rhetoric of sustainable urbanism, including energy-saving technologies and reusable resources, as a way to create an aura of authenticity. But these claims about ecological sustainability are largely justifications or legitimizations that often conceal as much as they reveal. At the end of the day, these instant cities are primarily corporate business platforms that promote entrepreneurialism and profit-making activities.[165]

In seeking alternative, radical, utopian futures freed from constraints of the present, instant cities have embraced the principles of modernist and hypermodernist city building. This commitment to leaving the past behind means that the construction of instant cities requires a ruthless break with all that has gone before.[166] By breaking decisively from traditionalism and historicism (and their nostalgic attachments to the past), modernist city building held firmly to a belief in progress, to the possibility of the rational planning of an ideal social order, and to the standardization of knowledge.[167] Instant cities conform to the developmentalist fantasy of pressing into the future with little or no resistance. They offer a vision of a hypermodernist city cleared of all unwanted encumbrances.[168] Instant cities follow the all-too-familiar modernist-utopian script of an ideal landscape, at once orderly and well-managed, sustainable and resilient. The fast-track approach favors abrupt rupture over incremental accretion—starting afresh (via creative destruction) rather than slow, piecemeal adaptation through a gradual process of organically overlaying usable building typologies onto disused or outmoded ones.[169] Measured against dystopian images of disorder and breakdown, instant cities appear to provide an oasis of stability and predictability in a surrounding sea of chaos.[170]

PLANNING FOR SUSTAINABLE URBANISM: THE SLOW CITY MOVEMENT AS MIRROR OPPOSITE OF INSTANT URBANISM

The stress on speed and mobility that defines the hypermodern metropolis of the twenty-first century has produced particular kinds of spatial landscapes. Rapid movement and uninterrupted circulation have become signature features of contemporary city building, fundamental ordering principles embodied in "fast urbanism." The physical placement of such infrastructural components as highways, high-speed railroads, and airports creates the urban fabric of contemporary metropolitan city-regions, in which transit points function as high-value crossroads tying together a loose-knit agglomeration of distinct spatial typologies (such as edge cities, shopping malls, big-box discount stores, tourism and entertainment districts, and office parks). Premised on the mantra "time is money" and a fixation on global competitiveness, this fast urbanism of the hypermodern city-region "structures a mobility regime" of speedy and convenient travel, inter- and intraurban accessibility, and high vehicular capacity at the expense of the "slow urbanism" of pedestrianized streetscapes, walkability, and relaxed pace.[171]

In the post-metropolitan world of sprawling city-regions, urban planners are often forced to choose between allying themselves with or against rapid mobility; that is, they gravitate between promoting the structural dimensions of fast urbanism or its opposite. They can use the ontological tension between the "space of places" (characterized by local spatial organization, physical proximity, and connectivity) and the "space of flows" (characterized by the accelerated pace of global flows of goods, people, and capital engendered through networks of telecommunications and fast mobility) to revalue the advantages of slow growth amid fast metropolitan expansion.[172] Such slow-world activities as the Italian-based Cittaslow movement, which offers relief not only from the repetitive sameness of fast urbanism but also from its accelerated tempo and ecological destructiveness, offers an alternative suggesting one of several anti-globalization utopias.[173] These "spaces of deceleration" refer not only to how particular sites slow the rapid pulse of cities trapped in the pressurized worlds of accelerated consumption, but also to how they assist in retarding the frenzied, market-driven cycles of property investment, urban obsolescence, and real estate speculation.[174] Champions of the "slow city" movement argue that place and identity in the contemporary city are in crisis, suggesting that the serial reproduction of generic, look-alike landscapes and hypermobility have undermined an authentic sense of place—a signature feature of what the "good city" should be. The Cittaslow movement offers a counternarrative to the "placelessness" (Relph) and "non-place" (Augé) landscapes of instant cities.[175]

The unprecedented scale of urban transformation on a global scale has engendered a "fast world" that values rapid movement over attachment to place and faceless networks over face-to-face encounters in lived communities. The fast world of instant urbanism depends upon a steady acceleration in flows of capital, commodities, ideas, and people. The convergence around speed, movement, and flows undermines the distinctiveness of place, devalues tradition, and subverts a durable sense of belonging. These peculiar circumstances of twenty-first-century urbanism present significant environmental and social challenges for cities built de novo in a single gesture as if by the wave of a magic wand. Prospects for long-term urban sustainability are increasingly hampered by the accelerated pace of construction, the profligate use of resources, and the overly speculative stance and myopic, short-term view of real estate capital.[176]

Yet for every rising trend in global urbanism there always seems to be a countermovement, an emergent current that points in another direction. The faster the pace of everyday life in search of profit and material consumption, the more the demand for local roots, predictability, and shared identities.[177] The mirror image of fast-track instant urbanism is the slow city (Cittaslow) movement.[178]

The Cittaslow movement seeks to facilitate an unhurried, decelerated pace of life without noise pollution and traffic congestion, along with environmentally unsustainable practices. As a deliberate and calculated response to the globalizing impulses of sameness, repetition, and mimicry, the Cittaslow movement calls for the creation of an authentic sense of place through the promotion of local distinctiveness. As a counterweight to the global diffusion and implantation of mass material culture, the Cittaslow movement promotes sustainable local economies, healthy environments, and the preservation of local uniqueness.[179]

The Cittaslow movement is closely related to the longer-established and better-known Slow Food movement.[180] The Cittaslow and Slow Food movements—the most established and globally connected "slow" organizations—are related institutionally through overlapping networks. The Slow Movement itself is a broad category of loosely connected organizations, social groups, and affiliates that seek to promote ways of living that offer an alternative to the unsustainable elements of modern urban life.[181]

Cittaslow and Slow Food are allied Slow Movement networks operating in more than 150 countries, including the United States. Member towns must meet the requirement of having a population of fewer than fifty thousand residents. The membership accreditation process requires towns to score at least 50 percent in a self-assessment process and inspection of approximately sixty "requirements for excellence." Taken together, these two networks promote an agenda that focuses on issues concerning quality of life and social-ecological resilience: local and regional food markets, recovery and protection of biodiversity and local ecologies, support for cultural traditions, agricultural policy reform, climate adaptation planning, economic development, social inclusion, livable urbanism in the built environment, and the politics of land development.[182] The aims of the two movements are similar but not identical. In broad terms, both seek to create an environmentally sustainable future by reconnecting people and their social and ecological environments.[183] Both are hostile to corporate business enterprise and the homogenizing effects of globalization, although their "driving motivation is not so much overtly political as it is ecological, humanistic, and aesthetic."[184]

Planning theorists remain divided over the extent to which the Cittaslow movement represents a model for sustainable urban development. Skeptics have suggested that the Cittaslow movement might "all too easily produce enervated, backward-looking, isolationist communities: living mausoleums where the puritanical zealotry of slowness displaces the fervent materialism of the fast world."[185] While the model may not be appropriate for direct transfer to towns in which the ways of life it proposes are not already in place, the Cittaslow framework offers small cities a platform for acknowledging authenticity and a way of formally validating the experiential uniqueness of place.[186] Others have argued that "in promoting the development of small towns," the Cittaslow movement can be most appropriately characterized as "defending enclaves of [local] interest, rather than offering plausible models for more general social transformation."[187]

In contrast, other planning theorists regard Cittaslow as a successful framework for sustainable urban development that at least tempers, if not indirectly resists, the encroachment of global corporate capitalism.[188] Heike Mayer and Paul Knox, for example, argue that the "Slow City agenda represents a viable model for alternative urban development that is especially sensitive and responsive to the complicated interdependencies between the goals for economic development, environmental protection, and social equity."[189] The Cittaslow movement constitutes, in their view, "what is arguably the broadest and most grassroots implementation of the principles associated with liveability, quality of life and sustainability."[190] The Cittaslow movement represents a workable alternative to traditional economic development planning with its stress on *laissez-faire* solutions, place marketing, and "smokestack chasing."[191] Those who conceived of the Cittaslow movement envisioned its core principles in opposition to the contemporary culture of speed, flows, and motion, with its "constant visual noise and unforgiving pace." The aim is to "lift us, if only momentarily," as Nicolai Ouroussoff argued in another context, "out of our increasingly frenetic lives—to slow us down and force us to look at the world around us, and at one another, more closely."[192]

PART III

The Future of Urbanism

CHAPTER 7

CONCLUSION

Urban Futures

Modernity, as Marshall Berman so eloquently suggested, thrives in a constant state of flux, "a maelstrom of relentless disintegration and renewal" that throws everything in its path into continuous upheaval and tumult.[1] The exemplary modern metropolis—as both the active agent of modernity and sometimes its unprepared recipient—occupies a precarious place of perpetual becoming, bedeviled by seemingly irreconcilable conflicts, internal contradictions, and irreducible ambiguities.[2] This trope of the provisional city or the unfinished city has been a persistent metaphor for the dynamics of urban transformation.[3] At the heart of the ongoing story of urban transformation are the conjoined processes of creative destruction and rejuvenation, or building and unbuilding.[4] Think of cities, as a number of prominent urban scholars have urged, as unfinished and incomplete (and always under construction)—embodiments of a kind of turbulent building and rebuilding, a process of constant erasure and reinscription that gestures toward resolution but never really ends. Constantly remade and recast, in myriad ways through monumental decisions as well as everyday encounters, cities are never-ending, expansive, and unbounded stories that unfold over time.[5] In short, they are never static, permanent, or fixed in place; instead, they reflect an inherent indeterminacy, instability, and open-endedness.[6]

At the center of the story of metropolitan modernity is a paradox: while cities converge around shared features and preoccupations, their morphological characteristics, pace of transformation, and scope of regulatory mechanisms vary considerably. As Nigel Thrift has suggested, "cities are not mirrors of modernity"; rather, their patterns of growth and transformation conform "not [to] a straight line but a curve."[7] Put in another way, rather than imagining a unified or singular logic of urban transformation, or a linear or teleological progression toward a shared goal, it is more fruitful to think in terms of multiple trajectories and circuitous routes that simultaneously produce alternative modernities. To think in terms of alternative modernities is to revise the distinction between modernization—defined as linear progress toward a single end point—and modernity, with its multiple pathways and divergent outcomes.[8] There are key historical conjunctures—decisive moments of upheaval and crisis—when relatively ossified and embedded social structures and institutional arrangements, established urban practices, and the ties that bind disparate pieces into the semblance of a whole experience breakdown and disintegration. There are times when new constellations of social forces appear to leap forward into unknown territory in search of a decisive rupture with the past just as easily as they look backward for dependable sources of stability. Similarly, there are fissures in the urban fabric that bring about the breakdown of existing mechanisms of urban governance and usher in the introduction of new modes of regulation. Rather than thinking of cities in terms of linear historical time ("Universal History") in relation to which the pace and direction of urban transformation—blocked and delayed or hastened and accelerated—receive their meaning, it is better to dispense altogether within a singular (and irresistible) narrative of modernization.[9] At the start of the twenty-first century, the multiple trajectories of urbanism on a global scale have broken away from the once regnant modernization impulse that projected an imagined linear or teleological progression through successive stages toward some common end point called development. Rethinking conventional urban theory enables us to treat urbanization as a complex, multifaceted, and sometimes contradictory process that proceeds along multiple pathways without a privileged, common end point.[10]

Uncertainty, provisionality, and contingency have always been key dimensions of urban life.[11] Uncertainty has long confounded efforts to plan, build, and govern (manage) cities. Architects, design specialists, and city builders indoctrinated in a modernist way of thinking have often sought remedial antidotes to those features of urban life that cannot easily be known or managed: the prototypical model that expresses an imagined vision of future possibility; the autonomous zone that overcomes ambiguous boundaries and fluid borders by partitioning territories into self-enclosed locations; the official census that enables rational calculation to measure and justify interventions; and the comprehensive master plan that offers the authoritative promise of a radiant city to come.[12]

However, uncertainty has been a problem not only for professional urbanists but also for city inhabitants. Early theorists of the modern urban experience were deeply concerned about the social and psychological effects of city life. In contrast to what they assumed to be the stable, familiar, and regular routines of rural life, the city was deemed to be a fundamentally volatile, capricious, and hence unknowable environment. What Georg Simmel called the "mental life of the metropolis" was a psychological reaction to the frenetic tempo, the unbounded multiplicity, and the infinite complexity of the modern city.[13] If one responded to all external stimuli, Simmel worried, "one would be completely atomized internally and come to an unimaginable psychic state." Along with others, he reasoned that a blasé attitude—a deliberative posture of impersonality, anonymity, and indifference—offered a coping mechanism, or personal defense, against the fundamental uncertainties of the modern metropolis.[14] This embedded duality in the historical legacy of urbanism—ambiguity as an inherent quality inseparable from urban living and unpredictability as a target of urban intervention—invites us to look into the circumstances that have given rise to indeterminacy and ambiguity and to analyze the various efforts to mitigate and manage them.[15]

Though unsettled circumstances have always accompanied the historical process of urbanization, uncertainty as an inherent condition of urban life has acquired new urgency at the start of the twenty-first century. There is a growing consensus in both scholarly writing and policy-making circles that cities on a global scale not only offer the best hope for humankind but also present the greatest challenges for dealing with world poverty, hunger, and deprivation. Such provocative proclamations as "the city is everywhere and in everything" epitomize what amounts to a "city-centric vision of the current geo-historical moment."[16] As Neil Brenner suggests, "urbanization has become one of the dominant metanarratives through which our current planetary situation is interpreted, both in academic circles and in the public sphere."[17] Yet there is disagreement at the most basic level over how to describe the global urban condition and how best to analyze and interpret it.[18] As the received wisdom of mainstream urban theories has come under intense scrutiny, there has been a proliferation of analytical frameworks competing for dominance as the best approach for understanding the contours of global urbanism.[19] If there is any agreement about the characteristics of global urbanism at the start of the twenty-first century, it is twofold: first, a widespread acknowledgment that cities "are now growing rapidly at locations all over the globe [and] that urbanization is moving ahead more forcefully than at any other time in human history," and second, a shared consensus around the impossibility of forecasting where urbanization may be heading and why.[20] The once confident vision that cities were converging, albeit at their own pace, on a common end point has come unglued as the conjoined ideologies of modernization and development have failed to deliver on their promise of linear progress.[21]

A parallel situation exists in the realm of urban policy and practice. Beyond the perfunctory projection that "the future is urban," there is no shared vision of how this future will unfold. As visions of urban futurity cede ground to tentative experiments in managing what cannot be confidently foreseen, urban policy makers and ordinary residents alike must orient themselves toward the uncertain and the unknown.[22] The widespread use of such concepts as emergency, disruption, crisis, vulnerability, precarity, hazardscapes, and even urbicide reflects a begrudging acceptance of future uncertainty as an existential condition of contemporary urbanism.[23] At the same time, the proliferation of such goal-oriented terms as sustainable urbanism, resilience, adaptability, smart cities, resourcefulness, preparedness, risk aversion, disaster mitigation, and healthy cities suggests that the utopian goal of the "good city" remains as elusive as ever.[24] With the widespread acknowledgment that "we now live in turbulent times" (sometimes simultaneously in the economic, political, and ecological spheres), uncertainty has become perhaps the key dimension of the contemporary urban condition, operating across "multiple levels and scales."[25]

At a theoretical level, the once all-consuming search for universally applicable paradigms and for general theories of urbanization has fallen into disfavor. What has filled the gap in the urban studies literature has been a withering array of analytical frameworks that offer piecemeal and fragmentary perspectives on cities and the contemporary urban condition. Without pretensions to the grand theories of yesteryear, these middle-range theoretical orientations and local idiographic perspectives yield insights on historically specific conditions while at the same time remain somewhat agnostic on world-historical trends of urbanization.[26] On a practical level, grand and comprehensive visions of urban futures have given way to tentative, localized experiments that focus either on place-specific circumstances that yield incremental remediation or on local practices that provide object lessons on what to avoid. Placing uncertainty, provisionality, and contingency close to the center of critical urban inquiry can yield insights into how contemporary cities are imagined, planned, built, governed, and inhabited.[27]

RETHINKING MAINSTREAM URBAN STUDIES

The hyperurbanism of the twenty-first century is both unprecedented in scale and scope and difficult to explain using the analytic tools of conventional urban studies. The urban excess of "supersized cities" elicits both awe and dread in almost equal measure. Seeking to make sense of the sheer velocity and scale of urban transformation, urban theorists, planning experts, and policy makers pondering the future oscillate between acclamations of human progress and forewarnings of a coming crisis. Once regarded as exemplars of urbanism, globalizing cities with world-class aspirations have become exceptional in many respects. City builders in struggling cities in distress have sought in vain to overcome the dysfunctional disorders associated with inadequate and broken-down infrastructure, overcrowded streetscapes, overburdened transit systems, social inequalities, absolute poverty, indecent housing, and inadequate structures of good governance. Similarly, the megacities of hypergrowth greatly amplify the problems of overcrowded and unhygienic slums associated with the late-nineteenth-century industrial city—predicaments that at the time seemed daunting indeed. Finally, the fast-track urbanism of "instant cities" offers the promise of starting afresh, bypassing the obstacles inherited from the past. All in all, building sustainable and just cities has become a popular slogan that often turns out to be an empty promise except for entrepreneurial elites who have successfully forged enclaves in the midst of deprivation.[28]

Global urbanism at the start of the twenty-first century is a dynamic field in which urban landscapes are both expanding and contracting, and growing and declining, in response to forces that operate at multiple scales. There is growing acknowledgment that cities like São Paulo, Shanghai, Seoul, Mumbai, Delhi, and Shenzhen are likely to challenge the preeminence of the recognized metropolitan centers of North America and Europe, "not just in size and density of economic activity—they have already done that—but primarily as leading incubators in the global economy, progenitors of new urban form, process, and identity."[29] Yet, as Ananya Roy, Edgar Pieterse, and others have stressed, even though the urban future lies elsewhere, in such cities of the Global South as Manila, Cairo, Lagos, Karachi, Mexico City, Rio de Janeiro, Dakar, and Kinshasa, mainstream theoretical work on global city-regions is still firmly located within the cities of the historic North American and European core of the capitalist world economy.[30] Rethinking conventional urban theory means using the experiences of these cities "off the map" to reconfigure the conceptual and analytic frameworks at the center of urban and metropolitan analysis.[31]

This growing recognition of the need to theorize the diversity of urban trajectories cannot be accomplished without unsettling and disrupting conventional approaches to urban growth and development. Rethinking the geographies of urban and regional theory involves destabilizing the key premises, concepts, and visions that have guided the understanding of urban transformation on a global scale over the past many decades. Much of the theoretical writing on global city-regions is firmly located in the urban experience of the leading cities of

North America and Western Europe. This kind of theoretical production is part of a canonical tradition whose key ideas have emerged in the crucible of a few "great" cities such as London, New York, Chicago, Paris, and Los Angeles. While critical urban scholarship has challenged the reigning conventional view that the top-down imposition of spatial order stands as a proxy for modernity, development, and progress, mainstream urban studies has remained largely impervious to these recent calls for epistemological renewal in urban theory production. By clinging to the prevailing normative view that rational use of space, efficient movement, and orderly environments constitute universal and all-encompassing characteristics that define "good urbanism," scholars still rooted in old paradigms consistently suggest that cities lacking these features are "not quite yet" modern, lagging behind in the competitive race toward development.[32]

As a general rule, the mainstream urban studies paradigm operates within the framework of grand narratives of progress, development, and modernization. In this conventional view, there is a singular logic that controls urbanization as a unidirectional historical process. Thus, mainstream urban studies assumes that urbanization as a global process unfolds in a more or less universal pattern along a continuum that stretches in linear time from less developed (or underdeveloped) to developing and finally to developed.[33] To accept the premises of modernization theory means to buy into a narrow narrative that naturalizes and normalizes this way of thinking about linear pathways to urban development.[34]

The effort to rethink mainstream urban theory contrasts sharply with the privileged lens that frames successful Euro-American cities in terms of the economic rationality and political reasoning of classical Western liberalism. Looking instead at the multiple trajectories of urban transformation that characterize global urbanism at the start of the twenty-first century offers a way of rejecting the self-referential, celebratory tone of conventional urban studies literature and deconstructing the neoliberal undertones behind the fanciful images and grand plans of aspiring global cities.[35]

The recurrent focus on growth has long been an integral feature of conventional theories of urban development. Theorizing about cities has never strayed too far from considerations about processes of growth—whether this thinking involves discussion of market-led regeneration or planned interventions to curb the negative impact of overgrowth and unfettered sprawl.[36] Yet for many cities and urban regions around the world, this seemingly natural connection between vibrant cities and steady growth has already been severed by the deleterious effects of deindustrialization, suburbanization, and sociodemographic decline.[37]

Within the framework of the contemporary age of globalization, scholars have identified three spatial restructuring tendencies as the main "causes" (or drivers) of urban shrinkage and decline.[38]Generally speaking, urban shrinkage and decline refer to a physical reduction in city size and density; unplanned, widespread losses in jobs and population; and a protracted socioeconomic downturn in a given (especially core) locality.[39] To a large extent, the term "shrinking cities" is used to describe older industrial urban centers in the core areas of the world economy that were typically characterized by relatively nondiversified local economies at the end of the World War II. For the most part, the socioeconomic stability of these modern industrial cities depended on mass-production industries, which accounted for the lion's share of stable employment.[40] Deindustrialization often went hand in hand with the inability of local economies to adjust to rapidly changing external conditions, such as the shift of markets away from manufacturing and toward services, the introduction of new technologies, and the corrosive impact of globalization. Postindustrial cities in decline have been slow or insufficiently able to adapt to the competitive demands of globalized markets. Consequently, corporate enterprises (with their power to absorb local labor supplies) have withdrawn from these cities at massive scales, leaving behind myriad socioeconomic, environmental, and political problems.[41]

THE INVERTED TELESCOPE

The convoluted nature of global urbanism at the start of the twenty-first century comes into sharper focus if we look outside Europe and North America as the privileged sites for the theoretical understanding of urbanization. Like a magnifying glass—or an "inverted telescope," to use Benedict Anderson's metaphorical expression—these other cities and metropolitan regions highlight what might too easily be overlooked closer to home.[42] The inverted telescope works to replace the privileged status of the great modern metropolises of North America and Europe as the main point of reference for theoretical thinking with a more nuanced and relational reading of global urbanism.[43] For Anderson, looking through the inverted telescope enables us to dispense with taking one perspective as matter-of-factly normal and expected, and instead see "simultaneously up close and from afar." For him, a singular vision exemplifies the "logic of seriality" in which peculiarities and particularities of local circumstances are lost in the rush to identify general patterns with universal application.[44]

In conventional thinking about urbanization, the modern, up-to-date city has always remained a universal ambition. Cities located in those peripheral terrains "beyond the West" are often disregarded for their insularity and backwardness, dismissed as derivative or imitative, or excluded as unexplained outliers that deviate from the norm. Yet they help define the center. Framing the successful "global city" has always required the "unsuccessful non-global city," to paraphrase Timothy Mitchell. The production of the global city involves the staging of differences. One current in the conventional urban studies literature proclaims that the envied position of the global city at the top of the urban hierarchy is reserved for only a few cities, while an equally influential current points to its enormous power of imitation and replication.[45] Yet dispensing with the universalist pretensions of conventional urban scholarship enables to see that ranked hierarchies are not inherently fixed, nor is urbanization-as-process unidirectional. New circumstances and conditions come into being, altering the appearance of global urbanism and its trajectories in unexpected ways. The variety of combinations ensures that cities will never be experienced in exactly the same way, look just alike, or benefit all residents in the same way.[46]

This viewpoint casts new light on various assumptions, both in terms of spatial imaginings and scholarly predispositions, that are usually taken for granted in conventional urban studies literature. Common axioms about the inherently homogenizing tendencies of emergent planetary urbanism, the contemporary interconnectedness of an increasingly globalizing world, and the normative value placed on the open, democratic public realm tend to discount and overlook the vast differences that exist in ordinary cities around the world. Why, for instance, should we assume that the flaneur so celebrated in Walter Benjamin's reading of the modern metropolis represents the universal essence of urban modernity? After all, the twentieth-century idea of "modern man," while intended to be progressive and cosmopolitan, unthinkingly ignored gender differences and denied the appeal of seemingly outdated forces such as religion and nationalism. The celebration of universal standards is never really a placeless gesture.[47] The failure to grasp historical particularities tends toward mystification rather than clarification. To reify urbanization as a process is to overlook, and hence downplay, its "ambiguities, inconsistencies, and repressed continuities."[48]

THE TIME/SPACE OF CITIES

The well-entrenched and taken-for-granted conceptions of a universal, homogeneous time have long constituted an ideal platform for various teleological interpretations of urban transformation. This mode of thinking has enabled mainstream urban theories to look upon cities as passing through distinct stages, (where each is dependent on what came before) in a single homogeneous time.[49] This historicist temptation to reduce all cities to expressions of a single temporal principle has greatly hindered our capacity to make sense of global urbanism. The starting point for overcoming this theoretical rigidity is to replace the ontology of copresence with an ontology of noncontemporaneity, thereby introducing an urban theory agenda that comes to terms with the incongruities, multiplicities, complexities, and unevenness of temporalities that accompany urban transformation.[50]

Seen in this light, all sorts of anomalies and oddities that mainstream urban studies literature typically treats as deviations from expected patterns of "normal" urban transformation can be reconceptualized as integral features of city building in places that fall outside the experience of globalizing cities with world-class aspirations. Thus, in challenging the idea that postindustrial cities in decline represent market failure and arrested development, Joshua Akers and others have argued that these struggling cities are active sites for renewed and reenergized bouts of capital accumulation. Rather than signifying an anomaly or a radical break with normalcy, the production of decline is part and parcel of a particular kind of property development that has become big business in its own right, as unbuilding replaces building as the primary strategy of accumulation. Parasitic investments, including speculation in abandoned property, demotion and removal, predatory mortgage lending, and rack-renting, mark the onset of new accumulation strategies rather than a withdrawal of capital. The introduction of new regulatory regimes that privilege private markets embeds market logic in urban governance.[51] In challenging what can be called growth fetishism, some scholars writing about postindustrial cities in decline have coined the term "degrowth machine" as an "implicit rejection of the single-minded pursuit of economic growth [with its alleged trickle-down benefits] at the expense of marginalized urban residents."[52]

When framed through the lens of conventional understandings of urban temporality, the four city types occupy distinct time frames, or time capsules. In short, they are distinct chronotopes (space-time entities). As a general rule, research and writing about globalizing cities with world-class aspirations are embedded in the circulation of self-affirming knowledge of a bright future. These vibrant, healthy cities with First World amenities occupy the radiant future, marking time as progress—i.e., the gradual realization of the ultimate goal of global recognition. These cities exist in linear time, marching expectantly forward toward the as yet unrealized goal of achieving world-class status. The conjoined discourses of creative destruction, provisionality, and cities-in-the-making accompany this expectant notion of forward-looking time.[53] Seen as vibrant, thriving, and successful, these cities are sites of innovation, setting the standards for laggard "loser" cities to emulate. Globalizing cities with world-class aspirations are imagined as occupying undifferentiated empty time, and all impediments and resistances are framed as archaic and backward—obsolete remnants of an ossified past that stand in the way of the arc of time.[54]

In contrast, postindustrial cities in decline occupy the past. While they suffer from neglect and abandonment, they can never escape from what they once were. They seem to have reversed direction, in full retreat, seemingly going backward in time. They mark time as before and after—a rupture, a break, that signals a turning point that divides a period of ascendancy (characterized by an exalted position of relative success) from one of decline (sometimes abrupt and sometimes slow). The arc of time passes from the apex of an ascendant past through a parabola of decline. Postindustrial cities generate a historical memory scripted as a fall from grace. Ruin, abandonment, and neglect evoke nostalgic remembrances of a former time of relative stability when the future looked bright—a paradise lost and what might have been. These cities are post-historical in that they are seen as outside of time, beyond their appointed time in the historical limelight, with no real prospects of ever returning to their glorious past. Their present is accompanied by nostalgia, melancholy, and regret.

The megacities of hypergrowth occupy a time warp, locked into an eternal present, a timeless time in which the future cruelly beckons but never arrives. They mark time as the emergency of permanent crisis. They may evolve, but they never really change. Trapped in a liminal state of stasis, these megacities of hypergrowth exist not in linear time but in cyclical time, trapped in an endless and repetitive loop. They lurch from crisis to crisis and, like Sisyphus, are condemned to repeated failures.[55]

What distinguishes the instant cities of the Asian Pacific Rim and the Persian Gulf is their sudden ascendancy in the rank order of worldliness. With their fast-track approach to city building, these instant cities occupy the surreal instantaneous realm of pure spontaneity. Coming into existence de novo and at breakneck speed endows them with a timeless quality that releases them from the normal conventions of life-cycle thinking about eventual decay and gradual ruination. They defy our conventional understanding of the evolving nature of linear urban time. Time starts in the present and is permanently attached to the endless now. These cities have grown at warp speed, leapfrogging over what came before and bypassing any presumption of staged linearity.[56] Defying both the linear time of modernism and the reversible time of postmodernism, they occupy the ahistorical time of contemporaneity—a synchronous time in which everything belongs to the same uniform conception of time, or to no time at all. Instant cities exist in a temporal condition that the contemporary art critic Nicolas Bourriaud has called "altermodern"—a form of time that bears no relation to the past or to chronology, while simultaneously signifying the modernity to come, pretending to embody the future in the present.[57] The featureless contemporaneity of instant cities gestures toward a kind of post-historical time, liberated from the tyranny of universal time and draining the passage of time of all significance and meaning whatsoever.[58]

Similarly, framed through the lens of conventional understandings of urban spatiality, these four city types occupy distinct space capsules. Globalizing cities with world-class aspirations form legible spatial wholes. They are centered—and figured—with recognizable boundaries. Clear demarcations between cores and peripheries give meaning and coherence to spatial arrangements. There is a place for everything, and correlatively, everything has a place. In contrast, postindustrial cities in decline are defined by voids, dead space, openness, and an underutilized built environment. They exist in empty space, with expansive voids and unimpeded sightlines. Their excess of derelict and ruined space signals a kind of arrested development in which not enough is happening.

Megacities of hypergrowth are formless and amoebalike, unruly, uncontrolled, and unfettered. Taking shape without logic, pattern, or rules, they consist of congested space, overcrowded streetscapes, and overused built environment where too much is happening. Finally, instant cities occupy the space of excess. They epitomize the time-space compression of hypermodernity. Speed transforms experiential space into spectacle.[59]

THE JANUS FACE OF GLOBAL URBANISM

Global urbanism at the start of the twenty-first century is characterized by increasing spatial polarization on a global scale where expansion and growth, on the one side, and contraction and decline, on the other, are intertwined in complex and dynamic webs. Growing cities and declining cities represent two faces of the same processes at work in the world economy. Taken together, the conjoined forces of globalization, deindustrialization, and neoliberalism have produced two types of postindustrial cities. On one hand, there are the resurgent cities that have successfully negotiated the difficult terrain from their once vibrant industrial base to newfound prosperity via information-age high-tech industries and service economies, culture-led redevelopment, gentrification, and tourist-entertainment sites. On the other hand, there are the dismal "loser" cities, those postindustrial places incapable of avoiding the downward spiral of neglect, abandonment, and decay. These postindustrial cities in decline are characterized, first and foremost, by their entanglements in a web of intractable problems from which there is no easy way out: pronounced demographic decline, high unemployment rates, low skill levels among remaining residents, and inordinately high levels of urban violence. The prosperity of these cities can be traced to their connection with the era of Fordist industrialization starting in the early twentieth century. The socioeconomic crisis of the 1970s and the relocations of industrial production sites linked to accelerating globalization affected them intensely. Deindustrialization, plant closings, and job losses lie at the root of the social problems encountered by their remaining residents. These postindustrial cities in decline represent one side of the global transformation of urban landscapes, whereas growing and expanding postindustrial cities represent the other.[60]

Cities are not only growing and expanding, but also stagnating or declining. Urban shrinkage—alternatively referred to as urban decline, abandonment, and neglect—is a distinct urban development pathway that has spread widely across old industrial regions of Western Europe, the post-socialist countries of Central and Eastern Europe, the American Rust Belt, northern Japan, and elsewhere around the world.[61] Whatever else, urban shrinkage needs to be understood not as a deviant outlier that somehow fails to conform to the expected norms of urban transformation, but as an object of inquiry in its own right. In its everyday usage, the term "shrinking cites" refers to metropolitan areas that feature economic decline and restructuring, population loss, increasing unemployment, fiscal austerity, environment degradation, along with the assortment of social problems associated with abandonment and decay, crumbling infrastructure, and vacant and derelict properties.[62] Urban shrinkage is the result of multiple parallel processes with manifold dimensions and effects.[63]

NOTES

PREFACE

1. Neil Brenner and Christian Schmid, "The 'Urban Age' in Question," *International Journal of Urban and Regional Research* 38, no. 3 (2014): 731–755; and Brendan Gleeson, "Critical Commentary: The Urban Age: Paradox and Prospect," *Urban Studies* 49, no. 5 (2012): 931–943.
2. Brenner and Schmid, "The 'Urban Age' in Question," 743.
3. Ashley Dawson and Brent Hayes Edwards, "Introduction: Global Cities of the South," *Social Text* 22, no. 4 (2004): 1.
4. Kees Koonings and Dirk Kruijt, eds., *Megacities: The Politics of Urban Exclusion and Violence in the Global South* (London: Zed, 2010); Ricky Burdettt and Philipp Rode, "Living in the Urban Age," in *Living in the Endless City*, ed. Ricky Burdett and Deyan Sudjic (London: Phaidon, 2011), 8–43; and Joel Kotkin and Wendell Cox, "The World's Fastest Growing Megacities," *Forbes*, April 8, 2013.
5. The use of the widely popular terms "Global South" and its polar opposite "Global North" is a convenient shorthand that does not suggest distinct geographic regions but instead indicates the differing positionalities associated with the structural asymmetries that characterize the contemporary world economy. For example, see Eric Sheppard and Richa Nagar, "From East–West to North–South," *Antipode* 36, no. 4 (2004): 557–563. In my view, the Global North / Global South distinction marks a step backward from the older bifurcation between core and peripheral zones of the world economy.
6. United Nations, *State of the World's Cities, 2002* (New York: United Nations Center for Human Settlement, 2002).
7. See, inter alia, Mike Davis, *Planet of Slums* (New York: Verso, 2006); Dirk Kruijt and Kees Koonings, "The Rise of Megacities and the Urbanization of Informality, Exclusion, and Violence," in Koonings and Kruijt, *Megacities*, 8–26; and Jennifer Robinson and Ananya Roy, "Debate on Global Urbanisms and the Nature of Urban Theory," *International Journal of Urban and Regional Research* 40, no. 1 (2016): 181–186.
8. Ivan Turok and Vlad Mykhnenko, "The Trajectories of European Cities, 1960–2005," *Cities* 24, no. 3 (2007): 165–182.
9. Witold Rybczynski and Peter Linneman, "How to Save Our Shrinking Cities," *Public Interest* 135 (1999): 30–44.
10. Philipp Oswalt and Tim Rienets, eds., *Atlas of Shrinking Cities* (Ostfildern, Germany: Hatje Cantz, 2006).

11. Theodore Gilman, *No Miracles Here: Fighting Urban Decline in Japan and the United States* (Albany: State University of New York Press, 2001).
12. Max Rousseau, "Re-imaging the City Centre for the Middle Classes: Regeneration, Gentrification and Symbolic Policies in 'Loser Cities'," *International Journal of Urban and Regional Research* 33, no. 3 (2009): 770–788; Richard Marshall, *Emerging Urbanity: Global Urban Projects in the Asia Pacific Rim* (New York: Spon, 2003), 201–204; and Fu-chen Lo and Peter Marcotullio, "Globalization and Urban Transformations in the Asia-Pacific Region: A Review," *Urban Studies* 37, no. 1 (2000): 77–111.
13. Jennifer Robinson, "Global and World Cities: A View from off the Map," *International Journal of Urban and Regional Research* 26, no. 3 (2002): 531.
14. AbdouMaliq Simone, *For the City Yet to Come: Changing African Life in Four Cities* (Durham, NC: Duke University Press, 2004), 1.
15. See contributions to Tahl Kaminer, Miguel Robles-Durán, and Heidi Sohn, eds., *Urban Asymmetries: Studies and Projects on Neoliberal Urbanization* (Rotterdam: 010 Publishers, 2011).
16. See Pedro Gadanho, "Mirroring Uneven Growth: A Speculation on Tomorrow's Cities Today," in *Uneven Growth: Tactical Urbanisms for Expanding Megacities*, ed. Pedro Gadanho (New York: Museum of Modern Art, 2015), 14–25. See also Martin J. Murray, "Large-Scale, Master-Planned Redevelopment Projects in Urbanizing Africa: Frictionless Utopias for the Contemporary Urban Age," in *Mega-Urbanization in the Global South: Fast Cities and New Urban Utopias in the Postcolonial State*, ed. Ayona Datta and Abdul Shaban (New York: Routledge, 2017), 31–53; Martin J. Murray, " 'City Doubles': Re-Urbanism in Africa," in *Cities and Inequalities in a Global and Neoliberal World*, ed. Faranak Miraftab, David Wilson, and Kenneth Salo (New York: Routledge, 2015), 92–109; and Ayona Datta, "Introduction: Fast Cities in an Urban Age," in *Mega-Urbanization in the Global South*, 1–28.
17. See Garth Myers and Martin J. Murray, "Introduction: Situating Contemporary Cities in Africa," in *Cities in Contemporary Africa*, ed. Martin J. Murray and Garth Myers (New York: Palgrave/Macmillan, 2008), 3.
18. David Smith, *Third World Cities in Global Perspective: The Political Economy of Uneven Urbanization* (Boulder, CO: Westview, 1996).
19. For "feral cities," see Richard Norton, "Feral Cities," *Naval War College Review* 56, no. 4 (2003): 97–106; and Robert J. Bunker, "The Emergence of Feral and Criminal Cities: U.S. Military Implications in a Time of Austerity," *Land Warfare Paper 99W* (Arlington, VA: Association of the United States Army, April 2014), 1–11, https://www.ausa.org/publications/emergence-feral-and-criminal-cities-us-military-implications-time-austerity.
20. Saskia Sassen, *The Global City: New York, London, Tokyo*, 2nd ed. (Princeton, NJ: Princeton University Press, 2001).
21. Nezar AlSayyad and Ananya Roy, "Medieval Modernity: On Citizenship and Urbanism in a Global Era," *Space and Polity* 10, no. 1 (2006): 2.
22. Neil Brenner and Christian Schmid, "Toward a New Epistemology of the Urban?" *City* 19, no. 2–3 (2015): 151–182.
23. See AbdouMaliq Simone, "The Urbanity of Movement: Dynamic Frontiers in Contemporary Africa," *Journal of Planning Education and Research* 31, no. 4 (2011): 379–391; and AbdouMaliq Simone, "On the Worlding of African Cities," *African Studies Review* 44, no. 2 (2001): 15–42.
24. Trevor Hogan and Julian Potter, "Big City Blues," *Thesis Eleven* 121, no. 1 (2014): 3.
25. See, inter alia, Max Page, *The Creative Destruction of Manhattan, 1900–1940* (Chicago: University of Chicago Press, 1999), 69–86.
26. Alan Mayne, *Slums: The History of a Global Injustice* (London: Reaktion, 2017).
27. See, for example, João Biehl, *Vita: Life in a Zone of Social Abandonment* (Berkeley: University of California Press, 2005); and Loïc Wacquant, "Class, Ethnicity, and the State in the Making of Marginality:

Revisiting Territories of Urban Relegation," in *Territories of Poverty: Rethinking North and South*, ed. Ananya Roy and Emma Shaw Crane (Athens: University of Georgia Press, 2015), 247–259.

28. Marcelo Balbo, "Beyond the City of Developing Countries: The New Urban Order of the 'Emerging City,'" *Planning Theory* 13, no. 3 (2014): 269–287.
29. See, for example, John Friedmann and Mike Douglass, *Cities for Citizens: Planning and the Rise of Civil Society in a Global Age* (New York: Wiley, 1998); and Peter Evans, *Livable Cities?* (Berkeley: University of California Press, 2002).
30. For his popular urban manifesto, see Edward Glaeser, *Triumph of the City: How Our Greatest Invention Makes Us Richer, Smarter, Greener, Healthier, and Happier* (New York: Penguin, 2011). For provocative critiques, see Jamie Peck, "Economic Rationality Meets Celebrity Urbanology: Exploring Edward Glaeser's City," *International Journal of Urban and Regional Research* 40, no. 1 (2016): 1–30; and Hillary Angelo and David Wachsmuth, "Why Does Everyone Think Cities Can Save the Planet?" *Urban Studies* 57, no. 11 (2020): 2201–2221.
31. George Packer, "The Megacity: Decoding the Chaos of Lagos," *New Yorker*, November 13, 2006, 62–75; and Eugene Linden, "The Exploding Cities of the Developing World," *Foreign Affairs* 75, no. 1 (1996): 52–65.
32. Mike Davis, *Planet of Slums* (New York: Verso, 2006); Jan Bremen, "Myth of the Global Safety Net," *New Left Review* 59 (2009): 29–38; and Jan Bremen, "Slumlands," *New Left Review* 40 (2006): 141–148.
33. Mike Davis, "Planet of Slums: Urban Involution and the Informal Proletariat," *New Left Review* 26 (2004): 28.
34. Davis, "Planet of Slums," 14, 16–17; Elliott Sclar, Pietro Garau, and Gabriella Carolini, "The 21st Century Health Challenge of Slums and Cities," *Lancet* 365, no. 9462 (2005): 901–903; and Tim Campbell and Alana Campbell, "Emerging Disease Burdens and the Poor in Cities of the Developing World," *Journal of Urban Health* 84, no. 1 (2007): 54–64.
35. See Martin J. Murray, *The Urbanism of Exception: The Dynamics of Global City Building in the Twenty-First Century* (New York: Cambridge University Press, 2017), 93–146.
36. See John Rundell, "Imagining Cities, Others: Strangers, Contingency and Fear," *Thesis Eleven* 121, no. 1 (2014): 10.
37. Engin Isin, "City.State: Critique of Scalar Thought," *Citizenship Studies* 11, no. 2 (2007): 211–228; and Engin Isin, "Historical Sociology of the City," in *Handbook of Historical Sociology*, ed. Gerard Delanty and Engin Isin (London: Sage, 2002), 312–325.
38. Rundell, "Imagining Cities, Others," 10–12.
39. Ananya Roy, "Conclusion: Postcolonial Urbanism: Speed, Hysteria, Mass Dreams," in *Worlding Cities: Asian Experiments and the Art of Being Global*, ed. Ananya Roy and Aiwha Ong (Malden, MA: Wiley-Blackwell, 2011), 307–335.
40. Robinson, "Global and World Cities," 531.
41. Dominique Malaquais, "Anti-Teleology: Re-Mapping the Imag(in)ed City," in *The African Cities Reader II: Mobilities and Fixtures*, ed. Ntone Edjabe and Edgar Pieterse (Cape Town: African Centre for Cities, 2010), 7.
42. See Allen Scott, "Emerging Cities of the Third Wave," *City* 15, no. 3–4 (2011): 289–321.
43. Many of the ideas expressed here are borrowed from Martin J. Murray, "Afterward: Re-engaging with Transnational Urbanism," in *Locating Right to the City in the Global South*, ed. Tony Samara, Shenjing He, and Guo Chen (New York: Routledge, 2013), 300–301.
44. For the source of some of the ideas expressed here, see Myers and Murray, "Introduction: Situating Cities in Africa." See also Jennifer Robinson, *Ordinary Cities: Between Modernity and Development* (New York: Routledge, 2006), 6–9.
45. Jennifer Robinson, "Cities in a World of Cities: The Comparative Gesture," *International Journal of Urban and Regional Research* 35, no. 1 (2011): 1–23; and Jennifer Robinson, "The Spaces of

Circulating Knowledge: City Strategies and Global Urban Governmentality," in *Mobile Urbanism: Cities and Policy-Making in the Global Age*, ed. Eugene McCann and Kevin Ward (Minneapolis: University of Minnesota Press, 2011), 15–40.

46. Roy, "Postcolonial Urbanism"; Ananya Roy, "Slumdog Cities: Rethinking Subaltern Urbanism," *International Journal of Urban and Regional Research* 35, no. 4 (2011): 223–238.
47. Patsy Healey, "The Universal and the Contingent: Some Reflections on the Transnational Flow of Planning Ideas and Practices," Planning Theory 11, no. 2 (2012): 188–207.
48. Richard Lloyd and Terry Nichols Clark, "The City as Entertainment Machine," in *Critical Perspectives on Urban Redevelopment: Volume 6*, ed. Kevin Fox Gotham (New York: Elsevier, 2001), 357.
49. Jennifer Robinson, "Global and World Cities," 549.
50. Helga Leitner and Eric Sheppard, "Provincializing Critical Urban Theory," *International Journal of Urban and Regional Research* 40, no. 1 (2016): 230.
51. Jamie Peck, "Cities Beyond Compare?" *Regional Studies* 49, no. 1 (2015): 162–163.
52. Ananya Roy, "Who's Afraid of Postcolonial Theory?" *International Journal of Urban and Regional Research* 40, no. 1 (2015): 200.
53. The ideas for this paragraph are derived from Partha Pukhopadhya, Marie-Hélène Zérah, and Eric Denis, "Subaltern Urbanization: Indian Insights for Urban Theory," *International Journal of Urban and Regional Research* 44, no. 4 (2020): 582; and Susan Parnell and Jennifer Robinson, "(Re)theorizing Cities from the Global South: Looking Beyond Neoliberalism," *Urban Geography* 33, no. 4 (2012): 593–617.
54. Vanessa Watson, "Seeing from the South: Refocusing Urban Planning on the Globe's Central Urban Issues," *Urban Studies* 46, no. 11 (2009): 2259–2275.
55. Eric Sheppard, Helga Leitner, and Anant Maringanti, "Provincializing Global Urbanism: A Manifesto," *Urban Geography* 34, no. 7 (2013): 893.
56. For "ordinary cities," see Ash Amin and Stephen Graham, "The Ordinary City," *Transactions of the Institute of British Geographers* 22, no. 4 (1997) 411–429; and Jennifer Robinson, *Ordinary Cities: Between Modernity and Development* (New York: Routledge, 2006). See Jennifer Robinson and Ananya Roy, "Debate on Global Urbanisms and the Nature of Urban Theory," *International Journal of Urban and Regional Research* 40, no. 1 (2016): 181–186; and Ananya Roy, "The 21st-Century Metropolis: New Geographies of Theory," *Regional Studies* 43, no. 6 (2009): 819–830.
57. Colin McFarlane, "The Comparative City: Knowledge, Learning, Urbanism," *International Journal of Urban and Regional Research* 34, no. 4 (2010): 725–742; and Roy, "Conclusion: Postcolonial Urbanism," in *Worlding Cities*, 307–335.
58. The title of this section is an idea borrowed from T. G. McGee, *Urbanization in the Third World: Explorations in Search of a Theory* (London: Bell, 1971).
59. In part, I take my inspiration (metaphorically at least) from Mike Davis when he suggests that he is excavating the future of Los Angeles. See Mike Davis, *City of Quartz: Excavating the Future of Los Angeles* (New York: Vintage, 1990).
60. See Saskia Sassen, "Does the City Have Speech?" *Public Culture* 25, no. 2 (2013): 209
61. See, for example, Tom Angotti, *The New Century of the Metropolis: Urban Enclaves and Orientalism* (New York: Routledge, 2013), 3–25.
62. See Jennifer Robinson, "In the Tracks of Comparative Urbanism: Difference, Urban Modernity and the Primitive," *Urban Geography* 25, no. 8 (2004): 709.
63. For an argument in favor of a universalizing theoretical framework, see Allen Scott and Michael Storper, "The Nature of Cities: The Scope and Limits of Urban Theory," *International Journal of Urban and Regional Research* 39, no. 1 (2015): 1–15. For an approach that gestures toward radical relativism, see AbdouMaliq Simone, "It's Just the City After All!" *International Journal of Urban and Regional Research* 39, no. 1 (2015): 210–218. For a critique of "radical uniqueness," see Robert

Beauregard, "Radical Uniqueness and the Flight from Urban Theory," in *The City, Revisited: Urban Theory from Chicago, Los Angeles, and New York*, ed. Dennis Judd and Dick Simpson (Minneapolis: University of Minnesota Press, 2013), 186–206.

64. Tim Edensor and Mark Jayne, "Introduction: Urban Theory Beyond the West," in *Urban Theory Beyond the West: A World of Cities*, ed. Tim Edensor and Mark Jayne (New York: Routledge, 2012), 1–27.

65. Robert Beauregard, *When America Became Suburban* (Minneapolis: University of Minnesota Press, 2006).

66. Edward Soja, "Inside Exopolis: Scenes from Orange County," in *Variations on a Theme Park*, ed. Michael Sorkin (New York: Hill & Wang, 1992), 94–122; Edward Soja, *Postmetropolis: Critical Studies of Cities and Regions* (New York: Wiley-Blackwell, 2000); Deyan Sudjic, *The 100-Mile City* (San Diego: Harcourt Brace, 1993); Jon Teaford, *Post-Suburbia: Government and Politics in the Edge Cities* (Baltimore: Johns Hopkins University Press, 1997); Peter Hall and Kathy Pain, "From Metropolis to Polyopolis," in *The Polycentric Metropolis: Learning from Mega-City Regions in Europe*, ed. Peter Hall and Kathy Pain (New York: Earthscan, 2006), 3–18; Robert Lang and Paul Knox, "The New Metropolis: Rethinking Megalopolis," *Regional Studies* 43, no. 6 (2009): 789–802; Paul Knox, *Metroburbia, USA* (New Brunswick: Rutgers University Press, 2008); Oliver Gillham, *The Limitless City: A Primer on the Urban Sprawl Debate* (London: Island, 1992); and Peter Gordon and Harry Richardson, "Beyond Polycentricity: The Dispersed Metropolis, Los Angeles, 1970–1990," *Journal of the American Planning Association* 62, no. 3 (1996): 289–295.

67. Neil Brenner, "Introduction: Urban Theory Without an Outside," in *Implosions/Explosions: Towards a Theory of Planetary Urbanization*, ed. Neil Brenner (Berlin: Jovis, 2014), 14–30; Christian Schmid, "Networks, Borders, and Difference: Towards a Theory of the Urban," in *Implosions/Explosions*, 67–79; and Christian Schmid, "Journeys Through Planetary Urbanization: Decentering Perspectives on the Urban," *Environment and Planning D: Society and Space* 36, no. 3 (2018): 591–610.

68. AbdouMaliq Simone, "(Non)Urban Humans: Questions for a Research Agenda (the Work the Urban *Could* Do)," *International Journal of Urban and Regional Research* 44, no. 4 (2020): 755.

69. Murray, *The Urbanism of Exception*, 108–110; and Roger Keil, "Extended Urbanization, 'Disjunct Fragments' and Global Suburbanisms," *Environment and Planning D: Society and Space* 3, no. 2 (2018): 494–511.

70. Solomon Benjamin, "Occupancy Urbanism: Radicalizing Politics and Economy Beyond Policy and Programs," *International Journal of Urban and Regional Research* 32, no. 3 (2008): 719.

71. These ideas are taken from Andrew Abbott, "Los Angeles and the Chicago School: A Comment on Michael Dear," *City and Community* 1, no. 1 (2002): 35–36; and Neil Brenner and Christian Schmid, "Toward a New Epistemology of the Urban?" *City* 19, no. 2–3 (2015): 151–182.

72. Ash Amin, "Spatialities of Globalization," *Environment & Planning A* 34, no. 3 (2002): 386.

73. See Kevin Ward, "Towards a Relational Comparative Approach to the Study of Cities," *Progress in Human Geography* 34, no. 4 (2010): 471–487; and McFarlane, "The Comparative City."

74. See Myers and Murray, "Introduction: Situating Contemporary Cities in Africa," 6.

75. See Edensor and Jayne, "Introduction: Urban Theory Beyond the West."

76. For journalistic accounts that fit this genre of poignant and colorful anecdotes, see George Packer, "The Megacity: Decoding the Chaos of Lagos," *New Yorker*, November 13, 2006, 62–75; and Robert Draper and Pascal Maitre, "Kinshasa, Urban Pulse of the Congo," *National Geographic* 224, no. 4 (2013): 100–123.

77. Jean Comaroff and John Comaroff, *Theory from the South: Or, How Euro-America Is Evolving Toward Africa* (New York: Paradigm, 2012), 1.

78. Achille Mbembe and Sarah Nuttall, "Writing the World from an African Metropolis," *Public Culture* 16, no. 3 (2004): 348, 366.

79. Robinson, *Ordinary Cities*, 116–140.

80. See David Wachsmuth, "City as Ideology: Reconciling the Explosion of the City Form with the Tenacity of the City Concept," *Environment & Planning D* 32, no. 1 (2014): 75–90.
81. Robinson and Roy, "Debate on Global Urbanisms and the Nature of Urban Theory," 182.
82. See Ilka Ruby and Andreas Ruby, "Forward," in *Urban Transformation*, ed. Ilka Ruby and Andreas Ruby (Berlin: Ruby, 2008), 11.
83. Jennifer Robinson, "Comparative Urbanism: New Geographies and Cultures of Theorizing the Urban," *International Journal of Urban and Regional Research* 40, no. 1 (2016): 188.
84. Robinson, "Comparative Urbanism," 194–195.
85. Saskia Sassen, "Cityness," in *Urban Transformation*, 84–87 (quotation from 84).
86. These ideas are borrowed from Ruby and Ruby, "Forward," 11–12. See also Stephen Graham, *Vertical: The City from Satellites to Bunkers* (New York: Verso, 2016).
87. See Murray, *The Urbanism of Exception*, 1–20.
88. Ruby and Ruby, "Forward," 12.
89. See Edgar Pieterse, "African Cities: Grasping the Unknowable: Coming to Grips with African Urbanisms," *Social Dynamics* 37, no. 1 (2011): 5–23.
90. Boanventura de Sousa Santos, *Toward a New Common Sense: Law, Science and Politics in the Paradigmatic Transition* (New York: Routledge, 1995), 7–11.
91. See Lisa Weinstein, *The Durable Slum: Dharavi and the Right to Stay Put in Globalizing Mumbai* (Minneapolis: University of Minnesota Press, 2014): 20–23.
92. These ideas are borrowed from Brodwyn Fischer, "Introduction," in *Cities from Scratch: Poverty and Informality in Urban Latin America*, ed. Brodwyn Fischer, Bryan McCann, and Javier Auyero (Durham, NC: Duke University Press, 2014), 1–8.

INTRODUCTION: RETHINKING GLOBAL URBANISM AT THE START OF THE TWENTY-FIRST CENTURY

1. For a clear assertion of this claim, see Michael Dear and Steven Flusty, "Postmodern Urbanism," *Annals of the Association of American Geographers* 88, no. 1 (1998): 50–72.
2. Brenda Yeoh, "Global/Globalizing Cities," *Progress in Human Geography* 23, no. 4 (1999): 612, 608. For an example of mainstream views of urban hierarchies, see Allen Scott, "Human Capital Resources and Requirements Across the Metropolitan Hierarchy of the United States," *Journal of Economic Geography* 9 (2009): 207–226.
3. See Martin J. Murray, "Ruination and Rejuvenation: Rethinking Growth and Decline Through an Inverted Telescope," *International Journal of Urban and Regional Research* 45, no. 2 (2021): 348–362.
4. Gavin Shatkin, "'Fourth World' Cities in the Global Economy: The Case of Phnom Penh, Cambodia," *International Journal of Urban and Regional Research* 22, no. 3 (1998): 378–393.
5. See, for example, Max Rousseau, "Re-imaging the City Centre for the Middle Classes: Regeneration, Gentrification and Symbolic Policies in 'Loser Cities,'" *International Journal of Urban and Regional Research* 33, no. 3 (2009): 772.
6. Ash Amin and Nigel Thrift, *Cities: Rethinking Urban Theory* (Cambridge: Polity, 2002); Jennifer Robinson, *Ordinary Cities: Between Modernity and Development* (London: Routledge, 2006); and Ash Amin, "The Urban Condition: A Challenge to Social Science," *Public Culture* 25, no. 2 (2013): 201–208. For another angle, see Robert Beauregard, "The Radical Break in Late Twentieth-Century Urbanization," *Area* 38, no. 2 (2006): 218–220.
7. See Richard Marshall, *Emerging Urbanity: Global Urban Projects in the Asia Pacific Rim* (London: Spon, 2003), 200–203.

8. See Shiuh-Shen Chien and Max Woodworth, "China's Urban Speed Machine: The Politics of Speed and Time in a Period of Rapid Urban Growth," *International Journal of Urban and Regional Research* 42, no. 4 (2018): 723–737.
9. See Neil Brenner, "Debating Planetary Urbanization: For an Engaged Pluralism," *Environment & Planning D* 36, no. 3 (2018): 575.
10. Tim Edensor and Mark Jayne, "Introduction: Urban Theory Beyond the West," in *Urban Theory Beyond the West: A World of Cities*, ed. Tim Edensor and Mark Jayne (New York: Routledge, 2012), 1–27; and Aiwha Ong, "Worlding of Cities, or the Art of Being Global," in *Worlding Cities: Asian Experiments and the Art of Being Global*, ed. Ananya Roy and Aihwa Ong (Malden, MA: Wiley-Blackwell, 2011), 1–26.
11. See Neil Brenner, "Theses on Urbanism," *Public Culture* 25, no. 1 (2013): 91; Edward Soja and Miguel Kanai, "The Urbanization of the World," in *The Endless City*, ed. Ricky Burdett and Deyan Sudjic (London: Phaidon, 2007), 54–68; and Neil Brenner and Christian Schmid, "Towards a New Epistemology of the Urban?" *City* 19, no. 2–3 (2015): 151–182.
12. See Martin J. Murray, *The Urbanism of Exception: The Dynamics of Global City Building in the Twenty-First Century* (New York: Cambridge University Press, 2017), 34–40, 49–54, 101–105.
13. See Deniz Ay, "Review of Cities of the Global South Reader," *Journal of Planning Education and Research* 38, no. 3 (2018): 380.
14. Jennifer Robinson, "Postcolonialising Geography: Tactics and Pitfalls," *Singapore Journal of Tropical Geography* 24, no. 3 (2003): 275.
15. These ideas are derived from Jennifer Robinson, "The Urban Now: Theorizing Cities Beyond the New," *European Journal of Cultural Studies* 16, no. 6 (2013): 660.
16. Murray, *The Urbanism of Exception*, 4–5.
17. See Josef Gugler, "World Cities in Poor Countries: Conclusions from Case Studies of the Principal Regional and Global Players," *International Journal of Urban and Regional Research* 27, no. 3 (2003): 707–712.
18. Jennifer Robinson, "World Cities, or a World of Cities," *Progress in Human Geography* 29, no. 6 (2005): 757–765.
19. Neil Brenner and Christian Schmid, "The 'Urban Age' in Question," *International Journal of Urban and Regional Research* 38, no. 3 (2014): 751.
20. Ash Amin and Stephen Graham, "The Ordinary City," *Transactions of the Institute of British Geographers* 22, no. 4 (1997): 416–417.
21. Jennifer Robinson, "Global and World Cities: A View from Off the Map," *International Journal of Urban and Regional Research* 26, no. 3 (2002): 531–554.
22. See Robinson, *Ordinary Cities*, 1–2, 15.
23. Saskia Sassen, "Reading the City in a Global Digital Age: Between Topographical Representation and Spatialized Power Projects," in *Future City*, ed. Stephen Read, Jürgen Rosemann, and Job van Eldijk (London: Spon, 2005), 148.
24. Rousseau, "Re-imaging the City Centre for the Middle Classes," 770–788. For "forgotten cities," see Lorene Hoyt and André Leroux, *Voices from Forgotten Cities: Innovative Revitalization Coalitions in America's Older Small Cities* (PolicyLink, Citizens' Housing and Planning Association, and MIT School of Architecture and Planning, 2007), 11–18.
25. Robinson, *Ordinary Cities*, 2.
26. Robinson, "Global and World Cities," 532.
27. Saskia Sassen, *Cities in a World Economy*, 4th ed. (Thousand Oaks, CA: Pine Forge, 2011).
28. See Ravi Sundaram, "Revisiting the Pirate Kingdom," *Third Text* 23, no. 3 (2009): 335–345.
29. See David Harvey, "From Managerialism to Entrepreneurialism: The Transformation in Urban Governance in Late Capitalism," *Geografiska Annaler: Series B, Human Geography* 71, no. 1 (1989): 3–17;

and Aihwa Ong, *Neoliberalism as Exception: Mutations in Citizenship and Sovereignty* (Durham, NC: Duke University Press, 2006).

30. Saskia Sassen, *Globalization and Its Discontents: Essays on the New Mobility of People and Money* (New York: New Press, 1999).
31. Peter Taylor, Ben Derudder, Pieter Saey, and Frank Witlox, "Introduction: Cities in Globalization," in *Cities in Globalization: Practices, Policies, and Theories*, ed. Peter Taylor, Ben Derudder, Pieter Saey, and Frank Witlox (New York: Routledge, 2007), 13–18; and David Etherington and Martin Jones, "City-Regions: New Geographies of Uneven Development and Inequality," *Regional Studies* 43, no. 2 (2009): 247–265.
32. John Friedmann, "The World-City Hypothesis," *Development and Change* 17, no. 1 (1986): 69–83; John Friedmann, "Where We Stand: A Decade of World City Research," in *World Cities in a World System*, ed. Paul Knox and Peter Taylor (Cambridge: Cambridge University Press, 1995), 21–47; and John Friedmann and Goetz Wolff, "World City Formation: An Agenda for Research and Action," *International Journal of Urban and Regional Research* 6, no. 3 (2002): 309–344.
33. Saskia Sassen, *The Global City: New York, London, Tokyo*, 2nd ed. (Princeton, NJ: Princeton University Press, 2001).
34. See, inter alia, Peter Taylor, "Specification of the World City Network," *Geographical Analysis* 33, no. 2 (2001):181–194; Peter Taylor, *World City Network: A Global Urban Analysis* (New York: Routledge, 2004); and Peter Taylor, "Leading World Cities: Empirical Evaluations of Urban Nodes in Multiple Networks," *Urban Studies* 42, no. 9 (2005): 1593–1608.
35. Jonathan Beaverstock, Richard Smith, and Peter Taylor, "A Roster of World Cities," *Cities* 16, no. 6 (1999): 445–458; and Jonathan Beaverstock, Richard Smith, and Peter Taylor, "World-City Network: A New Metageography?," *Annals of the Association of American Geographers* 90, no. 1 (2004): 123–134.
36. Tim Bunnell and Anant Marinanti, "Practising Urban and Regional Research Beyond Metrocentricity," *International Journal of Urban and Regional Research* 34, no. 2 (2010): 415–420.
37. Eugene McCann, "Urban Political Economy Beyond the 'Global City,' " *Urban Studies* 41, no. 12 (2004): 2315–2334; and Eugene McCann, "The Urban as an Object of Study in Global Cities Literatures: Representational Practices and Conceptions of Place and Scale," in G*eographies of Power: Placing Scale*, ed. Andrew Herod and Melissa Wright (Cambridge, MA: Blackwell, 2002), 61–84.
38. Robinson, "World Cities, or a World of Cities," 757; and Yeoh, "Global/Globalizing Cities."
39. Ash Amin and Nigel Thrift, *Seeing Like a City* (Cambridge: Polity, 2016).
40. This is the title of a 2011 Social Science Research Council (SSRC) Dissertation Proposal Development Workshop organized by Ananya Roy and Helga Leitner. See also Helga Leitner and Eric Sheppard, "Provincializing Critical Urban Theory," *International Journal of Urban and Regional Research* 40, no. 1 (2016): 228–235.
41. Edensor and Jayne, "Introduction: Urban Theory Beyond the West."
42. Amin and Graham, "The Ordinary City," 14–29; Robinson, *Ordinary Cities*; Jennifer Robinson, "Developing Ordinary Cities: City Visioning Processes in Durban and Johannesburg," *Environment and Planning A* 40, no. 1 (2008): 74–87; and Jennifer Robinson, "Cities in a World of Cities: The Comparative Gesture," *International Journal of Urban and Regional Research* 35, no. 1 (2011): 1–23.
43. Jennifer Wolch, "Radical Openness as Method in Urban Geography," *Urban Geography* 24, no. 8 (2003): 645–646.
44. Ananya Roy, "The 21st-Century Metropolis: New Geographies of Theory," *Regional Studies* 43, no. 6 (2009): 819–830.
45. Robinson, "Global and World Cities," 532.
46. Robinson, "World Cities, or a World of Cities," 757. See, for example, Richard Grant and Jan Nijman, "Globalization and the Corporate Geography of Cities in the Less-Developed World," *Annals of*

the Association of American Geographers 92, no. 2 (2002): 320–340; and Colin McFarlane, "Urban Shadows: Materiality, the 'Southern City' and Urban Theory," *Geography Compass* 2, no. 2 (2008): 340–358.

47. See, inter alia, Saskia Sassen, "Introduction: Deciphering the Global," in *Deciphering the Global, Its Scales, Spaces, and Subjects*, ed. Saskia Sassen (New York: Routledge, 2007), 1–18.
48. For the source of these ideas, see Ivonne Audirac, Emmanuèle Cunningham Sabot, Sylvie Fol, and Sergio Torres Moraes, "Declining Suburbs in Europe and Latin America," *International Journal of Urban and Regional Research* 36, no. 2 (2012): 229.
49. Adrián Aguilar and Peter Ward, "Globalization, Regional Development, and Mega-City Expansion in Latin America: Analyzing Mexico City's Peri-Urban Hinterland," *Cities* 20, no. 1 (2003): 3–21.
50. Stephen Graham and Simon Marvin, *Splintering Urbanism: Networked Infrastructures, Technological Mobilities and the Urban Condition* (New York: Routledge, 2001), 115.
51. Roy Jones and Brian Shaw, "Development, Postmodernism and the Postcolonial City in Southeast Asia," *Australian Development Studies Network: Development Bulletin* 45 (1998): 20.
52. This paragraph is derived from a reworking of Garth Myers and Martin J. Murray, "Introduction: Situating Contemporary Cities in Africa," in *Cities in Contemporary Africa*, ed. Martin J. Murray and Garth Myers (New York: Palgrave Macmillan, 2006), 8–9.
53. Mike Crang, "Grounded Speculations on Theories of the City and the City of Theory," *International Journal of Urban and Regional Research* 25, no. 3 (2001): 665–669.
54. Achille Mbembé and Sarah Nuttall, "Writing the World from an African Metropolis," *Public Culture* 16, no. 3 (2004): 348.
55. Jamie Peck, "Cities Beyond Compare?" *Regional Studies* 49, no. 1 (2015): 165.
56. Roy, "The 21st-Century Metropolis," 820.
57. Patricia Ehrkamp, "Internationalizing Urban Theory: Toward Collaboration," *Urban Geography* 32, no. 8 (2011): 1122–1128; Colin McFarlane, "The Comparative City: Knowledge, Learning, Urbanism," *International Journal of Urban and Regional Research* 34, no. 4 (2010): 725–742; and Robinson, *Ordinary Cities*, 1–12.
58. McFarlane, "Urban Shadows"; Robinson, "Urban Geography: World Cities, or a World of Cities," 757–765; and Roy, "The 21st-Century Metropolis."
59. Ananya Roy, "Slumdog Cities: Rethinking Subaltern Urbanism," *International Journal of Urban and Regional Research* 35, no. 4 (2011): 223–238; Abdoumaliq Simone, *For the City Yet to Come: Urban Life in Four African Cities* (Durham, NC: Duke University Press, 2006); Filip de Boeck and Marie-Françoise Plissard, *Kinshasa: Tales of the Invisible City* (Antwerp: Ludion, 2006); Ryan Bishop, John Phillips, and Wei Wei Yeo, eds., *Postcolonial Urbanism: Southeast Asian Cities and Global Processes* (New York: Routledge, 2003); and Brenda Yeoh, "Postcolonial Cities," *Progress in Human Geography* 25, no. 3 (2001): 456–468.
60. Vanessa Watson, "Seeing from the South: Refocusing Urban Planning on the Globe's Central Urban Issues," *Urban Studies* 46, no. 11 (2009): 2259–2275; and Edgar Pieterse, *City Futures: Confronting the Crisis of Urban Development* (London: Zed, 2008).
61. Ricky Burdett and Philipp Rode, "Living in the Urban Age," in *Living in the Endless City*, ed. Ricky Burdett and Deyan Sudjic (London: Phaidon, 2011), 8–43. For a more nuanced view, see Brenner and Schmid, "The 'Urban Age' in Question"; and Andy Merrifield, "The Urban Question Under Planetary Urbanization," *International Journal of Urban and Regional Research* 37, no. 3 (2013): 909–922.
62. Juan Miguel Kanai, "On the Peripheries of Planetary Urbanization: Globalizing Manaus and Its Expanding Impact," *Environment and Planning D: Society and Space* 32, no. 6 (2014): 1071–1087.
63. Chien and Woodworth, "China's Urban Speed Machine." Following the qualifications that Oren Yiftachel has adopted, my use of binary categories is aimed at sharpening arguments rather than at

describing any objective reality. Needless to say, there are no clear-cut analytic distinctions between such geographic characterizations as North and South, West and East. These categories are best understood as permeable zones in a fluid conceptual grid that attempts to draw attention to the main sources of power and difference within complex and dynamic social worlds. See Oren Yiftachel, "Essay: Re-engaging Planning Theory? Towards 'South-Eastern' Perspectives," *Planning Theory* 2006 5, no. 3 (2006): 212.

64. See Robinson, "Global and World Cities: A View from off the Map."
65. See AbdouMaliq Simone, "(Non)Urban Humans: Questions for a Research Agenda (the Work the Urban *Could* Do)," *International Journal of Urban and Regional Research* 44, no. 4 (2020): 756.
66. See, for example, Ananya Roy, "Why India Cannot Plan Its Cities: Informality, Insurgence and the Idiom of Urbanization," *Planning Theory* 8, no. 1 (2009): 76–87.
67. Carl Grodach and Anastasia Loukaitou-Sideris, "Cultural Development Strategies and Urban Revitalization: A Survey of US Cities," *International Journal of Cultural Policy* 13, no. 4 (2007): 349–370; and Malcolm Miles, "Interruptions: Testing the Rhetoric of Culturally Led Urban Development," *Urban Studies* 42, no. 5–6 (2005): 889–911.
68. Allen J. Scott, "Emerging Cities of the Third Wave," *City* 15, no. 3–4 (2011): 289–321; Allen J. Scott, *Social Economy of the Metropolis: Cognitive–Cultural Capitalism and the Global Resurgence of Cities* (Oxford: Oxford University Press, 2008); and Yann Moulier Boutang, *Cognitive Capitalism* (Cambridge: Polity, 2012).
69. H. V. Savitch and Paul Kantor, *Cities in the International Marketplace: The Political Economy of Urban Development in North America and Western Europe* (Princeton, NJ: Princeton University Press, 2002), 1–28; Nan Ellen, *Postmodern Urbanism*, rev. ed. (New York: Princeton Architectural Press, 1999), 154–204; Scott, "Emerging Cities of the Third Wave"; and Scott, *Social Economy of the Metropolis*, 1–18.
70. Savitch and Kantor, *Cities in the International Marketplace*, 1–28; Ellen, *Postmodern Urbanism*, 154–204; Scott, "Emerging Cities of the Third Wave," 289–321; and Scott, *Social Economy of the Metropolis*, 1–18.
71. Monica Degen, "Fighting for the Global Catwalk: Formalizing Public Life in Castlefield (Manchester) and Diluting Public Life in el Raval (Barcelona)," *International Journal of urban and Regional Research* 27, no. 4 (2003): 867–880.
72. Anna Klingmann, *Brandscapes: Architecture in the Experience Economy* (Cambridge, MA: MIT Press, 2007), 240.
73. Anne Moudon, "Proof of Goodness: A Substantive Basis for New Urbanism," *Places* 13, no. 2 (2000): 38–43; and John Friedman, "The Good City: Defense of Utopian Thinking," *International Journal of Urban and Regional Research* 24, no. 2 (2000): 460–472.
74. Katrin Großmann, Marco Bontje, Annegret Haase, and Vlad Mykhnenko, "Shrinking Cities: Notes for the Further Research Agenda," *Cities* 35 (2013): 221–225; and Annegret Haase, Dieter Rink, Katrin Grossmann, Matthias Bernt, and Vlad Mykhnenko, "Conceptualizing Urban Shrinkage," *Environment and Planning A* 46, no. 7 (2014): 1519–1534.
75. Robert Beauregard, "Representing Urban Decline: Postwar Cities as Narrative Objects," *Urban Affairs Review* 29, no. 2 (1993): 187–202; and Robert Beauregard, *Voices of Decline: The Postwar Fate of US Cities* (New York: Routledge, 2003).
76. Ivan Turok and Vlad Mykhnenko, "The Trajectories of European Cities, 1960–2005," *Cities* 24, no. 3 (2007): 165–182.
77. Rousseau, "Re-imaging the City Centre for the Middle Classes."
78. As a heuristic device, Max Rosseau prefers the term "loser cities" because it adds a social and cultural dimension—a symbolic element—to the simple economic and/or political characterization of terms such as "declining cities" or even just "(post)industrial cities." See Rousseau, "Re-imaging the City Centre for the Middle Classes."

79. Roy, "The 21st-Century Metropolis"; Edgar Pieterse, *City Futures: Confronting the Crisis of Urban Development* (London: Zed, 2008); and Mike Davis, *Planet of Slums* (New York: Verso, 2006).
80. Davis, *Planet of Slums.*
81. AbdouMaliq Simone and Vyjayanthi Rao, "Securing the Majority: Living Through Uncertainty in Jakarta," *International Journal of Urban and Regional Research* 36, no. 2 (2012): 315–335; AbdouMaliq Simone, "Cities of Uncertainty: Jakarta, the Urban Majority, and Inventive Political Technologies," *Theory, Culture and Society* 30 (2013): 243–263; and Carole Rakodi, "Order and Disorder in African Cities: Governance, Politics, and Land Development Processes," in *Under Siege: Four African Cities: Freetown, Johannesburg, Kinshasa, Lagos*, ed. Okwui Enwezor et al. (Ostfildern-Ruit, Germany: Hatje Cantz, 2002), 45–80.
82. Dirk Kruijt and Kees Koonings, "The Rise of Megacities and the Urbanization of Informality, Exclusion and Violence," in *Megacities: The Politics of Urban Exclusion and Violence in the Global South*, ed. Kees Koonings and Dirk Kruijt (London: Zed, 2009), 8–26.
83. Austin Zeiderman, *Endangered City: The Politics of Security and Risk in Bogotá* (Durham, NC: Duke University Press, 2016), 200.
84. Davis, *Planet of Slums*, 1–19; and Edgar Pieterse, "Grasping the Unknowable: Coming to Grips with African Urbanisms," in *Rogue Urbanism: Emergent African Cities*, ed. Edgar Pieterse and AbdouMaliq Simone (Cape Town: Jacana, 2013), 19–36.
85. See Teresa Caldeira, "Worlds Set Apart," in *Living in the Endless City*, ed. Ricky Burdett and Deyan Sudjic (London: Phaidon, 2011), 168–175.
86. Martin J. Murray, "Afterword: Re-engaging with Transnational Urbanism," in *Locating Right to the City in the Global South*, Tony Samara, Shenjing He, and Guo Chen (New York: Routledge, 2013), 287. See, inter alia, Darryl D'Monte, "A Matter of People," in *Living in the Endless City*, ed. Ricky Burdett and Deyan Sudjic (London: Phaidon, 2011), 94–101; and Rahul Mehrotra, "Negotiating the Static and Kinetic Cities: The Emergent Urbanism of Mumbai," in *Other Cities, Other Worlds: Urban Imaginaries in a Globalizing Age*, ed. Andreas Huyysen (Durham, NC: Duke University Press, 2008), 205–218.
87. Ayona Datta, "Introduction: Fast Cities in an Urban Age," in *Mega-Urbanization in the Global South: Fast Cities and New Urban Utopias of the Postcolonial State*, ed. Ayona Datta and Abdul Shaban (New York: Routledge, 2017), 8–9.
88. See Keller Easterling, "Zone," in *Urban Transformation*, ed. Ilka Ruby and Andreas Ruby (Berlin: Ruby, 2008), 30–45.
89. Tim Simpson, "Tourist Utopias: Biopolitics and the Genealogy of the Post–World Tourist City," *Current Issues in Tourism* 19, no. 1 (2016): 27–59.
90. Frederic Jameson, *Postmodernism, or, The Cultural Logic of Late Capitalism* (New York: Verso, 1984), 38, 44, 80, 83.
91. See James Scott, "The High-Modernist City: An Experiment and a Critique," in *Seeing Like a State: How Certain Schemes to Improve the Human Condition Have Failed* (New Haven, CT: Yale University Press, 1998), 103–146.
92. Datta, "Introduction: Fast Cities in an Urban Age," 8–9.
93. Fu-chen Lo and Peter Marcotullio, "Globalisation and Urban Transformations in the Asia-Pacific Region: A Review," *Urban Studies* 37, no. 1 (2000): 77–111.
94. See M. Christine Boyer, *The City of Collective Memory: Its Historical Imagery and Architectural Entertainments* (Cambridge, MA: MIT Press, 1996).
95. Tom Angotti, *The New Century of the Metropolis: Urban Enclaves and Orientalism* (New York: Routledge, 2013), 3–25.
96. Marshall, *Emerging Urbanity*, 191.
97. Max Page, *The Creative Destruction of Manhattan, 1900–1940* (Chicago: University of Chicago Press, 1999).

98. Brenner, "Debating Planetary Urbanization," 575.
99. Christian Schmid, "Journeys Through Planetary Urbanization: Decentering Perspectives on the Urban," *Environment and Planning D: Society and Space* 36, no. 3 (2018): 591–610; and Brenner and Schmid, "Towards a New Epistemology of the Urban?"
100. See Murray, *The Urbanism of Exception*, 51–54, 99–102.
101. Jennifer Robinson, "Cities in a World of Cities," 19.
102. Thomas Sigler, "Relational Cities: Doha, Panama City, and Dubai as 21st Century Entrepôts," *Urban Geography* 34, no. 5 (2013): 628.
103. Jane M. Jacobs, "A Geography of Big Things," *Cultural Geography* 13, no. 1 (2006): 1–27.
104. Pedro Gadanho, "Mirroring Uneven Growth: A Speculation on Tomorrow's Cities Today," in *Uneven Growth: Tactical Urbanisms for Expanding Megacities*, ed. Pedro Gadanho (New York: Museum of Modern Art, 2014), 14–25.
105. As an exception, see Stephen Marr, "Worlding and Wilding: Lagos and Detroit as Global Cities," *Race & Class* 57, no. 4 (2016): 3–21.
106. Ravi Sundaram, *Pirate Modernity: Delhi's Media Urbanism* (London: Routledge, 2010), 7–8.
107. Tahl Kaminer, Miguel Robles-Durán, and Heidi Sohn, "Introduction: Urban Asymmetries," in *Urban Asymmetries: Studies and Projects on Neoliberal Urbanization*, ed. Tahl Kaminer, Miguel Robles-Durán, and Heidi Sohn (Rotterdam: 011 Publishers, 2011), 11–20.
108. Edensor and Jayne, "Introduction," in *Urban Theory Beyond the West*, 10.
109. See Sundaram, *Pirate Modernity*, 7–8.

1. THE NARROW PREOCCUPATIONS OF CONVENTIONAL URBAN STUDIES

1. Tim Edensor and Mark Jayne, "Introduction," in *Urban Theory Beyond the West: A World of Cities*, ed. Tim Edensor and Mark Jayne (New York: Routledge, 2012), 1.
2. Eric Sheppard, Helga Leitner, and Anant Maringanti, "Provincializing Global Urbanism: A Manifesto," *Urban Geography* 34, no. 7 (2013): 893–900; Jamie Peck and Nik Theodore, "Mobilizing Policy: Models, Methods, and Mutations," *Geoforum* 41, no. 2 (2010): 169–174; and Jamie Peck, "Economic Rationality Meets Celebrity Urbanology: Exploring Edward Glaeser's City," *International Journal of Urban and Regional Research* 40, no. 1 (2016): 1–30.
3. Jennifer Robinson, *Ordinary Cities: Between Modernity and Development* (New York: Routledge, 2006), 2–3; and Jennifer Robinson, "Global and World Cities: A View from Off the Map," *International Journal of Urban and Regional Research* 26, no. 3 (2002): 531–554.
4. See AbdouMaliq Simone, *City Life from Jakarta to Dakar: Movements at the Crossroads* (New York: Routledge, 2011).
5. Richard Stren, "Local Development and Social Diversity in the Developing World: New Challenges for Globalizing City-Regions," in *Global City-Regions: Trends, Theory, Policy*, ed. Allen J. Scott (Oxford University Press, 2001), 193–213.
6. Ananya Roy, "The 21st-Century Metropolis: New Geographies of Theory," *Regional Studies* 43, no. 6 (2009): 820.
7. Mike Davis, "Planet of Slums: Urban Involution and the Informal Proletariat," *New Left Review* 26 (2004): 5, 23, 27.
8. For an example that received a great deal of publicity at the time, see Robert Kaplan, "The Coming Anarchy: How Scarcity, Crime, Overpopulation, Tribalism, and Disease Are Rapidly Destroying the Social Fabric of Our Planet," *Atlantic Monthly* 273, no. 2 (February 1994): 44–76.
9. Robinson, "Global and World Cities," 531. See also Roy, "The 21st-Century Metropolis," 820.

10. See Slavomíra Ferenćuhová, "Urban Theory Beyond the 'East/West Divide'? Cities and Urban Research in Postsocialist Europe," in *Urban Theory Beyond the West: A World of Cities*, ed. Tim Edensor and Mark Jayne (New York: Routledge, 2012), 65–74; and Tim Edensor and Mark Jayne, "Afterword: A World of Cities," in *Urban Theory Beyond the West*, 329–331.
11. Neil Brenner, "Stereotypes, Archetypes, and Prototypes: Three Uses of Superlatives in Contemporary Urban Studies," *City & Community* 2, no. 3 (2003): 205–216.
12. Walter Benjamin, *Illuminations* (London: Pimlico, 1999).
13. Thomas Bender, *The Unfinished City: New York and the Metropolitan Idea* (New York: New York University Press, 2002).
14. For the classical statement, see Robert Park, Ernest Burgess, Roderick McKenzie, *The City* (Chicago: University of Chicago Press, 1925).
15. Edward W. Soja, "It All Comes Together in Los Angeles," in *Postmodern Geographies: The Reassertion of Space in Critical Social Theory* (New York: Verso, 1989), 190–221; and Allen J. Scott and Edward W. Soja, "Los Angeles: Capital of the Late 20th Century," *Environment and Planning D* 4, no. 3 (1986): 249–254.
16. Some of the ideas developed here are derived from Garth Myers and Martin J. Murray, "Introduction: Situating Contemporary Cities in Africa," in *Cities in Contemporary Africa*, ed. Martin J. Murray and Garth Myers (New York: Palgrave MacMillan, 2006), 10–14. One current in the scholarly literature seems to cavalierly dismiss cities in Africa as representing a kind of "nowhere place" in relation to the "real action" of global cities; see Peter Taylor, "World Cities and Territorial States: The Rise and Fall of Their Mutuality," in *World Cities in a World System*, ed. Peter Taylor and Paul Knox (Cambridge: Cambridge University Press, 1995), 48–62.
17. See Robinson, "Global and World Cities," 531, 540; Howard Dick and Peter Rimmer, "Beyond the Third-World City: The New Urban Geography of Southeast Asia," *Urban Studies* 35, no. 12 (1998): 2303–2321; David Smith, *Third World Cities in Global Perspective: The Political Economy of Uneven Urbanization* (Boulder, CO: Westview, 1996); David Drakakis-Smith, *Third World Cities* (London: Routledge, 2000); and Izak van der Merve, "The Global Cities of Sub-Saharan African: Fact or Fiction?," *Urban Forum* 15, no. 1 (2004): 36–47.
18. Anthony King, *Global Cities: Post-Imperialism and the Internationalization of London* (London: Routledge and Kegan Paul, 1990).
19. See Myers and Murray, "Introduction: Situating Contemporary Cities in Africa," 1–21; and Robinson, "Global and World Cities," 531.
20. Allen J. Scott and Michael Storper, "The Nature of Cities: The Scope and Limits of Urban Theory," *International Journal of Urban and Regional Research* 39, no. 1 (2015): 12.
21. See Jennifer Robinson, "Urban Geography: World Cities, or a World of Cities," *Progress in Human Geography* 29, no. 6 (2005): 757. See also Dick and Rimmer, "Beyond the Third-World City."
22. Jennifer Robinson, " 'Arriving at' Urban Policies/the Urban: Traces of Elsewhere in Making Urban Futures," in *Critical Mobilities*, ed. Ola Söderström et al. (London: Routledge, 2013), 1–28; and Jennifer Robinson, "Cities in a World of Cities: The Comparative Gesture," *International Journal of Urban and Regional Research 35*, no. 1 (2010): 1–23.
23. This list includes Saskia Sassen, Manuel Castells, John Friedmann, and Peter Taylor but is in no way is restricted to them. See, inter alia, Saskia Sassen, *Global Cities: New York, London, Tokyo*, 2nd ed. (Princeton: Princeton University Press, 2001); Saskia Sassen, "New Frontiers Facing Urban Sociology at the Millennium," *British Journal of Sociology* 51, no. 1 (2000): 143–159; Saskia Sassen, ed., *Cities and Their Cross-Border Networks* (New York: Routledge, 2000); Manuel Castells, *The Informational City* (London: Blackwell, 1989); Manuel Castells, *The Rise of the Network Society* (London: Blackwell, 1996); John Friedmann, "The World-City Hypothesis," *Development and Change* 17, no. 1 (1986): 69–83; and Peter Taylor, "World Cities and Territorial States Under Conditions of Contemporary Globalisation," *Political Geography* 19, no. 1 (2000): 5–32.

24. From Myers and Murray, "Introduction: Situating Cities in Africa," 9. In a similar vein, numerous scholars have tended toward getting the cities they study recognized as bona fide global cities. As Jennifer Robinson has put it, "Global City as a concept becomes a regulating fiction. It offers an authorized image of city success (so people can buy into it) which also establishes an end point of development for ambitious cities" (Robinson, "Global and World Cities," 546).
25. Saskia Sassen, "The Urban Complex in the World Economy," *International Social Science Journal* 139 (1994): 42–63; and Saskia Sassen, *Cities in a World Economy* (London: Pine Forge, 1994).
26. Janet Abu-Lughod, *New York, Los Angeles, Chicago: America's Global Cities* (Minneapolis: University of Minnesota Press, 1999).
27. Andrew Jones, "The 'Global City' Misconceived: The Myth of 'Global Management' in Transnational Service Firms," *Geoforum* 33, no. 3 (2002): 335–350; Chris Hamnett, "Social Polarisation in Global Cities: Theory and Evidence," *Urban Studies* 31, no. 3 (1994): 410–424; James White, "Old Wine, Cracked Bottle?: Tokyo, Paris, and the Global City Hypothesis," *Urban Affairs Review* 33, no. 4 (1998): 451–477; Ben Derudder, "The Mismatch Between Concepts and Evidence in the Study of a Global Urban Network," in *Cities in Globalization: Practices, Policies, and Theories*, ed. Peter Taylor, Ben Derudder, Pieter Saey, and Frank Witlox (New York: Routledge, 2007), 261–275; and Richard G. Smith, "Beyond the Global City Concept and the Myth of 'Command and Control,'" *International Journal of Urban and Regional Research* 38, no. 1 (2014): 98–115.
28. Robinson, "Global and World Cities"; and Robinson, *Ordinary Cities*, 1–7.
29. For the source of these ideas, see Rob Shields, "The Urban Question as Cargo Cult: Opportunities for a New Urban Pedagogy," *International Journal of Urban and Regional Research* 32, no. 3 (2008): 712.
30. Shields, "The Urban Question as Cargo Cult," 712, 714.
31. Shields, "The Urban Question as Cargo Cult," 712, 714, 717.
32. Myers and Murray, "Introduction: Situating Cities in Africa," 9. See also Jack Burgers and Sako Musterd, "Understanding Urban Inequality: A Model Based on Existing Theories and an Empirical Illustration," *International Journal of Urban and Regional Research* 26, no. 2 (2002): 403.
33. See, for example, David Smith and Michael Timberlake, "Hierarchies of Dominance Among World Cities: A Network Approach," in *Global Networks, Linked Cities*, ed. Saskia Sassen (New York: Routledge, 2002), 117–141; and Arthur Alderson and Jason Beckfield, "Power and Position in the World City System," *American Journal of Sociology* 109 (2004): 811–851.
34. Carl Nordlund, "A Critical Comment on the Taylor Approach for Measuring World City Interlock Linkages," *Geographical Analysis* 36, no. 3 (2004): 290–296; Zackery Neal, "Structural Determinism in the Interlocking World City Network," *Geographical Analysis* 44, no. 2 (2012): 162–170; and Xingjian Liu and Ben Derudder, "Two-Mode Networks and the Interlocking World City Network Model: A Reply to Neal," *Geographical Analysis* 44, no. 2 (2012): 171–173.
35. Shields, "The Urban Question as Cargo Cult, 713.
36. Robinson, *Ordinary Cities*, 35–42. See Michael Hoyler and John Harrison, "Global Cities Research and Urban Theory Making," *Environment and Planning A* 49, no. 12 (2017): 2853–2858.
37. Robinson, "Global and World Cities," 536, 546.
38. Austin Zeiderman, "Beyond the Enclave of Urban Theory?" *International Journal of Urban and Regional Research* (2018): 1114.
39. Zeiderman, "Beyond the Enclave of Urban Theory?," 1114. See Ananya Roy, "Worlding the South: Towards a Post-Colonial Urban Theory," in *The Routledge Handbook on Cities of the Global South*, ed. Sue Parnell and Sophie Oldfield (New York: Routledge, 2014), 9–20; Martin J. Murray, *The Urbanism of Exception: Global City Making in the 21st Century* (New York: Cambridge University Press, 2017); and Neil Brenner, ed., *Implosions/Explosions: Towards a Study of Planetary Urbanization* (Berlin: Jovis, 2014).

40. Zeiderman, "Beyond the Enclave of Urban Theory?," 1115.
41. See, inter alia, Jennifer Robinson and Ananya Roy, "Global Urbanisms and the Nature of Urban Theory," *International Journal of Urban and Regional Research* 40, no. 1 (2016): 181–186; Ananya Roy, "Who's Afraid of Postcolonial Theory?," *International Journal of Urban and Regional Research* 40, no. 1 (2016): 200–209; and Ananya Roy, "What Is Urban About Critical Urban Theory?," *Urban Geography* 37, no. 6 (2016): 810–823.
42. Jennifer Robinson, "In the Tracks of Comparative Urbanism: Difference, Urban Modernity and the Primitive," *Urban Geography* 25, no. 8 (2004): 709.
43. Scott and Storper, "The Nature of Cities," 10.
44. See Scott and Storper, "The Nature of Cities," 1.
45. Michael Storper and Allen J. Scott, "Current Debates in Urban Theory: A Critical Assessment," *Urban Studies* 53, no. 16 (2016): 1115.
46. Scott and Storper, "The Nature of Cities," 12.
47. Scott and Storper, "The Nature of Cities," 1, 11.
48. Scott and Storper, "The Nature of Cities," 4.
49. Scott and Storper, "The Nature of Cities," 1.
50. Scott and Storper, "The Nature of Cities," 1, 12.
51. Wilhem Windelband, "History and Natural Science," *Theory and Psychology* 8, no. 1 (1894/1998): 5–22.
52. For a reinterpretation of this debate, see Derek Gregory, "Areal Differentiation and Post-Modern Human Geography," in *Horizons in Human Geography*, ed. Derek Gregory and Rex Walford (New York: Palgrave, 1989), 67–96.
53. Storper and Scott, "Current Debates in Urban Theory," 1132.
54. Storper and Scott, "Current Debates in Urban Theory," 1132.
55. See Richard Walker, "Why Cities? A Response," *International Journal of Urban and Regional Research* 40, no. 1 (2016): 176.
56. Jennifer Robinson and Ananya Roy, "Debate on Global Urbanisms and the Nature of Urban Theory," *International Journal of Urban and Regional Research* 40, no. 1 (2016): 181. For a critique of universalizing impulses in theory building, see Mary Lawhon, Jonathan Silver, Henrik Ernstson, and Joseph Pierce, "Unlearning (Un)Located Ideas in the Provincialization of Urban Theory," *Regional Studies* 50, no. 9 (2016): 1611–1622.
57. Robinson, "In the Tracks of Comparative Urbanism," 720.
58. Walker, "Why Cities?" 176, 164.
59. Matthias Bernt, "The Limits of Shrinkage: Conceptual Pitfalls and Alternatives in the Discussion of Urban Population Loss," *International Journal of Urban and Regional Research* 40, no. 2 (2016): 449.
60. Oli Mould, "A Limitless Urban Theory? A Response to Scott and Storper's 'The Nature of Cities: The Scope and Limits of Urban Theory,' " *International Journal of Urban and Regional Research* 39, no. 1 (2015): 157, 161, 162.
61. Roy, "Conclusion: Postcolonial Urbanism: Speed, Mass Hysteria, and Dreams," in *Worlding Cities: Asian Experiments and the Art of Being Global*, ed. Ananya Roy and Aihwa Ong (Malden, MA: Wiley-Blackwell, 2011), 307.
62. Robert Beauregard, "Radical Uniqueness and the Flight from Urban Theory," in *The City Revisited: Urban Theory from Chicago, Los Angeles, New York*, ed. Dennis Judd and Dick Simpson (Minneapolis: University of Minnesota Press, 2011), 186.
63. Beauregard, "Radical Uniqueness," 186.
64. See Larry Bourne, "On Schools of Thought, Comparative Research, and Inclusiveness: A Commentary," *Urban Geography* 29, no. 2 (2008): 177–186; and Robert Beauregard, "City of Superlatives," *City & Community* 2, no. 3 (2003): 183–199.

65. Juan Obarrio, "Symposium Theory from the South," *Johannesburg Salon 5* (2012): 5; Arjun Appadurai, *Modernity at Large: Cultural Dimensions of Globalization* (Minneapolis: University of Minnesota Press, 1996); and Jean and John Comaroff, *Theory from the South, or How Euro-America Is Evolving Toward Africa* (New York: Paradigm, 2012).
66. See Helga Leiter and Eric Sheppard, "Provincializing Critical Urban Theory: Extending the Ecosystem of Possibilities," *International Journal of Urban and Regional Research* 39, no. 1 (2015): 228, 231.
67. Neil Brenner and Christian Schmid, "The 'Urban Age' in Question," *International Journal of Urban and Regional Research* 38, no. 3 (2014): 751. For an important intervention, see Roger Keil, "The Empty Shell of the Planetary: Re-rooting the Urban in the Experience of the Urbanites," *Urban Geography* 39, no. 10 (2018): 1589–1602.
68. Martin J. Murray, *The Urbanism of Exception: The Dynamics of Global City Building in the Twenty-first Century* (New York: Cambridge University Press, 2017), 106, 108, 109.
69. Andy Merrifield, "Planetary Urbanisation: *Une affaire de perception*," *Urban Geography* 39, no. 10 (2018): 1605.
70. Hillary Angelo and David Wachsmuth, "Urbanizing Urban Political Ecology: A Critique of Methodological Cityism," *International Journal of Urban and Regional Research* 39, no. 1 (2015): 20; Creighton Connolly, "Urban Political Ecology Beyond Methodological Cityism," *International Journal of Urban and Regional Research* 39, no. 1 (2015): 63–75; and Richardo Martinez, Tim Bunnell, and Michelle Acuto, "Productive Tensions? The 'City' Across Geographies of Planetary Urbanization and the Urban Age," *Urban Geography*, October 20, 2020, https://doi.org/10.1080/02723638.2020.1835128(2020).
71. Murray, *The Urbanism of Exception*, 106.
72. Linda Peake, Darren Patrick, Rajyashree Reddy, Gökbörü Sarp Tanyildiz, Sue Ruddick, and Roza Tchoukaleyska, "Placing Planetary Urbanization in Other Fields of Vision," *Environment and Planning D: Society and Space* 36, no. 3 (2018): 381.
73. Natalie Oswin, "Planetary Urbanization: A View from Outside," *Environment and Planning D: Society and Space* 36, no. 3 (2018): 542; Heather McLean, "In Praise of Chaotic Research Pathways: A Feminist Response to Planetary Urbanization," *Environment and Planning D: Society and Space* 36, no. 3 (2018): 547–555; Linda Peake, "On Feminism and Feminist Allies in Urban Geography," *Urban Geography* 37, no. 6 (2016): 830–838; Linda Peake, "The Twenty-First Century Quest for Feminism and the Global Urban," *International Journal of Urban and Regional Research* 40, no. 1 (2016): 219–227; and Rajyashree Reddy, "The Urban Under Erasure: Towards a Postcolonial Critique of Planetary Urbanization," *Environment and Planning D: Society and Space* 36, no. 3 (2018): 529–539.
74. Neil Brenner, "Debating Planetary Urbanization: For an Engaged Pluralism," *Environment and Planning D: Society and Space* 36, no. 3 (2018): 571.
75. Hillary Angelo and Kian Goh, "Out in Space: Difference and Abstraction in Planetary Urbanization," *International Journal of Urban and Regional Research*, 45, 4 (2021), 733–734, 738, 734. On the matter of "difference," see Kanishka Goonewardena, "Planetary Urbanization and Totality," *Environment and Planning D: Society and Space* 36, no. 3 (2018): 459–460.
76. See Angelo and Goh, "Out in Space," 744; Kate Derickson, "Masters of the Universe," *Environment and Planning D: Society and Space* 36, no. 3 (2018): 356–362; and Michelle Buckley and Kendra Strauss, "With, Against and Beyond Lefebvre: Planetary Urbanization and Epistemic Plurality," *Environment and Planning D: Society and Space* 34, no. 4 (2016): 617–636.
77. Sue Ruddick, Linda Peake, Gökbörü Tanyildiz, and Darren Patrick, "Planetary Urbanization: An Urban Theory for Our Time?," *Environment and Planning D: Society and Space* 36, no. 3 (2018): 387–404.

78. Goonewardena, "Planetary Urbanization and Totality."
79. Brenner, "Debating Planetary Urbanization," 575, 578.
80. Brenner, "Debating Planetary Urbanization," 576.
81. Christian Schmid, "Journeys Through Planetary Urbanization: Decentering Perspectives on the Urban," *Environment and Planning D: Society and Space* 36, no. 3 (2018): 591.
82. Brenner, "Debating Planetary Urbanization," 573, 575, 581; and Neil Brenner and Christian Schmid, "Towards a New Epistemology of the Urban," *City* 19, no. 2–3 (2015): 151–182.
83. See Jan Nijman, "Comparative Urbanism—An Introduction," *Urban Geography* 28, no. 1 (2007): 1–6.
84. Robinson and Roy, "Debate on Global Urbanisms," 182.
85. Robinson, "Global and World Cities," 548–549.
86. Nijman, "Comparative Urbanism—An Introduction."
87. Dennis Rodgers, "An Illness Called Managua: 'Extraordinary' Urbanization and 'Mal-development' in Nicaragua," in *Urban Theory Beyond the West*, ed. Tim Edensor and Mark Jayne (New York, Routledge, 2012), 134,.
88. Nijman, "Comparative Urbanism—An Introduction."
89. Robinson and Roy, "Debate on Global Urbanisms," 183.
90. See Jane M. Jacobs, "Commentary: Comparing Comparative Urbanism," *Urban Geography* 33, no. 6 (2012): 904–914.
91. Beauregard, "Radical Uniqueness," 196.
92. Beauregard, "Radical Uniqueness," 196.
93. James Fraser, "Globalization, Development and Ordinary Cities: A Review Essay," *Journal of World-Systems Research* 12, no. 1 (2006): 189.
94. Doreen Massey, *For Space* (Thousand Oaks, CA: Sage, 2005), 87.
95. Fraser, "Globalization, Development and Ordinary Cities," 190.
96. Massey, *For Space*, 87.
97. Fraser, "Globalization, Development and Ordinary Cities," 190.
98. For debate on this theme, see Volker Schmidt, "Multiple Modernities or Varieties of Modernity?," *Current Sociology* 54, no. 1 (2006): 77–97.
99. Bernt, "The Limits of Shrinkage," 447.
100. Jennifer Robinson, "The Urban Now: Theorising Cities Beyond the New," *European Journal of Cultural Studies* 16, no. 6 (2013): 659–677; and Jennifer Robinson, "Thinking Cities Through Elsewhere: Comparative Tactics for a More Global Urban Studies," *Progress in Human Geography* 40, no. 1 (2015): 3–29.
101. Ash Amin, "The Good City," *Urban Studies* 43, no. 5–6 (2006): 1009–1023.
102. Robinson, "Cities in a World of Cities"; and Jennifer Robinson, "Comparisons: Cosmopolitan or Colonial?," *Singapore Journal of Tropical Geography* 32, no. 2 (2011): 125–140.
103. Robinson, " 'Arriving at' Urban Policies/the Urban."
104. Robinson, *Ordinary Cities*, 141–143, 154–155; and Jennifer Robinson, "Living in Dystopia: Past, Present and Future," in *Noir Urbanisms*, ed. Gyan Prakash (Princeton, NJ: Princeton University Press, 2010), 218–240.
105. This and the following paragraphs are adopted from Myers and Murray, "Introduction: Situating Contemporary Cities in Africa," 3.
106. See Robinson, "Global and World Cities"; and Robinson, *Ordinary Cities*, 2–3.
107. See Colin McFarlane, "Urban Shadows: Materiality, the 'Southern City' and Urban Theory," *Geography Compass* 2, no. 2 (2008): 341; Murray, *The Urbanism of Exception*', 1–45, 102–103.
108. Stephen Read, "The Form of the Future," in *Future City*, ed. Stephen Read, Jürgen Rosemann, and Job van Eldijk (London: Spon, 2005), 3–17. For an argument that does not erase the nation-state, see

Göran Therborn, *Cities of Power: The Urban, the National, the Popular, the Global* (London: Verso, 2017). For an insightful critique, see Owen Hatherley, "Comparing Capitals," *New Left Review* 105 (2017): 107–132.

109. See Stefan Krätke, "Cities in Contemporary Capitalism," *International Journal of Urban and Regional Research* 38, no. 5 (2014): 1661; and Stefan Krätke, "How Manufacturing Industries Connect Cities Across the World: Extending Research on 'Multiple Globalizations,' " *Global Networks* 14, no. 2 (2013): 121–147.
110. See Robinson, *Ordinary Cities*, 1–11.
111. Davis, "Planet of Slums," 5.
112. See, for example, Leonie Pearson, Peter Newton, and Peter Roberts, eds., *Resilient Sustainable Cities: A Future* (New York: Routledge, 2014).
113. See Sean Fox, "Urbanization as a Global Historical Process: Theory and Evidence from Sub-Saharan Africa," *Population and Development Review* 38, no. 2 (2012): 285; and Douglas Gollin, Remi Jedwan, and Dietrich Vollrath, "Urbanization with and Without Industrialization," *Journal of Economic Growth* 21, no. 1 (2016): 35–70.
114. Mike Davis, "The Urbanization of Empire: Megacities and the Laws of Chaos," *Social Text* 22, no. 4 (2004): 9–15.
115. Abidin Kusno, Melani Budianta, and Hilmar Farid, "Editorial Introduction: Runaway City / Leftover Spaces," *Inter-Asia Cultural Studies* 12, no. 4 (2011): 473–481; and Carole Rakodi, "Global Forces, Urban Change, and Urban Management in Africa," in *The Urban Challenge in Africa: Growth and Management of its Largest Cities*, ed. Carole Rakodi (Tokyo: United Nations University Press, 1997), 17–73.
116. See Paul Knox and Sallie Marston, *Human Geography: Places and Regions in Global Context* (Upper Saddle River, NJ: Prentice-Hall, 2000); and Josef Gugler, "Over-Urbanization Reconsidered," in *Cities in the Developing World: Issues, Theory and Policy*, ed. Josef Gugler (Oxford: Oxford University Press, 1997), 114–123. Manchester, Birmingham, Liverpool, and Leeds were the great "shock cities" of the industrial age of the early nineteenth century—a term coined by Platt in his pioneering work on the environmental impact of industrialization. See Harold Platt, *Shock Cities: The Environmental Transformation and Reform of Manchester and Chicago* (Chicago: University of Chicago Press, 2005).
117. See Myers and Murray, "Introduction: Situating Cities in Africa," 5.
118. Richard Grant and Jan Nijman, "Globalization and the Corporate Geography of Cities in the Less-Developed World," *Annals of the Association of American Geographers* 92, no. 2 (2002): 320–340.
119. Michael Watts, "Development II: The Privatization of Everything?," *Progress in Human Geography* 18, no. 3 (1994): 371–384.
120. AbdouMaliq Simone, "Between Ghetto and Globe: Remaking Urban Life in Africa," in *Associational Life in African Cities: Popular Responses to the Urban Crisis*, ed. Arne Tostensen, Inge Tvedten, and Mariken Vaa (Stockholm: Nordiska Afrikainstitutet, 2001), 46.
121. AbdouMaliq Simone, "Straddling the Divides: Remaking Associational Life in the Informal City," *International Journal of Urban and Regional Research* 25, no. 1 (2001): 102–117; and AbdouMaliq Simone, *For the City Yet to Come: Changing African Life in Four Cities* (Durham, NC: Duke University Press, 2004).
122. See Myers and Murray, "Introduction: Situating Cities in Africa," 6.
123. Simone, *For the City Yet to Come*, 1–2. See also AbdouMaliq Simone, "It's Just the City After All," *International Journal of Urban and Regional Research* 40, no. 1 (2016): 210–218.
124. Myers and Murray, "Introduction: Situating Cities in Africa," 7.
125. Simone, *For the City Yet to Come*, 2.
126. Simone, *For the City Yet to Come*, 15–16.

2. THE UNIVERSALIZING PRETENSIONS OF MAINSTREAM URBAN STUDIES: GENERIC CITIES AND THE CONVERGENCE THESIS

1. See Gavin Shatkin, "Global Cities of the South: Emerging Perspectives on Growth and Inequality," *Cities* 24, no. 1 (2007): 1.
2. See Michael Cohen, "The Hypothesis of Urban Convergence: Are Cities in the North and South Becoming More Alike in an Age of Globalization?," in *Preparing for the Urban Future: Global Pressures and Local Forces*, ed. Michael Cohen, Blair Ruble, Joseph Tulchin and Allison Garland (Washington, DC: Woodrow Wilson Center Press, 1996), 25–38.
3. Howard Dick and Peter Rimmer, "Beyond the Third World City: The New Urban Geography of South-East Asia," *Urban Studies* 35, no. 12 (1998): 2303–2321.
4. Loretta Lees, Hyun Bang Shin, and Ernesto Lopez-Morales, *Urban Futures: Planetary Gentrification* (Cambridge: Polity, 2016), 111–139; and Loretta Lees, Tom Slater, and Elvin Wyly, *Gentrification* (New York: Routledge, 2008), 129–162.
5. Cohen, "The Hypothesis of Urban Convergence."
6. For some notable examples of the convergence thesis, see Michael Peter Smith, *Transnational Urbanism: Locating Globalization* (Malden, MA: Blackwell, 2001); Georg Glasze, Chris Webster, and Klaus Franz, eds., *Private Cities: Global and Local Perspectives* (London: Routledge, 2006); Peter Marcuse and Ronald van Kempen, "Introduction," in *Globalizing Cities: A New Spatial Order*? ed. Peter Marcuse and Ronald van Kempen (London: Blackwell, 2000), 1–21; and Allen J. Scott, John Agnew, Edward W. Soja, and Michael Storper, "Global City-Regions," in *Global City-Regions: Trends, Theory, Policy*, ed. Allan J. Scott (Oxford: Oxford University Press, 2002), 11–30.
7. Gavin Shatkin, "The City and the Bottom Line: Urban Megaprojects and the Privatization of Planning in Southeast Asia," *Environment and Planning A* 40, no. 2 (2008): 383. In similar fashion, the reduction of globalization to McDonaldization, neocolonialism, or even cultural imperialism reflects the underlying assumption of convergence.
8. Dick and Rimmer, "Beyond the Third World City," 2319.
9. Peter Rimmer and Howard Dick, *The City in Southeast Asia: Patterns, Processes and Policy* (Singapore: Nation University of Singapore Press, 2009). This book explores ways of moving beyond outmoded paradigms of the Third World city.
10. Dick and Rimmer, "Beyond the Third World City," 2303.
11. Rem Koolhass, "The Generic City," in Rem Koolhaas and Bruce Mau, *S,M,L,XL* (New York: Monacelli, 1995), 1248.
12. Richard Prouty, "Buying Generic: The Generic City in Dubai," *London Consortium Static* 8 (2009): 1–8.
13. Rem Koolhaas, "What ever Happened to Urbanism?" in Koolhaas and Mau, *S,M,L,XL*, 959.
14. Koolhaas, "The Generic City," in Koolhaas and Mau, *S,M,L,XL*, 1239–1264. For a critical exploration of the generic city, see Siddharth Puri, "Specifying the Generic: A Theoretical Unpacking of Rem Koolhaas' 'Generic City' " (PhD diss., University of Cincinnati, 2007).
15. David Ley, "Transnational Spaces and Everyday Lives," *Transactions of the Institute of British Geographers*, n.s., 29, no. 2 (2004): 151–164.
16. Lila Leontidou, "Alternatives to Modernism in (Southern) Urban Theory: Exploring In-Between Spaces," *International Journal of Urban and Regional Research* 30, no. 2 (1996): 180.
17. Neil Smith, "New Globalism, New Urbanism: Gentrification as Global Urban Strategy," *Antipode* 34, no. 3 (2002): 427.
18. Rowland Atkinson and Gary Bridge, "Introduction," in *Gentrification in a Global Context: the New Urban Colonialism*, ed. Rowland Atkinson and Gary Bridge (New York: Routledge, 2005), 1–17.

19. See, inter alia, Peter Marcuse and Ronald van Kempen, "Conclusion: A Changed Spatial Order," in *Globalizing Cities*, ed. Peter Marcuse and Ronald van Kempen (London: Blackwell, 2000), 249–275.
20. Caglar Keyder, "Globalization and Social Exclusion in Istanbul," *International Journal of Urban and Regional Research* 29, no. 1 (2005): 124–134.
21. James Simmie, "Trading Places: Competitive Cities in the Global Economy," *European Planning Studies* 10, no. 2 (2002): 201–214; Simon Anholt, *Competitive Identity: The New Brand Management for Nations, Cities and Regions* (New York: Palgrave Macmillan, 2007); Erik Swyngedouw, Frank Moulaert, and Arantxa Rodriguez, " 'The World in a Grain of Sand': Large-Scale Urban Development Projects and the Dynamics of 'Glocal' Transformations," in *The Globalized City: Economic Restructuring and Social Polarization in European Cities*, ed. Frank Moulaert, Arantxa Rodriguez, and Erik Swyngedouw (Oxford: Oxford University Press, 2005), 9–28; and Sako Musterd and Alan Murie, "Making Cities Competitive: Challenges and Debates," in *Making Competitive Cities*, ed. Sako Musterd and Alan Murie (Oxford: Wiley-Blackwell, 2007), 3–16.
22. Elizabeth Currid, "New York as a Global Creative Hub: A Competitive Analysis of Four Theories on World Cities," *Economic Development Quarterly* 204 (2006): 330–350.
23. Jennifer Robinson, "Global and World Cities: A View from Off the Map," *International Journal of Urban and Regional Research* 26, no. 3 (2002): 548.
24. Amy Ellen Schwartz and Ingrid Gould Ellen, "No Easy Answers: Cautionary Notes for Competitive Cities," *Brookings Review* 18, no. 3 (2000): 44–47; and Iain Begg, " 'Investability': The Key to Competitive Regions and Cities?," *Regional Studies* 36, no. 2 (2002): 187–193.
25. Jamie Peck, "Struggling with the Creative Class," *International Journal of Urban and Regional Research* 24, no. 4 (2005): 740. See also Allen J. Scott, "Cultural Economy and the Creative Field of the City," *Geografiska Annaler: Series B, Human Geography* 92 (2010): 115–130; and Allen J. Scott, "Capitalism and Urbanization in a New Key? The Cognitive-Cultural Dimension," *Social Forces* 85 (2007): 1465–1482.
26. Elizabeth Currid-Halkett and Allen J. Scott, "The Geography of Celebrity and Glamour: Reflections on Economy, Culture, and Desire in the City," *City, Culture and Society* 4 (2013): 3. See also Alberto Vanolo, "The Image of the Creative City: Some Reflections on Urban Branding in Turin," *Cities* 25, no. 6 (2008): 370.
27. Nick Buck, Ian Gordon, Alan Harding, and Ivan Turok, "Conclusion: Moving Beyond the Conventional Wisdom," in *Changing Cities: Rethinking Urban Competitiveness, Cohesion and Governance*, ed. Ian Gordon, Nick Buck, Alan Harding, and Ivan Turok (New York: Palgrave Macmillan, 2005), 265–282. See also Ivan Turok, "Cities, Regions, and Competitiveness," *Regional Studies* 38, no. 9 (2004): 1069–1083.
28. See, for example, Charles Landry, *The Creative City: A Toolkit for Urban Innovation* (London: Earthscan, 2000).
29. Paul Chatterton, "Will the Real Creative City Please Stand Up?," *City* 4, no. 3 (2000): 392.
30. Jamie Peck, "Creative Moments: Working Culture, Through Municipal Socialism and Neoliberal Urbanism," in *Mobile Urbanism: Cities and Policymaking in the Global Age*, ed. Eugene McCann and Kevin Ward (Minneapolis: University of Minnesota Press, 2011), 41–70.
31. Chatterton, "Will the Real Creative City Please Stand Up?," 391–392; Graeme Evans, "Creative Cities, Creative Spaces and Urban Policy," *Urban Studies* 46, no. 5–6 (2009): 1003–1040; and Graeme Evans, "From Cultural Quarters to Creative Clusters—Creative Spaces in the New City Economy," in *The Sustainability and Development of Cultural Quarters: International Perspectives*, ed. Mattias Legner (Stockholm: Institute of Urban History, 2009), 32–59.
32. See, for example, Yusuf Shahid and Kaoru Nabeshima, "Creative Industries in East Asia," *Cities* 22, no. 2 (2005): 109–122; and Allen J. Scott, "Emerging Cities of the Third Wave," *City* 15 (2011): 289–321.

33. Peck, "Creative Moments," 41. See Allen J. Scott, "Creative Cities: Conceptual Issues and Policy Questions," *Journal of Urban Affairs* 28, no. 1 (2006): 1–17.
34. Jinna Tay, "Creative Cities," in *Creative Industries*, ed. John Hartley (Malden, MA: Blackwell, 2005), 220–232.
35. Peck, "Creative Moments," 41; and Darrin Bayliss, "The Rise of the Creative City: Culture and Creativity in Copenhagen," *European Planning Studies* 15, no. 7 (2007): 889–903.
36. Richard Florida, *The Rise of the Creative Class* (New York: Basic Books, 2002); Richard Florida, "Cities and the Creative Class," *City & Community* 2, no. 1 (2003): 3–19; and Richard Florida, *Cities and the Creative Class* (New York: Routledge, 2004). For a critique, see Peck, "Struggling with the Creative Class," 740–770.
37. Vivek Chibber, "Reviving the Developmental State? The Myth of the National Bourgeoisie," in *The Empire Reloaded: The Socialist Register*, ed. Leo Panitch and Colin Leys (New York: Monthly Review Press, 2005), 146.
38. See Gordon MacLeod, "From Urban Entrepreneurialism to a 'Revanchist City'? On the Spatial Injustices of Glasgow's Renaissance," *Antipode* 34, no. 3 (2002): 602–614.
39. Florida, *The Rise of the Creative Class*, 1–10; Dan Eugen Ratiu, "Creative Cities and/or Sustainable Cities: Discourses and Practices," *City, Culture and Society* 4 (2013): 125–135.
40. Jeffrey Zimmerman, "From Brew Town to Cool Town: Neoliberalism and the Creative City Development Strategy in Milwaukee," *Cities* 25, no. 4 (2008): 230–242.
41. Zimmerman, "From Brew Town to Cool Town," 230.
42. Peck, "Creative Moments," 43.
43. Eugene McCann and Kevin Ward, "Introduction. Urban Assemblages: Territories, Relations, Practices, and Power," in *Mobile Urbanism*, xiv.
44. Jennifer Robinson, "The Spaces of Circulating Knowledge: City Strategies and Global Urban Governmentality," in *Mobile Urbanism*, 15.
45. Peck, "Creative Moments," 43.
46. Andy Pratt, "Creative Cities: The Cultural Industries and the Creative Class," *Geografiska Annaler: Series B, Human Geography* 90, no. 2 (2008): 107–117.
47. See Max Rousseau, "Re-imaging the City Centre for the Middle Classes: Regeneration, Gentrification and Symbolic Policies in 'Loser Cities,' " *International Journal of Urban and Regional Research* 33, no. 3 (2009): 770–788.
48. Peck, "Creative Moments," 42.
49. Landry, *The Creative City*, 10–15; and Charles Landry, *The Art of City Making* (London: Earthscan, 2006).
50. See Ivan Turok, "The Distinctive City: Pitfalls in the Pursuit of Differential Advantage," *Environment and Planning A* 41 (2009): 13–30; and Graeme Evans, "Cultural Industry Quarters: From Pre-Industrial to Post-Industrial Production," in *City of Quarters: Urban Villages in the Contemporary City*, ed. Duncan Bell and Mark Jayne (Aldershot, UK: Ashgate, 2004), 71–92.
51. Jamie Peck, "Creativity Fix," *Eurozine*, June 28, 2007, https://www.eurozine.com/the-creativity-fix/. See also Stefan Krätke, " 'Creative Cities' and the Rise of the Dealer Class: A Critique of Richard Florida's Approach to Urban Theory," *International Journal of Urban and Regional Research* 34, no. 4 (2010): 835–853.
52. Peck, "Creativity Fix."
53. Jamie Peck and Adam Tickell, "Neoliberalizing Space," *Antipode* 34, no. 3 (2002): 380–404.
54. Peck, "Creativity Fix."
55. Ratiu, "Creative Cities and/or Sustainable Cities," 134.
56. Chatterton, "Will the Real Creative City Please Stand Up?," 391–392.
57. Davide Ponzini and Ugo Rossi, "Becoming a Creative City: The Entrepreneurial Mayor, Network Politics and the Promise of an Urban Renaissance," *Urban Studies* 47, no. 5 (2010): 1037.

58. Gert-Jan Hospers and Cees-Jan Pen, "A View on Creative Cities Beyond the Hype," *Creativity and Innovation Management* 17, no. 4 (2008): 259–270.
59. Chatterton, "Will the Real Creative City Please Stand Up?," 392.
60. Pierre Bourdieu, *The State Nobility: Elite Schools in the Field of Power* (Stanford, CA: Stanford University Press, 1996), 180.
61. Chatterton, "Will the Real Creative City Please Stand Up?," 392.
62. Mary Donegan and Nichola Lowe, "Inequality in the Creative City: Is There Still a Place for "Old-Fashioned" Institutions?," *Economic Development Quarterly* 22, no. 1 (2008): 46–62; Hans Pruijt, "Culture Wars, Revanchism, Moral Panics, and the Creative City: A Reconstruction of a Decline of Tolerant Public Policy: The Case of Dutch Anti-Squatting Legislation," *Urban Studies* 50, no. 6 (2013): 1114–1129; and John Paul Catungal, Deborah Leslie, and Yvonne Hii, "Geographies of Displacement in the Creative City: The Case of Liberty Village, Toronto," *Urban Studies* 46, no. 5–6 (2009): 1095–1114.
63. Chatterton, "Will the Real Creative City Please Stand Up?," 392.
64. This quotation, taken in a different context, comes from Mattias Bernt, "The Limits of Shrinkage: Conceptual Pitfalls and Alternatives in the Discussion of Urban Population Loss," *International Journal of Urban and Regional Research* 40, no. 2 (2016): 441, 447.
65. Shatkin, "Global Cities of the South," 1.
66. Shatkin, "The City and the Bottom Line," 386.
67. Shatkin, "Global Cities of the South," 1; and Shatkin, "The City and the Bottom Line," 386.
68. Shatkin, "The City and the Bottom Line," 396, 398.
69. H. V. Savitch and Paul Kantor, *Cities in the International Marketplace: The Political Economy of Urban Development in North America and Western Europe* (Princeton, NJ: Princeton University Press, 2002).
70. Brenda Yeoh, "Global/Globalizing Cities," *Progress in Human Geography* 23 (1999): 607–616.
71. Robert Beauregard and Anne Haila, "The Unavoidable Continuities of the City," in *Globalizing Cities: A New Spatial Order?*, 22–36; Richard Grant and Jan Nijman, "Globalization and the Corporate Geography of Cities in the Less-Developed World," *Annals of the Association of American Geographers* 92 (2002): 320–340.
72. George Lin, "Chinese Urbanism in Question: State, Society, and the Reproduction of Urban Spaces," *Urban Geography* 28 (2007): 7–29.
73. Lawrence Ma and Fulong Wu, "Restructuring the Chinese City: Diverse Processes and Reconstituted Spaces," in *Restructuring the Chinese City: Changing Society, Economy, and Space*, ed. Lawrence Ma and Fulong Wu (New York: Routledge, 2005), 10.
74. Shatkin, "The City and the Bottom Line."
75. Maarten Hajer, "Review of *S,M,L,XL*, by Rem Koolhaas and Bruce Mau," *Theory, Culture & Society* 16, no. 4 (1999): 137–144.
76. Jeroen de Kloet and Lena Scheen, "Pudong: the Shanzhai Global City," *European Journal of Cultural Studies* 16, no. 6 (2013): 692–709.
77. Atkinson and Bridge, "Introduction," 6; and Smith, "New Globalism, New Urbanism," 447.
78. Andrew Harris, "The Metonymic Urbanism of Twenty-First-Century Mumbai," *Urban Studies* 49, no. 13 (2012): 2961.
79. See Shatkin, "Global Cities of the South," 4–7.
80. Richard Child Hill, "Cities and Nested Hierarchies," *International Social Science Journal* 56, no. 181 (2004): 374.
81. Nezar AlSayyad, "Hybrid Culture / Hybrid Urbanism: Pandora's Box of the 'Third Place,' " in *Hybrid Urbanism*, ed. Nezar AlSayyad (Westport, CT: Praeger, 2001), 1–20; and Abidin Kusno, *Behind the Postcolonial* (London: Routledge, 2000).

82. Joe Nasr and Mercedes Volait, "Introduction: Transporting Planning," in *Urbanism Imported or Exported? Native Aspirations and Foreign Plans*, ed. Joe Nasr and Mercedes Volait (Chichester, UK: Wiley, 2003), xx.
83. Shatkin, "Global Cities of the South," 2; and Cathy Liu, "From Los Angeles to Shanghai: Testing the Applicability of Five Urban Paradigms," *International Journal of Urban and Regional Research* 36, no. 6 (2012): 1127–1145.
84. Shatkin, "Global Cities of the South," 1.
85. Shatkin, "The City and the Bottom Line," 384.
86. Jennifer Robinson and Anya Roy, "Global Urbanisms and the Nature of Urban Theory," *International Journal of Urban and Regional Research* 40, no. 1 (2016): 181.
87. Neil Brenner and Christian Schmid, "Towards a New Epistemology of the Urban?," *City* 19, no. 2–3 (2015): 151–182.
88. Ash Amin and Stephen Graham, "The Ordinary City," *Transactions of the Institute of British Geographers* 22, no. 4 (1997): 411–429.
89. These ideas are derived, almost verbatim, from Jamie Peck, "Cities Beyond Compare?," *Regional Studies* 49, no. 1 (2015): 165.
90. Robinson and Roy, "Global Urbanisms and the Nature of Theory," 181–186.
91. Peck, "Cities Beyond Compare?," 161, 170–171.
92. Ananya Roy, "Slumdog Cities: Rethinking Subaltern Urbanism," *International Journal of Urban and Regional Research* 35, no. 2 (2011): 223–238.
93. Peck, "Cities Beyond Compare?," 161. See also Raewyn Connell, *Southern Theory: The Global Dynamics of Knowledge in Social Science* (Sydney: Allen & Unwin, 2007); Jean Comaroff and John Comaroff, *Theory from the South: Or, How Euro-America Is Evolving Toward Africa* (Boulder, CO: Paradigm, 2011); and Tim Edensor and Mark Jayne, "Introduction: Urban Theory Beyond the West," in *Urban Theory Beyond the West: A World of Cities*, ed. Tim Edensor and Mark Jayne (New York: Routledge, 2012), 1–27.
94. Vanessa Watson, "Seeing from the South: Refocusing Urban Planning on the Globe's Central Urban Issues," *Urban Studies* 46, no. 11 (2009): 2259–2275; and Susan Parnell and Jennifer Robinson, "(Re) theorizing Cities from the Global South: Looking Beyond Neoliberalism," *Urban Geography* 33, no. 4 (2012): 593–617.
95. Josef Gugler, "World Cities in Poor Countries: Conclusions from Case Studies of the Principal Regional and Global Players," *International Journal of Urban and Regional Research* 27, no. 3 (2003): 707–712; and Edensor and Jayne, "Introduction: Urban Theory Beyond the West."
96. Raewyn Connell, "Using Southern Theory: Decolonizing Social Thought in Theory, Research and Application," *Planning Theory* 13, no. 2 (2014): 210–223.
97. Austin Zeiderman, "Cities of the Future? Megacities and the Space/Time of Urban Modernity," *Critical Planning* 15 (2008): 23–39.
98. Harris, "The Metonymic Urbanism of Twenty-First-Century Mumbai," 2960.
99. See, inter alia, Ananya Roy, "Worlding the South: Toward a Post-Colonial Theory," in *The Routledge Handbook on Cities of the Global South*, ed. Susan Parnell and Sophie Oldfield (New York: Routledge, 2014), 9–20; and Alan Mabin, "Grounding Southern City Theory in Time and Place," in *The Routledge Handbook on Cities of the Global South*, 21–36. For a particularly acerbic statement that actually reinforces binary thinking, see Sarah Nuttall and Achille Mbembé, "A Blasé Attitude: A Response to Michael Watts," *Public Culture* 17, no. 1 (2005): 193–202.
100. Srinivas Aravamudan, "Surpassing the North: Can the Antipodean Avantgarde Trump Postcolonial Belatedness?," *Society for Cultural Anthropology*, February 25, 2012, https://culanth.org/fieldsights/surpassing-the-north-can-the-antipodean-avantgarde-trump-postcolonial-belatedness.

101. Ashley Dawson and Brent Hayes Edwards, "Global Cities of the South," *Social Text* 22, no. 4 (2004): 1–7; Watson, "Seeing from the South"; and Vanessa Watson, "Planning and the 'Stubborn Realities' of Global South-East Cities: Some Emerging Ideas," *Planning Theory* 12, no. 1 (2013): 81–100.
102. Michael Hoyler and John Harrison, "Global Cities Research and Urban Theory Making," *Environment and Planning A* 49, no. 12 (2017): 2854. See also Mark Jayne and Kevin Ward, "A Twenty-First Century Introduction to Urban Theory," in *Urban Theory: New Critical Perspectives*, ed. Mark Jayne and Kevin Ward (London: Routledge, 2016), 1–18; Donald McNeill, *Global Cities and Urban Theory* (London: Sage, 2017); and Phil Hubbard, *City* (London: Routledge, 2006), 57.
103. Michiel van Meeteren, Ben Derudder, and David Bassens, "Can the Straw Man Speak? An Engagement with Postcolonial Critiques of 'Global Cities Research,' " *Dialogues in Human Geography* 6, no. 3 (2016): 250. For a careful rejoinder, see Jennifer Robinson, "Theorizing the Global Urban with 'Global and World Cities Research': Beyond Cities and Synechdoche," *Dialogues in Human Geography* 6, no. 3 (2016): 268–272.
104. Ann El Khoury, "Pluralizing Global Urbanism: Replacing the God-Trick with Goddess Tactics," *Dialogues in Human Geography* 6, no. 3 (2016): 292.
105. Michael Hoyer and John Harrison, "Global Cities Research and Urban Theory Making," *Environment and Planning A* 49, no. 12 (2017): 2853–2858; Jamie Peck, "Transatlantic City, Part 2: Late Entrepreneurialism," *Urban Studies* 54, no. 2 (2017): 327–363; and Maarten van Meeteren, David Bassens, and Ben Derudder, "Doing Global Urban Studies: On the Need for Engaged Pluralism, Frame Switching, and Methodological Cross-Fertilization," *Dialogues in Human Geography* 6, no. 3 (2016): 296–301.
106. Shatkin, "Global Cities of the South," 3; and David Bassens and Michiel van Meeteren, "World Cities Under Conditions of Financialized Globalization: Towards an Augmented World City Hypothesis," *Progress in Human Geography* 39, no. 6 (2015): 752–775.
107. See, inter alia, Peter Marcuse and Ronald Van Kempen, "Introduction," in *Globalizing Cities: A New Spatial Order?*, 1–21; and Martin J. Murray, *Urbanism of Exception: Global City Building at the Start of the Twenty-First Century* (New York: Cambridge University Press, 2017), 93–146.
108. Simon Marvin and Steven Graham, *Splintering Urbanism: Networked Infrastructures, Technological Mobilities and the Urban Condition* (New York: Routledge, 2001); and Stephen Graham, "Constructing Premium Network Spaces: Reflections on Infrastructure Networks and Contemporary Urban Development," *International Journal of Urban and Regional Research* 24, no. 1 (2000): 183–200.
109. Jennifer Robinson, *Ordinary Cities: Between Modernity and Development* (London: Routledge, 2006). For a stylized and polemical assault on the "ordinary cities" argument, see Richard Smith, "The Ordinary City Trap," *Environment and Planning A* 45, no. 10 (2013): 2290–2304.
110. Ananya Roy, "Conclusion: Postcolonial Urbanism: Speed, Hysteria, Mass Dreams," in *Worlding Cities: Asian Experiments and the Art of Being Global*, ed. Ananya Roy and Aihwa Ong (Malden, MA: Wiley-Blackwell, 2011), 306–335.
111. Susan Parnell and Sophie Oldfield, " 'From the South,' " in *The Routledge Handbook on Cities of the Global South*, 4; Edensor and Jayne, "Introduction: Urban Theory Beyond the West"; and Watson, "Seeing from the South," 2259–2275.
112. See Christopher Harker, "Theorizing the Urban from the 'South'?," *City* 15, no. 1 (2011): 120, 122; and Seth Schindler, "Towards a Paradigm of Southern Urbanism," *City* 21, no. 1 (2017): 47–64.
113. See Loretta Lees, "The Geography of Gentrification: Thinking Through Comparative Urbanism," *Progress in Human Geography* 36, no. 2 (2012): 156.
114. Leontidou, "Alternatives to Modernism in (Southern) Urban Theory," 180.
115. See James Sidaway, Tim Bunnell, and Brenda Yeoh, "Geography and Postcolonialism," *Singapore Journal of Tropical Geography* 24, no. 3 (2003): 269.

116. Gavin Shatkin, "Contesting the Indian City: Global Visions and the Politics of the Local," *International Journal of Urban and Regional Research* 38, no. 1 (2014): 1–13.
117. AbdouMaliq Simone, *City Life from Jakarta to Dakar: Movements at the Crossroads* (New York: Routledge, 2010), 280.
118. Smith, "New Globalism, New Urbanism."
119. Harris, "The Metonymic Urbanism of Twenty-First-Century Mumbai," 2960.
120. Harris, "The Metonymic Urbanism of Twenty-First-Century Mumbai," 2960.
121. Mabin, "Grounding Southern City Theory in Time and Place," 21–36.
122. Ananya Roy, "The 21st-Century Metropolis: New Geographies of Theory," *Regional Studies* 43, no. 6 (2009): 820.
123. Harris, "The Metonymic Urbanism of Twenty-First-Century Mumbai," 2969; and Robert Beauregard, "City of Superlatives," *City and Community* 2, no. 3 (2003): 183–199.
124. See Eric Sheppard, Helga Leitner, and Anant Maringanti, "Urban Pulse—Provincializing Global Urbanism: A Manifesto," *Urban Geography* 34, no. 7 (2013): 893–900.
125. Gregor McLennan, "Postcolonial Critique: The Necessity of Sociology," in *Postcolonial Sociology*, ed. Julian Go (Bingley, UK: Emerald, 2013), 119–144.
126. Christine Hentschel, "Postcolonizing Berlin and the Fabrication of the Urban," *International Journal of Urban and Regional Research* 39, no. 1 (2015): 79–91.
127. Peck, "Cities Beyond Compare?," 171.
128. This framing is derived from a reading of Aravamudan, "Surpassing the North."
129. Schindler, "Towards a Paradigm of Southern Urbanism," 52.
130. Peck, "Cities Beyond Compare?," 170–171.
131. See, for example, Aravamudan, "Surpassing the North."
132. Jean Comaroff and John Comaroff, "Theory from the South: A Rejoinder," in *The Johannesburg Salon: Volume Five*, ed. Achille Mbembe (Johannesburg: Johannesburg Workshop in Theory and Criticism, 2012), 31.
133. Comaroff and Comaroff, "Theory from the South: A Rejoinder," 31.
134. Roy, "Worlding the South," 15.
135. Rajesh Vengopal, "Neoliberalism as a Concept," *Economy and Society* 44, no. 2 (2015): 170, 171; and Bill Dunn, "Against Neoliberalism as a Concept," *Capital & Class* 41, no. 3 (2017): 437.
136. Peck, "Cities Beyond Compare?," 165.
137. Peck, "Cities Beyond Compare?," 160.
138. Peck, "Cities Beyond Compare?," 160.
139. Ann Varley, "Feminist Perspectives on Urban Poverty," in *Rethinking Feminist Interventions Into the Urban*, ed. Linda Peake and Martina Rieker (London: Routledge, 2013), 128, 126.
140. Peck, "Cities Beyond Compare?," 171.
141. Peck, "Cities Beyond Compare?," 171.
142. Varley, "Feminist Perspectives on Urban Poverty," 126.
143. Hentschel, "Postcolonizing Berlin and the Fabrication of the Urban," 79.

3. GLOBALIZING CITIES WITH WORLD-CLASS ASPIRATIONS: THE EMERGENCE OF THE POSTINDUSTRIAL TOURIST-ENTERTAINMENT CITY

1. Sam Bass Warner and Lawrence Vale, "Introduction: Cities, Media, and Imaging," in *Imaging the City: Continuing Struggles and New Directions*, ed. Lawrence Vale and Sam Bass Warner (New Brunswick, NJ: Center for Urban Policy Research, 2001), xxiii.

2. Sharon Zukin, *The Cultures of Cities* (Oxford: Blackwell, 1995); Scott Salmon, "Imagineering the Inner City? Landscapes of Pleasure and the Commodification of Cultural Spectacle in the Postmodern City," in *Popular Culture: Production and Consumption*, ed. C. Lee Harrington and Denise Bielby (Malden, MA: Blackwell, 2001), 106–120; and Sharon Zukin, "Cultural Strategies for Economic Development and the Hegemony of Vision," in *Urbanization of Injustice*, ed. Erik Swyngedouw and Andrew Merrifield (London: Lawrence and Wishart, 1996), 223–243.
3. Allen Cunningham, "The Modern City Revisited—Envoi," in *The Modern City Revisited*, ed. Thomas Deckker (London: Spon, 2000), 247.
4. David Morley and Kevin Robins, *Spaces of Identity: Global Media, Electronic Landscapes and Cultural Boundaries* (New York: Routledge, 1995), 26.
5. Richard Lloyd and Terry Nichols Clark, "The City as Entertainment Machine," in *Critical Perspectives on Urban Redevelopment: Volume 6*, ed. Kevin Fox Gotham (New York: Elsevier, 2001), 357. See also Terry Nichols Clark, ed., *The City as an Entertainment Machine*, rev. ed. (Lanham, MD: Lexington, 2011).
6. Lloyd and Clark, "The City as Entertainment Machine."
7. Susan Fainstein and Dennis Judd, eds., *The Tourist City* (New Haven, CT: Yale University Press, 1999); and Dennis Judd, ed., *The Infrastructure of Play: Building the Tourist City* (Armonk, NY: Sharp, 2004).
8. John Hannigan, *Fantasy City: Pleasure and Profit in the Postmodern Metropolis* (New York: Routledge, 1998).
9. M. Christine Boyer, *The City of Collective Memory: Its Historical Imagery and Architectural Entertainments* (Cambridge, MA: MIT Press, 1994), 320–321, 369.
10. M. Christine Boyer, "The City of Illusion: New York's Public Places," in *The Restless Urban Landscape*, ed. Paul Knox (Englewood Cliffs, NJ: Prentice Hall, 1993), 111–126.
11. Lloyd and Clark, "The City as Entertainment Machine"; and Richard Lloyd, *Neo-Bohemia: Art and Commerce in the Post-Industrial City* (New York: Routledge, 2006).
12. Tim Hall and Phil Hubbard, "The Entrepreneurial City: New Urban Politics, New Urban Geography?," *Progress in Human Geography* 20, no. 2 (1996): 160.
13. Susan Fainstein and Dennis Judd, "Cities as Places to Play," in *The Tourist City*, ed. Dennis Judd and Susan Fainstein (New Haven, CT: Yale University Press, 1999), 261–272.
14. See Max Rousseau, "Re-imaging the City Centre for the Middle Classes: Regeneration, Gentrification and Symbolic Policies in 'Loser Cities,' " *International Journal of Urban and Regional Research* 33, no. 3 (2009): 770–788.
15. Zukin, *The Cultures of Cities*, 2.
16. Peter Eisinger, "The Politics of Bread and Circuses: Building the City for the Visitor Class," *Urban Affairs Review* 35, no. 3 (2000): 317.
17. David Harvey, "From Managerialism to Entrepreneurialism: The Transformation of Urban Governance in Late Capitalism," *Geografiska Annaler: Series B, Human Geography* 71, no. 1 (1989): 3–17; Neil Brenner and Nik Theodore, "Neoliberalism and the Urban Condition," *City* 9, no. 1 (2005): 101–107; and Jamie Peck, "Entrepreneurial Urbanism: Between Uncommon Sense and Dull Compulsion," *Geografiska Annaler: Series B, Human Geography* 96, no. 4 (2014): 396–401.
18. See, inter alia, Hall and Hubbard, "The Entrepreneurial City," 153–174; Bob Jessop, "The Entrepreneurial City: Re-imaging Localities, Redesigning Economic Governance, or Restructuring Capital?," in *Transforming Cities: Contested Governance and New Spatial Divisions*, ed. Nick Jewson and Susanne MacGregor (London: Routledge, 1997), 28–41; Bob Jessop, "The Narrative of Enterprise and the Enterprise of Narrative," in *The Entrepreneurial City: Geographies of Politics, Regime, and Representation*, ed. Tim Hall and Phil Hubbard (Chichester, NY: Wiley, 1998), 77–99; Gordon MacLeod, "From Urban Entrepreneurialism to a 'Revanchist City'? On the Spatial Injustices of Glasgow's Renaissance," *Antipode* 34, no. 3 (2002): 602–623; Phil Hubbard, "Urban Design and

City Regeneration: Social Representations and Entrepreneurial Landscapes," *Urban Studies* 33, no. 8 (1996): 1441–1461; Gwyndaf Williams, "Rebuilding the Entrepreneurial City: The Master Planning Response to the Bombing of Manchester," *Environment and Planning B* 27, no. 4 (2000): 485–505; and Kevin Ward, "Entrepreneurial Urbanism, State Restructuring and Civilising 'New' East Manchester," *Area* 35, no. 2 (2003): 116–127.

19. Katharyne Mitchell, "Transnationalism, Neo-Liberalism, and the Rise of the Shadow State," *Economy & Society* 30, no. 2 (2001): 165–189.
20. Brendan Bartley and Kasey Treadwell Shine, "Competitive City: Governance and the Changing Dynamics of Urban Regeneration in Dublin," in *The Globalized City: Economic Restructuring and Social Polarization in European Cities*, ed. Frank Moulaert, Arantxa Rodríguez, and Erik Swyngedouw (Oxford: Oxford University Press, 2003), 164; and Salmon, "Imagineering the Inner City?," 108.
21. Patrick Loftman and Brendan Nevin, "Going for Growth: Prestige Projects in Three British Cities," *Urban Studies* 33, no. 6 (1996): 991–1019; and Hall and Hubbard, "The Entrepreneurial City," 153–174.
22. Harvey, "From Managerialism to Entrepreneurialism," 3–17.
23. Patrick Loftman and Brendan Nevin, "Prestige Projects and Urban Regeneration in the 1980s and 1990s: A Review of the Benefits and Limitations," *Planning Practice and Research* 10, no. 3–4 (1995): 299–315; Hall and Hubbard, "The Entrepreneurial City"; Ivor Turok, "Cities, Regions and Competitiveness," *Regional Studies* 38, no. 9 (2004): 1061–1075; and Ivan Turok, "Cities, Competition and Competitiveness: Identifying New Connections," in *Changing Cities: Rethinking Urban Competitiveness, Cohesion and Governance*, ed. Nick Buck, Ian Gordon, Alan Harding, and Ivan Turok (New York: Palgrave MacMillan, 2005), 25–43.
24. Keith Bassett, "Partnerships, Business Elites and Urban Politics: New Forms of Governance in an English City?," *Urban Studies* 33, no. 3 (1996): 539–555; Philip Booth, "Partnerships and Networks: The Governance of Urban Regeneration in Britain," *Journal of Housing and the Built Environment* 20, no. 3 (2005): 257–269; Hartmut Häußermann and Katja Simons, "Facing Fiscal Crisis: Urban Flagship Projects in Berlin," in *The Globalized City*, 107–124; and Rachel Weber, "Extracting Value from the City: Neoliberalism and Urban Redevelopment," *Antipode* 34, no. 3 (2002): 519–539.
25. Tony Gilmour, *Sustaining Heritage: Giving the Past a Future* (Sydney: Sydney University Press, 2007); and Andrew MacClaran and Pauline McGuirk, "Planning the City," in *Making Space*, 63–94.
26. Andrew MacClaran, "Masters of Space: The Property Development Sector," in *Making Space*, 7–62.
27. Lorene Hoyt, "Planning Through Compulsory Commercial Clubs: Business Improvement Districts," *Economic Affairs* 25, no. 4 (2005): 24–27; Jerry Mitchell, "Business Improvement Districts and the 'New' Revitalization of Downtown," *Economic Development Quarterly* 15, no. 2 (2001): 115–123; Stephen Osborne, *Public–Private Partnerships: Theory and Practice in International Perspective* (London: Routledge, 2000); and Kevin Ward, "Entrepreneurial Urbanism and Business Improvement Districts in the State of Wisconsin: A Cosmopolitan Critique," *Annals of the Association of American Geographers*, 100, no. 5 (2010): 1177–1196.
28. Monica Degen, "Fighting for the Global Catwalk: Formalizing Public Life in Castlefield (Manchester) and Diluting Public Life in el Raval (Barcelona)," *International Journal of Urban and Regional Research* 27, no. 4 (2003): 867–868.
29. Eugene McCann, "Collaborative Visioning or Urban Planning as Therapy? The Politics of Public-Private Policy Making," *Professional Geographer* 53, no. 2 (2001): 207–218.
30. Neil Brenner and Nik Theodore, "Cities and the Geographies of 'Actually Existing Neoliberalism,' " in *Spaces of Neoliberalism: Urban Restructuring in North America and Western Europe*, ed. Neil Brenner and Nik Theodore (Oxford: Wiley-Blackwell, 2002), 23.

31. Gordon MacLeod, "Urban Politics Reconsidered: Growth Machine to Post-Democratic City?," *Urban Studies* 48, no. 12 (2011): 2629–2660.
32. Hall and Hubbard, "The Entrepreneurial City."
33. Kevin Fox Gotham, "Theorizing Urban Spectacles: Festivals, Tourism, and the Transformation of Urban Space," *City* 9, no. 2 (2005): 225–246; Briavel Holcomb, "Revisioning Place: De- and Re-constructing the Image of the Industrial City," in *Selling Places: The City as Cultural Capital, Past and Present*, ed. Gerry Kearns and Chris Philo (New York: Pergamon, 1993), 133–144; and Briavel Holcomb, "Marketing Cities for Tourism," in *The Tourist City*, ed. Dennis Judd and Susan Fainstein (New Haven, CT: Yale University Press, 1999), 54–70.
34. Boyer, *The City of Collective Memory*, 1–8; Boyer, "The City of Illusion: New York's Public Places"; M. Christine Boyer, "The Great Frame-Up: Fantastic Appearances in Contemporary Spatial Politics," in *Spatial Practices*, ed. Helen Liggett and David Perry (London: Sage, 1995), 81–109; M. Christine Boyer, "Twice Told Stories: The Double Erasure of Times Square," in *The Unknown City: Contesting Architecture and Social Space*, ed. Iain Borden, Joe Kerr, Jane Rendell, with Alicia Pivaro (Cambridge, MA: MIT Press, 2001), 30–53; Stephanie Hemelryk Donald, Eleonore Kofman, and Catherine Kevin, eds., *Branding Cities: Cosmopolitanism, Parochialism, and Social Change* (New York: Routledge, 2008).
35. Charles Rutheiser, *Imagineering Atlanta: The Politics of Place in the City of Dreams* (New York: Verso, 1996). For a less than positive view, see Peggy Teo, "The Limits of Imagineering: A Case Study of Penang," *International Journal of Urban and Regional Research* 27, no. 3 (2003): 545–563; and Kevin Archer, "The Limits to the Imagineered City: Sociospatial Polarization in Orlando," *Economic Geography* 73, no. 3 (1997): 322–336.
36. Bart Eeckhout, "The 'Disneyfication' of Times Square: Back to the Future?," in *Critical Perspectives on Urban Redevelopment*, 379–428.
37. Theming, or the styling of urban space, is a concept that far exceeds the normal purview of both architecture and urban planning. Themed spaces are elements of city landscapes that seek to attach themselves to something other than what they are by imitating bygone eras or faraway places. Theming creates illusions and images of nonexistent realities or of other times and places. See T. C. Chang, "Theming Cities, Taming Places: Insights from Singapore," *Geografiska Annaler: Series B, Human Geography* 82, no. 1 (2000): 35–54.
38. Claire Colomb, *Staging the New Berlin: Place Marketing and the Politics of Urban Reinvention Post-1980* (New York: Routledge, 2012).
39. Miriam Greenberg, *Branding New York: How a City in Crisis Was Sold to the World* (New York: Routledge, 2008); Graeme Evans, "Hard-Branding the Cultural City—From Prado to Prada," *International Journal of Urban and Regional Research* 27, no. 2 (2003): 417–440; Michalis Kavaratzis, "From City Marketing to City Branding: Towards a Theoretical Framework for Developing City Brands," *Place Branding* 1, no. 1 (2004): 58–73; and Michalis Kavaratzis and Gregory Ashworth, "City Branding: An Effective Assertion of Identity or a Transitory Marketing Trick?," *Tijdschrift voor Economische en Sociale Geografie* 96, no. 5 (2005): 506–514.
40. Christopher Mele, *Selling the Lower East Side: Culture, Real Estate, and Resistance in New York City* (Minneapolis: University of Minnesota Press, 2000); M. Christine Boyer, "Cities for Sale: Merchandising History at South Street Seaport," in *Variations on a Theme Park: The New American City and the End of Public Space*, ed. Michael Sorkin (New York: Hill and Wang, 1992), 181–204; Boyer, "Twice Told Stories"; and Phil Hubbard, "Urban Design and Local Economic Development: A Case Study in Birmingham," *Cities* 12, no. 4 (1995): 243–251.
41. Boyer, *The City of Collective Memory*, 47–48.
42. Richard Scherr, "The Synthetic City: Excursions Into the Real–Not Real," *Places* 18, no. 2 (2008): 8. See Boyer, *The City of Collective Memory*, 47–48, 59.
43. Boyer, *The City of Collective Memory*, 47, 448–449, 51.

44. Anne-Marie Broudehoux, "Image Making, City Marketing and the Aesthetization of Social Inequality in Rio de Janeiro," in *Consuming Tradition, Manufacturing Heritage: Global Norms and Urban Forms in the Age of Tourism*, ed. Nezar AlSayyad (New York: Routledge, 2001), 274.
45. Boyer, "The City of Illusion," 113.
46. Broudehoux, "Image Making," 276.
47. David Harvey, *The Urban Experience* (New York: Blackwell, 1989), 270–271.
48. Aspa Gospodini, "European Cities in Competition and the New 'Uses' of Urban Design," *Journal of Urban Design* 7, no. 1 (2002): 60.
49. Guy Julier, "Urban Designscapes and the Production of Aesthetic Consent," *Urban Studies* 42, no. 5–6 (2005): 869–887.
50. Craig Young, Martina Diep and Stephanie Drabble, "Living with Difference? The 'Cosmopolitan City' and Urban Reimaging in Manchester, UK," *Urban Studies* 43, no. 10 (2006): 1687–1714.
51. Paul Knox and Kathy Pain, "Globalization, Neoliberalism and International Homogeneity in Architecture and Urban Development," *Informationen zur Raumentwicklung* 5, no. 6 (2010): 423.
52. Rachel Webber, "The Metropolitan Habitus: Its Manifestations, Locations and Consumption Profiles," *Environment and Planning A* 39 (2007): 182–207.
53. Knox and Pain, "Globalization, Neoliberalism and International Homogeneity," 421. See also Martin J. Murray, "The Quandary of Post-Public Space: New Urbanism, Melrose Arch and the Rebuilding of Johannesburg After Apartheid," *Journal of Urban Design* 18, no. 1 (2013): 119–144.
54. Leslie Sklair, "Commentary: From the Consumerist/Oppressive City to the Functional/Emancipatory City," *Urban Studies* 46, no. 12 (2009): 2703–2711.
55. Anna Klingmann, *Brandscapes: Architecture in the Experience Economy* (Cambridge, MA: MIT Press, 2007), 83.
56. Jim McGuigan, "Neo-Liberalism, Culture and Policy," *International Journal of Cultural Policy* 11, no. 3 (2005): 229–241.
57. David Harvey, *The Condition of Postmodernity: An Enquiry Into the Origins of Cultural Change* (Cambridge, MA: Blackwell, 1989), 52–54, 62–63, 75–76.
58. Monica Degen, *Sensing Cities: Regenerating Urban Life in Barcelona and Manchester* (New York: Routledge 2008); and Monica Degen, Caitlin De Silvey, and Gillian Rose, "Experiencing Visualities in Designed Urban Environments: Learning from Milton Keynes," *Environment & Planning A* 40, no. 8 (2008): 1901–1920.
59. Edward Relph, *Place and Placelessness* (London: Pion, 1976); and Marc Augé, *Non-Places: Introduction to an Anthropology of Supermodernity*, trans. John Howe (New York: Verso, 1995).
60. Nan Ellen, *Postmodern Urbanism*, rev. ed. (New York: Princeton Architectural Press, 1999), 15, 16, 22–24, 182, 224–225, 289–291, 324, 334.
61. See, for example, Robert Venturi, *Complexity and Contradiction in Architecture* (New York: Museum of Modern Art, 1996).
62. Harvey, *The Condition of Postmodernity*, 41–2, 54, 66–98.
63. Ash Amin and Nigel Thrift, *Cities: Reimaging the Urban* (Cambridge: Polity, 2002), 51–77.
64. H. Lim, "Cultural Strategies for Revitalizing the City: A Review and Evaluation," *Regional Studies* 27, no. 6 (1993): 589–595; Miguel Kanai and Iliana Ortega-Alcázar, "The Prospects for Culture-Led Urban Regeneration in Latin America: Cases from Mexico City and Buenos Aires," *International Journal of Urban and Regional Research* 33, no. 2 (2009): 483–501; Ute Lehrer, "Willing the Global City: Berlin's Cultural Strategies of Interurban Competition After 1989," in *The Global Cities Reader*, ed. Neil Brenner and Roger Keil (New York: Routledge, 2006), 332–338; Elizabeth Strom, "From Pork to Porcelain: Cultural Institutions and Downtown Development," *Urban Affairs Review* 38, no. 1 (2002): 3–21; Judith Kenny, "Making Milwaukee Famous: Cultural Capital, Urban Image and the Politics of Place," *Urban Geography* 16 (1995): 440–458; Paul Knox, *Cities and Design* (New York: Routledge, 2011); and Graeme Evans, "Cultural Industry Quarters: From Pre-Industrial to

Post-Industrial Production," in *City of Quarters: Urban Villages in the Contemporary City*, ed. Duncan Bell and Mark Jayne (Aldershot, UK: Ashgate, 2004), 71–92.

65. Ian Gordon and Nick Buck, "Introduction: Cities in the New Conventional Wisdom," in *Changing Cities: Rethinking Urban Competitiveness, Cohesion, and Governance*, 6.
66. Greenberg, *Branding New York*, 253–260; Gospodini, "European Cities in Competition and the New 'Uses' of Urban Design," 59–73; Miriam Greenberg, "Branding Cities: A Social History of the Urban Lifestyle Magazine," *Urban Affairs Review* 36, no. 2 (2000): 228–263; and Keith Bassett, "Urban Cultural Strategies and Urban Regeneration: A Case Study and Critique," *Environment and Planning A* 25, 12 (1993): 1773–1788.
67. Terry Nichols Clark, Richard Lloyd, Kenneth Wong, and Pushpam Jain, "Amenities Drive Urban Growth," *Journal of Urban Affairs* 24, no. 5 (2002): 493–515; Allen J. Scott, *The Cultural Economy of Cities: Essays on the Geography of Image Producing Industries* (London: Sage, 2000); and Elizabeth Currid-Halkett and Allen J. Scott, "The Geography of Celebrity and Glamour: Reflections on Economy, Culture, and Desire in the City," *City, Culture, and Society* 4, no. 1 (2013): 2–11; and Davorka Mikulić and Lidija Petrić, "Can Culture and Tourism Be the Foothold of Urban Regeneration? A Croatian Case Study," *Tourism* 62, no. 4 (2014): 377–395.
68. Michael Ian Borer, "The Location of Culture: The Urban Culturalist Perspective," *City & Community* 5, no. 2 (2006): 173–197; John McCarthy, "Regeneration of Cultural Quarters: Public Art for Place Image or Place Identity?," *Journal of Urban Design* 11, no. 2 (2006): 243–262; and Allen J. Scott, "Cultural-Products Industries and Urban Economic Development: Prospects for Growth and Market Contestation in a Global Context," *Urban Affairs Review* 39, no. 4 (2004): 461–490.
69. Bernadette Quinn, "Arts Festivals and the City," *Urban Studies* 42, no. 5–6 (2005): 927–943; Elizabeth Strom, "Let's Put on a Show: Performing Arts and Urban Revitalization in Newark, New Jersey," *Journal of Urban Affairs* 21, no. 4 (1999): 423–435; Catherine Cameron, "The Marketing of Tradition: The Value of Culture in American Life," *City & Society* 1, no. 2 (1987): 162–174; Catherine Cameron, "The Marketing of Heritage: From the Western World to the Global Stage," *City & Society* 20, no. 2 (2008): 160–168; Elizabeth Currid, "How Art and Culture Happen in New York: Implications for Urban Economic Development," *Journal of the American Planning Association* 73, no. 4 (2007): 454–467; and Elizabeth Currid and Sarah Williams, "The Geography of Buzz: Art, Culture and the Social Milieu in New York and Los Angeles," *Journal of Economic Geography* 10, no. 3 (2010): 423–451.
70. Joshua Sapotichne, "Rhetorical Strategy in Stadium Development Politics," *City, Culture and Society* 3, no. 1 (2012): 169–180.
71. Loretta Lees, "The Ambivalence of Diversity and the Politics of Urban Renaissance: The Case of Growth in Downtown Portland, Maine," *International Journal of Urban and Regional Research* 27, no. 3 (2003): 614.
72. Graeme Evans, "Branding the City of Culture—The Death of City Planning?," in *Culture, Urbanism and Planning*, ed. Javier Monclús and Manuel Guardia (Aldershot, UK: Ashgate, 2006), 197–214; and Graeme Evans, "Creative Cities, Creative Spaces and Urban Policy," *Urban Studies* 46, no. 5–6 (2009): 1003–1040.
73. Christopher Jencks, *What Is Postmodernism?* (New York: St. Martin's Press, 1986); Ellen, *Postmodern Urbanism*, 20–24; Diane Ghirardo, *Architecture After Modernism* (London: Thames & Hudson, 1996), 7–27.
74. Salmon, "Imagineering the Inner City?," 109. See Ernest Sternberg, "What Makes Buildings Catalytic? How Cultural Facilities Can Be Designed to Spur Surrounding Development," *Journal of Architectural and Planning Research* 19, no. 1 (2002): 30–43.
75. Klingmann, *Brandscapes*, 273–281.
76. Close to a half-century ago, Robert Venturi, Denise Scott Brown, and Steven Izenour distinguished between buildings whose architectural image was primarily the result of surface ornament applied to structures shaped by their functions and buildings whose image was the largely the result of their

unusual form. They named the former "decorated sheds" and the latter "ducks" (a tongue-in-cheek reference to a roadside stand on Long Island that sold poultry and was shaped like a duck). Italian Renaissance palazzi, for example, which are essentially straightforward structures with exquisitely ornamented exteriors and interiors, are exemplary expressions of decorated sheds. In contrast, Gothic cathedrals, with their flying buttresses, pinnacles, and soaring steeples, are examples of ducks. See Robert Venturi, Denise Scott Brown, and Steven Izenour, *Learning from Las Vegas: The Forgotten Symbolism of Architectural Form* (Cambridge, MA: MIT Press, 1972), 6–7, 17, 90–101.

77. Ivan Turok, "The Distinctive City: Pitfalls in the Pursuit of Differential Advantage," *Environment and Planning A* 41, no. 1 (2009): 24; and Donald Crilley, "Architecture as Advertising: Constructing the Image of Redevelopment," in *Selling Places: The City as Cultural Capital, Past and Present*, ed. Gerry Kearns and Chris Philo (Oxford: Pergamon, 1993), 233–234.
78. Arantxa Rodríguez and Elena Martinez, "Restructuring Cities: Miracles and Mirages in Urban Revitalization in Bilbao," in *The Globalized City*, 181; and Lorenzo Vicario and Manuel Martínez Monje, "Another 'Guggenheim Effect'? The Generation of a Potentially Gentrifiable Neighbourhood in Bilbao," *Urban Studies* 40, no. 12 (2003): 2383–2400.
79. Klingmann, *Brandscapes*, 240.
80. María Gomez, "Reflective Images: The Case of Urban Regeneration in Glasgow and Bilbao," *International Journal of Urban and Regional Research* 22, no. 1 (1998): 106–121; Beatriz Plaza, "The Guggenheim-Bilbao Museum Effect: A Reply to María Gomez," *International Journal of Urban and Regional Research* 23, no. 3 (1999): 589–592; Beatriz Plaza, "Evaluating the Influence of a Large Cultural Artifact in the Attraction of Tourism: The Guggenheim-Bilbao Museum Case," *Urban Affairs Review* 36, no. 2 (2000): 264–274; María Gomez and Sara Gonzalez, "A Reply to Beatriz Plaza's 'The Guggenheim-Bilbao Museum Effect,'" *International Journal of Urban and Regional Research* 25, no. 4 (2001): 898–900; Beatriz Plaza, "The Return on Investment of the Guggenheim Museum, Bilbao," *International Journal of Urban and Regional Research* 30, no. 2 (2006): 452–467; Sara González Ceballos, "The Role of the Guggenheim Museum in the Development of Urban Entrepreneurial Practices in Bilbao," *International Journal of Iberian Studies* 16, no. 3 (2004): 177–186.
81. Matthew Wansborough and Andrea Mangeean, "The Role of Urban Design in Cultural Regeneration," *Journal of Urban Design* 5, no. 2 (2000): 181–197; Javier Martinez, "Selling Avant-Garde: How Antwerp Became a Fashion Capital (1990–2002)," *Urban Studies* 44, no. 2 (2007): 2449–2464; Monika De Frantz, "From Cultural Regeneration to Discursive Governance: Constructing the Flagship of the 'Museumsquarter Vienna' as a Plural Symbol of Change," *International Journal of Urban and Regional Research* 29, no. 1 (2005): 50–66; Kevin Hetherington, "Manchester's Urbis: Urban Regeneration, Museums and Symbolic Economies," *Cultural Studies* 21, no. 4–5 (2007): 630–649; and Elizabeth Currid, *The Warhol Economy: How Fashion, Art, and Music Drive New York City* (Princeton, NJ: Princeton University Press, 2007).
82. See Ash Amin and Nigel Thrift, "Neo-Marshallian Nodes in Global Networks," *International Journal of Urban and Regional Research* 16, no. 4 (1992): 571–587; and Harvey Molotch, "Place in Product," *International Journal of Urban and Regional Research* 26, no. 4 (2002): 678.
83. Carl Grodach, "Museums as Urban Catalysts: The Role of Urban Design in Flagship Cultural Development," *Journal of Urban Design* 13, no. 2 (2008): 196.
84. Irina Van Aalst and Inez Boogaarts, "From Museum to Mass Entertainment: The Evolution of the Role of Museums in Cities," *European Urban and Regional Studies* 9, no. 3 (2002): 197.
85. Knox, *Cities and Design*, 192; and Graeme Evans, "Tourism, Creativity and the City," in *Tourism, Creativity and Development*, ed. Greg Richards and Julie Wilson (London: Routledge, 2007), 57–72.
86. John Urry coined the term "spectacle-ization" for the process through which places enter the global order, motivated by the desire for recognition as being on the global stage. See John Urry, "The Power of Spectacle," in *Visionary Power, Producing the Contemporary City*, ed. Christine de Baan, Joachim Declerck, and Veronique Patteeuw (Rotterdam: NAi, 2007), 134.

87. Roger Sherman, "Opinion: Risky Business," *Planetzen*, September 18, 2007, http://www.planetizen.com/node/27120.
88. Laurice Taitz, "Shock and Awe," *Sunday Times Lifestyle* [Johannesburg], November 25, 2007, 12.
89. Sharon Zukin, *Naked City: The Death and Life of Authentic Urban Places* (New York: Oxford University Press, 2010), 232.
90. Knox and Pain, "Globalization, Neoliberalism and International Homogeneity," 421.
91. Luis Fernandez-Galiano, "Spectacle and Its Discontents; or, the Joys of Architainment," in *Commodification and Spectacle in Architecture: A Harvard Design Magazine Reader*, ed. William Saunders (Minneapolis: University of Minnesota Press, 2005), 1–7.
92. Jane M. Jacobs, "A Geography of Big Things," *Cultural Geographies* 13, no. 1 (2006): 1–27; and Deyan Sudjic, *The Edifice Complex: How the Rich and Powerful—and Their Architects—Shape the World* (New York: Penguin, 2005), 296.
93. Alessandro Angelini, "Book Review of *Spaces of Global Cultures*," *Future Anterior* 2, no. 1 (2005): 69.
94. Yasser Elsheshtawy, "Navigating the Spectacle: Landscapes of Consumption in Dubai," *Architectural Theory Review* 13, no. 2 (2008): 164.
95. Jacobs, "A Geography of Big Things"; Yasser Elsheshtawy, *Dubai: Behind an Urban Spectacle* (New York: Routledge, 2010), 152–154; Anthony King, "Worlds in the City: Manhattan Transfer and the Ascendance of Spectacular Space," *Planning Perspectives* 11, no. 2 (1996): 87–114; and Anthony King, "(Post)Colonial Geographies: Material and Symbolic," *Historical Geography* 27, no. 1 (1999): 99–118.
96. Susan Fainstein, "Mega-Projects in New York, London and Amsterdam," *International Journal of Urban and Regional Research* 32, no. 4 (2008): 867–880; Ute Lehrer and Jennefer Laidy, "Old Mega-Projects Newly Packaged? Waterfront Redevelopment in Toronto," *International Journal of Urban and Regional Research* 32, no. 4 (2008): 786–803; Erik Swyngedouw, Frank Moulaert, and Arantxa Rodriguez, "Neoliberal Urbanization in Europe: Large-Scale Urban Development Projects and the New Urban Policy," *Antipode* 34, no. 3 (2002): 542–577; and Julian Bolleter, "Charting a Changing Waterfront: A Review of Key Schemes for Perth's Foreshore," *Journal of Urban Design* 19, no. 5 (2014): 569–592.
97. Jon Goss, "Disquiet on the Waterfront: Reflections on Nostalgia and Utopia in the Urban Archetypes of Festival Market Places," *Urban Geography* 17, no. 3 (1996): 221–247; Tim Butler, "Re-urbanizing London Docklands: Gentrification, Suburbanization or New Urbanism?" *International Journal of Urban and Regional Research* 31, no. 4 (2007): 759–781; and Leonie Sandercock and Kim Dovey, "Pleasure, Politics and the 'Public Interest': Melbourne's Riverscape Revitalization," *Journal of the American Planning Association* 68, no. 2 (2002): 151–164.
98. J.-K. Seo, "Re-urbanisation in Regenerated Areas of Manchester and Glasgow," *Cities* 19, no. 2 (2002): 113–121; Brendan Bartley and Kasey Treadwell Shine, "Competitive City: Governance and the Changing Dynamics of Urban Regeneration in Dublin," in *The Globalized City*, 145–166; and João Cabral and Berta Rato, "Urban Development for Competitiveness and Cohesion: The Expo 98 Urban Project in Lisbon," in *The Globalized City*, 209–228.
99. Tim Edensor, *Industrial Waste: Space, Aesthetics and Materiality* (London: Berg, 2005); Tim Edensor, "The Ghosts of Industrial Ruins: Ordering and Disordering Memory in Excessive Space," *Environment and Planning D* 23, no. 6 (2005): 830–849; and Tim Edensor, "Waste Matter: The Debris of Industrial Ruins and the Disordering of the Material World," *Journal of Material Culture* 10, no. 3 (2000): 311–332; and Andreas Huyssen, "Authentic Ruins: Products of Modernity," in *Ruins of Modernity*, ed. Julia Hell and Andreas Schönle (Durham, NC: Duke University Press, 2010), 17–28.
100. Degen, "Fighting for the Global Catwalk," 871.
101. Boyer, "The City of Illusion," 121.
102. See, inter alia, Keith Jacobs, "Waterfront Redevelopment: A Critical Discourse Analysis of the Policy-Making Process Within the Chatham Maritime Project," *Urban Studies* 41, no. 4 (2004):

817–832; Aspa Gospodini, "Urban Waterfront Redevelopment in Greek Cities: A Framework for Redesigning Space," *Cities* 18, no. 5 (2001): 285–295; Brian Doucet, "Variations of the Entrepreneurial City: Goals, Roles and Visions in Rotterdam's *Kop van Zuid* and the Glasgow Harbour Megaprojects," *International Journal of Urban and Regional Research* 37, no. 6 (2013): 2035–2051; David Gordon, "Planning, Design and Managing Change in Urban Waterfront Redevelopment," *Town Planning Review* 67, no. 3 (1996): 261–290; Brian Hoyle, "Global and Local Change on the Port-City Waterfront," *Geographical Review* 90, no. 3 (2000): 395–417; Susan Oakley, "Working Port or Lifestyle Port? A Preliminary Analysis of the Port Adelaide Waterfront Redevelopment," *Geographical Research* 43, no. 3 (2005): 319–326; Richard Marshall, "Contemporary Urban Space-Making at the Water's Edge," in *Waterfronts in Post-Industrial Cities*, ed. Richard Marshall (New York: Spon, 2001): 3–14; Martin Millsbaugh, "Waterfronts as Catalysts for City Renewal," in *Waterfronts in Post-Industrial Cities*, 74–85; and Daniel Galland and Carsten Hansen, "The Roles of Planning in Waterfront Redevelopment: From Plan-Led and Market-Driven Styles to Hybrid Planning?," *Planning Practice and Research* 27, no. 2 (2012): 203–225.

103. Dikmen Bezmez, "The Politics of Urban Waterfront Regeneration: The Case of Haliç (the Golden Horn), Istanbul," *International Journal of Urban and Regional Research* 32, no. 4 (2008): 815–840; and Gene Desfor and John Jørgensen, "Flexible Urban Governance: The Case of Copenhagen's Recent Waterfront Development," *European Planning Studies* 12, no. 4 (2004): 479–496.
104. Sharon Zukin, *Landscapes of Power: From Detroit to Disneyworld* (Berkeley: University of California Press, 1991), 20.
105. Klingmann, *Brandscapes*, 280.
106. Lloyd and Clark," City as Entertainment Machine," 372.
107. Boyer, "Cities for Sale," 181–204; Ann Breen and Dick Rigby, *Waterfronts: Cities Reclaim Their Edge* (New York: McGraw Hill, 1993); Ann Breen and Dick Rigby, *The New Waterfront: A Worldwide Success Story* (London: Thames & Hudson, 1996); Patrick Malone (ed.), *City, Capital and Water* (London: Routledge, 1996); Stephen McGovern, "Evolving Visions of Waterfront Development in Postindustrial Philadelphia: The Formative Role of Elite Ideologies," *Journal of Planning History* 7, no. 4 (2008): 295–326; Glen Norcliffe, Keith Bassett, and Tony Hoare, "The Emergence of Postmodernism on the Urban Waterfront: Geographical Perspectives on Changing Relationships," *Journal of Transport Geography* 4, no. 2 (1996): 123–134; and Sanette Ferreira and Gustav Visser, "Creating an African Riviera: Revisiting the Impact of the Victoria and Alfred Waterfront Development in Cape Town," *Urban Forum* 18, no. 3 (2007): 227–246.
108. Quintin Stevens and Kim Dovey, "Appropriating the Spectacle: Play and Politics in a Leisure Landscape," *Journal of Urban Design* 9, no. 3 (2004): 362.
109. Ron Griffiths, "Making Sameness: Place Marketing and the New Urban Entrepreneurialism," in *Cities, Economic Competition and Urban Policy*, ed. Nick Oatley (London: Paul Chapman, 1998), 43.
110. Klingmann, *Brandscapes*, 280.
111. Zukin, *Landscapes of Power*, 42.
112. Klingmann, *Brandscapes*, 281. See Darryll Kilian and Belinda Dodson, "Between the Devil and the Deep Blue Sea: Functional Conflicts in Cape Town's Victoria and Alfred Waterfront," *Geoforum* 27, no. 4 (1996): 495–507.
113. Weber, "Extracting Value from the City."
114. Andrew Jones, "Issues in Waterfront Regeneration: More Sobering Thoughts—A UK Perspective," *Planning Practice & Research* 13, no. 4 (1998): 433–442.
115. Neil Brenner and Nik Theodore, "Preface: From the 'New Localism' to the Spaces of Neoliberalism," *Antipode* 34, no. 3 (2002): 341–347.
116. Erik Swyngedouw, Frank Moulaert, and Arantxa Rodriguez, " 'The World in a Grain of Sand': Large-Scale Urban Development Projects and the Dynamics of 'Glocal' Transformations," in *The Globalized City*, 9–28.

117. Brian Doucet, Ronald van Kempen, and Jan van Weesep, " 'We're a Rich City with Poor People': Municipal Strategies of New-Build Gentrification in Rotterdam and Glasgow," *Environment and Planning A* 43, no. 6 (2011): 1438–1454.
118. Franco Bianchini, Jon Dawson, and Richard Evans, "Flagship Projects in Urban Regeneration," in *Rebuilding the City: Property-Led Urban Regeneration*, ed. Patsey Healey, Simin Davoudi, Mo O'Toole, David Usher, and Solmaz Tavsanoglu (London: Spon, 1992), 245–255; Holcomb, "Marketing Cities for Tourism"; and Herbert Smyth, *Marketing the City: The Role of Flagship Developments in Urban Regeneration* (London: Spon, 1994).
119. Fernando Diaz Orueta and Susan Fainstein, "The New Mega-Projects: Genesis and Impacts," *International Journal of Urban and Regional Research* 32, no. 4 (2008): 759–767.
120. Deike Peters, "The Renaissance of Inner-City Railway Station Areas: A Key Element in the Contemporary Dynamics of Urban Restructuring," *Critical Planning* 15, no. 1 (2009): 162–185; and Matti Siemiatycki, "The Making of a Mega Project in the Neoliberal City: The Case of Mass Rapid Transit Infrastructure Investment in Vancouver, Canada," *City* 9, no. 1 (2005): 67–83.
121. Swyngedouw, Moulaert, and Rodriguez, "Neoliberal Urbanization in Europe," 542–543.
122. See Idalina Baptista, "Practices of Exception in Urban Governance: Reconfiguring Power Inside the State," *Urban Studies* 50, no. 1 (2013): 43; and Martin J. Murray, *The Urbanism of Exception: The Dynamics of Global City-Building in the Twenty-First Century* (New York: Cambridge University Press, 2017), 277–300.
123. Swyngedouw, Moulaert, and Rodriguez, "Neoliberal Urbanization in Europe," 572, 556.
124. Bent Flyvbjerg, "Machiavellian Megaprojects," *Antipode* 37, no. 1 (2005): 18–22; Malcolm Miles, "Interruptions: Testing the Rhetoric of Culturally Led Urban Development," *Urban Studies* 42, no. 5–6 (2005): 889–911; Graeme Evans, "Measure for Measure: Evaluating the Evidence of Culture's Contribution to Regeneration," *Urban Studies* 42, no. 5–6 (2005): 959–983; Johannes Novy and Deike Peters, "Railway Station Mega-Projects as Public Controversies: The Case of Stuttgart 21," *Built Environment* 38, no. 1 (2012): 128–145; Thomas Gunton, "Megaprojects and Regional Development: Pathologies in Project Planning," *Regional Studies* 37, no. 5 (2003): 505–519; and Paul Gellert and Barbara Lynch, "Mega-Projects as Displacements," *International Social Science Journal* 55, no. 175 (2003): 15–25.
125. Lehrer and Laidley, "Old Mega-Projects Newly Packaged?," 788.
126. Flyvbjerg, "Machiavellian Megaprojects," 18. See also Bent Flyvbjerg, "Design by Deception: The Politics of Megaproject Approval," *Harvard Design Magazine* 22 (Spring–Summer 2005): 50–59.
127. Flyvbjerg, "Machiavellian Megaprojects," 22; Bent Flyvbjerg, Nils Bruzelius, and Werner Rothengatter, *Megaprojects and Risk: An Anatomy of Ambition* (Cambridge: Cambridge University Press, 2003); and Alan Altshuler and David Luberoff, *Mega-Projects: The Changing Politics of Urban Public Investment* (Washington, DC: Brookings Institution Press, 2003).
128. Flyvbjerg, "Machiavellian Megaprojects."
129. Fainstein, "Mega-Projects in New York, London and Amsterdam," 768; Erik Swyngedouw, Frank Moulaert, and Arantxa Rodriguez, "The Contradictions of Urbanizing Globalization," in *The Globalized City*, 247–265; and Rachel Weber, *Why We Overbuild* (Chicago: University of Chicago Press, 2015).
130. Patrick Loftman and Brendan Nevin, "Prestige Projects in Three British Cities," *Urban Studies* 33, no. 6 (1996): 991–1019; and Rachel Weber, "Selling City Futures: The Financialization of Urban Redevelopment Policy," *Economic Geography* 86, no. 3 (2010): 251–274.
131. Michael Friedman, Jacob Bustad, and David Andrews, "Feeding the Downtown Monster: (Re) developing Baltimore's 'Tourist Bubble,' " *City, Culture and Society* 3, no. 3 (2012): 209.
132. Mary McCarthy, *Venice Observed* (New York: Harcourt Brace, 1963), 7–8.
133. Anne Tyler, *The Accidental Tourist* (New York: Berkley, 1986), 11.

134. See Patrick Mullins, "Tourism Urbanization," *International Journal of Urban and Regional Research* 15, no. 3 (1991): 326–342; and John Urry, *The Tourist Gaze: Leisure and Travel in Contemporary Societies* (Thousand Oaks, CA: Sage, 1990).
135. Mark Gottdiener, *Theming of America: Dreams, Visions, and Commercial Spaces* (Boulder, CO: Westview, 1997); Mark Gottdiener, ed., *New Forms of Consumption: Consumers, Culture, and Commodification* (New York: Rowman & Littlefield, 2000; Saskia Sassen and Frank Roost, "The City: Strategic Site for the Global Entertainment Industry," in *The Tourist City*, ed. Susan Fainstein and Dennis Judd (New Haven, CT: Yale University Press, 1999), 143–154; Robert Hollands and Paul Chatterton, "Producing Nightlife in the New Urban Entertainment Economy: Corporatization, Branding, and Market Segmentation," *International Journal of Urban and Regional Research* 27, no. 2 (2003): 361–385; and Kevin Fox Gotham, "Tourism from Above and Below: Globalization, Localization and New Orleans's Mardi Gras," *International Journal of Urban and Regional Research* 29, no. 2 (2005): 309–326.
136. Stephen Britton, "Tourism, Capital, and Place: Towards a Critical Geography of Tourism," *Environment and Planning D* 9, no. 4 (1991): 470.
137. For a wider view, see Vanessa Schwartz, *Spectacular Realities: Early Mass Culture in Fin-de-Siecle Paris* (Berkeley: University of California Press, 1998), 39, 46.
138. The classic text is Dean MacCannell, *The Tourist: A New Theory of the Leisure Class* (London: Macmillan, 1976).
139. Urry, *The Tourist Gaze*, 140–141; and Jon Goss, "Placing the Market and Marketing the Place: Tourist Advertising of the Hawaiian Islands, 1972–1992," *Environment and Planning D* 11, no. 6 (1993): 671–672.
140. Barbara Kirshenblatt-Gimblett, *Destination Culture: Tourism, Museums, and Heritage* (Berkeley: University of California Press, 1998), 194.
141. Kevin Methan, *Tourism in Global Society: Place, Culture, and Consumption* (New York: Palgrave, 2001); and Susan Fainstein, Lily Hoffman, and Dennis Judd, "Introduction," in *Cities and Visitors: Regulating People, Markets, and City Space*, ed. Lily Hoffman, Susan Fainstein, and Dennis Judd (New York: Blackwell, 2003), 1–20.
142. Harvey, *The Condition of Postmodernity*, 294–295.
143. Gotham, "Tourism from Above and Below," 309–310.
144. Molotch, "Place in Product," 677.
145. Stephen Britton, "Tourism, Capital, and Place."
146. See Mimi Sheller and John Urry, eds., *Tourism Mobilities: Places to Play, Places in Play* (New York: Routledge, 2004).
147. Tim Coles, "Urban Tourism, Place Promotion and Economic Restructuring: The Case of Post-Socialist Leipzig," *Tourism Geographies* 5, no. 2 (2003): 190–219.
148. Dean MacCannell, "Staged Authenticity: Arrangements of Social Space in Tourist Settings," *American Journal of Sociology* 79, no. 3 (1973): 589–603; Dean MacCannell, *Empty Meeting Grounds* (New York: Routledge, 1992); Erik Cohen, "Authenticity and Commodification in Tourism," *Annals of Tourism Research* 15, no. 3 (1988): 371–386; Nezar AlSayyad, "Global Norms and Urban Forms in the Age of Tourism: Manufacturing Heritage, Consuming Tradition," in *Consuming Tradition, Manufacturing Heritage: Global Norms and Urban Forms in the Age of Tourism*, ed. Nezar AlSayyad (London: Routledge, 2001), 1–33; and David Howes, "Introduction: Commodities and Cultural Borders," in *Cross-Cultural Consumption: Global Markets, Local Realities*, ed. David Howes (New York: Routledge, 1996), 1–16.
149. Robert David Sack, *Place, Modernity, and the Consumer's World: A Relational Framework for Geographical Analysis* (Baltimore: Johns Hopkins University Press, 1992), 161–162; Erik Cohen, "Contemporary Tourism—Trends and Challenges: Sustainable Authenticity or Contrived Post-

Modernity?," in *Change in Tourism: People, Places, Processes*, ed. Richard Butler and Douglas Pearce (New York: Routledge, 1995), 12–29.

150. George Ritzer, *The Globalization of Nothing* (New York: Pine Forge, 2004), 105.
151. Greg Richards and Julie Wilson, "The Impact of Cultural Events on City Image: Rotterdam, Cultural Capital of Europe 2001," Urban Studies 41, no. 10 (2004): 1931–1951; Andrew Smith, " 'Borrowing' Public Space to Stage Major Events: The Greenwich Park Controversy," *Urban Studies* 51, no. 2 (2014): 247–263; Kevin Fox Gotham, "Marketing Mardi Gras: Commodification, Spectacle and the Political Economy of Tourism in New Orleans," *Urban Studies* 39, no. 10 (2002): 1735–1756; Kevin Fox Gotham, "(Re)Branding the Big Easy: Tourism Rebuilding in Post-Katrina New Orleans," *Urban Affairs Review* 42, no. 6 (2007): 823–850; and Kevin Fox Gotham, *Authentic New Orleans: Tourism, Culture, and Race in the Big Easy* (New York: New York University Press, 2007).
152. Scarlett Cornelissen, "Crafting Legacies: The Changing Political Economy of Global Sport and the 2010 FIFA World Cup™," *Politikon* 34, no. 3 (2007): 241–259; Scarlett Cornelissen, "Scripting the Nation: Sport, Mega-Events and State-Building in Post-Apartheid South Africa," *Sport in Society* 11, no. 4 (2008): 481–493; Scarlett Cornelissen and Kamilla Swart, "The 2010 World Cup as a Political Construct: The Challenge of Making Good on an African Promise," *Sociological Review* 54, no. 2 (2006): 108–123; and C. M. Hall, "Urban Entrepreneurship, Corporate Interests and Sports Mega-Events: The Thin Policies of Competitiveness Within the Hard Outcomes of Neoliberalism," *Sociological Review* 54, no. 2 (2006): 59–70.
153. Greg Adranovich, Matthew Burbank, and Charles Heying, "Olympic Cities: Lessons Learned from Mega-Event Politics," *Journal of Urban Affairs* 23, no. 2 (2001): 113–132; Harry Hiller, "Mega-Events, Urban Boosterism and Growth Strategies: An Analysis of the Objectives and Legitimations of the Cape Town 2004 Olympic Bid," *International Journal of Urban and Regional Research* 24, no. 2 (2000): 449–458; and Elias Beriatos and Aspa Gospodini, "Glocalizing Urban Landscapes: Athens and the 2004 Olympics," *Cities* 21, no. 3 (2004): 187–202.
154. See Juan Miguel Kanai, "Buenos Aires, the Capital of Tango: Tourism, Redevelopment and the Cultural Politics of Neoliberal Urbanism," *Urban Geography* 35, no. 8 (2014): 1111–1117; and Juan Miguel Kanai, "Buenos Aires Beyond (Homo)Sexualized Urban Entrepreneurialism: Queer Geographies of Tango," *Antipode* 47, no. 3 (2015): 652–670.
155. Ian Taylor, "European Ethnoscapes and Urban Redevelopment: The Return of Little Italy in 21st Century Manchester," *City* 4, no. 1 (2000): 27–42; and Stephen Shaw, Susan Bagwell, and Joanna Karmowska, "Ethnoscapes as Spectacle: Reimaging Multicultural Districts as New Destinations for Leisure and Tourism Consumption," *Urban Studies* 41, no. 10 (2004): 1983–2000.
156. Allen J. Scott, "The Cultural Economy of Cities," *International Journal of Urban and Regional Research* 21, no. 2 (1997): 323–339; and Scott, "Cultural-Products Industries and Urban Economic Development."
157. See, for example, Christopher Law, "Urban Tourism and Its Contribution to Economic Regeneration," *Urban Studies* 29, no. 3–4 (1992): 599–618.
158. Jason Hackworth, *The Neoliberal City: Governance, Ideology, and Development in American Urbanism* (Ithaca, NY: Cornell University Press, 2007).
159. See, inter alia, Brenner and Theodore, "Cities and the Geographies of 'Actually Existing Neoliberalism"; Jamie Peck and Adam Tickell, "Neoliberalizing Space," *Antipode* 34, no. 3 (2002): 380–404; and David Harvey, "Neoliberalism and the City," *Studies in Social Justice* 1, no. 1 (2007): 1–13.
160. Akhil Gupta and James Ferguson, "Spatializing States: Toward an Ethnography of Neo-Liberal Governmentality," *American Ethnologist* 29, no. 4 (2000): 981–1002.
161. Jamie Peck, "Review of Jason Hackworth, *The Neoliberal City*," *Annals of the Association of American Geographers* 97, no. 4 (2007): 807.
162. See Harvey, "From Managerialism to Entrepreneurialism."

163. Jeremy Németh, "Defining a Public: The Management of Privately Owned Public Space," *Urban Studies* 46, no. 1 (2009): 2463–2490; Jeremy Németh and Justin Hollander, "Security Zones and New York City's Shrinking Public Space," *International Journal of Urban and Regional Research* 34, no. 1 (2010): 20–34; Jeremy Németh and Stephan Schmidt, "Toward a Methodology for Measuring the Security of Publicly Accessible Spaces," *Journal of the American Planning Association* 73, no. 3 (2007): 283–297; Jeremy Németh, "Control in the Commons: How Public Is Public Space?," *Urban Affairs Review* 48, no. 6 (2012): 814–838.
164. Roger Wettenhall, "The Rhetoric and Reality of Public-Private Partnerships," *Public Organization Review* 3, no. 1 (2003): 77–107; and Gerry Stoker, "Public-Private Partnerships and Urban Governance," in *Partnerships in Urban Governance: European and American Experience*, ed. Jon Pierre (London: Macmillan, 1998), 34–51.
165. Lorene Hoyt, "Do Business Improvement District Organizations Make a Difference? Crime in and Around Commercial Areas in Philadelphia," *Journal of Planning Education and Research* 25, no. 2 (2005): 185–199.
166. Chris Gibson, "Economic Geography, to What Ends? From Privilege to Progressive Performances of Expertise," *Environment and Planning A: Economy and Space* 51, no. 3 (2019): 807.
167. Dallas Rogers and Chris Gibson, "Unsolicited Urbanism: Development Monopolies, Regulatory-Technical Fixes and Planning-as-Deal-Making," *Environment and Planning A: Society and Space* 53, no. 3 (2020): 2, 3.
168. Rogers and Gibson, "Unsolicited Urbanism," 3.
169. Murray, *The Urbanism of Exception*, 280.
170. Rogers and Gibson, "Unsolicited Urbanism," 2
171. Baptista, "Practices of Exception in Urban Governance," 39–54.
172. Murray, *The Urbanism of Exception*, 299–300.
173. Rogers and Gibson, "Unsolicited Urbanism," 2.
174. Tridib Banerjee, "The Future of Public Space: Beyond Invented Streets and Reinvented Places," *Journal of the American Planning Association* 67, no. 1 (2001): 12–13.
175. Scherr, "The Synthetic City."
176. Boyer, "The City of Illusion."
177. Zukin, *The Cultures of Cities*, 49–78; Sharon Zukin, "Urban Lifestyles: Diversity and Standardization in Spaces of Consumption," *Urban Studies* 35, no. 5–6 (1998): 825–839; Zukin, *Naked City*, 35–75; Peter Jackson, "Domesticating the Street," in *Images of the Street: Planning, Identity and Control in Public Space*, ed. Nicholas Fyfe (London: Routledge, 1998), 176–191; and Rowland Atkinson, "Domestication by Cappuccino or a Revenge on Urban Space? Control and Empowerment in the Management of Public Spaces," *Urban Studies* 40, no. 9 (2003): 1829–1843.
178. For example, Anna Minton argues that "the privatisation of the public realm, through the growth of 'private-public' space, produces over-controlled, sterile places that lack connection to the reality and diversity of the local environment, with the result that they all tend to look the same. They also raise serious questions about democracy and accountability. But perhaps most worrying of all are the effects on cohesion, battered by the creation of atomized enclaves of private space which displace social problems into neighbouring districts." Anna Minton, *The Privatisation of Public Space: What Kind of World Are We Building?* (London: Royal Institution of Chartered Surveyors, 2006), 3.
179. See Margaret Kohn, *Brave New Neighborhoods: The Privatization of Public Space* (New York: Routledge, 2004); and Setha Low, "How Private Interests Take Over Public Space," in *The Politics of Public Space*, ed. Setha Low and Neil Smith ((New York: Routledge, 2006), 81–104.
180. Anastasia Loukaitou-Sideris and Tridip Banerjee, *Urban Design Downtown: Poetics and Politics of Form* (Berkeley: University of California Press, 1998), 183–185.
181. Mike Davis, "Fortress Los Angeles: The Militarization of Urban Space," in *Variations on a Theme Park*, ed. Michael Sorkin (New York: Noonday, 1992), 154–180; and Steven Flusty, "The Banality

of Interdiction: Surveillance, Control and the Displacement of Diversity," *International Journal of Urban and Regional Research* 25, no. 3 (2001): 658–664.

182. Neil Smith, *The New Urban Frontier: Gentrification and the Revanchist City* (London: Routledge, 1996); and Jon Coaffee, "Urban Renaissance in the Age of Terrorism: Revanchism, Automated Social Control or the End of Reflection?" *International Journal of Urban and Regional Research* 29, no. 2 (2005): 447–454; Don Mitchell, "The End of Public Space? People's Park, Definitions of the Public Democracy," *Annals of the Association of American Geographers* 85, no. 1 (1995): 108–133; and Steven Flusty, *Building Paranoia: The Proliferation of Interdictory Space and the Erosion of Spatial Justice* (Los Angeles: Los Angeles Forum for Architecture and Urban Design, 1994).
183. Davis, "Fortress Los Angeles"; and Susan Christopherson, "The Fortress City: Privatized Spaces, Consumer Citizenship," in *Post-Fordism: A Reader*, ed. Ash Amin (Oxford: Blackwell, 1994), 409–427.
184. John Allen, "Ambient Power: Berlin's Potsdamer Platz and the Seductive Logic of Public Spaces," *Urban Studies* 43, no. 2 (2006): 443.
185. Jon Coaffee, "Rings of Steel, Rings of Concrete and Rings of Confidence: Designing Out Terrorism in Central London Pre and Post September 11th," *International Journal of Urban and Regional Research* 28, no. 1 (2004): 201–211; Jon Coaffee, "Recasting the 'Ring of Steel': Designing Out Terrorism in the City of London?," in *Cities, War and Terrorism: Towards an Urban Geopolitics*, ed. Stephen Graham (Oxford: Blackwell, 2004), 276–296; and Don Mitchell, "The S.U.V. Model of Citizenship: Floating Bubbles, Buffer Zones and the Rise of the Purely Atomic Individual," *Political Geography* 24, no. 1 (2004): 77–100.
186. Allen, "Ambient Power," 443.
187. Allen, "Ambient Power," 443.
188. Ada Louise Huxtable, *The Unreal America: Architecture and Illusion* (New York: New Press, 1997).
189. Klingmann, *Brandscapes*, 35–36, 280–281. See Murray, "The Quandary of Post-Public Space, 119–144.
190. Atkinson, "Domestication by Cappuccino or a Revenge on Urban Space?"
191. Allen, "Ambient Power," 442–443.
192. Zukin, *The Cultures of Cities*, 28.
193. Allen, "Ambient Power," 449; and Klingmann, *Brandscapes*, 16–17, 313.
194. Allen, "Ambient Power," 449, 441.
195. Zukin, *The Culture of Cities*, xiv; Atkinson, "Domestication by Cappuccino or a Revenge on Urban Space?"; and Regan Koch and Alan Latham, "On the Hard Work of Domesticating a Public Space," *Urban Studies* 50, no. 1 (2013): 6–21.
196. David Madden, "Revisiting the End of Public Space: Assembling the Public in an Urban Park," *City & Community* 9, no. 2 (2010): 187–207.
197. Allen, "Ambient Power," 454.
198. See Christine Hentschel, "City Ghosts: The Haunted Struggles for Downtown Durban and Berlin *Neukölln*," in *Locating Right to the City in the Global South*, ed. Tony Samara, Shenjing He, and Guo Chen (New York: Routledge, 2013), 195–217.

4. STRUGGLING POSTINDUSTRIAL CITIES IN DECLINE

1. Mitch McEwen, "Interview with Keller Easterling About Subtraction," *Archinect*, August 26, 2014, www.https://archinect.com/another/interview-with-keller-easterling-about-subtraction.
2. Ivonne Audirac, "Introduction: Shrinking Cities from Marginal to Mainstream: Views from North America and Europe," *Cities* 75 (2018): 1.

3. Ian Gordon and Nick Buck, "Introduction: Cities in the New Conventional Wisdom," in *Changing Cities: Rethinking Urban Competitiveness, Cohesion and Governance*, ed. Nick Buck, Ian Gordon, Alan Harding, and Ivan Turok (New York: Palgrave Macmillan, 2005), 1–24; and Paul Hardin Kapp and Paul Armstrong, eds., *SynergiCity: Reinventing the Postindustrial City* (Urbana: University of Illinois Press, 2012).
4. Ivan Turok, "The Distinctive City: Pitfalls in the Pursuit of Differential Advantage," *Environment and Planning A* 41, no. 1 (2009): 13. For a critique of this literature, see Joshua Leon, "Global Cities at Any Cost: Resisting Municipal Mercantilism," *City* 21, no. 1 (2017): 6–24.
5. Saskia Sassen, *The Global City: New York, London, Tokyo*, 2nd ed. (Princeton, NJ: Princeton University Press, 2001).
6. See, for example, Hazel Duffy, *Competitive Cities: Succeeding in the Global Economy*, 2nd ed. (London: Spon, 2004); Sako Musterd and Alan Murie, "The Idea of the Creative or Knowledge-Based City," in *Making Competitive Cities*, ed. Sako Musterd and Alan Murie (Oxford: Wiley-Blackwell, 2010), 17–32; and James Simmie and Peter Wood, "Innovation and Competitive Cities in the Global Economy: Introduction to the Special Issue," *European Planning Studies* 10, no. 2 (2002): 149–151.
7. See Martin Boddy and Michael Parkinson, eds., *City Matters: Competitiveness, Cohesion, and Urban Governance* (Bristol, UK: Policy Press, University of Bristol, 2004).
8. See Martin J. Murray, *Urbanism of Exception: City Building at the Start of the Twenty-First Century* (New York: Cambridge University Press, 2017), 40–41; Asher Gherner, *Rule by Aesthetics: World-Class City Making in Delhi* (New York: Oxford University Press, 2015); and Maryam Nastar, "The Quest to Become a World City: Implications for Access to Water," *Cities* 41 (2014): 1–9.
9. Edward Glaeser and William Kerr, "What Makes a City Entrepreneurial?" *Harvard Policy School Policy Briefs*, February 2010, 1–4; Charles Landry and Franco Bianchini, *The Creative City* (London: Demos Comedia, 1995); Richard Florida, *Cities and the Creative Class* (New York: Routledge, 2004); Allen J. Scott, "Creative Cities: Conceptual Issues and Policy Questions," *Journal of Urban Affairs* 28, no. 1 (2006): 1–17; Sam Allwinkle and Peter Cruickshank, "Creating Smarter Cities: An Overview," *Journal of Urban Technology* 18, no. 2 (2011): 1–16; and Alberto Vanolo, "Smartmentality: The Smart City as Disciplinary Strategy," *Urban Studies* 51, no. 5 (2014): 883–898.
10. See William Lever and Ivan Turok, "Competitive Cities: Introduction to the Review," *Urban Studies* 36, no. 5–6 (1999): 791–793; and Iain Begg, "Cities and Competitiveness," *Urban Studies* 36, no. 5–6 (1999): 795–809.
11. Tim Hall and Phil Hubbard, "The Entrepreneurial City: New Urban Politics, New Urban Geography?," *Progress in Human Geography* 20, no. 2 (1996): 153–174; Bob Jessop, "The Entrepreneurial City: Re-imaging Localities, Redesigning Economic Governance, or Restructuring Capital?," in *Transforming Cities: Contested Governance and New Spatial Divisions*, ed. Nick Jewson and Susanne MacGregor (London: Routledge, 1997), 28–41; Bob Jessop, "The Narrative of Enterprise and the Enterprise of Narrative," in *The Entrepreneurial City: Geographies of Politics, Regime, and Representation*, ed. Tim Hall and Phil Hubbard (Chichester, NY: Wiley, 1998), 77–99; and Gordon MacLeod, "From Urban Entrepreneurialism to a 'Revanchist City'? On the Spatial Injustices of Glasgow's Renaissance," *Antipode* 34, no. 3 (2002): 602–623.
12. Jason Hackworth, *The Neoliberal City: Governance, Ideology, and Development in American Urbanism* (Ithaca, NY: Cornell University Press, 2007); and Jamie Peck, Nik Theodore, and Neil Brenner, "Neoliberal Urbanism Redux?," *International Journal of Urban Regional Research* 37, no. 3 (2013): 1091–1099.
13. Monica Degen, "Fighting for the Global Catwalk: Formalizing Public Life in Castlefield (Manchester) and Diluting Public Life in el Raval (Barcelona)," *International Journal of Urban and Regional Research* 27, no. 4 (2003): 867–868.

14. See Manuel Wolff and Thorsten Wiechmann, "Urban Growth and Decline: Europe's Shrinking Cities in a Comparative Perspective 1990–2010," *European Urban and Regional Studies* 25, no. 2 (2018): 122–139; Emmanuèle Cunningham Sabot and Sylvie Fol, "Schrumpfende Stäedte in Westeuropa: Fallstudien aus Frankreich und Grossbritannien," *Berliner Debatte Initial* 18, no. 1 (2007): 22–35. Other shrinking cities in Eastern/Central Europe include Ostrava (Czech Republic), Sosnowiec/Bytom (Poland), Timisoara (Romania), and Donetsk/Makiivka (Ukraine).
15. See, for example, Mike Davis, *Dead Cities and Other Tales* (New York: New Press, 2002), 1–20; and Robert Mark Silverman, Li Yin, and Kelly Patterson, "Dawn of the Dead City: An Exploratory Analysis of Vacant Addresses in Buffalo, NY 2008–2010," *Journal of Urban Affairs* 35, no. 2 (2013): 131–152.
16. Ivonne Audirac, Emmanuèle Cunningham Sabot, Sylvie Fol, and Sergio Torres Moraes, "Declining Suburbs in Europe and Latin America," *International Journal of Urban and Regional Research* 36, no. 2 (2012): 229–230. See Marcel Langner and Wilfried Endlicher, eds., *Shrinking Cities: Effects on Urban Ecology and Challenges for Urban Development* (Frankfurt: Peter Lang, 2007).
17. Robert Beauregard, *Voices of Decline: The Postwar Fate of U.S. Cities*, 2nd ed. (Oxford: Blackwell, 2003); and Robert Beauregard, "The Radical Break in Late Twentieth-Century Urbanization," *Area* 38, no. 2 (2006): 218–220.
18. Nate Millington, "Post-Industrial Imaginaries: Nature, Representation and Ruin in Detroit, Michigan," *International Journal of Urban and Regional Research* 37, no. 1 (2013): 279–296.
19. Keller Easterling, "The Wrong Story," *Perspecta* 41 (2008): 81.
20. Marco Bontje, "Facing the Challenge of Shrinking Cities in East Germany: The Case of Leipzig," *Geojournal* 61 (2004): 13–21.
21. Sylvie Fol and Emmanuèle Cunningham Sabot, "Urban Decline and Shrinking Cities: A Critical Assessment of Approaches to Urban Shrinkage," *Annales de géographie* 119 (2010): 364.
22. Max Rousseau, *"Re-imaging the City* Centre for the Middle Classes: Regeneration, Gentrification and Symbolic Policies in 'Loser Cities,' " *International Journal of Urban and Regional Research* 33, no. 3 (2009): 770–788.
23. Rem Koolhaas, "What Ever Happened to Urbanism?," in *S,M,L,XL*, by Rem Koolhaas and Bruce Mau (New York: Monacelli, 1995), 961.
24. See Lorene Hoyt and André Leroux, *Voices from Forgotten Cities: Innovative Revitalization Coalitions in America's Older Small Cities* (PolicyLink, Citizens' Housing and Planning Association, and MIT School of Architecture and Planning, 2007).
25. Robert Beauregard, "Urban Population Loss in Historical Perspective: United States, 1820–2000," *Environment and Planning A* 41, no. 3 (2009): 514–528.
26. Karina Pallagst, *Growth Management in the U.S.: Between Theory and Practice* (Aldershot, UK: Ashgate, 2007); Karina Pallagst, "Shrinking Cities: Planning Challenges from an International Perspective," in *Cities Growing Smaller*, ed. Steve Rugare and Terry Schwarz (Cleveland: Kent State University and Cleveland Urban Design Collaborative, 2008), 6–16; and Karina Pallagst et al., *The Future of Shrinking Cities: Problems, Patterns and Strategies of Urban Transformation in a Global Context* (Berkeley: Center for Global Metropolitan Studies and the Institute of Urban and Regional Development and the Shrinking Cities International Research Network, 2009). For an excellent effort at explaining decline in the U.S. Rust Belt, see Jason Hackworth, *Manufacturing Decline: How Racism and the Conservative Movement Crush the Rust Belt* (New York: Columbia University Press, 2019).
27. Thorsten Wiechmann and Karina Pallagst, "Urban Shrinkage in Germany and the USA: A Comparison of Transformation Patterns and Local Strategies," *International Journal of Urban and Regional Research* 36, no. 2 (2012): 261–280; and Cristina Martinez-Fernandez, Ivonne Audirac, Sylvie Fol, and Emmanuèle Cunningham Sabot, "Shrinking Cities: Urban Challenges of Globalization," *International Journal of Urban and Regional Research* 36, no. 2 (2012): 213–225.

28. Ivonne Audirac, "Introduction: Shrinking Cities from Marginal to Mainstream: Views from North America and Europe," *Cities* 75 (2018): 1.
29. Caru Browns, "Shrinkage Happens . . . in Small Towns Too! Responding to De-population and Loss of Place in Susquehanna River Towns," *Urban Design International* 18, no. 1 (2013): 61–77; and Paul Knox and Heike Mayer, *Small Towns Sustainability* (Basel, Switzerland: Birkhauser, 2009).
30. Sabrina Deitrick and Robert Beauregard, "From Front-Runner to Also-Ran: The Transformation of a Once Dominant Industrial Region: Pennsylvania, USA," in *The Rise of the Rustbelt: Revitalizing Older Industrial Regions*, ed. Philip Cooke (London: UCL Press, 1995), 52–71.
31. Fernando Ortiz-Moya and Nieves Moreno, "The Incredible Shrinking Japan: Cinematic Representations of Urban Decline," *City* 20, no. 6 (2017): 880–903; and Andreas Luescher and Sujata Shetty, "An Introductory Review to the Special Issue: Shrinking Cities and Towns: Challenge and Responses," *Urban Design International* 18, no. 1 (2013): 1.
32. Sassen, *The Global City*, xvii–xxiv.
33. See, for example, Paul Kantor, H. V. Savitch, and Serena Vicari Haddock, "The Political Economy of Urban Regimes: A Comparative Perspective," *Urban Affairs Review* 32, no. 3 (1997): 348–377; Karina Pallagst, "Shrinking Cities in the United States of America: Three Cases, Three Planning Stories," in *The Future of Shrinking Cities: Problems, Patterns and Strategies of Urban Transformation in a Global Context*, ed. Karina Pallagst et al. (Berkeley: Center for Global Metropolitan Studies and the Institute of Urban and Regional Development and the Shrinking Cities International Research Network, 2009), 81–88.
34. Wolff and Wiechmann, "Urban Growth and Decline" ; and Manuel Wolff, Sylvie Fol, Hélène Roth, and Emmanuèle Cunningham Sabot, "Is Planning Needed? Shrinking Cities in the French Urban System," *Transnational Planning Review* 88, no. 1 (2017): 131–145.
35. William Tabb, "The Wider Context of Austerity Urbanism," *City* 18, no. 2 (2014): 87–100; Jason Hackworth, "Rightsizing as Spatial Austerity in the American Rust Belt," *Environment and Planning A* 47, no. 4 (2015): 766–782; and Jamie Peck, "Austerity Urbanism," *City* 16, no. 6 (2012): 626–655.
36. Hall and Hubbard, "The Entrepreneurial City," 153–174; and Jessop, "The Entrepreneurial City."
37. MacLeod, "From Urban Entrepreneurialism to a 'Revanchist City'?"; and Neil Brenner and Nik Theodore, "Cities and the Geographies of 'Actually Existing Neoliberalism,' " in *Spaces of Neoliberalism: Urban Restructuring in North America and Western Europe*, ed. Neil Brenner and Nik Theodore (Oxford: Wiley-Blackwell, 2002), 23.
38. Ivonne Audirac, "Shrinking Cities: An Unfit Term for American Urban Policy?," *Cities* 75 (2018): 12–19; Mattias Bernt, Annegret Haase, Katrin Großmann, Matthew Cocks, Chris Couch, Caterina Cortese, and Robert Krzysztofik, "How Does(n't) Urban Shrinkage Get Onto the Agenda? Experiences from Leipzig, Liverpool, Genoa and Bytom," *International Journal of Urban and Regional Research* 38, no. 5 (2014): 1749–1766; and Sílvia Sousa and Paulo Pinho, "Planning for Shrinkage: Paradox or Paradigm," *European Planning Studies* 23, no. 1 (2015): 12–32.
39. See Brent Ryan, *Design After Decline: How America Rebuilds Shrinking Cities* (Philadelphia: University of Pennsylvania Press, 2012), ix–xiv.
40. Joshua Akers, "Making Markets: Think Tank Legislation and Private Property in Detroit," *Urban Geography* 34, no. 8 (2013): 1070–1095; Matthias Bernt, "Partnerships for Demolition: The Governance of Urban Renewal in East Germany's Shrinking Cities," *International Journal of Urban and Regional Research* 33, no. 3 (2009): 754–769; Alice Mah, *Industrial Ruination, Community, and Place: Landscapes and Legacies of Urban Decline* (Toronto: University of Toronto Press, 2012); Mathieu Hikaru Desan, "Bankrupted Detroit," *Thesis Eleven* 121, no. 1 (2014): 122–130; and Mathieu Hikaru Desan and George Steinmetz, "The Spontaneous Sociology of Detroit's Hyper-Crisis," in *Reinventing Detroit: The Politics of Possibility*, ed. Michael Peter Smith and L. Owen Kirkpatrick (New Brunswick, NJ: Transaction, 2015), 15–36.

41. Manuel Castells, *The Rise of the Network Society*, 2nd ed. (New York: Blackwell, 2000), 168.
42. W. Brian Arthur, "Increasing Returns and the New World of Business," *Harvard Business Review*, July–August 1996, 100–109.
43. See, for example, Rafael Longoria and Susan Rogers, "Exodus Within an Expanding City: The Case of Houston's Historic African-American Communities," *Urban Design International* 18, no. 1 (2012): 24–42; Ryan, *Design After Decline*, 37–83; and Renia Ehrenfuecht and Marla Nelson, "Recovery in a Shrinking City: Challenges to Right-Sizing Post-Katrina New Orleans," in *The City After Abandonment*, ed. Margaret Dewar and June Manning Thomas (Philadelphia: University of Pennsylvania Press, 2012), 133–150.
44. Michelle Wilde Anderson, "Dissolving Cities," *Yale Law Journal* 121 (2011): 1364; and Michelle Wilde Anderson, "The New Minimal Cities," *Yale Law Journal* 123 (2014): 1118–1227.
45. Anderson, "Dissolving Cities," 1366; and Mike Maciag, "Are Cities That Lost Population Making a Comeback?," *Governing*, May 23, 2013.
46. See Richard Meegan, Patricia Kennett, Gerwyn Jones, and Jacqui Croft, "Global Economic Crisis, Austerity and Neoliberal Urban Governance in England," *Cambridge Journal of Regions, Economy and Society* 7, no. 1 (2014): 137–153.
47. See Daniel Hummel, "Right-Sizing Cities in the United States: Defining Its Strategies," *Journal of Urban Affairs* 37, no. 4 (2015): 397–409; and Thorsten Wiechmann and Marco Bontje, "Responding to Tough Times: Policy and Planning Strategies in Shrinking Cities," *European Planning Studies* 23, no. 1 (2015): 1–11; and Witold Rybczynski and Peter Linneman, "How to Save Our Shrinking Cities," *Public Interest* 135 (1999): 30–44.
48. Kate Stohr, "Shrinking City Syndrome," *New York Times*, February 5, 2004, D8.
49. Ivonne Audirac, Sylvie Fol, and Cristina Martinez-Fernandez, "Shrinking Cities in a Time of Crisis," *Berkeley Planning Journal* 23, no. 1 (2010): 51–57.
50. Philipp Oswalt, "Shrinking Cities: Hypotheses on Urban Shrinking in the 21st Century," accessed September 23, 2013, http://www.shrinkingcities.com/hypothesen.0.html%3F&L=1.html. See also Philipp Oswalt, ed., *Shrinking Cities: Volume 1, International Research* (Ostfildern-Ruit, Germany: Hatje Cantz, 2006).
51. Audirac, Fol, and Martinez-Fernandez, "Shrinking Cities in a Time of Crisis." See also Annegret Haase, Dieter Rink, Katrin Grossmann, Matthias Bernt, Vlad Mykhnenko, "Conceptualizing Urban Shrinkage," *Environment and Planning A* 46, no. 7 (2014): 1519–1534; and Bernt et al., "How Does(n't) Urban Shrinkage Get Onto the Agenda?"
52. John Rennie Short, Lisa Benton, William Luce, and Judith Walton, "The Reconstruction of the Image of a Postindustrial City," *Annals of Association of American Geographers* 83, no. 2 (1993): 222. See also Justin Hollander, "Can a City Successfully Shrink? Evidence from Survey Data on Neighborhood Quality," *Urban Affairs Review* 47, no. 1 (2011): 129–141.
53. Margaret Dewar and Matt Weber, "City Abandonment," in *The Oxford Handbook of Urban Planning*, ed. Rachel Weber and Randall Crane (New York: Oxford University Press, 2011), 564.
54. Timothy Chapin, "From Growth Controls, to Comprehensive Planning, to Smart Growth: Planning's Emerging Fourth Wave," *Journal of the American Planning Association* 78, no. 1 (2012): 5–15; and Nancey Green Leigh and Nathanael Hoelzel, "Smart Growth's Blind Side," *Journal of the American Planning Association* 78, no. 1 (2012): 87–103.
55. Sílvia Sousa and Paulo Pinho, "Shrinkage in Portuguese National Policy and Regional Spatial Plans: Concern or Unspoken Word?" *Journal of Spatial and Organizational Dynamics* 2, no. 4 (2014): 260–275.
56. Hartmut Häußermann and Walter Siebel, *Neue Urbanität* (Frankfurt: Suhrkamp, 1987).
57. See, for example, Dieter Rehfeld, "Disintegration and Reintegration of Production Clusters in the Ruhr Area," in *The Rise of the Rustbelt: Revitalizing Older Industrial Regions*, ed. Philip Cooke (London: UCL Press, 1995), 85–102.

58. Häussermann and Siebel, *Neue Urbanität*, 1–10. Robert Beauregard suggests that the economist Mabel Walker used the term "shrinkage" for the first time in 1947 as a metaphor to draw attention to cities in North America. See, inter alia, Robert Beauregard, *When America Became Suburban* (Minneapolis: University of Minnesota Press, 2006).
59. Hartmut Häußermann and Walter Siebel, "Die Schrumpfende Stadt und die Stadtsoziologie" [The shrinking city and urban sociology], in *Soziologische Stadtforschung*, ed. Jürgen Friedrichs (Opladen, Germany: Westdeutscher Verlag, 1998), 80, 83.
60. An early (rather loose) use of the term can be found in Robert Thornbury, *The Changing Urban School* (London: Taylor & Francis, 1978); Robert Weaver, "The Suburbanization of America or the Shrinking of the Cities," *Civil Rights Digest* 9, no. 3 (1977): 2–11; and Gurney Breckenfeld, "Coping with City Shrinkage," *Civil Engineering* 48, no. 11 (1978): 112–113.
61. Häußermann and Siebel, "Die Schrumpfende Stadt und die Stadtsoziologie," 78–94.
62. Katrin Großmann, Annegret Haase, Dieter Rink, and Annett Steinführer, "Urban Shrinkage in East Central Europe? Benefits and Limits of a Cross-National Transfer of Research Approaches," in *Declining Cities/Developing Cities: Polish and German Perspectives*, ed. Marek Nowak and Michał Nowosielski (Poznan, Poland: Instytut Zachodni, 2008), 77–99; Vlad Mykhnenko, and Ivan Turok, "East European Cities: Patterns of Growth and Decline, 1960–2005," *International Planning Studies* 13, no. 4 (2008): 311–342.
63. Matthias Bernt, "The Limits of Shrinkage: Conceptual Pitfalls and Alternatives in the Discussion of Urban Population Loss," *International Journal of Urban and Regional Research* 40, no. 2 (2016): 442.
64. See Philipp Oswalt, ed., *Shrinking Cities: Volume 2, Interventions* (Ostfildern, Germany: Hatje Cantz, 2006).
65. See Oswalt, *Shrinking Cities: Volume 1.*
66. Philipp Oswalt and Tim Rieniets, eds., *Atlas of Shrinking Cities* (Ostfildern, Germany: Hatje Cantz, 2006).
67. Miriam Axel-Lute, "Small Is Beautiful—Again," Shelterforce, July 23, 2007, http://www.nhi.org/online/issues/150/smallisbeautiful.html.
68. Martinez-Fernandez, Audirac, Fol, and Cunningham Sabot, "Shrinking Cities: Urban Challenges of Globalization," 213, 214. For an early theory of decline, see Jiirgen Friedrichs, "A Theory of Urban Decline: Economy, Demography and Political Elites," *Urban Studies* 30, no. 6 (1993): 907–917.
69. See Robert Beauregard, "Aberrant Cities: Urban Population Loss in the United States, 1820–1930," *Urban Geography* 24, no. 8 (2003): 672–690; and Bernt, "Partnerships for Demolition."
70. See, inter alia, Stephanie Ryberg-Webster and J. Rosie Tighe, "The Legacy of Legacy Cities," in *Legacy Cities: Continuity and Change Amid Decline and Revival*, ed. Stephanie Ryberg-Webster and J. Rosie Tighe (Pittsburghh: University of Pittsburg Press, 2019), 3–17.
71. Stephen V. Ward, "Spatial Planning Approaches to Growth and Shrinkage: A Historical Perspective," in *Parallel Patterns of Shrinking Cities and Urban Growth: Spatial Planning for Sustainable Development of City Regions and Rural Areas*, ed. Robin Ganser and Rocky Piro (Burlington, VT: Ashgate, 2012), 7–26.
72. Beauregard, *Voices of Decline*, 150–178; Anthony Downs, "The Challenge of Our Declining Big Cities," *Housing Policy Debate* 8 (1997): 359–408; and Robert Ebel, "Urban Decline in the World's Developed Economies: An Examination of the Trends," *Research in Urban Economics* 5 (1985): 1–19.
73. Fol and Cunningham Sabot, "Urban Decline and Shrinking Cities," 373–375.
74. Marco Bontje and Sako Musterd, "Understanding Shrinkage in European Regions," *Built Environment* 38, no. 2 (2012): 153–161; and Ivan Turok and Vlad Mykhnenko, "The Trajectories of European Cities, 1960–2005," *Cities* 24, no. 3 (2007): 165–182.
75. Oswalt and Rieniets, *Atlas of Shrinking Cities.*
76. Wolff and Wiechmann, "Urban Growth and Decline."
77. Turok and Mykhnenko, "The Trajectories of European Cities, 1960–2005."

78. Eduardo López Moreno (principal author), *State of the World's Cities 2008/9: Harmonious Cities—UN-HABITAT* (London: Earthscan, 2008), 40.
79. Karina Pallagst, "Das Ende der Wachstumsmaschine," *Berliner Debatte Initial* 18, no. 1 (2007): 4–13.
80. Justin Hollander, "Moving Toward a Shrinking Cities Metric: Analyzing Land Use Changes Associated with Depopulation in Flint, Michigan," *Cityscape: A Journal of Policy Development and Research* 12, no. 1 (2010): 133–151.
81. For Buffalo, see Silverman, Yin, and Patterson, "Dawn of the Dead City"; Susan Hansen, Carolyn Ban, and Leonard Huggins, "Explaining the Brain Drain from Older Industrial Regions: The Pittsburgh Region," *Economic Development Quarterly* 17, no. 2 (2003): 132–147. For Flint, see Victoria Morckel, "Why the Flint, Michigan, USA Water Crisis Is an Urban Planning Failure," *Cities* 62 (2017): 23–27. For Detroit, see Brent Ryan, "The Restructuring of Detroit: City Block Form Change in a Shrinking City, 1900–2000," *Urban Design International* 13 (2008): 156–168.
82. See Rybczynski and Linneman, "How to Save Our Shrinking Cities?"; Alan Mallach, *Facing the Urban Challenge: The Federal Government and America's Older Distressed Cities* (Washington, DC: Metropolitan Policy Program at Brookings Institution, 2010); and Dean Stansel, "Why Some Cities Are Growing and Others Are Shrinking," *Cato Journal* 31, no. 2 (2011): 285–303.
83. Kate Allen, "Shrinking Cities: Population Decline in the World's Rust-Belt Areas," *Financial Times*, June 16, 2017.
84. Jon Teaford, *The Twentieth-Century American City* (Baltimore: Johns Hopkins University Press, 1986); and Robert Beauregard, "Federal Policy and Postwar Urban Decline: A Case of Government Complicity?," *Housing Policy Debate* 12, no. 1 (2001): 129–151. For Flint, see Hollander, "Moving Toward a Shrinking Cities Metric"; and Christine Hannemann, "The Industrial City as a Shrinking City and the Special Case of Flint, MI," in *Industrial Cities: History and Future*, ed. Clemens Zimmerman (Frankfurt-on-Main: Campus Verlag, 2013), 142–164.
85. Michael Gecan, "On Borrowed Time: Urban Decline Moves to the Suburbs," *Boston Review*, March 8, 2008.
86. Matt Woolsey, "America's Fastest-Dying Towns," *Forbes*, December 9, 2008, https://www.forbes.com/2008/12/08/towns-ten-economy-forbeslife-cx_mw_1209dying_slide.html?sh=7e96672214f0; and Joshua Zumbrun, "America's Fastest-Dying Cities," *Forbes*, August 5, 2008, https://www.forbes.com/2008/08/04/economy-ohio-michigan-biz_cx_jz_0805dying.html?sh=2b57f68b180d.
87. Alan Mallach, *Facing the Urban Challenge: The Federal Government and America's Older Distressed Cities* (Washington, DC: Metropolitan Policy Program at Brookings, 2010), 1–13; and Fol and Cunningham Sabot, "Urban Decline and Shrinking Cities," 374–375.
88. Rybczynski and Linneman, "How to Save Our Shrinking Cities."
89. Fol and Cunningham Sabot, "Urban Decline and Shrinking Cities," 373–375.
90. Tim Rieniets, "Shrinking Cities: Causes and Effects of Urban Population Losses in the Twentieth Century," *Nature & Culture* 4 (2009): 251.
91. Oswalt and Rieniets, *Atlas of Shrinking Cities*; and Cristina Martinez-Fernandez, Noya Antonella, and Weyman Tamara, eds., *Demographic Change and Local Development Shrinkage, Regeneration and Social Dynamics* (Paris: OECD, Local and Employment Development, 2012)..
92. Marek Nowak and Michał Nowosielski, eds., *Declining Cities/Developing Cities: Polish and German Perspectives* (Poznan, Poland: Instytut Zachodny, 2008); Ján Buček and Branislav Bleha, "Urban Shrinkage as a Challenge to Local Development Planning in Slovakia," *Morovian Geographical Reports* 21, no. 1 (2013): 5–15; Ján Buček, "Slovakia," in *Urban Issues and Urban Policies in the New EU Countries*, ed. Ronald Van Kempen, Marcel Vermeulen, and Ad Baan (Aldershot, UK: Ashgate, 2005), 79–108; Marieke Maes, Maarten Loopmans, and Christian Kesteloot, "Urban Shrinkage and Everyday Life in Post-Socialist Cities: Living with Diversity in Hrušov, Ostrava, Czech Republic," *Built Environment* 38, no. 2 (2012): 229–243; and Gerd Lintz, Bernhart Müller, and Karl Schmude,

"The Future of Industrial Cities and Regions in Central and Eastern Europe," *Geoforum* 38, no. 3 (2007): 512–519.

93. Lintz, Müller, and Schmude, "The Future of Industrial Cities and Regions in Central and Eastern Europe"; and Šerý Ondřej, Svobodová Hana, Šilhan Zdeněk, and Szczyrba Zdeněk, "Shrinking of Cities in the Czech Republic and Its Reflection on Society: Case Study of Karviná City," *Geographica Pannonica* 22, no. 1 (2018): 68–80.
94. Britt Dale, "An Institutionalist Approach to Local Restructuring: The Case of Four Norwegian Mining Towns," *European Urban and Regional Studies* 9, no. 1 (2002): 5–20; Tadeusz Stryjakiewicz, Przemysław Ciesiółka, and Emilia Jaroszewska, " 'Urban Shrinkage and the Post-Socialist Transformation: The Case of Poland," *Built Environment* 38, no. 2 (2012): 196–213; and Ilinca Păun Constantinescu, "Shrinking Cities in Romania: Former Mining Cities in Valea Jiului," *Built Environment* 38, no. 2 (2012): 214–228.
95. Emmanuèle Cunningham Sabot and Sylvie Fol, "Shrinking Cities in France and Great Britain: A Silent Process," in *The Future of Shrinking Cities:* 24–35. See also Wolff et al., "Is Planning Needed?"
96. Turok and Mykhnenko, "The Trajectories of European Cities, 1960–2005"; and Simón Sánchez-Moral, Ricardo Méndez, and José Prada-Trigo, "Resurgent Cities: Local Strategies and Institutional Networks to Counteract Shrinkage in Avilés (Spain)," *European Planning Studies* 23, no. 1 (2015): 33–52.
97. Rieniets, "Shrinking Cities," 231–254.
98. Benno Brandstetter, Thilo Lang, and Anne Pfeifer, "Umgang mit der schrumpfenden Stadt—ein Debattenüberblick," *Berliner Debatte Initial* 16, no. 6 (2005): 55–68.
99. For example, in the Ruhr Valley, Duisburg has declined more than Dortmund, and the larger cities have declined less than the smaller ones. See Wiechmann and Pallagst, "Urban Shrinkage in Germany and the USA, 261–280.
100. Häußermann, and Siebel, "Die Schrumpfende Stadt und die Stadtsoziologie."
101. Shrinkage and decline are highly uneven in spatial terms. In the post-socialist east, Halle-Leipzig and Dresden, for example, have shrunk less than, say, Chemnitiz. The southern former GDR regions (Saxony and Thuringia) have suffered less abandonment and decline than the northern regions (Mecklenburg-Vorpommern, Pomerania, and Brandenburg). For eastern Germany, see Thilo Lang, "Shrinkage, Metropolization and Peripheralization in East Germany," *European Planning Studies* 20, no. 10 (2012): 1747–1754; Heike Liebmann and Thomas Kuder, "Pathways and Strategies of Urban Regeneration—Deindustrialized Cities in Eastern Germany," *European Planning Studies* 20, no. 7 (2012): 1155–1172; and Eric Erbacher and Sina Nitzsche, "Performing the Double Rupture: Kraftklub, Popular Music and Post-Socialist Urban Identity in Chemnitz, Germany," *International Journal of Cultural Studies* 20, no. 4 (2017): 437–455. Thanks to Sina Nitzsche for pointing these conditions out to me.
102. Birgit Glock and Hartmut Häußermann, "New Trends in Urban Development and Public Policy in Eastern Germany: Dealing with the Vacant Housing Problem at the Local Level," *International Journal of Urban and Regional Research* 28, no. 4 (2004): 919–929.
103. Timothy Moss, " 'Cold Spots' of Urban Infrastructure: 'Shrinking' Process in Eastern Germany and the Modern Infrastructure Ideal," *International Journal of Urban and Regional Research* 32, no. 2 (2007): 437; and Leinhard Lötsher, "Shrinking East German Cities?," *Geographia Polonica* 78, no. 1 (2005): 79–98.
104. Bontje, "Facing the Challenge of Shrinking Cities in East Germany."
105. Cunningham Sabot and Fol, "Shrinking Cities in France and Great Britain."
106. Nadja Kabisch, Dagmar Haase, and Annegret Haase, "Urban Population Development in Europe, 1991–2008: The Examples of Poland and the UK," *International Journal of Urban and Regional Research* 36, no. 6 (2012): 1326–1348.

107. Philip Boland, "Unpacking the Theory-Policy Interface of Local Economic Development: An Analysis of Cardiff and Liverpool," *Urban Studies* 4, no. 5–6 (2007): 1019–1039.
108. Philipp Oswalt, "The Ephemeral," in *Urban Ecology: Detroit and Beyond*, ed. Kyong Park (Hong Kong: Map Book Publishers, 2005), 115–120; and Mark Boyle, Christopher McWilliams, and Gareth Rice, "The Spatialities of Actually Existing Neoliberalism in Glasgow, 1977 to Present," *Geografiska Annaler: Series B, Human Geography* 90, no. 4 (2008): 313–325.
109. Cunningham Sabot and Fol, "Schrumpfende Staedte in Westeuropa."
110. Cunningham Sabot and Fol, "Shrinking Cities in France and Great Britain."
111. Cunningham Sabot and Fol, "Shrinking Cities in France and Great Britain."
112. Chris Couch and Matthew Cocks, "Housing Vacancy and the Shrinking City: Trends and Policies in the UK and the City of Liverpool," *Housing Studies* 28, no. 3 (2013): 499–519; and Matthew Cocks and Chris Couch, "The Governance of a Shrinking City: Housing Renewal in the Liverpool Conurbation, UK," *International Planning Studies* 17, no. 3 (2012): 277–301 See also Ed Ferrari and Jonathan Roberts, "Liverpool—Changing Urban Form," in *Shrinking Cities: Working Papers—Manchester/Liverpool II*, ed. Philipp Oswalt (Berlin: Office Philipp Oswalt, 2004), 130–136.
113. Aidan While, Andrew Jonas, and David Gibbs, "The Environment and the Entrepreneurial City: Searching for the Urban 'Sustainability Fix' in Manchester and Leeds," *International Journal of Urban and Regional Research* 28, no. 3 (2004): 549–569; Michael Edema Leary, "A Lefebvrian Analysis of the Production of Glorious, Gruesome Public Space in Manchester," *Progress in Planning* 85 (2013): 1–52; and Kevin Ward, "Entrepreneurial Urbanism, State Restructuring and Civilizing 'New' East Manchester," *Area* 35, no. 2 (2003): 116–127.
114. Anonymous, "Rustbelt Britain: The Urban Ghosts," *Economist*, October 12, 2013, 67.
115. Steve Hawkes, "'Failing' Cities and Towns," *Telegraph*, October 11, 2013.
116. Kate Allen, "Shrinking Cities: Population Decline in the World's Rust-Belt Areas," *Financial Times*, June 16, 2017; Staff Reporter, "Britain's Decaying Towns: City Sicker," *Economist*, October 12, 2013; and Frances Perraudin, "Ten of Top 12 Most Declining UK Cities are in North of England—Report," *Guardian*, February 29, 2016.
117. David Edgington, "Restructuring Japan's Rustbelt: The Case of Muroran, Hokkaido, 1985–2010," *Urban Affairs Review* 49, no. 4 (2012): 475–524; Winfried Flüchter, "Shrinking Cities in Japan: Between Megalopolises and Rural Peripheries," *Electronic Journal of Contemporary Japanese Studies* (2008), 3–8; Winfried Flüchter, "Urbanisation, City, and City System in Japan Between Development and Shrinking: Coping with Shrinking Cities in Times of Demographic Change," in *Urban Spaces in Japan: Cultural and Social Perspectives*, ed. Christoph Brumann and Evelyn Schulz (London: Routledge, 2012), 15–36; Winfried Flüchter, "Megapolises and Rural Peripheries: Shrinking Cities in Japan," in *Shrinking Cities: Volume 1*, 83–95; and Yasuyuki Fuji, "Shrinkage in Japan," in *Shrinking Cities: Volume 1*, 96–100.
118. Allen, "Shrinking Cities."
119. Sophie Buhnik, "The Dynamics of Urban Degrowth in Japanese Metropolitan Areas: What Are the Outcomes of Urban Recentralisation Strategies?," *Town Planning Review* 88, no. 1 (2017): 79–92; Sophie Buhnik, "The Uneven Impacts of Demographic Decline in a Japanese Metropolis: A Three-Scale Approach to Urban Shrinkage Patterns in the Osaka Metropolitan Area," in *Globalization and New Intra-Urban Dynamics in Asian Cities*, ed. Natacha Aveline-Dubach, Sue-Ching Jou, and Hsin-Huang Michael Hsiao (Taiwan: National Taiwan University Press, 2014), 199–238; and Sophie Buhnik, "From Shrinking Cities to Toshi no Shukushō: Identifying Patterns of Urban Shrinkage in the Osaka Metropolitan Area," *Berkeley Planning Journal* 23, no. 1 (2010): 132–155.
120. Jae Seung Lee, Sehyung Won, and Saehoon Kim, "Describing Changes in the Built Environment of Shrinking Cities: Case Study of Incheon, South Korea," *Journal of Urban Planning and Development* 142, no. 2 (2016).

121. Sukumar Ganapati, "The Paradox of Shrinking Cities in India," *Shrinking Cities: A Global Perspective*, ed. in Harry W. Richardson and Chang Woon Nam (New York: Routledge, 2014), 169–181.
122. See Karina Pallagst, Cristina Martinez-Fernandez, and Thorsten Wiechmann, "Introduction," in *Shrinking Cities: International Perspectives and Policy Implications*, ed. Karina Pallagst, Thorsten Wiechmann, and Cristina Martinez-Fernandez (New York: Routledge, 2014), 1–13; and Ivonne Audirac and Jesús Arroyo Alejandre, "Introduction," in *Shrinking Cities South/North*, ed. Ivonne Audirac and Jesús Arroyo Alejandre (Tallahassee: Florida State University, 2010), 9–36.
123. Cristina Martinez-Fernandez, "De-industrialization and De-innovation in Shrinking Cities," in *Shrinking Cities South/North*, 51–68.
124. Brian Alexander, "What America Is Losing As Its Small Towns Struggle: To Erode Small-Town Culture Is to Erode the Culture of the Nation," *Atlantic*, October 18, 2017; and Paul Overberg, "The Divide Between America's Prosperous Cities and Struggling Small Towns—in 20 Charts," *Wall Street Journal*, December 29, 2017. For the countertrend of small-town demographic growth, see Christiana Brennan, Darlenne Hackler, and Christopher Hoene, "Demographic Change in Small Cities, 1990 to 2000," *Urban Affairs Review* 40, no. 3 (2005): 342–361.
125. See, for example, Juha Kotilainen, Ilkka Eisto, and Eero Vatanen, "Uncovering Mechanisms for Resilience: Strategies to Counter Shrinkage in a Peripheral City in Finland," *European Planning Studies* 23, no. 1 (2015): 53–68; Kadri Leetmaa, Agnes Kriszan, Mari Nuga, and Joachim Burdack, "Strategies to Cope with Shrinkage in the Lower End of the Urban Hierarchy in Estonia and Central Germany," *European Planning Studies* 23, no. 1 (2015), 147–165; and Elena Batunova and Maria Gunko, "Urban Shrinkage: An Unspoken Challenge of Spatial Planning in Russian Small and Medium-Sized Cities," *European Planning Studies* 26, no. 8 (2018): 1580–1597.
126. Fol and Cunningham Sabot, "Urban Decline and Shrinking Cities," 375. See also Turok and Mykhnenko, "The Trajectories of European Cities, 1960–2005."
127. Alan Mallach, "What We Talk About When We Talk About Shrinking Cities: The Ambiguity of Discourse and Policy Response in the United States," *Cities* 69 (2017): 109–115.
128. Wiechmann and Pallagst, "Urban Shrinkage in Germany and the USA"; Thorsten Wiechmann, "Errors Expected—Aligning Urban Strategy with Demographic Uncertainty in Shrinking Cities," *International Planning Studies* 13, no. 4 (2008): 431–446; and contributions to Pallagst, Wiechmann, and Martinez-Fernandez, eds., *Shrinking Cities: International Perspectives and Policy Implications*.
129. See Bernt, "The Limits of Shrinkage."
130. Cunningham Sabot and Fol, "Shrinking Cities in France and Great Britain."
131. Thorsten Wiechmann, "Conversion Strategies Under Uncertainty in Post-Socialist Shrinking Cities: The Example of Dresden in Eastern Germany," in *The Future of Shrinking Cities*, 5–16.
132. Oswalt and Rieniets, eds., *Atlas of Shrinking Cities*; Turok and Mykhnenko, "The Trajectories of European Cities, 1960–2005"; and Karina Pallagst, "Shrinking Cities in the United States of America: Three Cases, Three Planning Stories," in *The Future of Shrinking Cities*, 81–88.
133. Pallagst, "Shrinking Cities: Planning Challenges from an International Perspective."
134. Cunningham Sabot and Fol, "Shrinking Cities in France and Great Britain."
135. Dirk Van de Kaa, "Europe's Second Demographic Transition," *Population Bulletin* 42 (1987): 1–57; and Sophie Buhnik, "From Shrinking Cities to Toshi no Shukushō."
136. Maxell Hartt and Jason Hackworth, "Shrinking Cities, Shrinking Households, or Both?," *International Journal of Urban and Regional Research* 44, no. 6 (2020): 1083–1095.
137. Stefan Siedentop and Stefan Fina, "Who Sprawls Most? Exploring the Patterns of Urban Growth Across 26 European Countries," *Environment and Planning A* 44, no. 11 (2012): 2765–2784; Maxwell Douglas Hartt, "How Cities Shrink: Complex Pathways to Population Decline," *Cities* 75 (2018): 38–49; and Annegret Haase, Anja Nelle, and Alan Mallach, "Representing Urban Shrinkage:

The Importance of Discourse as a Frame for Understanding Conditions and Policy," *Cities* 69 (2017): 95–101.

138. Allan Mallach, Annegret Haase, and Keiro Hattori, "The Shrinking City in Comparative Perspective: Contrasting Dynamics and Responses to Urban Shrinkage," *Cities* 69 (2017): 102–108; Haase, et al., "Conceptualizing Urban Shrinkage"; Wolff and Wiechmann, "Urban Growth and Decline"; and Josje Hoekveld, "Time-Space Relations and the Differences Between Shrinking Regions," *Built Environment* 38, no. 2 (2012): 179–195.

139. See Andy Pike, Danny MacKinnon, Mike Coombes, Tony Champion, David Bradley, Andrew Cumbers, Liz Robson, and Colin Wymer, "Unequal Growth: Tackling Declining Cities," *Report from the Joseph Roundtree Foundation* (Newcastle, UK: Newcastle University, 2016).

140. Sousa and Pinho, "Planning for Shrinkage: Paradox or Paradigm."

141. See Robert Beauregard, "Growth and Depopulation in the United States," in *Rebuilding America's Legacy Cities: New Directions for the Industrial Heartland*, ed. Allan Mallach (New York: American Assembly, Columbia University, 2012), 1–24.

142. See, for example, David Wilson, "Metaphors, Growth Coalition Discourses and Black Poverty Neighborhoods in a U.S. City," *Antipode* 28, no. 1 (1996): 72–96.

143. David Wilson, "Performative Neoliberal-Parasitic Economies and the Making of Political Realities: The Chicago Case," *International Journal of Urban and Regional Research* 35, No. 4 (2011): 691–711; and David Wilson, Dean Beck, and Adrian Bailey, "Neoliberal-Parasitic Economies and Space Building: Chicago's Southwest Side," *Annals of the Association of American Geographers* 99, no. 2 (2009): 301–324.

144. David Harvey, *Spaces of Global Capitalism: A Theory of Uneven Geographical Development* (New York: Verso, 2006).

145. See Raju Das, "David Harvey's Theory of Uneven Geographical Development: A Marxist Critique," *Capital & Class* 41, no. 3 (2017): 511–536.

146. Keller Easterling, *Subtraction* (Berlin: Sternberg, 2014), 1–5, 14–17.

147. Paul Dobraszczyk, *The Dead City: Urban Ruins and the Spectacle of Decay* (London: Taurus, 2017), 3–4.

148. Diana Reckien and Cristina Martinez-Fernandez, "Why Do Cities Shrink?," *European Planning Studies* 19, no. 8 (2011): 1375–1397.

149. Beauregard, *Voices of Decline*, 36–38.

150. Peter Hall, "Balancing European Territory: The Challenge of the Post-Liberal Era," in *Parallel Patterns of Shrinking Cities and Urban Growth: Spatial Planning for Sustainable City Regions and Rural Areas*, ed. Robin Ganser and Rocky Piro (Burlington, VT: Ashgate, 2012), 27–44.

151. See Pallagst, "Shrinking Cities in the United States of America"; and Annett Steinführer and Annegret Haase, "Demographic Change as Future Challenge for Cities in East Central Europe," *Geografiska Annaler B* 89, no. 2 (2007): 183–195.

152. Bernt, "Partnerships for Demolition"; and Moss, " 'Cold Spots' of Urban Infrastructure," 437.

153. Glock and Häußermann, "New Trends in Urban Development and Public Policy in Eastern Germany"; Manfred Kühn Matthias Bernt, and Laura Colini, "Power, Politics and Peripheralization: Two Eastern German Cities," *European Urban and Regional Studies* 24, no. 3 (2017): 258–273; Sarah Dubeaux and Emmanuèle Cunningham Sabot, "Maximizing the Potential of Vacant Spaces Within Shrinking Cities, a German Approach," *Cities* 75 (2018): 6–11; and Dieter Rink, "Wilderness: The Nature of Urban Shrinkage? The Debate on Urban Restructuring and Restoration in Eastern Germany," *Nature and Culture* 4, no. 3 (2009): 275–292.

154. Dagmar Haase, Sven Lautenbach, and Ralf Seppelt, "Modeling and Simulating Residential Mobility in a Shrinking City Using an Agent-Based Approach," *Environmental Modeling and Software* 25, no. 10 (2010): 1225–1240; Wiechmann and Bontje, "Responding to Tough Times"; Bernhard Müller and Stefan Siedentop, "Growth and Shrinkage in Germany: Trends, Perspectives and Challenges for Spatial Planning and Development," *German Journal of Urban Studies* 44, no. 1 (2004): 1–12;

and Peter Franz, "Suburbanization and the Clash of Urban Regimes: Developmental Problems of East Germany Cities in a Free Market Environment," *European Urban and Regional Studies* 7, no. 2 (2000): 135–146.

155. Wiechmann and Pallagst, "Urban Shrinkage in Germany and the USA"; Pallagst, "Shrinking Cities: Planning Challenges from an International Perspective"; and Wiechmann, "Errors Expected."

156. See Justin Hollander, *Polluted and Dangerous: America's Worst Abandoned Properties and What Can Be Done About Them* (Burlington: University of Vermont Press, 2009), 176–202.

157. Cristina Martinez-Fernandez, Chung-Tong Wu, Laura Schatz, Nobuhisa Taira, and José Vargas-Hernandez, "The Shrinking Mining City: Urban Dynamics and Contested Territory," *International Journal of Urban and Regional Research* 36, no. 2 (2012): 245–260; and Dong-Hung Chin, "A Cluster of the Four Coal Mining Cities in Korea from a Global Perspective: How Did the People Overcome a Crisis After a Massive Closure of Mines?," in *Shrinking Cities: International Perspectives and Policy Implications*, 186–204.

158. Allen, "Shrinking Cities." See Chang Yu, Martin de Jong, and Baodong Cheng, " Getting Depleted Resource-Based Cities Back on Their Feet Again—the Example of Yichun in China," *Journal of Cleaner Production* 134 (part A), no. 15 (2016): 42–50; Hu Yinan, "Tough Time for Towns Where Mines Run Dry," *China Daily*, April 16, 2010; and Dan Yang, "Tourism Development and Transformation of Resource-Exhausted City: A Case Study of Wansheng District, Chongqing, China," *International Journal of Humanities and Social Science* 4, no. 10 (2014): 194–198.

159. Ben Marsh, "Continuity and Decline in the Anthracite Towns of Pennsylvania," *Annals of the Association of American Geographers* 77, no. 3 (1987): 337–352; James Randall and R. Geoff Ironside, "Communities on the Edge: An Economic Geography of Resource-Dependent Communities in Canada," *Canadian Geographer* 40, no. 1 (1996): 17–35.

160. Josje Hoekveld, "Understanding Spatial Differentiation in Urban Decline Levels," *European Planning Studies* 22, no. 2 (2014): 362–382; Josje Hoekveld, "Spatial Differentiation of Population Development in a Declining Region: The Case of Saarland," *Geografiska Annaler: Series B, Human Geography* 97, no. 1 (2015): 47–68; and Sylvie Fol, "Urban Shrinkage and Socio-Spatial Disparities: Are the Remedies Worse Than the Disease?," *Built Environment* 38, no. 2 (2012): 259–275.

161. Deborah Popper and Frank Popper, "Small Can Be Beautiful: Coming to Terms with Decline," *Planning* 68, no. 7 (2002): 20–23; and Karina Pallagst and Thorsten Wiechmann, "Shrinking Smart—Städtische Schrumpfungsprozesse in den USA," in *Jahrbuch Stadtregion 2004/05, Schwerpunkt: Schrumpfende Staedte*, ed. Norbert Gestring et al. (Wiesbaden, Germany: V. S. Verlag fuer Sozialwissenschaften, 2005), 105–127.

162. Oleg Golubchikov, Anna Badyina, and Alla Makhrova, "The Hybrid Spatialities of Transition: Capitalism, Legacy and Uneven Urban Economic Restructuring," *Urban Studies* 51, no. 4 (2014): 617–633.

163. Friedrichs, "A Theory of Urban Decline" ; Johann Jessen, "Conceptualizing Shrinking Cities—A Challenge for Planning Theory," in *Parallel Patterns of Shrinking Cities and Urban Growth: Spatial Planning for Sustainable City Regions and Rural Areas*, ed. Robin Ganser and Rocky Piro (Burlington, VT: Ashgate, 2012), 45–58.

164. See Andreas Luescher and Sujata Shetty, "An Introductory Review to the Special Issue: Shrinking Cities and Towns: Challenge and Responses," *Urban Design International* 18, no. 1 (2013): 1.

165. Annegret Haase, Matthias Bernt, Katrin Großmann, Vlad Mykhnenko, and Dieter Rink, "Varieties of Shrinkage in European Cities," *European Urban and Regional Studies* 23, no. 1 (2016): 86–102; and Gert-Jan Hospers, "Coping with Shrinkage in Europe's Cities and Towns," *Urban Design International* 18, no. 1 (2013): 78–89.

166. See, for example, Rūta Ubarevičienė and Maarten van Ham, "Population Decline in Lithuania: Who Lives in Declining Regions and Who Leaves?" Discussion Paper No. 10160 (Bonn: Institute for the Study of Labor, 2016).

167. Tiadla Haartsen and Viktor Venhorst, "Planning for Decline: Anticipating on Population Decline in the Netherlands," *Tijdschrift Voor Economische En Sociale Geografie* 101, no. 2 (2010): 218–227; Annegret Haase, Alexandra Athanasopoulou, and Dieter Rink, "Urban Shrinkage as an Emerging Concern for European Policymaking," *European Urban and Regional Studies* 23, no. 1 (2016): 103–107; Hospers, "Coping with Shrinkage in Europe's Cities and Towns"; and David Reher, "The Demographic Transition Revisited as a Global Process," *Population, Space and Place* 10, no. 1 (2004): 19–41.
168. Manfred Fuhrich and Robert Kaltenbrunner, "Der Osten—jetzt auch im Westen? Stadtumbau-West und Stadtumbau Ost—zwei ungleiche Geschwister," *Berliner Debatte Initial* 16, no. 6 (2005): 41–54; Großmann et al., "Urban Shrinkage in East Central Europe?"; and Annegret Haase, Antonin Vaishar, and Grzegorz Węcławowicz, "The Post-Socialist Condition and Beyond: Framing and Explaining Urban Change in East Central Europe," in *Residential Change and Demographic Challenge: The Inner City of East Central Europe in the 21st Century*, ed. Annegret Haase, Annett Steinführer, Sigrun Kabisch, Katrin Großmann, and Ray Hall (Burlington, VT: Ashgate, 2011), 63–83.
169. For post-socialist cities in eastern Germany, see Bontje, "Facing the Challenge of Shrinking Cities in East Germany"; and Bernt, "Partnerships for Demolition."
170. See, for example, Annett Steinführer, Adam Bierzynski, Katrin Großmann, Annegret Haase, and Sigrun Kabish, Nadja Kabisch, "Population Decline in Polish and Czech Cities During Post-Socialism? Looking Behind the Official Statistics," *Urban Studies* 47, no. 11 (2010): 2325–2346.
171. For "demographic shock," see Annett Steinfuhrer and Annagret Haase, "Demographic Change as Future Challenge for Cities in East Central Europe," *Geografiska Annaler: Series B, Human Geography* 89, no. 2 (2007): 183–195. For demographic decline and low fertility rates, see Sefan Buzar, Philip Ogden, and Ray Hall, "Households Matter: The Quiet Demography of Urban Transformation," *Progress in Human Geography* 29, no. 4 (2005): 413–436; and Stefan Buzar, Philip Ogden, Ray Hall, Annegret Haase, Sigrun Kabisch, and Annett Steinführer, "Splintering Urban Populations: Emergent Landscapes of Reurbanisation in Four European Cities," *Urban Studies* 44, no. 4 (2007): 651–677.
172. Douglas Booth, "Regional Long Waves and Urban Policy," *Urban Studies* 24, no. 6 (1987): 447–459.
173. Beauregard, "The Radical Break in the Late Twentieth-Century Urbanization." For Camden, see Neil Smith, Paul Caris, and Elvin Wyly, "The 'Camden Syndrome' and the Menace of Suburban Decline: Residential Disinvestment and the Discontents in Camden County, New Jersey," *Urban Affairs Review* 36, no. 4 (2001): 497–531.
174. Justin Hollander, *Sunburnt Cities: The Great Recession, Depopulation, and Urban Planning in the American Sunbelt* (New York: Routledge, 2011).
175. Jason Hackworth, "Why There Is No Detroit in Canada," *Urban Geography* 37, no. 2 (2016): 275; Hackworth, *Manufacturing Decline*, 3, 42–53, 61, 75, 117, 227; and Colin Gordon, *Mapping Decline: St. Louis and the Fate of the American City* (Philadelphia: University of Pennsylvania Press, 2008). For natural disaster metaphor, see Laura Reese, "Economic Versus Natural Disasters: If Detroit Had a Hurricane," *Economic Development Quarterly* 20, no. 3 (2006): 219–231.
176. Hackworth, "Why There Is No Detroit in Canada," 27; Bernadette Hanlon, "The Decline of Older, Inner Suburbs in Metropolitan America," *Housing Policy Debate* 19, no. 3 (2008): 423–456; Bernadette Hanlon, " A Typology of Inner-Ring Suburbs: Class, Race, and Ethnicity in U.S. Suburbia," *City & Community* 8, no. 3 (2009): 221–246; and John Rennie Short, Bernadette Hanlon, and Thomas Vicino, "The Decline of Inner Suburbs: The New Suburban Gothic in the United States," *Geography Compass* 1, no. 3 (2007): 641–656.
177. For suburbanization as cause for decline, see Anthony Downs, "Some Realities About Sprawl and Urban Decline," *Housing Policy Debate* 10, no. 4 (1999): 955–974; Jan Bruekner, "Urban Sprawl: Diagnosis and Remedies," *International Regional Science Review* 23, no. 2 (2000): 160–171; Glen Atkinson and Ted Oleson, "Urban Sprawl as a Path Dependent Process," *Journal of Economic Issues*

30, no. 2 (1996): 609–615; and Steven High, *Industrial Sunset: The Making of North America's Rust Belt, 1969–1984* (Toronto: University of Toronto Press, 2003).

178. Kenneth Jackson, *Crabgrass Frontier: The Suburbanization of the United States* (New York: Oxford University Press, 1985).
179. Beauregard, "Federal Policy and Postwar Urban Decline"; and Downs, "The Challenge of Our Declining Big Cities."
180. Yael Allweil, "Shrinking Cities: Like a Slow-Motion Katrina," *Places* 9, no. 1 (2007): 191.
181. Pallagst, "Das Ende der Washstumsmaschine."
182. Hoekveld, "Time-Space Relations and the Differences Between Shrinking Regions"; and Bontje and Musterd, "Understanding Shrinkage in European Regions."
183. Silverman, Yin, and Patterson, "Dawn of the Dead City"; and Allen Goodman, "Central Cities and Housing Supply: Growth and Decline in US Cities," *Journal of Housing Economics* 14, no. 4 (2005): 315–335.
184. Karina Pallagst and Thorsten Wiechmann, "Shrinking Smart—Städtische Schrumpfungsprozesse in den USA," in *Jahrbuch Stadtregion 2004/05, Schwerpunkt: Schrumpfende Staedte*, ed. Norbert Gestring et al. (Wiesbaden, Germany: V. S. Verlag fuer Sozialwissenschaften, 2005), 105–127; and Pallagst, "Shrinking Cities: Planning Challenges from an International Perspective." For Glasgow, see Cunningham Sabot and Fol, "Schrumpfende Städte in Westeuropa: Fallstudien aus Frankreich und Grossbritannien." For Detroit and St. Louis, respectively, see Thomas Sugrue, *The Origins of the Urban Crisis* (Princeton, NJ: Princeton University Press, 1996); and Gordon, *Mapping Decline*. For Baltic cities, see Matas Cirtautas, "Urban Sprawl of Major Cities in the Baltic States," *Architecture and Urban Planning* 7 (2013): 72–79.
185. See Anonymous, "Cities: The Doughnut Effect," *Economist*, January 17, 2002, for the confusing use of this metaphor.
186. Adrián Aguilar and Peter Ward, "Globalization, Regional Development, and Mega-City Expansion in Latin America: Analyzing Mexico City's Peri-Urban Hinterland," *Cities* 20, no. 1 (2003): 3–21; Bryan Roberts, "Globalization and Latin American Cities," *International Journal of Urban and Regional Research* 29, no. 1 (2006): 110–123; and Audirac et al., "Declining Suburbs in Europe and Latin America," 235.
187. Ivonne Audirac, "Shrinking Cities in Latin America: An Oxymoron?," in *Shrinking Cities: A Global Perspective*, ed. Harry W. Richardson and Chang Woon Nam (New York: Routledge, 2014), 28.
188. Cunningham Sabot and Fol, "Schrumpfende Städte in Westeuropa: Fallstudien aus Frankreich und Grossbritannien"; Marie-Fleur Albecker, "The Effects of Globalization in the First Suburbs of Paris: From Decline to Revival?," *Berkeley Planning Journal* 23, no. 1 (2010): 102–131; and Marie-Fleur Albecker and Sylvie Fol, "The Restructuring of Declining Suburbs in the Paris Region," in *Shrinking Cities: International Perspectives and Policy Implications*, 78–98.
189. Pallagst, "Shrinking Cities: Planning Challenges from an International Perspective"; Fuhrich and Kaltenbrunner, "Der Osten—jetzt auch im Westen?"; Henning Nuissl and Dieter Rink, "The 'Production' of Urban Sprawl: Urban Sprawl in Eastern Germany as a Phenomenon of Post-Socialist Transformation," *Cities* 22, no. 2 (2005): 123–134; Daniel Florentin, "The 'Perforated City': Leipzig's Model of Urban Shrinkage Management," *Berkeley Planning Journal* 23, no. 1 (2010): 83–101; Stephan Schmidt, "Sprawl Without Growth in Eastern Germany," *Urban Geography* 32, no. 1 (2011): 105–128; and Alan Mace, Peter Hall, and Nick Gallert, "New East Manchester: Urban Renaissance or Urban Opportunism?," *European Planning Studies* 15, no. 1 (2007): 51–65.
190. Moss, " 'Cold Spots' of Urban Infrastructure"; Timothy Moss, "Utilities, Land-Use Change and Urban Development: Brownfield Sites as 'Cold-Spots' of Infrastructure Networks in Berlin," *Environment and Planning A* 35, no. 3 (2003): 511–529; and Tobias Armborst, Daniel D'Oca, and Georgeen Theodore, "Improve Your Lot," in *Cities Growing Smaller*, ed. Steve Rugare and Terry Schwarz (Cleveland: Kent State University and Cleveland Urban Design Collaborative, 2008), 47–64.

191. See, for example, Daniel Clement, *The Spatial Injustice of Crisis-Driven Neoliberal Urban Restructuring in Detroit* (PhD diss., University of Miami, 2013); Kimberley Kinder, *DIY Detroit: Making Do in a City Without Services* (Minneapolis: University of Minnesota Press, 2016); and Kimberley Kinder, "Guerrilla-Style Defensive Architecture in Detroit: A Self-Provisional Strategy in a Neoliberal Space of Disinvestment," *International Journal of Urban and Regional Research* 38, no. 5 (2014): 1767–1784.
192. J. Rosie Tighe and Joanna Ganning, "The Divergent City: Unequal and Uneven Development in St. Louis," *Urban Geography* 36, no. 5 (2015): 654–673.
193. Nuissl and Rink, "The 'Production' of Urban Sprawl in Eastern Germany," 123.
194. Nuissl and Rink, "The 'Production' of Urban Sprawl in Eastern Germany," 123, 124–127.
195. Kornelia Ehrlich, Agnes Kriszan, and Thilo Lang, "Urban Development in Central and Eastern Europe—Between Peripheralization and Centralization?" *disP—The Planning Review* 48, no. 2 (2012): 77–92; and Simin Davoudi, "Polycentricity in European Spatial Planning: From an Analytical Tool to a Normative Agenda," *European Planning Studies* 11, no. 8 (2003): 979–990.
196. Günter Herfert, "Berlin: The Metropolitan Periphery Between Boom and Shrinkage," *Revue European Spatial Research Policy* 13, no. 2 (2006): 5–22.
197. Betka Zakirova, "Shrinkage at the Urban Fringe: Crisis or Opportunity?," *Berkeley Planning Journal* 23, no. 1 (2010): 58–82. See Miriam Fritsche, Marcel Langner, Hadia Köhler, Anke Ruckes, Daniela Schüler, Betka Zakirova, Katharina Appel, Valeska Contardo, Ellen Diermayer, Mathias Hofmann, Christoph Kulemeyer, Peter Meffert, and Janneke Westermann, "Shrinking Cities: A New Challenge for Research in Urban Ecology," in *Shrinking Cities: Effects on Urban Ecology and Challenges for Urban Development*, ed. Marcell Langner and Wilfried Endlicher (Frankfurt: Peter Lang, 2007), 17–34.
198. In offering a broad yet bland and generic description, Cristina Martinez-Fernandez, Ivonne Audirac, Sylvie Fol, and Emmanuèle Cunningham Sabot, for example, define "shrinking cities" as urban areas—whether a city, part of a city, or an entire metropolitan region—"that have experienced population loss, economic downturn, employment decline and social problems as symptoms of a structural crisis." See Martinez-Fernandez, Audirac, Fol, and Cunningham Sabot, "Shrinking Cities," 213.
199. Aksel Olsen, "Shrinking Cities: Fuzzy Concept or Useful Framework?," *Berkeley Planning Journal* 26, no. 1 (2013): 107.
200. Haase, Rink, Großmann, and Mykhnenko, "Conceptualizing Urban Shrinkage"; and Wiechmann and Pallagst, "Urban Shrinkage in Germany and the USA."
201. Bernt, "The Limits of Shrinkage," 443.
202. Joseph Schilling and John Logan, "Greening the Rust Belt: A Green Infrastructure Model for Right-sizing America's Shrinking Cities," *Journal of the American Planning Association* 74, no. 4 (2008): 451–466.
203. See Philipp Oswalt and Tim Rieniets, "Introduction", in *Atlas of Shrinking Cities*, 6–7.
204. Haase, Bernt, Großmann, Mykhnenko, and Rink, "Varieties of Shrinkage in European Cities."
205. Bernt, "The Limits of Shrinkage," 443.
206. Hollander, *Sunburnt Cities*; Pallagst, "Shrinking Cities in the United States of America"; and Wiechmann and Pallagst, "Urban Shrinkage in Germany and the USA."
207. Olsen, "Shrinking Cities: Fuzzy Concept or Useful Framework?," 107.
208. Hackworth, *Manufacturing Decline*, 35–62. See also Bernt, "Partnerships for Demolition," 757–759.
209. Bernt, "Limitations of Shrinkage," 443.
210. Ilinca Păun Constantinescu, "Shrinking Cities in Romania: Former Mining Cities in Valea Jiului," *Built Environment* 38, no. 2 (2012): 214–228; and Hoekveld, "Spatial Differentiation of Population Development in a Declining Region".
211. Hoekveld, "Understanding Spatial Differentiation in Urban Decline Levels"; Hoekveld, "Time-Space Relations and the Differences between Shrinking Regions"; Wolff and Wiechmann, "Urban

Growth and Decline"; and Haase, Bernt, Großmann, and Mykhnenko, "Varieties of Shrinkage in European Cities."

212. Chris Couch, Jay Karecha, Henning Nuissl, and Dieter Rink, "Decline and Sprawl: An Evolving Type of Urban Development—Observed in Liverpool and Leipzig," *European Planning Studies* 13, no. 1 (2005): 117–136.
213. Hoekveld, "Understanding Spatial Differentiation in Urban Decline Levels," 362–363, 369–370, 374–375; and Reckien and Martinez-Fernandez, "Why Do Cities Shrink?"
214. Olsen, "Shrinking Cities: Fuzzy Concept or Useful Framework?," 107.
215. Beauregard, *Voices of Decline*, 35. See also Hoekveld, "Understanding Spatial Differentiation in Urban Decline Levels," 363–364.
216. Beauregard, *Voices of Decline*, 36.
217. There are exceptions. For example, some scholars have begun to investigate the cumulative and self-reinforcing process of urban shrinkage. See, for example, Nina Schwarz and Dagmar Haase, "Urban Shrinkage: A Vicious Circle for Residents and Infrastructure? Coupling Agent-Based Models on Residential Location Choice and Urban Infrastructure Development," *Proceedings of the Fifth Biennial Conference of the International Environmental Modelling and Software Society* (*iEMSs*), Ottawa, July 5–8, 2010, 2:817–824.
218. Bernard Müller and Stefan Siedentop, "Growth and Shrinkage in Germany: Trends, Perspectives and Challenges for Spatial Planning and Development," *German Journal of Urban Studies* 44, no. 1 (2004): 14–32; and Ivonne Audirac, "Urban Shrinkage Amid Fast Metropolitan Growth (Two Faces of Contemporary Urbanism)," in *The Future of Shrinking Cities*, 69–80.
219. Wolff and Wiechmann, "Urban Growth and Decline," 122–123; and Hoekveld, "Time-Space Relations and the Differences Between Shrinking Regions"; and Katrin Großmann, Marco Bontje, Annegret Haase, and Vlad Mykhnenko, "Shrinking Cities: Notes for the Further Research Agenda," *Cities* 35 (2013): 221–225.
220. Hoekveld, "Time-Space Relations and the Differences Between Shrinking Regions," 179–180.
221. José Reis, Elisabete Silva, and Paulo Pinho, "Measuring Space: A Review of Spatial Metrics for Urban Growth and Shrinkage," in *The Routledge Handbook of Planning Research Methods*, ed. Elisabete Silva, Patsey Healey, Neil Harris, and Pieter Van den Broeck (New York: Routledge, 2014), 279.
222. George Bentley, Priscilla McCutcheon, Robert Cromley, and Dean Hanink, "Race, Class, Unemployment, and Housing Vacancies in Detroit: An Empirical Analysis," *Urban Geography* 37, no. 5 (2016): 785–800; and Siobhan Gregory, "Detroit Is a Blank Slate: Metaphors in the Journalistic Discourse of Art and Entrepreneurship in the City of Detroit," *EPIC [Ethnographic Praxis Industry Conference] 2012 Proceedings* (2012): 217–233.
223. See, for example, José Reis, Elisabete Silva, and Paulo Pinho, "Spatial Metrics to Study Urban Patterns in Growing and Shrinking Cities," *Urban Geography* 37, no. 2 (2016): 246–271; Reis, Silva, and Pinho, "Measuring Space"; and Dieter Rink and Sigrun Kabisch, "Introduction: The Ecology of Shrinkage," *Nature and Culture* 4, no. 3 (2009): 223–230.
224. Bernt, "Limits of Shrinkage," 445.
225. Rink and Kabisch, "Introduction: The Ecology of Shrinkage."
226. Sousa and Pinho, "Planning for Shrinkage: Paradox or Paradigm."
227. Olsen, "Shrinking Cities: Fuzzy Concept or Useful Framework?," 107; Hollander, Pallagst, Schwartz, and Popper, "Planning Shrinking Cities"; and Bernt, "The Limits of Shrinkage," 441–442.
228. Olsen, Shrinking Cities: Fuzzy Concept or Useful Framework?" 107.
229. Bernt, "The Limits of Shrinkage."
230. See, for example, Margaret Dewar and June Manning Thomas, "Introduction," in *The City After Abandonment*, 1–14.
231. Bernt, "The Limits of Shrinkage," 441.
232. Bernt, "The Limits of Shrinkage," 444.

233. Haase, Bernt, Großmann, Mykhnenko, and Rink, "Varieties of Shrinkage in European Cities," 86; and Großmann, Bontje, Haase, and Mykhnenko, "Shrinking Cities: Notes for the Further Research Agenda."
234. Sylvie Fol, "Urban Shrinkage and Socio-Spatial Disparities: Are the Remedies Worse Than the Disease?," *Built Environment* 38, no. 2 (2012): 259–275; and Tighe and Ganning, "The Divergent City."
235. Chris Hamnett, *Unequal City: London in the Global Arena* (New York: Routledge, 2003); Peter Marcuse and Ronald van Kempen, eds., *Globalizing Cities: A New Spatial Order?* (London: Blackwell, 2000); Paul Cheshire, "Resurgent Cities, Urban Myths and Policy Hubris," *Urban Studies* 43, no. 8 (2006): 1231–1246; and Trudi Bunting and Piere Filion, "Uneven Cities: Addressing Rising Inequality in the Twenty-First Century," *Canadian Geographer* 45, no. 1 (2001): 126–141.
236. See Russell Weaver, Sharmistha Bagchi-Sen, Jamie Knight, and Amy Frazier, *Shrinking Cities: Understanding Urban Decline in the United States* (New York: Routledge, 2017).
237. Hoekveld, "Understanding Spatial Differentiation in Urban Decline Levels"; and Hoekveld, "Time-Space Relations and the Differences Between Shrinking Regions."
238. Davia Downey and Laura Reese, "Sudden Versus Slow Death of Cities: New Orleans and Detroit," *Du Bois Review: Social Science Research on Race* 14, no. 1 (2017): 219–243.
239. Wolff and Wiechmann, "Urban Growth and Decline," 132.
240. See, for example, Couch et al., "Decline and Sprawl."
241. Cristina Martinez-Fernandez, Tamara Weyman, Sylvie Fol, Ivonne Audirac, Emmanuèle Cunningham Sabot, Thorsten Wiechmann, and Hiroshi Yahagi, "Shrinking Cities in Australia, Japan, Europe and the USA: From a Global Process to Local Policy Responses," *Progress in Planning* 105 (2016): 1–48.
242. Moss, "'Cold Spots' of Urban Infrastructure," 437.

5. SPRAWLING MEGACITIES OF HYPERGROWTH: THE UNPLANNED URBANISM OF THE TWENTY-FIRST CENTURY

1. UN-Habitat, *The State of the World's Cities* (London: Earthscan, 2006); and Ash Amin, "Telescopic Urbanism and the Poor," *City* 17, no. 4 (2013): 476.
2. Dirk Kruijt and Kees Koonings, "The Rise of Megacities and the Urbanization of Informality, Exclusion and Violence," in *Megacities: The Politics of Urban Exclusion and Violence in the Global South*, ed. Kees Koonings and Dirk Kruijt (London: Zed, 2009), 8–26; Edward Soja and Miguel Kanai, "The Urbanization of the World," in *Implosions/Explosions: Towards a Study of Planetary Urbanization*, ed. Neil Brenner (Berlin: Jovis, 2014), 142–159; and Ricky Burdett and Philipp Rode, "Living in an Urban Age," in *Living in the Endless City*, ed. Ricky Burdett and Deyan Sudjic (London: Phaildon, 2011), 8–43. See also contributions to Frauke Kraas, Surinder Aggarwal, Martin Coy, Günter Mertins, eds., *Megacities: Our Global Urban Future* (New York: Springer, 2012); and Andre Sorensen and Junichiro Okata, eds., *Megacities: Urban Form, Governance, and Sustainability* (New York: Springer, 2011).
3. Eugenie Birch and Susan Wachter, "World Urbanization: The Critical Issue of the Twenty-First Century," in *Global Urbanization*, ed. Eugenie Birch and Susan Wachter (Philadelphia: University of Pennsylvania Press, 2011), 3–23. See also Austin Zeiderman, "Cities of the Future? Megacities and the Space/Time of Urban Modernity," *Critical Planning* (2008): 26.
4. Zeiderman, "Cities of the Future?," 27.
5. Zeiderman, "Cities of the Future?," 26.
6. Tahl Kaminer, Miguel Robles-Durán, and Heidi Sohn, "Introduction: Urban Asymmetries," in *Urban Asymmetries: Studies and Projects on Neoliberal Urbanization*, ed. Tahl Kaminer, Miguel

Robles-Durán, and Heidi Sohn (Rotterdam: 010 Publishers, 2011), 10–20; and Martin J. Murray, *The Urbanism of Exception: The Dynamics of Global City Building in the Twenty-First Century* (New York: Cambridge University Press, 2016).

7. AbdouMaliq Simone, "The Urban Poor and Their Ambivalent Exceptionalities: Some Notes from Jakarta," *Current Anthropology* 56, no. S11 (2015): 515–523; Javier Auyero, "Taking Bourdieu to the Shantytown," *International Journal of Urban and Regional Research* 45, no. 1 (2021): 176–185; and Gabriel Giorgi, "Improper Selves: Cultures of Precarity," *Social Text* 31, no. 2 [115](2013): 69–81.
8. Edgar Pieterse, *City Futures: Confronting the Crisis of Urban Development* (London: Zed, 2008), 1–15.
9. Zygmunt Bauman, *Wasted Lives: Modernity and Its Outcasts* (Cambridge: Polity, 2004); and Jan Breman, "Slumlands," *New Left Review* 40 (2006): 141–148.
10. Abidin Kusno, Melani Budianta, and Hilmar Faride, "Editorial Introduction: Runaway City/Leftover Spaces," *Inter-Asia Cultural Studies* 12, no. 4 (2011): 474; Timir Angin, "Jakarta Leftover Spaces," *Inter-Asia Cultural Studies* 12, no. 4 (2011): 568–583; and Danny Hoffman, *Monrovia Modern: Urban Form and Political Imagination in Liberia* (Durham, NC: Duke University Press, 2017).
11. Mike Davis, *Planet of Slums* (New York: Verso, 2006), 20–49. See also Jeremy Seabrook, *In the Cities of the South: Some Scenes from a Developing World* (New York: Verso, 1996); and Robert Neuwirth, *Shadow Cities: A Billion Squatters, a New World* (New York: Routledge, 2005).
12. See Tim Bunnell and Andrew Harris, "Re-Viewing Informality: Perspectives from Urban Asia," *International Development Planning Review* 34, no. 4 (2012): 339–340.
13. See, for example, Romola Sanyal, "Slum Tours as Politics: Global Urbanism and Representations of Poverty," *International Political Sociology* 9, no. 1 (2015): 93–96; Vyjayanthi Rao, "Slum as Theory," *International Journal of Urban and Regional Research* 30, no. 1 (2006): 225–232; and Ananya Roy, "Slumdog Cities: Rethinking Subaltern Urbanism," *International Journal of Urban and Regional Research* 35, no. 2 (2011): 223–238.
14. Gareth A. Jones and Romola Sanyal, "Spectacle and Suffering: The Mumbai Slum as a Worlded Space," *Geoforum* 65 (2015): 437.
15. "Representing slum life as a capitalist libertarian dream free from an oppressive state, in which dynamic individuals prosper through hard work while those less fortunate display stoic resilience bolstered by the strength of community relies on other potential misreadings" (Jones and Sanyal, "Spectacle and Suffering," 438). See also Julia Meschkank, "Investigations Into Slum Tourism in Mumbai: Poverty Tourism and the Tensions Between Different Constructions of Reality," *GeoJournal* 76, no. 1 (2011): 47–62; Peter Dyson, "Slum Tourism: Representing and Interpreting 'Reality' in Dharavi, Mumbai," *Tourism Geographies* 14, no. 2 (2012): 254–274; Thomas Frisch, "Glimpses of Another World: The *Favela* as a Tourist Attraction," *Tourism Geographies* 14, no. 2 (2012): 320–338; Kim Dovey and Ross King, "Informal Urbanism and the Taste for Slums," *Tourism Geographies* 14, no. 2 (2012): 275–293; and Fabian Frebzel, Ko Koens, Malke Steinbrink, and Christian Rogerson, "Slum Tourism: State of the Art," *Tourism Review International* 18, no. 4 (2015): 237–252.
16. For an argument along these lines, see Amin, "Telescopic Urbanism and the Poor"; and Ash Amin, "The Urban Condition: A Challenge to Social Science," *Public Culture* 25, no. 2 (2013): 201–208.
17. AbdouMaliq Simone, "Cities of Uncertainty: Jakarta, the Urban Majority, and Inventive Political Technologies," *Theory, Culture & Society* 30, no. 7–8 (2013): 243–263; AbdouMaliq Simone, "The Visible and Invisible: Remaking Cities in Africa," in *Under Siege: Four African Cities: Freetown, Johannesburg, Kinshasa, Lagos*, ed. Okwui Enwezor et al. (Ostfildern-Ruit, Germany: Hatje Cantz, 2002), 23–44; and Vyjayanthi Rao, "Proximate Distances: The Phenomenology of Density in Mumbai," *Built Environment* 33, no. 2 (2007): 227–248.
18. Ravi Sundaram, *Pirate Modernity: Delhi's Media Urbanism* (New York: Routledge, 2010), 113–124; Ravi Sundaram, "Recycling Modernity: Pirate Electronic Cultures in India," *Third Text* 13, no. 47

(1999): 59–65; and Ravi Sundaram, "Revisiting the Pirate Kingdom," *Third Text* 23, no. 3 (2009): 335–345.

19. Teddy Cruz, "Tijuana Case Study: Tactics of Invasion: Manufactured Sites," *Architectural Design* 75, no. 5 (2005): 32–37; Rahul Mehrotra, "Negotiating the Static and Kinetic Cities: The Emergent Urbanism of Mumbai," in *Other Cities, Other Worlds: Urban Imaginaries in a Globalizing Age*, ed. Andreas Huyssen (Durham, NC: Duke University Press, 2008), 205–218; and Garth Myers, "Remaking the Edges: Surveillance and Flows in Sub-Saharan Africa's New Suburbs," in *The Design of Frontier Spaces: Control and Ambiguity*, ed. Carolyn Loeb and Andreas Luescher (Burlington, VT: Ashgate, 2015), 45–64.
20. AbdouMaliq Simone, "On the Worlding of African Cities," *African Studies Review* 44, no. 2 (2001: 15–41; AbdouMaliq Simone, *City Life from Jakarta to Dakar* (Routledge, New York, 2010); AbdouMaliq Simone, "Straddling the Divides: Remaking Associational Life in the Informal African City," *International Journal of Urban and Regional Research* 25, no. 1 (2001): 102–117; and AbdouMaliq Simone, "Resource of Intersection: Remaking Social Collaboration in Urban Africa," *Canadian Journal of African Studies* 37, no. 2–3 (2003): 513–538.
21. Garth Myers and Martin J. Murray, "Introduction: Situating Contemporary Cities in Africa," in *Cities in Contemporary Africa*, ed. Martin J. Murray and Garth Myers (New York: Palgrave/Macmillan), 1–31.
22. See, inter alia, Lansana Gberie, "Review Article: Africa—The Troubled Continent," *African Affairs* 104, no. 415 (2005): 337–342; George Packer, "The Megacity: Decoding the Chaos of Lagos, *New Yorker*, November 13, 2006, 62–75; and Eugene Linden, "The Exploding Cities of the Developing World," *Foreign Affairs* 75, no. 1 (1996): 52–65.
23. Peter Adey, "Air/Atmospheres of the Megacity," *Theory, Culture & Society* 30, no. 7–8 (2013): 291–308.
24. Eeva Berglund, "Troubled Landscapes of Change: Limits and Natures in Grassroots Urbanism," in *Dwelling in Political Landscapes: Contemporary Anthropological Perspectives*, ed. Ann Lounela, Eeva Berglund, and Timo Kallinen (Helsinki: Finnish Literature Society, 2019), 196.
25. Jennifer Robinson, "Global and World Cities: A View from Off the Map," *International Journal of Urban and Regional Research* 26, no. 3 (2002): 531; Jennifer Robinson, "Urban Geography: World Cities, or a World of Cities," *Progress in Human Geography* 29, no. 6 (2005): 757–765; and Jennifer Robinson, "Living in Dystopia: Past, Present and Future," in *Noir Urbanisms*, ed. Gyan Prakash (Princeton, NJ: Princeton University Press, 2010), 218–240.
26. See Jennifer Robinson, *Ordinary Cities: Between Modernity and Development* (New York: Routledge, 2006).
27. Colin McFarlane, "Urban Shadows: Materiality, the 'Southern City' and Urban Theory," *Geography Compass* 2, no. 2 (2008): 341.
28. For an example of this approach, see Andrew Hamer, "Economic Impacts of Third World Mega-Cities: Is Size the Issue?," in *Mega-City Growth and the Future*, ed. Roland Fuchs, Ellen Brennan, Joseph Chamie, Fuchen Lo, and Joha Uitto (Tokyo: United Nations University Press, 1994), 172–191.
29. Ananya Roy, "Urban Informality: Toward an Epistemology of Planning," *Journal of the American Planning Association* 71, no. 2 (2005): 147; and Robinson, "Global and World Cities," 547–548.
30. Doreen Massey, "Cities in the World," in *City Worlds*, ed. Doreen Massey, John Allen and Steve Pile (London: Routledge, 1999), 115.
31. See AbdouMaliq Simone, *For the City Yet to Come: Changing African Life in Four Cities* (Durham, NC: Duke University Press, 2004); Pieterse, *City Futures*, 16–38; Zeiderman, "Cities of the Future?," 27; and Hans Westlund, "Urban Futures in Planning, Policy and Regional Science: Are We Entering a Post-Urban World?" *Built Environment* 40, no. 4 (2014): 447–457.
32. Davis, *Planet of Slums*, 19.
33. Neuwirth, *Shadow Cities*, 1–23.

34. Davis, *Planet of Slums*, 19.
35. Matthew Gandy, "Learning from Lagos," *New Left Review* 33 (2005): 38–40.
36. Edward Glaeser, *Triumph of the City: How Our Greatest Invention Makes Us Richer, Smarter, Greener, Healthier, and Happier* (New York: Penguin, 2011).
37. Paul Webster and Jacob Burke, "How the Rise of the Megacity Is Changing the Way We Live," *Observer*, January 21, 2012. See Jo Beall, "Cities, Terrorism, and Development," *Journal of International Development* 18, no. 1 (2006): 105–120; Dennis Rodgers, "Slum Wars of the 21st Century: Gangs, Mano Dura, and the New Urban Geography of Conflict in Central America," *Development and Change* 40, no. 5 (2009): 949–976; Dennis Rodgers, "The State as a Gang: Conceptualising the Governmentality of Violence in Contemporary Nicaragua," *Critique of Anthropology* 26, no. 3 (2006): 315–330; and Kess Koonings and Dirk Kruijt, "Fractured Cities, Second-Class Citizenship, and Urban Violence," in *Fractured Cities: Social Exclusion, Urban Violence and Contested Spaces in Latin America*, ed. Kess Koonings and Dirk Kruijt (London: Zed, 2007), 7–22.
38. Zeiderman, "Cities of the Future?," 26.
39. Gandy, "Learning from Lagos," 38.
40. Tayyab Mahmud, " 'Surplus Humanity' and the Margins of Legality: Slums, Slumdogs, and Accumulation by Dispossession," *Chapman Law Reviewno.* 4, 1 (2010): 4; and Denise Ferreira da Silva, "Towards a Critique of the Socio-Logos of Justice: The Analytics of Raciality and the Production of Universality," *Social Identities* 7, no. 3 (2001): 421–454.
41. See, for example, Richard Norton, "Feral Cities," *Naval War College Review* 56, no. 4 (2003): 97; Robert Bunker and John Sullivan, "Integrating Feral Cities and Third Phase Cartels/Third Generation Gangs Research: The Rise of Criminal (Narco) City Networks and BlackFor," *Small Wars & Insurgencies* 22, no. 5 (2011): 764–786; Richard J. Norton, "Feral Cities: Problems Today, Battlefields Tomorrow?," *Marine Corps University Journal* 1, no. 1 (2010): 50–77; Matt Carr, "Slouching Towards Dystopia: The New Military Futurism," *Race & Class* 51, no. 3 (2010): 13–32; and Mathieu Atkins, "Gangs of Karachi: Meet the Mobsters Who Run the Show in One of the World's Deadliest Cities," *Harpers* 331, no. 1984 (September 2015): 34–46.
42. Packer, "The Megacity," 62–63.
43. Robert Kaplan, "The Coming Anarchy: How Scarcity, Crime, Overpopulation, Tribalism, and Disease Are Rapidly Destroying the Social Fabric of Our Planet," *Atlantic Monthly*, February 1994, 46.
44. See David Harvey, *Spaces of Global Capitalism: A Theory of Uneven Geographical Development* (New York: Verso, 2006), 69–116.
45. Zeiderman, "Cities of the Future?," 26; Davis, *Planet of Slums*, 199–206; and Kaplan, "The Coming Anarchy," 46.
46. Zeiderman, "Cities of the Future?," 27.
47. Linden, "The Exploding Cities of the Developing World," 53.
48. Matthew Power, "The Magic Mountain: Trickle-Down Economics in a Philippine Garbage Dump," *Harper's Magazine*, December 2006, 57.
49. Kaplan, "The Coming Anarchy," 46.
50. Davis, *Planet of Slums*, 20; Zeiderman, "Cities of the Future?," 27; and Liza Weinstein, *The Durable Slum: Dharavi and the Right to Stay Put in Globalizing Mumbai* (Minneapolis: University of Minnesota Press, 2014), 9–15.
51. Rem Koolhaas suggests that rather than viewing the conditions of dysfunctionality as a distinctive way of becoming modern, it can just as easily be argued that "Lagos represents a developed, extreme paradigmatic case-study of a city at the forefront of globalizing modernity. This is to say that Lagos is not catching up with us. Rather we may be catching up with Lagos." Rem Koolhaas and Harvard Project on the City, "Lagos," in *Mutations*, ed. Stefano Boeri, Sanford Kwinter, Nadia Tazi, and Hans Ulrich Obrist (Barcelona: Actar, 2001), 653. This hyperbolic gesture, as Okwui Enwezor contends, can be construed as a "celebration of the pathological . . . the unstable and the culture of make-do."

Okwui Enwezor, "Terminal Modernity: Rem Koolhaas' Discourse on Entropy," in *What Is OMA: Considering Rem Koolhaas and the Office of Metropolitan Architecture*, ed. V. Patteeuw (Rotterdam: NAi, 2003), 116.

52. Zeiderman, "Cities of the Future?," 27; Breman, "Slumlands"; and Paul Mason, "Slumlands—Filthy Secret of the Modern Mega-City," *New Statesman*, August 8, 2011.
53. Much of the analysis in the following paragraphs is derived from a critical reading of Zeiderman, "Cities of the Future?," 26; and Robinson, "Global and World Cities."
54. Austin Zeiderman, *Endangered City: The Politics of Security and Risk in Bogotá* (Durham, NC: Duke University Press, 2016), 200; and James Ferguson, *Global Shadows: Africa in the Neoliberal World* (Durham, NC: Duke University Press, 2006), 190–191.
55. Zeiderman, "Cities of the Future?," 27.
56. See Timothy Mitchell, "The Stage of Modernity," in *Questions of Modernity*, ed. Timothy Mitchell (Minneapolis: University of Minnesota Press, 2000), 8–16.
57. Zeiderman, "Cities of the Future?," 27; and Zeiderman, *Endangered City*, 200–202.
58. Zeiderman, "Cities of the Future?," 28; Mitchell, "The Stage of Modernity," 9; and Zeiderman, *Endangered City*, 200, 201.
59. Rem Koolhaas and Harvard Project on the City, "Lagos," 653.
60. Gandy, "Learning from Lagos," 42.
61. See Neil Smith, "The Satanic Geographies of Globalization: Uneven Development in the 1990s," *Public Culture* 10, no. 1 (1997): 174–178.
62. Robinson, "Global and World Cities," 531.
63. Rao, "Slum as Theory," 227.
64. Pieterse, *City Futures*, 130–135.
65. Edgar Pieterse, "Exploratory Notes on African Urbanism," paper presented at the Third European Conference on African Studies, Leipzig, June 4–7, 2009, http://www.stellenboschheritage.co.za/wp-content/uploads/Exploratory-Notes-on-African-Urbanism.pdf, 2.
66. Mitchell, "The Stage of Modernity," 9; Robinson, "Global and World Cities"; and Ferguson, *Global Shadows*, 188–192.
67. Zeiderman, "Cities of the Future?," 29.
68. Zeiderman, *Endangered City*, 201.
69. Robinson, "Urban Geography"; and Jennifer Robinson, "Cities in a World of Cities: The Comparative Gesture," *International Journal of Urban and Regional Research* 35, no. 1 (2011): 1–23.
70. Ferguson, *Global Shadows*, 186.
71. Zeiderman, "Cities of the Future?," 29.
72. Ferguson, *Global Shadows*, 182, 183.
73. Zeiderman, "Cities of the Future?," 29.
74. Zeiderman, "Cities of the Future?," 29; Ferguson, *Global Shadows*, 189–192; and Davis, *Planet of Slums*, 199–206.
75. Rao, "Slum as Theory"; and Davis, *Planet of Slums*, 20–49.
76. Rao, "Slum as Theory," 228; and Roy, "Slumdog Cities," 224.
77. My thinking deviates somewhat from Roy, "Slumdog Cities," 224, 234. See also Andrew Harris, "The Metonymic Urbanism of Twenty-First-Century Mumbai," *Urban Studies* 49, no. 13 (2012), 2955–2973; and Alan Gilbert, "The Return of the Slum: Does Language Matter?," *International Journal of Urban and Regional Research* 31, no. 4 (2007): 697–713.
78. See Franz Fanon, *The Wretched of the Earth* (New York: Grove, 1993). See also Alan Mayne, *Slums: The History of a Global Injustice* (London: Reaktion, 2017), 7–15.
79. Robert Fishman, *Bourgeois Utopias: The Rise and Fall of Suburbia* (New York: Basic Books, 1987), 62.

80. Richard Harris and Charlotte Vorms, "Introduction," in *What's in a Name? Talking About Urban Peripheries*, ed. Richard Harris and Charlotte Vorms (Toronto: University of Toronto Press, 2017), 17. See also Mayne, *Slums*, 26–39.
81. See, for example, McFarlane, "Urban Shadows."
82. Pieterse, *City Futures*, 108; and Françoise Lieberherr-Gardiol, "Slums Forever? Globalisation and Its Consequences," *European Journal of Development Research* 18, no. 2 (2006): 275–283.
83. Pieterse, *City Futures*, 108.
84. See, for example, Clara Olmedo and Martin J. Murray, "The Formalization of Informal/Precarious Labor in Contemporary Argentina," *International Sociology* 17, no. 2 (2002): 421–443.
85. Roy, "Slumdog Cities," 224.
86. Rao, "Slum as Theory," 228, 229; and Mayne, *Slums*, 7–15.
87. Sarah Nuttall and Achille Mbembé, "A Blasé Attitude: A Response to Michael Watts," *Public Culture* 17, no. 1 (2005): 194.
88. Vyjayanthi Rao, "Slum as Theory," *Editoriale Lotus* 143 (2010): 11. See also Vyjayanthi Rao, "Slum as Theory: Mega-Cities and Urban Models," in *The Sage Handbook of Architectural Theory*, ed. C. Greig Crysler, Stephen Cairns, and Hilde Heynen (Thousand Oaks, CA: Sage, 2012), 671–686.
89. Rao, "Slum as Theory," *Editoriale Lotus*, 12; and Solomon Benjamin, "Occupancy Urbanism: Radicalizing Politics and Economy Beyond Policy and Programs," *International Journal of Urban and Regional Research* 32, no. 3 (2008): 719–729.
90. Marie Huchzermeyer, "Troubling Continuities: Use and Utility of the Term 'Slum,' " in *The Routledge Handbook on Cities of the Global South*, ed. Susan Parnell and Sophie Oldfield (New York: Routledge, 2014), 86.
91. Kaminer, Robles-Durán, and Sohn, "Introduction: Urban Asymmetries," 12.
92. Benjamin Marx, Thomas Stoker, and Tavneet Suri, "The Economics of Slums in the Developing World," *Journal of Economic Perspectives* 27, no. 4 (2013): 187.
93. Mayne, *Slums*, 16–39.
94. Gareth A. Jones, "Slumming About," *City* 15, no. 6 (2011): 698.
95. AbdouMaliq Simone, "What You See Is Not Always What You Know: Struggles Against Re-containment and the Capacities to Remake Urban Life in Jakarta's Majority World," *South East Asia Research* 23, no. 2 (2015): 227–244; and João Marcelo Melo, "Aesthetics and Ethics in *City of God*: Content Fails, Form Talks," *Third Text* 18, no. 5 (2004): 475–481.
96. Pushpa Arabindoo, "Beyond the Return of the 'Slum'," *City* 15, no. 6 (2011): 633; and Huchzermeyer, "Troubling Continuities."
97. See, for example, Tom Angotti, "Apocalyptic Anti-Urbanism: Mike Davis and His Planet of Slums," *International Journal of Urban and Regional Research* 30, no. 4 (2006): 961–967; and Mayne, *Slums*, 9–10. For an earlier commentary, see Janice Perlman, *The Myth of Marginality* (Berkeley: University of California Press, 1976).
98. Pushpa Arabindoo, "Rhetoric of the Slum: Rethinking Urban Poverty," *City* 15, no. 6 (2011): 643. See also Angotti, "Apocalyptic Anti-Urbanism."
99. Gilbert, "The Return of the Slum," 701; Mayne, *Slums*, 9–11; and Arabindoo, "Beyond the Return of the 'Slum.' "
100. Arabindoo, "Beyond the Return of the 'Slum'," 633.
101. Arabindoo, "Beyond the Return of the 'Slum'," 633.
102. Gilbert, "The Return of the Slum," 701–706.
103. Gilbert, "The Return of the Slum," 701. See also Allan Gilbert, "Slums, Tenants and Home-Ownership: On Blindness to the Obvious," *International Development Planning Review* 30, no. 2 (2008): i–x; Allan Gilbert, "Extreme Thinking About Slums and Slum Dwellers: A Critique," *SAIS Review* 29, no. 1 (2009): 35–48; Alan Gilbert, "Epilogue," *City* 15, no. 6 (2011): 722–726.

104. Peter Brand and Julio Dávila, "Mobility Innovation at the Urban Margins," *City* 15, no. 6 (2011): 657; and Arabindoo, "Rhetoric of the 'Slum'," 636.
105. Deborah Potts, "Shanties, Slums, Breeze Blocks and Bricks," *City* 15, no. 6 (2011): 709–721.
106. David Simon, "Situating Slums," *City* 15, no. 6 (2011): 678.
107. Rao, "Slum as Theory," 228, 231.
108. Roy, "Slumdog Cities," 224, 227.
109. Roy, "Slumdog Cities," 231.
110. Uwe Altrock, "Conceptualising Informality: Some Thoughts on the Way Towards Generalization," in *Urban Informalities: Reflections on the Formal and Informal*, ed. Colin McFarlane and Michael Waibel (London: Routledge, 2016), 171–194.
111. Tayyab Mahmud, " 'Surplus Humanity' and Margins of Legality: Slums, Slumdogs, and Accumulation by Dispossession," *Chapman Law Review* 14, no. 1 (2010): 1–3.
112. Frank Snowden, *Naples in the Time of Cholera* (New York: Cambridge University Press, 1995), 35–36; quoted in Davis, *Planet of Slums*, 175.
113. Mahmud, " 'Surplus Humanity' and the Margins of Legality," 11. See also Mike Davis, "Planet of Slums: Urban Involution and the Informal Proletariat," *New Left Review* 26 (2004): 5–34.
114. Mahmud, " 'Surplus Humanity' and the Margins of Legality," 11.
115. Derek Gregory, "The Black Flag: Guantánamo Bay and the Space of Exception," *Geografiska Annaler: Series B, Human Geography* 88, no. 4 (2006): 405–427; Claudio Minca, "The Return of the Camp," *Progress in Human Geography* 29 (2004): 405–412; and Michel Algier, "Between War and the City: Towards an Urban Ethnography of Refugee Camps," *Ethnography* 3, no. 3 (2002): 317–341.
116. On problems associated with treating terms like "slums" as paradigms, see Andrew Norris, "The Exemplary Exception: Philosophical and Political Decisions in Giorgio Agamben's *Homo Sacer*," in *Politics, Metaphysics and Death: Essays on Giorgio Agamben's Homo Sacer*, ed. Andrew Norris (Durham, NC: Duke University Press, 2005), 262–283.
117. Davis, *Planet of Slums*, 151–173; Mahmud, " 'Surplus Humanity' and the Margins of Legality," 22–23; and Tayyab Mahmud, "Slums, Slumdogs, and Resistance," *American University Journal of* Gender Social Policy *and* Law 18, no. 3 (2010): 685–710.
118. Mahmud, " 'Surplus Humanity' and Margins of Legality," 2; and Saskia Sassen, *Expulsions: Brutality and Complexity in the Global Economy* (Cambridge, MA: Belknap, 2014), 16–17, 18, 211.
119. Martin J. Murray, "Fire and Ice: Unnatural Disasters and the Disposable Urban Poor in Post-Apartheid Johannesburg," *International Journal of Urban and Regional Research* 33, no. 1 (2009): 174.
120. UN-Habitat, "The Challenge of Slums: Global Report on Human Settlements 2003," *Management of Environmental Quality* 15, no. 3 (2004): 337–338; Carlos Brillembourg, "The New Slum Urbanism of Caracas, Invasions and Settlements, Colonialism, Democracy, Capitalism and Devil Worship," *AD: Architectural Design* 74, no. 2 (2004): 77–81; and Breman, "Slumlands."
121. Jan Nijman, "A Study of Space Inside Mumbai's Slums," *Tijdschrift voor economische en sociale geografie* 101, no. 1 (2010): 4–17.
122. For an earlier intervention into the mainstream scholarly literature, see Charles Stokes, "A Theory of Slums," *Land Economics* 38, no. 3 (1962): 187–197. See also Mike Davis, "The Urbanization of Empire: Megacities and the Laws of Chaos," *Social Text* 22, no. 4 (2004): 9–15; and Gandy, "Learning from Lagos," 52.
123. Kruijt and Koonings, "The Rise of Megacities."
124. Roy, "Slumdog Cities, 224.
125. Robinson, "Global and World Cities," 531. See also Robinson, *Ordinary Cities*; and Robinson, "Cities in a World of Cities."
126. Roy, "Slumdog Cities," 224.
127. Potts, "Shanties, Slums, Breeze Blocks and Bricks."
128. Davis, *Planet of Slums*, 39.

129. Eileen Stillwaggon, *Stunted Lives, Stagnant Economies: Poverty, Disease and Underdevelopment* (New Brunswick, NJ: Rutgers University Press, 1998), 67.
130. Murray, "Fire and Ice."
131. Martin J. Murray, *Taming the Disorderly City: The Spatial Landscape of Johannesburg After Apartheid* (Ithaca, NY: Cornell University Press, 2008), 189–224.
132. Amos Rappaport, "Spontaneous Settlements as Vernacular Design," in *Spontaneous Shelter: International Perspectives and Prospects*, ed. Carl Patton (Philadelphia: Temple University Press, 1988), 52.
133. Rosa Flores Fernandez, "Physical and Spatial Characteristics of Slum Territories Vulnerable to Natural Disasters," *Les Cahiers de l'Afrique de l'Est* 44 (2011): 5–22.
134. Raul Lejano and Corinna Del Bianco, "The Logic of Informality: Pattern and Process in a São Paulo Favela," *Geoforum* 91 (2018): 205, 197.
135. Larissa Lomnitz, *Networks and Marginality: Life in a Mexican Shantytown* (New York: Academic, 1977), 91.
136. Françoise Lieberherr-Gardiol, "Slums Forever?," Globalization and its Consequences," *European Journal of Development Research* 18, no. 2 (2006) 275; and Ronak Patel and Thomas Burke, "Urbanization—An Emerging Humanitarian Disaster," *New England Journal of Medicine* 361 (2009): 741–743.
137. Lieberherr-Gardiol, "Slums Forever?," 275, 276, 279.
138. See Lieberherr-Gardiol, "Slums Forever?," 275–276.
139. Colin McFarlane, "Sanitation in Mumbai's Informal Settlements: State, 'Slum,' and Infrastructure," *Environment and Planning A* 40, no. 1 (2008): 91.
140. Sapana Doshi, "The Politics of the Evicted: Redevelopment, Subjectivity and Difference in Mumbai's Slum Frontier," *Antipode* 45, no. 3 (2013): 1–22.
141. Tom Angotti, "Apocalyptic Anti-Urbanism," 961. See also David Cunningham, "Slumming It: Mike Davis's Grand Narrative of Urban Revolution," *Radical Philosophy* 142 (2007): 8–18; and Richard Pithouse, "Review of Mike Davis, *Planet of Slums*," *Journal of Asian and African Studies* 43 (2008): 567–574.
142. James Holston, "Insurgent Citizenship in an Era of Global Urban Peripheries," *City & Society* 21, no. 2 (2009): 245–267; and James Holston and Teresa Caldeira, "Urban Peripheries and the Invention of Citizenship," *Harvard Design Magazine* 28 (2008): 18.
143. Jeremy Harding, "It Migrates to Them," *London Review of Books* 29, no. 5 (March, 8, 2007): 25–27; Colin McFarlane, "The Entrepreneurial Slum: Civil Society, Mobility and the Co-production of Urban Development," *Urban Studies* 49, no. 13 (2012): 2795–2816; and Simone, "Cities of Uncertainty."
144. AbdouMaliq Simone, "Emergency Democracy and the 'Governing Composite,' " *Social Text* 95, no. 26 (2) (2008): 13–33; Weinstein, *The Durable Slum*, 115–140; and Jonathan Silver, "Incremental Infrastructures: Material Improvisation and Social Collaboration Across Post-Colonial Accra," *Urban Geography* 35, no. 6 (2014): 788–804.
145. Alexander Vasudevan, "The Makeshift City: Towards a Global Geography of Squatting," *Progress in Human Geography* 39, no. 3 (2015): 341.
146. See Giorgio Agamben, *Homo Sacer: Sovereign Power and Bare Life*, trans. Daniel Heller-Roazen (Palo-Alto, CA: Stanford University Press, 1998), 170–171; and Jaão Biehl, *Vita: Life in a Zone of Social Abandonment* (Berkeley: University of California Press, 2005), 35–45.
147. Davis, *Planet of Slums*, 121–150; and Ayona Datta, *The Illegal City: Space, Law and Gender in a Delhi Squatter Settlement* (London: Ashgate, 2012), 67–87.
148. Lindsay Sawyer, "Piecemeal Urbanisation at the Peripheries of Lagos," *African Studies* 7, no. 2 (2014): 282, 283.
149. Janice Perlman, "Six Misconceptions About Squatter Settlements," *Development* 4 (1986): 40–44.
150. Vasudevan, "The Makeshift City," 342, 343.

151. AbouMaliq Simone, "The Ineligible Majority: Urbanizing the Postcolony in Africa and Southeast Asia," *Geoforum* 42 (2011): 266–270; and Colin McFarlane, "The City as Assemblage: Dwelling and Urban Space," *Environment and Planning D: Society and Space* 29, no. 4 (2011): 649–671. These ideas are derived from Vasudevan, "The Makeshift City," 344.
152. AbouMaliq Simone, "The Ineligible Majority," 269.
153. Neema Kudva, "The Everyday and the Episodic: The Spatial and Political Impacts of Urban Informality," *Environment and Planning A* 41, no. 7 (2009): 1624.
154. Frank Gaffikin and David Perry, "The Contemporary Urban Condition: Understanding the Globalizing City as Informal, Contested, and Anchored," *Urban Affairs Review* 48, no. 5 (2012): 705.
155. Gaffikin and Perry, "The Contemporary Urban Condition," 706; and Partha Chatterjee, "The Rights of the Governed," *Identity, Culture and Politics* 3, no. 2 (2002): 66.
156. Chatterjee, "The Rights of the Governed."
157. Simone, "Straddling the Divides," 102.
158. See, inter alia, Mona Fawaz, "An Unusual Clique of City-Makers: Social Networks in the Production of a Neighborhood in Beirut (1950–75)," *International Journal of Urban and Regional Research* 32, no. 3 (2008): 565–585; Jonathan Shapiro Anjaria, *The Slow Boil: Street Food, Rights and Public Space in Mumbai* (Stanford, CA: Stanford University Press, 2016), 176–180; and Colin McFarlane, "The City as Assemblage: Dwelling and Urban Space," *Environment and Planning D: Society and Space* 29, no. 4 (2011): 649–671.
159. Asef Bayat, "Radical Religion and the Habitus of the Dispossessed: Does Islamic Militancy Have an Urban Ecology?," *International Journal of Urban and Regional Research* 31, no. 3 (2007): 579; and Asef Bayat, *Life as Politics: How Ordinary People Changed the Middle East* (Stanford, CA: Stanford University Press, 2010).
160. Asef Bayat, "Un-Civil Society: The Politics of the Informal People," *Third World Quarterly* 18, no. 1 (1997): 56); AbdouMaliq Simone, "People as Infrastructure: Intersecting Fragments in Johannesburg," *Public Culture* 16, no. 3 (2004): 407–429; Vasudevan, "The Makeshift City"; and Alexander Vasudevan, "The Autonomous City: Towards a Critical Geography of Occupation," *Progress in Human Geography* 39, no. 3 (2015): 313–334.
161. Kudva, "The Everyday and the Episodic," 1617. For a thoughtful challenge to the alleged resourcefulness of the urban poor, see Mercedes González de la Rocha, "The Myth of Survival," *Development & Change* 38, no. 1 (2007): 45–66.
162. Bayat, "Radical Religion and the Habitus of the Dispossessed," 579. See, inter alia, Anjaria, *The Slow Boil*, 161–182; Vasudevan, "The Makeshift City"; Roy, "Slumdog Cities"; and James Holston, *Insurgent Citizenship: Disjunctions of Democracy and Modernity in Brazil* (Princeton, NJ: Princeton University Press, 2009).
163. Asef Bayat follows in the footsteps of James Scott, who coined the phrase "weapons of the weak" to capture everyday forms of peasant resistance. See James C. Scott, *Weapons of the Weak: Everyday Forms of Peasant Resistance* (New Haven, CT: Yale University Press, 1985). See also Bayat, "Un-Civil Society"; and Asef Bayat, "From 'Dangerous Classes' to 'Quiet Rebels': Politics of the Urban Subaltern in the Global South," *International Sociology* 15, no. 3 (2000): 269–278.
164. Kudva, "The Everyday and the Episodic," 1617.
165. Bayat, "Un-Civil Society."
166. Kudva, "The Everyday and the Episodic," 1615; and Tom Gillespie, "From Quiet to Bold Encroachment: Contesting Dispossession in Accra's Informal Sector," *Urban Geography* 38, no. 7 (2017): 974–992.
167. Holston and Caldeira, "Urban Peripheries and the Invention of Citizenship," 18.
168. González de la Rocha, "The Construction of the Myth of Survival," 62–63.
169. Holston and Caldeira, "Urban Peripheries and the Invention of Citizenship," 18–19.

170. See, inter alia, Karen Tranberg Hansen and Mariken Vaa, eds., *Reconsidering Informality: Perspectives from Urban Africa* (Uppsula, Sweden: Nordiska Afrikainstitutet, 2004); Colin McFarlane, "Rethinking Informality: Politics, Crisis, and the City," *Planning Theory and Practice* 13, no. 1 (2012): 89–108; and Ann Varley, "Postcolonialising Informality?," *Environment & Planning D: Society and Space* 31, no. 1 (2013): 4–22.
171. Louis Ferman, Stuart Henry, and Michelle Hoyman, "Issues and Prospects for the Study of Informal Economies: Concepts, Research Strategies, and Policy," *Annals of the American Academy of Political and Social Science* 493 (1987): 154–172.
172. Keith Hart, "Informal Income Opportunities and Urban Employment in Ghana," *Modern African Studies* 11, no. 1 (1973): 61–89.
173. See Olmedo and Murray, "The Formalization of Informal/Precarious Labor in Contemporary Argentina."
174. Colin Marx and Emily Kelling, "Knowing Urban Informalities," *Urban Studies* 56, no. 3 (2019): 494–509.
175. Jenny Mbaye and Cecilia Dinardi, "Ins and Outs of the Cultural Polis: Informality, Culture and Governance in the Global South," *Urban Studies* 56, no. 3 (2009): 578–593; Colin McFarlane, "Thinking with and Beyond the Informal–Formal Relation in Urban Thought," *Urban Studies* 56, no. 3 (2009): 620–623; and Monika Streule, Ozan Karaman, Lindsay Sawyer, and Christian Schmid, "Popular Urbanization: Conceptualizing Urbanization Processes Beyond Informality," *International Journal of Urban and Regional Research* 44, no. 4 (2020): 652–672.
176. For a mainstream view, see Rafael La Porta and Andrei Shleifer, "Informality and Development," *Journal of Economic Perspectives* 28, no. 3 (2014): 109–110. See also Marx, Stoker, and Suri, "The Economics of Slums in the Developing World."
177. Bryan Roberts, Ruth Finnegan, and Duncan Gallie, "Introduction," in *New Approaches to Economic Life*, ed. Bryan Roberts, Ruth Finnegan, and Duncan Gallie (Manchester, UK: Manchester University Press, 1985), 8–12.
178. Dzodzi Tsikata, "Informalization, the Informal Economy, and Urban Women's Livelihoods in Sub-Saharan Africa Since the 1990s," in *The Gendered Impacts of Liberalization: Towards "Embedded Liberalism"?*, ed. Shahra Razavi (New York: Routledge, 2009), 131.
179. The thinking behind this way of framing "informality" is borrowed from Dean Tipps, "Modernization Theory and the Comparative Study of Societies: A Critical Perspective," *Comparative Studies in Society and History* 15, no. 2 (1973): 199.
180. Gaffikin and Perry, "The Contemporary Urban Condition," 702.
181. Colin McFarlane and Michael Waibel, "Introduction: The Formal-Informal Divide in Context," in *Urban Informalities: Reflections on the Formal and Informal*, ed. Colin McFarlane and Michael Waibel (New York: Routledge, 2012), 1.
182. Barbara Misztal, *Informality: Social Theory and Contemporary Practice* (New York: Routledge, 2000), 47–67; Tatiana Thieme, "The 'Hustle' Amongst Youth Entrepreneurs in Mathare's Informal Waste Economy," *Journal of Eastern African Studies* 7 (2013): 389–412; and Tatiana Thieme, "The Hustle Economy: Informality, Uncertainty and the Geographies of Getting By," *Progress in Human Geography* 42, no. 4 (2018): 529–548.
183. Volker Kreibich, "The Mode of Informal Urbanisation: Reconciling Social and Statutory Regulation in Urban Land Management," in *Urban Informalities*, 149.
184. Roy, "Urban Informality," 147.
185. Martha Alter Chen, "Rethinking the Informal Economy: Linkages with the Formal Economy and Formal Regulatory Environment," in *Linking the Formal and Informal Economy: Concepts and Politics*, ed. Basudeb Guha-Khasnobis, Ravi Kanbur, and Elinor Ostrom (New York: Oxford University Press, 2007), 75–92.

186. Ananya Roy, "The 21st-Century Metropolis: New Geographies of Theory," *Regional Studies* 43, no. 6 (2009): 819–830; Roy, "Urban Informality"; and Roy, "Slumdog Cities," 224.
187. Nezar AlSayyad, "Urban Informality as a 'New' Way of Life," in *Urban Informality: Transnational Perspectives from the Middle East, Latin America, and South Asia*, ed. Ananya Roy and Nezar AlSayyad (Lanham, MD: Lexington, 2004), 7–30.
188. Kudva, "The Everyday and the Episodic," 1614.
189. Kudva, "The Everyday and the Episodic," 1614–1615, 1625.
190. Ananya Roy, "Why India Cannot Plan Its Cities: Informality, Insurgence and the Idiom of Urbanization," *Planning Theory* 8, no. 1 (2009): 82.
191. Timothy Mitchell, "The Properties of Markets: Informal Housing and Capitalism's Mystery," Cultural Political Economy Working Paper Number 2 (Lancaster, UK: Institute for Advanced Studies in Social and Management Studies, University of Lancaster, 2007), 1. See also Timothy Mitchell, "The Properties of Markets," in *Do Economists Make Markets? On the Performativity of Economics*, ed. Donald MacKenzie, Fabian Muniesa, and Lucia Siu (Princeton, NJ: Princeton University Press, 2007), 245, 253–254.
192. Timothy Mitchell, "Fixing the Economy," *Cultural Studies* 12, no. 1 (1998): 80–101.
193. Carmen Gonzalez, "Squatters, Pirates, and Entrepreneurs: Is Informality the Solution to the Urban Housing Crisis?," *University of Miami Inter-American Law Review* 40, no. 2 (2008): 244.
194. Ragui Assaad, "Formal and Informal Institutions in the Labor Market, with Applications to the Construction Sector in Egypt," *World Development* 21, no. 6 (1993): 926.
195. Kirsten Hackenbroch and Shahadat Hossain, " 'The Organised Encroachment of the Powerful'—Everyday Practices of Public Space and Water Supply in Dhaka, Bangladesh," *Planning Theory & Practice* 13, no. 3 (2012): 399.
196. Bunnell and Harris, "Re-Viewing Informality," 339–340.
197. For an early iteration of this mainstream thinking, see Paul Bairoch, *Urban Unemployment in Developing Countries: The Nature of the Problem and Proposals for Its Solution* (Geneva: International Labour Office, 1973).
198. Kim Dovey, "Informal Urbanism and Complex Adaptive Assemblage," *International Development Planning Review* 34, no. 4 (2012): 349–367; and Mara Ferreri, "The Seductions of Temporary Urbanism," *Ephemera* 15, no. 1 (2015): 181–191.
199. Bunnell and Harris, "Re-Viewing Informality"; and McFarlane, "Rethinking Informality."
200. Dovey, "Informal Urbanism and Complex Adaptive Assemblage," 363.
201. Monique Nuijten and Gerhard Anders, "Corruption and the Secret of Law: An Introduction," in *Corruption and the Secret of Law: A Legal Anthropological Perspective*, 2nd ed., ed. Monique Nuijten and Gerhard Anders (New York: Routledge, 2017), 1–26.
202. Hackenbroch and Hossain, " 'The Organised Encroachment of the Powerful,' " 399.
203. Hackenbroch and Hossain, " 'The Organised Encroachment of the Powerful,' " 399.
204. Roy, "Urban Informality," 148.
205. Roy, "Urban Informality," 148.
206. Roy, "Slumdog Cities," 233.
207. See Liza Weinstein, "Mumbai's Development Mafias: Globalization, Organized Crime and Land Development," *International Journal of Urban and Regional Research* 32, no. 1 (2008): 22–39.
208. See Koenraad Bogaert, *Globalized Authoritarianism: Megaprojects, Slums, and Class Relations in Urban Morocco* (Minneapolis: University of Minnesota Press, 2018), 15–19, 248–250.
209. Roy, "Slumdog Cities," 233.
210. Roy, "Slumdog Cities," 233.
211. Roy, "Slumdog Cities," 233.
212. Nezar AlSayyad and Ananya Roy, "Medieval Modernity: On Citizenship and Urbanism in a Global Era," *Space and Polity* 10, no. 1 (2006): 8.
213. Nezar AlSayyad and Ananya Roy, "Urban Informality: Crossing Borders," in *Urban Informality*, 5.

214. AlSayyad and Roy, "Medieval Modernity," 8.
215. Roy, "Urban Informality," 149.
216. See Bunnell and Harris, "Re-Viewing Urban Informality"; Kudva, "The Everyday and the Episodic," 1614–1615, 1625; and AlSayyad, "Urban Informality as a 'New' Way of Life."
217. Susan Marlow, Scott Taylor, and Amanda Thompson, "Informality and Formality in Medium-Sized Companies: Contestation and Synchronization," *British Journal of Management* 21, no. 4 (2010): 954–966.
218. Janice Perlman, *Favela: Four Decades of Living on the Edge in Rio de Janeiro* (New York: Oxford University Press, 2010), 29–30.
219. Iris Marilyn Young, *Justice and the Politics of Difference* (Princeton, NJ: Princeton University Press, 1990), 99.
220. Gonzalez, "Squatters, Pirates, and Entrepreneurs," 252.
221. Roy, "Urban Informality."
222. See Ryan Devlin, " 'An Area That Governs Itself': Informality, Uncertainty and the Management of Street Vending in New York City," *Planning Theory* 10, no. 1 (2011): 61; and Patricia Fernandez-Kelly and Anna Garcia, "Informalization at the Core: Hispanic Women, Homework, and the Advanced Capitalist State," in *The Informal Economy: Studies in Advanced and Less Developed Countries*, ed. Alejandro Portes, Manuel Castells, and Laura Benton (Baltimore: Johns Hopkins University Press, 1989), 254.
223. Devlin, " 'An Area That Governs Itself,' " 64.
224. See Ananya Roy, "The Gentleman's City: Urban Informality in the Calcutta of New Communism," in *Urban Informality*, 159.
225. Devlin, " 'An Area That Governs Itself,' " 54; and Thomas Aguilera and Alan Smart, "Squatting, North, South and Turnabout: A Dialogue Comparing Illegal Housing Research," in *Public Goods Versus Economic Interests: Global Perspectives on the History of Squatting*, ed. Freia Anders and Alexander Sedlmaier (New York: Routledge, 2017), 42.
226. Devlin, " 'An Area That Governs Itself,' " 61.
227. Alejandro Portes, "The Informal Economy and its Paradoxes," in *The Handbook of Economic Sociology*, ed. Neil Smelser and Richard Swedberg (Princeton, NJ: Princeton University Press, 1994), 426–450.
228. Roy, "Why India Cannot Plan Its Cities," 81.
229. Martijn Koster and Monique Nuijten, "Coproducing Urban Space: Rethinking the Formal/Informal Dichotomy," *Singapore Journal of Tropical Geography* 37, no. 3 (2016): 282–294.
230. Ray Bromley, "The Informal Sector: Why Is It Worth Discussing?," *World Development* 6 (1978): 1033–1039; and Francesco Chiodelli and Erez Tzfadia, "The Multifaceted Relation Between Formal Institutions and the Production of Informal Urban Spaces: An Editorial Introduction," *Geography Research Forum* 36 (2016): 1–14.
231. Roy, "Why India Cannot Plan Its Cities"; and Libby Porter, "Informality, the Commons and the Paradoxes for Planning: Concepts and Debates for Informality and Planning," *Planning Theory & Practice* 12 (2011): 115–153.
232. Porter, "Informality, the Commons and the Paradoxes for Planning," 116.
233. Roy, "Why India Cannot Plan Its Cities," 82.
234. Roy, "Why India Cannot Plan Its Cities," 81. See also Jacqueline Groth and Eric Corijn, "Reclaiming Urbanity: Indeterminate Spaces, Informal Actors and Urban Agenda Setting," *Urban Studies* 42, no. 3 (2005): 503–520.
235. James Holston, *Insurgent Citizenship: Disjunctions of Democracy and Modernity in Brazil* (Princeton, NJ: Princeton University Press, 2009), 228, 207.
236. Roy, "Why India Cannot Plan Its Cities," 80.
237. Roy, "Why India Cannot Plan Its Cities," 82, 83, 84, 86.

238. Roy, "Why India Cannot Plan Its Cities," 83. Veronica Crossa makes this point as well. See Veronica Crossa, "Reading for Difference on the Street: De-homogenising Street Vending in Mexico City," *Urban Studies* 53, no. 2 (2016): 287–301.
239. Jonathan Shapiro Anjaria, "Street Hawkers and Public Space in Mumbai," *Economic and Political Weekly*, May 27, 2006, 2140–2146.
240. Oren Yiftachel, "Theoretical Notes on 'Gray Cities': The Coming of Urban *Apartheid*," *Planning Theory* 8, no. 1 (2009): 88.
241. Oren Yiftachel, "(Un)Settling Colonial Presents," *Political Geography* 27, no. 3 (2008): 365–370; and Oren Yiftachel, "Critical Theory and Gray Space: Mobilization of the Colonized," *City* 13, no. 2–3 (2009): 246–263.
242. Hackenbroch and Hossain, " 'The Organised Encroachment of the Powerful,' " 400.
243. Hackenbroch and Hossain, " 'The Organised Encroachment of the Powerful,' " 399.
244. Kate Meagher, "Crisis, Informalization and the Urban Informal Sector in Sub-Saharan Africa," *Development and Change* 26, no. 2 (1995): 279, 259; and Kate Meagher, "The Tangled Web of Associational Life: Urban Governance and the Politics of Popular Livelihoods in Nigeria," *Urban Forum* 21 (2010): 299–313.
245. Rahul Mehrotra, "Forward," in *Rethinking the Informal City: Critical Perspectives from Latin America*, ed. Filipe Hernández, Peter Kellett, and Lea Allen (New York: Berghahn, 2010), xiii–xiv.
246. Mehrotra, "Forward," xii; and Mehrotra, "Negotiating the Static and Kinetic Cities," 206, 213; and Kudva, "The Everyday and the Episodic," 1625.
247. Mehrotra, "Forward," xiii; Michele Acuto, Cecelia Dinardi, and Colin Marx, "Transcending (In) Formal Urbanism," *Urban Studies* 56, no. 3 (2019): 475–487; and Dovey, "Informal Urbanism and Complex Adaptive Assemblage."
248. Kim Dovey, "Incremental Urbanisms: The Emergence of Informal Settlements," in *Emergent Urbanism: Urban Planning & Design in Times of Structural and Systemic Change*, ed. Kigran Haas and Krister Olsson (New York: Routledge, 2014), 45.
249. Idalina Baptista, "Electricity Services Always in the Making: Informality and the Work of Infrastructure Maintenance and Repair in an African City," *Urban Studies* 56, no. 3 (2019): 521.
250. Andy Pratt, "Formality as Exception," *Urban Studies* 56, no. 3 (2019): 612.
251. Mehrotra, "Forward," xiii; and Kim Dovey and Ronald King, "Forms of Informality: Morphology and Visibility of Informal Settlements," *Built Environment* 37, no. 1 (2011): 11–29.
252. Mitchell, "The Properties of Markets."
253. Kim Dovey, "Uprooting Critical Urbanism," *City* 15, no. 3–4 (2011): 347–354.
254. Kudva, "The Everyday and the Episodic," 1614, 1615, 1625.
255. Kudva, "The Everyday and the Episodic," 1615, 1624–1625.
256. Ravi Sundaram, "Revisiting the Pirate Kingdom," *Third Text* 23, no. 3 (2009): 336, 345.

6. BUILDING CITIES ON A GRAND SCALE: THE INSTANT URBANISM OF THE TWENTY-FIRST CENTURY

1. Marshall Berman, *All That Is Solid Melts Into Air* (New York: Simon & Schuster, 1982), 232.
2. David Scobey, *Empire City: The Making and Meaning of the New York City Landscape* (Philadelphia: Temple University Press, 2003); Max Paige, *The Creative Destruction of Manhattan, 1900–1940* (Chicago: University of Chicago Press, 1999); and Thomas Bender, *The Unfinished City: New York and the Metropolitan Idea* (New York: New York University Press, 2002).
3. Tom Angotti, *The New Century of the Metropolis: Urban Enclaves and Orientalism* (New York: Routledge, 2013), 3–25.

4. Gavin Shatkin and Sanjeev Vidyarthi, "Introduction: Contesting the Indian City: Global Visions and the Politics of the Local," in *Contesting the Indian City: Global Visions and the Politics of the Local*, ed. Gavin Shatkin (Malden, MA: Wiley Blackwell, 2014), 1–38; Gavin Shatkin, *Cities for Profit: The Real Estate Turn in Asia's Urban Politics* (Ithaca, NY: Cornell University Press, 2017); and Richard Marshall, *Emerging Urbanity: Global Urban Projects in the Asia Pacific Rim* (New York: Spon, 2003).
5. See Arjen Oosterman, "Editorial: Notes from the Tele-Present," January 29, 2013," in *Volume #34: City in a Box*, ed. Arjen Oosterman, Michelle Provoost, and Paul Kroese (Amsterdam: Stichting Archis, 2013), archis.org/volume/notes-from-the-tele-present/. See also Greg Lindsay, "Cities-in-a-Box," *Slate*, November 26, 2011.
6. See Martin J. Murray, "Large-Scale, Master-Planned Redevelopment Projects in Urbanizing Africa: Frictionless Utopias for the Contemporary Urban Age," in *Mega-Urbanization in the Global South: Fast Cities and New Urban Utopias in the Postcolonial State*, ed. Ayona Datta and Abdul Shaban (New York: Routledge, 2017), 31–53; and Martin J. Murray, " 'City Doubles': Re-Urbanism in Africa," in *Cities and Inequalities in a Global and Neoliberal World*, ed. Faranak Miraftab, David Wilson, and Kenneth Salo (New York: Routledge, 2015), 92–109.
7. Marshall, *Emerging Urbanity*, 191.
8. Page, *The Creative Destruction of Manhattan, 1900–1940*, Dana Cuff, *Provisional City: Los Angeles Stories of Architecture and Urbanism* (Cambridge, MA: MIT Press, 2001); David Harvey, *Spaces of Hope* (Berkeley: University of California Press, 2000), 133–181; and David Harvey, "Neo-Liberalism as Creative Destruction," *Geografiska Annaler: Series B, Human Geography* 88, no. 2 (2006): 145–158.
9. Peter Hall, *Cities in Civilization: Culture, Innovation, and Urban Order* (London: Weindenfield & Nicolson, 1998), 989; Marshall, *Emerging Urbanity*, 191.
10. See, for example, John Tomlinson, *The Culture of Speed: The Coming of Immediacy* (London: Sage, 2007).
11. The following sections are derived from Martin J. Murray, "Cities on a Grand Scale: Instant Urbanism at the Start of the Twenty-First Century," in *Routledge Handbook on Urban Spaces*, ed. David Wilson, Byron Miller, Kevin Ward, and Andrew Jonas (New York: Routledge, 2018), 184–196. See also Alessandro Gubitosi, "Fast Track Cities—Dubai's Time Machine," in *Instant Cities: Emergent Trends in Architecture and Urbanism in the Arab World: The Third International Conference of the Center for the Study of Architecture in the Arab Region*, ed. Amer Moustafa, Jamal Al-Qawasmi, and Kevin Mitchell (Sharaj, UAE: CSAAR Press, 2008), 89–102; and Isaac Lerner, "Instant City: City as Recreation/Re-Creation," in *Instant Cities*, 33–44.
12. Tim Bunnell, Daniel Goh, Chee-Kien Lai, and Choon-Piew Pow, "Introduction: Global Urban Frontiers? Asian Cities in Theory, Practice and Imagination," *Urban Studies* 49, no. 13 (2012): 2786.
13. See Orit Halpern, Jesse LeCavalier, Nerea Calvillo, and Wolfgang Pietsch, "Test Bed Urbanism," *Public Culture* 25, no. 2 [70] (2013): 272–306; and John Boudreau, "Cisco Systems Helps Build 'Cities-in-a-Box,' " *Washington Post*, June 9, 2010.
14. Samer Bagaeen, "Brand Dubai: The Instant City; or the Instantly Recognizable City," *International Planning Studies* 12, no. 1 (2007): 174.
15. Gavin Shatkin, "The City and the Bottom Line: Urban Megaprojects and the Privatization of Planning in Southeast Asia," *Environment and Planning A* 40, no. 2 (2008): 383–401; and Allan Irving, "The Modern/Postmodern Divide in Urban Planning," *University of Toronto Quarterly* 62, no. 4 (1993): 474–488.
16. Jung In Kim, "Making Cities Global: The New City Development of Songdo, Yujiapu and Lingang," *Planning Perspectives* 29, no. 3 (2014): 329–356; and Jonathan Barnett, *City Design: Modernist, Traditional, Green and Systems Perspectives*, 2nd ed. (New York: Routledge, 2016).
17. Nan Ellin, *Postmodern Urbanism*, rev. ed. (New York: Princeton Architectural Press, 1999), 132–133, 160–165.

18. Ellin, *Postmodern Urbanism*, 164–166; and Gubitosi, "Fast Track Cities—Dubai's Time Machine." See also Sarah Moser, Marian Swain, and Mohammed Alkhabbaz, "King Abdullah Economic City: Engineering Saudi Arabia's Post-Oil Future," *Cities* 45 (2015): 71–80; and Pamila Gupta, "Futures, Fakes and Discourses of the Gigantic and Miniature in 'The World' Islands, Dubai," *Island Studies Journal* 10, no. (2015): 181–196.
19. Murray, " 'City Doubles': Re-Urbanism in Africa"; Thomas Campanella, *The Concrete Dragon: China's Urban Revolution and What It Means for the World* (New York: Princeton Architectural Press, 2008); Pierre-Arnaud Barthel, "Arab Mega-Projects: Between the Dubai Effect, Global Crisis, Social Mobilization and a Sustainable Shift," *Built Environment* 36, no. 2 (2010): 136; Willem Paling, "Planning a Future for Phnom Penh: Mega Projects, Aid Dependence and Disjointed Governance," *Urban Studies* 49, no. 13 (2012): 2889–2912; Howard Dick and Paul Rimmer, "Beyond the Third World City: The New Urban Geography of South East Asia," *Urban Studies* 35, no. 12 (1998): 2303–2321; and Danielle Labbé and Julie-Ann Boudreau, "Understanding the Causes of Urban Fragmentation in Hanoi: The Case of New Urban Areas," *International Development Planning Review* 33, no. 3 (2011): 273–291.
20. Tom Percival and Peter Waley, "Articulating Intra-Asian Urbanism: The Production of Satellite Cities in Phnom Penh," *Urban Studies* 49, no. 13 (2012): 2872–2888; Mike Douglass and Liling Huang, "Globalizing the City in Southeast Asia: Utopia on the Urban Edge—The Case of Phu My Hung, Saigon," *International Journal of Asia-Pacific Studies* 3, no. 2 (2007): 1–42.
21. Gavin Shatkin, "Planning Privatopolis: Representation and Contestation in the Development of Urban Integrated Mega-Projects," in *Worlding Cities: Asian Experiments and the Art of Being Global*, ed. Ananya Roy and Aihwa Ong (Malden, MA: Wiley-Blackwell, 2011), 77–97; and Shatkin, "The City and the Bottom Line."
22. Shatkin, "Planning Privatopolis." See Martin J. Murray, *The Urbanism of Exception: The Dynamics of Global City Building in the Twenty-first Century* (New York: Cambridge University Press, 2017), 147-198.
23. Shatkin, "Planning Privatopolis," 77.
24. Shatkin, "Planning Privatopolis," 77.
25. Chad Haines, "Cracks in the Façade: Landscapes of Hope and Desire in Dubai," in *Worlding Cities*, 167; and Federico Cugurullo, "How to Build a Sandcastle: An Analysis of the Genesis and Development of Masdar City," *Journal of Urban Technology* 20, no. 1 (2013): 23–37.
26. Ola Söderström, Till Paasche, and Francisco Klauser, "Smart Cities as Corporate Storytelling," City 18, no. 3 (2014): 307–320; Federico Cugurullo, "The Business of Utopia: Estidama and the Road to the Sustainable City," *Utopian Studies* 24, no. 1 (2013): 66–88; Dennis Hardy, "Dubai—Planning Miracle or Desert Mirage?," *Town and Country Planning* 77, no. 3 (2008): 143; and Melodena Balakrisnan, "Dubai—A Star in the East: A Case Study in Strategic Destination Branding," *Journal of Place Management and Development* 1, no. 1 (2008): 62–91.
27. See Labbé and Boudreau, "Understanding the Causes of Urban Fragmentation in Hanoi," 286.
28. Martin J. Murray, "Waterfall City (Johannesburg): Privatized Urbanism in Extremis," *Environment & Planning A* 47, no. 3 (2015): 503–520; and Claire Herbert and Martin J. Murray, "Building New Cities from Scratch: Privatized Urbanism and the Spatial Restructuring of Johannesburg After Apartheid," *International Journal of Urban and Regional Research* 39, no. 3 (2015): 471–494.
29. Richard Lloyd and Terry Nichols Clark, "The City as Entertainment Machine," in *Critical Perspectives on Urban Redevelopment: Volume 6*, ed. Kevin Fox Gotham (New York: Elsevier, 2001), 357–378. See also Natalie Koch, "Urban 'Utopias': The Disney Stigma and Discourses of 'False Modernity,' " Environment and Planning A 44, no. 10 (2012): 2445–2462; and Natalie Koch, "Why Not a World City? Astana, Ankara, and Geopolitical Scripts in Urban Networks," *Urban Geography* 34, no. 1 (2013): 109–130.

30. Ruth Eaton, *Ideal Cities—Utopianism and the (Un)Built Environment* (London: Thames & Hudson, 2002); David Harvey, *Spaces of Hope* (Berkeley: University California, 2000); Bruce Mazlish, "A Tale of Two Enclosures: Self and Society as a Setting for Utopias," *Theory, Culture & Society* 20, no. 1 (2003): 43–60; Krishan Kumar, "Aspects of the Western Utopian Tradition," *History of the Human Sciences* 16, no. 1 (2003): 63–77; and Ellen Shoshkes, "East-West: Interactions Between the United States and Japan and Their Effect on Utopian Realism," *Journal of Planning History* 3, no. 3 (2004): 215–240.
31. Examples of master-planned company towns include the proto-industrial Werksiedlungsbau in the German Ruhr area (housing estates built for workers employed by Krupp industries), the planned settlements constructed by Pullman (Pullman Town in Chicago, 1880), Carnegie Steel Company (McDonald, Ohio, near Youngstown), Cadbury (Bourneville, near Birmingham, 1880), and Lever (Port Sunlight, near Liverpool, 1887).
32. Margaret Crawford, *Building the Workingman's Paradise: The Design of American Company Towns* (New York: Verso, 1995), 1–2.
33. See contributions to Oliver Dinius and Angela Vergara, eds., *Company Towns in the Americas: Landscape, Power, and Working-Class Communities* (Athens: University of Georgia Press, 2011); John Garner, ed., *The Company Town: Architecture and Society in the Early Industrial Age* (Oxford: Oxford University Press, 1992); John Garner, ed., *The Model Company Town: Urban Design Through Private Enterprise in Nineteenth-Century New England* (Amherst: University of Massachusetts Press, 1984); Hardy Green, *The Company Town: The Industrial Edens and Satanic Mills That Shaped the American Economy* (New York: Basic Books, 2010); Charles Dellheim, "The Creation of Company Culture: Cadburys, 1861–1931," *American Historical Review* 92, no. 1 (1987): 13–44; Martin Gaskell, "Model Industrial Villages in S. Yorkshire/N. Derbyshire and the Early Town Planning Movement," *Town Planning Review* 50, no. 4 (1979): 437–458; and Rupert Hebblethwaite, "The Municipal Housing Programme in Sheffield Before 1914," *Architectural History* 30 (1987): 143–179.
34. James Ferguson, "Seeing Like an Oil Company: Space, Security, and Global Capital in Neoliberal Africa," *American Anthropologist* 107, no. 3 (2005): 377–382; and Hannah Appel, "Offshore Work: Oil, Modularity, and the How of Capitalism in Equatorial Guinea," *American Ethnologist* 39, no. 4 (2012): 692–709.
35. See Standish Meacham, *Regaining Paradise: Englishness and the Early Garden City Movement* (New Haven, CT: Yale University Press, 1999); Peter Hall, *Cities of Tomorrow: An Intellectual History of Urban Planning and Design in the Twentieth Century* (Oxford: Blackwell, 2002); Gordon MacLeod and Kevin Ward, "Spaces of Utopia and Dystopia: Landscaping the Contemporary City," *Geografiska Annaler: Series B, Human Geography* 84, no. 3–4 (2002): 153–170; and David Pinder, "In Defence of Utopian Urbanism: Imaging Cities After the End of Utopia," *Geografiska Annaler: Series B, Human Geography* 84, no. 3–4 (2002): 229–241.
36. Ebenezer Howard, *The Garden Cities of To-morrow* (London: Swan Sonnenschein, 1902).
37. Stanley Buder, "Ebenezer Howard: The Genesis of a Town Planning Movement," *Journal of the American Institute of Planners* 35, no. 6 (1969): 390–398; Stanley Buder, *Visionaries and Planners: The Garden City Movement and the Modern Community* (Oxford: Oxford University Press, 1990); Peter Batchelor, "The Origin of the Garden City Concept of Urban Form," *Journal of the Society of Architectural Historians* 28, no. 3 (1969): 184–200; Patricia Burgess, "City Planning and the Planning of Cities: The Recent Historiography," *Journal of Planning Literature* 7, no. 4 (1993): 314–327; and Louise Mozingo, *Pastoral Capitalism: A History of Corporate Suburban Landscapes* (Cambridge, MA: MIT Press, 2011).
38. Liora Bigon, *A History of Urban Planning in Two West African Colonial Capitals: Residential Segregation in British Lagos and French Dakar (1850–1930)* (Lewiston, NY: Mellen, 2009); Liora Bigon, "'Garden City' in the Tropics? French Dakar in Comparative Perspective," *Journal of Historical Geography* 38, no. 1 (2012): 35–44; Garth Andrew Myers, *Verandahs of Power: Colonialism and Space*

in *Urban Africa* (Syracuse, NY: Syracuse University Press, 2003); Robert Home, *Of Planting and Planning: The Making of British Colonial Cities* (London: Spon, 1997); Alan Mabin, "Comparative Segregation: The Origins of the Group Areas Act and Its Planning Apparatuses," *Journal of Southern African Studies* 18, no. 2 (1992): 405–429; Alan Mabin and Dan Smith, "Reconstructing South Africa's Cities? The Making of Urban Planning 1900—2000," *Planning Perspectives* 12, no. 2 (1997): 193–223; Gwendolyn Wright, *The Politics of Design in French Colonial Urbanism* (Chicago: University of Chicago Press, 1991); Robert Home, "Town Planning and Garden Cities in the British Colonial Empire, 1910–1940," *Planning Perspectives* 5, no. 1 (1990): 23–37; and John Collins, "Lusaka: Urban Planning in a British Colony, 1931–1964," in *Shaping an Urban World*, ed. G. E. Cherry (London: Mansell, 1980), 227–241.

39. For the origins of the notion of dual cities, see Janet Abu-Lughod, *Rabat: Urban Apartheid in Morocco* (Princeton, NJ: Princeton University Press, 1980).
40. Liora Bigon, "Garden Cities in Colonial Africa: A Note on Historiography," *Planning Perspectives* 28, no. 3 (2013): 477, 482.
41. Anthony Alofsin, "Broadacre City: Nemesis," *American Art* 25, no. 2 (2011): 21–25.
42. Seth Schindler, "Governing the Twenty-First Century Metropolis and Transforming Territory," *Territory, Politics, Governance* 3, no. 1 (2015): 18.
43. Annapurna Shaw, "Town Planning in Postcolonial India, 1947–1965: Chandigarh Re-examined," *Urban Geography* 30, no. 8 (2009): 857–878; and James Holston, *The Modernist City: An Anthropological Critique of Brasilia* (Chicago: University of Chicago Press, 1989).
44. Mark Clapson, *Invincible Green Suburbs, Brave New Towns: Social Change and Urban Dispersal in Post-War England* (Manchester, UK: Manchester University Press, 1998); Alan Simson, "The Post-Romantic Landscape of Telford New Town," *Landscape and Urban Planning* 52, no. 2–3 (2001): 189–197; and Anthony Alexander, *Britain's New Towns: From Garden Cities to Sustainable Communities* (New York: Routledge, 2009).
45. Schindler, "Governing the Twenty-First Century Metropolis and Transforming Territory," 18–19. For New York City, Robert Caro chronicles the grandiose ambitions of Robert Moses, perhaps the quintessential modernist city builder of the twentieth century. See Robert Caro, *The Power Broker: Robert Moses and the Fall of New York* (New York: Knopf, 1974).
46. Louis Mumford, *The Story of Utopia* (New York: Boni & Liveright, 1922); and Charles Turner, "Mannheim's Utopia Today," *History of the Human Sciences* 16, no. 1 (2002): 27–47.
47. Robert Fishman, *Bourgeois Utopias: The Rise and Fall of Suburbia* (New York: Basic Books, 1987), 3.
48. Douglass and Huang, "Globalizing the City in Southeast Asia," 9. See Jeffrey Alexander, "Robust Utopias and Civil Repairs," *International Sociology* 16, no. 4 (2001): 579–591.
49. Jane Lumumba, "Why Africa Should Be Wary of Its 'New Cities,'" The Rockefeller Foundation's Informal City Dialogues, February 5, 2013, http://nextcity.org/informalcity/entry/why-africa-should-be-wary-of-its-new-cities. See also Vanessa Watson, "African Urban Fantasies: Dreams or Nightmares?," *Environment and Urbanization* 26, no. 1 (2014): 215–231; and Gautam Bhan, "The Real Lives of Urban Fantasies," *Environment and Urbanization* 26, no. 1 (2014): 232–235.
50. Natalie Koch, "The Violence of Spectacle: Statist Schemes to Green the Desert and Constructing Astana and Ashgabat as Urban Oases," *Social and Cultural Geography* 16, no. 6 (2015): 675–697; and Natalie Koch, *The Geopolitics of Spectacle: Space, Synecdoche, and the New Capitals of Asia* (Ithaca, NY: Cornell University Press, 2018).
51. Robert Hughes, *Trouble in Utopia*, episode 4, "The Shock of the New" (BBC and Time-Life Films, 1980; DVD, New York: Ambrose Video, 2001). See James Ferguson, *Global Shadows: Africa in the Neoliberal World Order* (Durham, NC: Duke University Press, 2006), 17.
52. Aihwa Ong, "Hyperbuilding: Spectacle, Speculation, and the Hyperspace of Sovereignty," in *Worlding Cities*, 205–226; and James Sidaway, "Enclave Space: A New Metageography of Development?," Area 39, no. 3 (2007): 331–339.

53. Walter Benjamin, *The Arcades Project*, trans. Howard Eiland and Kevin McLaughlin (Cambridge, MA: Belknap, 1999), 546.
54. Susan Buck-Morss, "The City as Dreamworld and Catastrophe," *October* 73 (1995): 3–26; Tim Edensor, *Industrial Ruins: Space, Aesthetics, and Materiality* (London: Berg, 2005); and Rebecca Kinney, *Beautiful Wasteland: The Rise of Detroit as America's Postindustrial Frontier* (Minneapolis: University of Minnesota Press, 2016).
55. Charles Piot, *Nostalgia for the Future: West Africa After the Cold War* (Chicago: University of Chicago Press, 2010).
56. See Ayona Datta, "Introduction: Fast Cities in an Urban Age," in *Mega-Urbanism in the Global South*, 1–28.
57. Erik Swyngedouw, "Exit 'Post'—The Making of 'Glocal' Urban Modernities," in *Future City*, ed. Stephen Read, Jürgen Rosemann, and Job van Eldijk (London: Spon, 2005), 127–128.
58. Michelle Acuto, "High-Rise Dubai Urban Entrepreneurialism and the Technology of Symbolic Power," *Cities* 27, no. 4 (2010): 272–284; Angotti, *The New Century of the Metropolis*, 3–25; Ayona Datta, "India's Ecocity? Environment, Urbanisation, and Mobility in the Making of Lavasa," *Environment and Planning C* 30, no. 6 (2012): 982–996; Ayona Datta, "A 100 Smart Cities, a 100 Utopias," *Dialogues in Human Geography* 5, no. 1 (2015): 49–53; Trevor Hogan and Christopher Houston, "Corporate Cities: Urban Gateways of Gated Communities Against the City: The Case of Lippo, Jakarta," in *Critical Reflections on Cities in Southeast Asia*, ed. Tim Bunnell, Lisa Drummond, and K.C. Ho (Singapore: Times Academic, 2002), 243–264; and Murray, "Waterfall City (Johannesburg)."
59. Bagaeen, "Brand Dubai"; Michael Cameron Dempsey, *Castles in the Sand: An Urban Planner in Abu Dhabi* (Jefferson, NC: MacFarland, 2014); Yasser Elsheshtawy, "Cities of Sand and Fog: Abu Dhabi's Global Ambitions," in *The Evolving Arab City: Tradition, Modernity and Urban Development*, ed. Yasser Elsheshtawy (New York: Routledge, 2008), 258–304; Natalie Koch, "The Monumental and the Miniature: Imagining 'Modernity' in Astana," *Social and Cultural Geography* 11, no. 8 (2010): 769–787; and Shiuh-shen Chien and Max Woodworth, "China's Urban Speed Machine: The Politics of Speed and Time in a Period of Rapid Urban Growth," *International Journal of Urban and Regional Research* 42, no. 4 (2018): 723–737.
60. Christopher Marcinkoski, *The City That Never Was: Reconsidering the Speculative Nature of Contemporary Urbanization* (New York: Princeton Architectural Press, 2016).
61. Linda Samuels, "Book Review: *The City That Never Was*, by Christopher Marcinkoski," *Journal of Architectural Education*, June 2016, 1–11, https://www.jaeonline.org/articles/review/city-never-was#/page1/. See also Rachel Weber, *From Boom to Bubble: How Finance Built the New Chicago* (Chicago: University of Chicago Press, 2015).
62. Femke van Noorloos and Marjan Kloosterboer, "Africa's New Cities: The Contested Future of Urbanization," *Urban Studies* 55, no. 6 (2018): 1223–1241; Laurence Côté-Roy and Sarah Moser, " 'Does Africa Not Deserve Shiny New Cities?' The Power of Seductive Rhetoric Around New Cities in Africa," *African Studies* 56, no. 12 (2019): 2391–2407; and Tom Goodfellow, "Urban Fortunes and Skeleton Cityscapes: Real Estate and Late Urbanization in Kigali and Addis Ababa," *International Journal of Urban and Regional Research* 41, no. 5 (2017): 786–803.
63. Robert Fishman, *Urban Utopias in the Twentieth Century: Ebenezer Howard, Frank Lloyd Wright, Le Corbusier* (Cambridge, MA: MIT Press, 1982); and Hall, *Cities of Tomorrow*, 1–25.
64. Elizabeth Rapoport, "Utopian Visions and Real Estate Dreams: The Eco-City Past, Present and Future," *Geography Compass* 8, no. 2 (2014): 137–149.
65. See Datta, "Introduction: Fast Cities in an Urban Age." See also Alexander Cuthbert, "Alphaville and Masdar: The Future of Urban Space and Form," in *Emergent Urbanism: Urban Planning & Design in Times of Structural and Systematic Change*, 2nd ed., ed. Tigran Haas and Krister Olsson (New York: Routledge, 2016), 9–18.

66. Norma Evenson, *Chandigarh* (Berkeley: University of California Press, 1966); Ravi Kalia, *Chandigarh: The Making of an Indian City* (New York: Oxford University Press, 2002); Shaw, "Town Planning in Postcolonial India, 1947–1965"; and Lydia Polgreen, "In a Dream City, a Nightmare for the Common Man," *New York Times*, December 23, 2006.
67. Datta, "Introduction: Fast Cities in an Urban Age." For the Asia-Pacific region, see Mike Douglass, "Mega-Urban Regions and World City Formation: Globalization, the Economic Crisis and Urban Policy Issues in Pacific Asia," *Urban Studies* 37, no. 12 (2000): 2315–2335.
68. Kim, "Making Cities Global," 23.
69. Rem Koolhaas and AMO, *The Gulf* (Baden, Switzerland: Lars Muller, 2007), 7.
70. The following paragraphs are revised versions of earlier published material in Martin J. Murray, "Afterward: Re-engaging with Transnational Urbanism," in *Locating Right to the City in the Global South*, ed. Tony Samara, Shenjing He, and Guo Chen (New York: Routledge, 2013), 285–310. For Dubai, see Yasser Elsheshtawy, "Redrawing Boundaries: Dubai, an Emerging Global City," in *Planning Middle Eastern Cities: An Urban Kaleidoscope in a Globalizing World*, ed. Yasser Elsheshtawy (London: Routledge, 2004), 169–199; Yasser Elsheshtawy, *Dubai: Behind an Urban Spectacle* (New York: Routledge, 2010); and Yasser Elsheshtawy, "Navigating the Spectacle: Landscapes of Consumption in Dubai," *Architectural Theory Review* 13, no. 2 (2008): 164–187.
71. Acuto, "High-Rise Dubai"; and Khaled Adham, "Rediscovering the Island: Doha's Urbanity from Pearls to Spectacle," in *The Evolving Arab City: Tradition, Modernity and Urban Development*, ed. Yasser Elsheshtawy (London: Routledge, 2008), 218–256.
72. Christopher Dickey, "Is Dubai's Party Over?," *Newsweek*, December 15, 2008.
73. Richard Pithouse, "Review of Mike Davis, *Planet of Slums*," *Journal of Asian and African Studies* 43, no. 5 (2008): 567.
74. Florian Wiedmann, Ashraf Salama, and Alain Thierstein, "A Framework for Investigating Urban Qualities in Emerging Knowledge Economies: The Case of Doha," *Archnet-IJAR, International Journal of Architectural Research* 6, no. 1 (2012): 48–49; Alain Thierstein and Elisabeth Schein, "Emerging Cities on the Arabian Peninsula: Urban Space in the Knowledge Economy Context," *International Journal of Architectural Research* 2, no. 2 (2008): 178–195; and Agatino Rizzo, "Metro Doha," *Cities* 31 (2013): 533–543.
75. Christopher Davidson, *Dubai: The Vulnerability of Success* (London: Hurst, 2008); Ahmed Kanna, "Flexible Citizenship in Dubai: Neoliberal Subjectivity in the Emerging 'City-Corporation,' " *Cultural Anthropology* 25, no. 1 (2010): 100–129; Mike Davis, "Fear and Money in Dubai," *New Left Review* 41 (2006): 47–68; Elsheshtawy, "Redrawing Boundaries"; Elsheshtawy, *Dubai: Behind an Urban Spectacle*; and Elsheshtawy, "Navigating the Spectacle."
76. Ahmed Kanna, "Dubai in a Jagged World," *Middle East Report* 243 (2007): 22–29; and Yasser Elsheshtawy, "Transitory Sites: Mapping Dubai's 'Forgotten' Urban Spaces," *International Journal of Urban and Regional Research* 32, no. 4 (2008): 668–688.
77. Hal Rothman and Mike Davis, "Introduction: The Many Faces of Las Vegas," in *The Grit Beneath the Glitter: Tales from the Real Las Vegas*, ed. Hal Rothman and Mike Davis (Berkeley: University of California Press, 2002), 1–14.
78. Todd Reisz, "Making Dubai: A Process in Crisis," [in "Post-Traumatic Urbanism," Special Issue], *Architectural Design* 80, no. 5 (2010): 38–43; and Michelle Buckley, "From Kerala to Dubai and Back Again: Construction Migrants and the Global Economic Crisis," *Geoforum* 43, no. 2 (2012): 250–259.
79. Richard Prouty, "Buying Generic: The Generic City in Dubai," *London Consortium Static* 8 (2009): 1–8. See also Non Arkaraprasertkul, "Power, Politics, and the Making of Shanghai," *Journal of Planning History* 9, no. 4 (2010): 232–259.

80. Rem Koolhaas, "The Generic City," in *S,M,L,X*, ed. Rem Koolhaas and Bruce Mau (New York: Monacelli, 1998), 1239–1264; and Maarten Hajer, "The Generic City," *Theory, Culture & Society* 16, no. 4 (1999), 137–144.
81. Campanella, *The Concrete Dragon*, 1–10; Marshall, *Emerging Urbanity*, 9–30; Herbert Wright, *Instant Cities* (London: Black Dog, 2008); and Elsheshtawy, "Navigating the Spectacle."
82. See, for example, Umberto Eco, *Travels in Hyperreality* (San Diego: Harcourt Brace Jovanovich, 1983), 30–31, 40–44.
83. Luis Fernandez-Galiano, "Spectacle and Its Discontents; or, the Joys of Architainment," in *Commodification and Spectacle in Architecture: A Harvard Design Magazine Reader*, ed. William Saunders (Minneapolis: University of Minnesota Press, 2005), 1–7.
84. See, inter alia, Donald McNeil, "Skyscraper Geography," *Progress in Human Geography* 29, no. 1 (2005): 41–55; Mona Domosh, "The Symbolism of the Skyscraper: Case Studies of New York's First Tall Buildings," *Journal of Urban History* 14, no. 3 (1988): 321–345; Deyan Sudjic, *The Ediface Complex: How the Rich and Powerful—and Their Architects—Shape the World* (New York: Penguin, 2005); and Larry Ford, "World Cities and Global Change: Observations on Monumentality in Urban Design," *Eurasian Geography and Economics* 49, n. 3 (2008): 237–262.
85. Stephen Graham, *Vertical: The City from Satellites to Bunkers* (New York: Verso, 2016), 149–173; and Stephen Graham and Lucy Hewitt, "Getting Off the Ground: On the Politics of Urban Verticality," *Progress in Human Geography* 37 (2013): 72–92.
86. Andrew Harris, "Vertical Urbanisms: Opening Up Geographies of the Three-Dimensional City," *Progress in Human Geography* 39, no. 5 (2015): 601–620; and Martine Drozdz, Manuel Appert, and Andrew Harris, "High-Rise Urbanism in Contemporary Europe," *Built Environment* 43, no. 4 (2017): 469–480.
87. Alessandro Angelini, "Book Review of *Spaces of Global Cultures*," *Future Anterior* 2, no. 1 (2005): 69.
88. Elsheshtawy, "Navigating the Spectacle," 164; Leslie Sklair, "Iconic Architecture and Capitalist Globalization," *City* 10, no. 1 (2006): 21–47; and Anthony King, "Worlds in a City: Manhattan Transfer and the Ascendance of Spectacular Space," *Planning Perspectives* 11 (1996): 87–114.
89. Graham, *Vertical*, 149–173.
90. Elsheshtawy, *Dubai: Behind an Urban Spectacle*, 152–154; Stephen Graham, "Vanity and Violence: On the Politics of Skyscrapers," *City* 20, no. 5 (2016): 755–771; and Jane Jacobs, "A Geography of Big Things," *Cultural Geographies* 13, no. 1 (2006): 1–27.
91. Steven Velegrinis and Richard Weller, "The 21st Garden City? The Metaphor of the Garden in Contemporary Singaporean Urbanism," *Journal of Landscape Architecture* 2, no. 2 (2007): 30–41; C. M. L. Wong, "The Developmental State in Ecological Modernization and the Politics of Environmental Framings: The Case of Singapore and Implications for East Asia," *Nature and Culture* 7, no. 1 (2012): 95–119; Belinda Yuen, "Creating the Garden City: The Singapore Experience," *Urban Studies* 33, no. 6 (1996): 955–970; and Puay Yok Tan, James Wang, and Angela Sia, "Perspectives on Five Decades of the Urban Greening of Singapore," *Cities* 32 (2013): 24–32.
92. Chua Beng Huat, "Singapore as Model: Planning Innovations, Knowledge Experts," in *Worlding Cities*, 29–54; and Heejin Han, "Singapore, a Garden City: Authoritarian Environmentalism in a Developmental State," *Journal of Environment & Development* 26, no. 1 (2017): 3–24.
93. Aihwa Ong, "Introduction: Worlding Cities, or the Art of being Global," in *Worlding Cities*, 15.
94. Acuto, "High-Rise Dubai"; Bagaeen, "Brand Dubai"; Ahmed Kanna, "Flexible Citizenship in Dubai: Neoliberal Subjectivity in the Emerging 'City-Corporation,' " *Cultural Anthropology* 25, no. 1 (2010): 100–129; Alessandro Petti, "Dubai Offshore Urbanism," in *Heterotopia and the City*, ed. Michiel Dehaene and Lieven De Cauter (London: Routledge. 2008), 287–296; Elsheshtawy, "Redrawing Boundaries"; Elsheshtawy, *Dubai: Behind an Urban Spectacle*, 1–16 ; Jim Krane, *Dubai: The Story of the World's Fastest City* (London: Atlantic, 2009); Michael Pacione, "City Profile Dubai," *Cities* 22,

no. 3 (2005): 255–265; Stephen Ramos, *Dubai Amplified: The Engineering of a Port Geography* (New York: Routledge, 2010); and Alamira Reem Bani Hashim, Clara Irazábal, and Greta Byrum, "The Scheherazade Syndrome: Fiction and Fact in Dubai's Quest to Become a Global City," *Architectural Theory Review* 15, no. 2 (2010): 210–231.

95. Shatkin, "Planning Privatopolis," 78.
96. Pierre-Arnaud Barthel and Sabine Planel, "Tangier-Med and Casa-Marina, Prestige Projects in Morocco: New Capitalist Frameworks and Local Context," *Built Environment* 36, no. 2 (2010): 176–191; Koenraad Bogaert, "New State Space Formation in Morocco: The Example of the Bouregreg Valley," *Urban Studies* 49, no. 2 (2012): 255–270; and Murray, *The Urbanism of Exception*, 11–18, 221–227.
97. Timothy Mitchell, "Dreamland," in *Evil Paradises: Dreamworlds of Neoliberalism*, ed. Mike Davis and Dan Monk (New York: New Press, 2007), 1–33.
98. Stefano Moroni and David Emanuel Andersson, "Introduction: Private Enterprise and the Future of Urban Planning," in *Cities and Private Planning: Property Rights, Entrepreneurship and Transaction Costs*, ed. David Emanuel Andersson and Stefano Moroni (Northampton, MA: Edward Elgar, 2014), 1–16.
99. Marshall, *Emerging Urbanity*, 3–4.
100. Marshall, *Emerging Urbanity*, 4.
101. Ong, "Hyperbuilding," 209.
102. Ong, "Hyperbuilding," 209.
103. Marshall, *Emerging Urbanity*, 9–28.
104. Douglass and Huang, "Globalizing the City in Southeast Asia"; Percival and Waley, "Articulating Intra-Asian Urbanism"; Tim Bunnell, "Antecedent Cities and Inter-referencing Effects: Learning from and Extending Beyond Critiques of Neoliberalisation," *Urban Studies* 52, no. 11 (2015): 1983–2000; Paling, "Planning a Future for Phnom Penh"; Trevor Hogan and Christopher Houston, "Corporate Cities: Urban Gateways of Gated Communities Against the City: The Case of Lippo, Jakarta," in *Critical Reflections on Cities in Southeast Asia*, ed. Tim Bunnell, Lisa Drummond, and K.C. Ho (Singapore: Times Academic, 2002), 243–264; Abidin Kusno, "Runaway City: Jakarta Bay, the Pioneer and the Last Frontier," *Inter-Asia Cultural Studies* 12, no. 4 (2011): 513–531; and Napong Rugkhapan and Martin J. Murray, "Songdo IBD (International Business District): Experimental Prototype for the City of Tomorrow?," *International Planning Studies* 24, no. 3–4 (2019): 272–292.
105. Zygmunt Bauman, "Seeking Shelter in Pandora's Box," *City* 9, no. 2 (2005): 161.
106. Nick Osbaldiston, "Consuming Space Slowly: Reflections on Authenticity, Place and the Self," in *Culture of the Slow: Social Deceleration in an Accelerated World*, ed. Nick Osbaldiston (New York: Palgrave Macmillan, 2013), 71–93.
107. See Mike Crang, "The Calculus of Speed: Accelerated Worlds, Worlds of Acceleration," *Time & Society* 19, no. 3 (2010): 404.
108. Crang, "The Calculus of Speed," 405.
109. See Tomlinson, *The Culture of Speed*, 72–93.
110. Siegfried Kracauer, "Ahasuerus, or the Riddle of Time," in *History: The Last Things Before the Last* (New York: Oxford University Press, 1969), 139.
111. Stefan Tanaka, "History Without Chronology," *Public Culture* 28, no. 1 (2015): 162.
112. See Reinhart Koselleck, *Futures Past: On the Semantics of Historical Time*, trans. Keith Tribe (Cambridge, MA: MIT Press, 1985).
113. Koch, "Urban 'Utopias,'" 2447, 2459.
114. Richard Scherr, "The Synthetic City: Excursions Into the Real–Not Real," *Places* 18, no. 2 (2006): 10.
115. Koch, "Urban 'Utopias,'" 2447, 2459; Elsheshtawy, "Cities of Sand and Fog," 258; and Phil Hubbard and Keith Lilley, "Pacemaking the Modern City: The Urban Politics of Speed and Slowness," *Environment and Planning D* 22, no. 2 (2004): 273–294.

116. Schindler, "Governing the Twenty-First Century Metropolis and Transforming Territory," 19.
117. Zack Lee, "Eco-Cities as an Assemblage of Worlding Practices," *International Journal of Built Environment and Sustainability* 2, no. 3 (2015): 183–191.
118. Helena Mattssen, "Staging a Milieu," in *Foucault, Biopolitics, and Governmentality*, ed. Jakob Nilsson and Sven-Olov Wallenstein (Stockholm: Södertörn Philosophical Studies, 2013), 124.
119. Eugene McCann, "Urban Policy Mobilities and Global Circuits of Knowledge: Toward a Research Agenda," *Annals of the Association of American Geographers* 101, no. 1 (2011): 107–130; and Jamie Peck and Nic Theodore, "Mobilizing Policy: Models, Methods, and Mutations," *Geoforum* 41, no. 2 (2010): 169–174.
120. See, for example, Anthony Sorensen and Richard Day, "Libertarian Planning," *Town Planning Review* 52, no. 4 (1981): 390–402.
121. Michael Storper, "Governing the Large Metropolis," *Territory, Politics, Governance* 2, no. 2 (2014): 117.
122. See Casey Lynch, " 'Vote with Your Feet': Neoliberalism, the Democratic Nation-State, and Utopian Enclave Libertarianism," *Political Geography* 59 (2017): 82.
123. See Murray, *The Urbanism of Exception*, 232–248.
124. Christopher Freiman, "Cosmopolitanism Within Borders: On Behalf of Charter Cities," *Journal of Applied Philosophy* 30, no. 1 (2013): 40–52.
125. Brandon Fuller and Paul Romer, "Cities from Scratch," *City Journal* 20, no. 4 (2010), http://city-journal.org/2010/20_4_charter-cities.html.
126. Paul Romer, "Why the World Needs Charter Cities," *TEDGlobal* 2009, http://www.ted.com/talks/paul_romer.html.
127. Jonathan Bach, "Modernity and Urban Imagination in Economic Zones," *Theory, Culture, & Society* 28, no. 5 (2011): 98–122; Andrew Barry, "Technological Zones," *European Journal of Social Theory* 9, no. 2 (2006): 239–253; and Keller Easterling, "The Corporate City Is the Zone," in *Visionary Power: Producing the Contemporary City*, ed. Christine de Baan, Joachim Declerck, Veronique Patteeuw (Rotterdam: NAi, 2007), 75–85.
128. See Anonymous, "Honduras Shrugged," *Economist*, December 10, 2011; and Anonymous, "Charter Cities: Unchartered Territory," *Economist*, October 6, 2012. See also Luis Guillermo Pineda Rodas, "The Making of a Free City: The Foundation of Laissez-Faire Capitalist Free Cities in Honduras in the Juncture of Globalization," Unpublished Paper. Department of Society and Globalization, Roskilde Universitet [Denmark], 2013, 7–10, 36–38, 50.
129. Storper, "Governing the Large Metropolis," 117.
130. Romina Ruiz-Goiriena, "*Paseo Cayala*: Guatemala Builds Private 'Cayala City' for Rich to Escape Crime," *Yahoo! News*, January 8, 2013, https://news.yahoo.com/guatemala-builds-private-city-escape-crime-180518632.html; and Katherine Davies, "The Architecture of Appearance: Arendt's Feminism and Guatemala's Private City," *Arendt Studies* 4 (2020): 53–82, https://doi.org/10.5840/arendtstudies20209825.
131. Rahul Sagar, "Are Charter Cities Legitimate?," *Journal of Political Philosophy* 24, no. 4 (2016): 510.
132. Andrew Swift, "The FP Top 100 Global Thinkers," *Foreign Policy*, November 29, 2010.
133. Aditya Chakrabortty, "Paul Romer Is a Brilliant Economist—but His Idea for Charter Cities Is Bad," *Guardian*, July 27, 2010; and Kee-Cheok Cheong, "Charter Cities: An Idea Whose Time Has Come or Should Have Gone?," *Malaysian Journal of Economic Studies* 47, no. 2 (2010): 165–168.
134. Koch, "The Monumental and the Miniature"; Koch, "Why Not a World City?"; Bernhard Köppen, "The Production of a New Eurasian Capital on the Kazakh Steppe: Architecture, Urban Design, and Identity in Astana," *Nationalities Papers: The Journal of Nationalism and Ethnicity* 41, no. 4 (2013): 590–605; Edward Schatz, "What Capital Cities Say About State and Nation Building," *Nationalism and Ethnic Politics* 9, no. 4 (2004): 111–140; and Mateusz Laszezkowski, "Shrek Meets the President:

Magical Authoritarianism in a 'Fairy Tale' City," in *Kazakhstan in the Making: Legitimacy, Symbols, and Social Changes*, ed. Marlene Laruelle (New York: Lexington, 2016), 63–88.

135. Oosterman, "Notes from the Tele-Present."
136. Oosterman, "Notes from the Tele-Present." See also Greg Lindsay, "Cisco's Big Bet on New Songdo: Creating Cities from Scratch," *Fast Company*, February 1, 2010.
137. Tim Bunnell, "Smart City Returns," *Dialogues in Human Geography* 5, no. 1 (2015): 45–48; Ayona Datta, "New Urban Utopias of Postcolonial India: 'Entrepreneurial Urbanization' in Dholera Smart City, Gujarat," *Dialogues in Human Geography* 5, no. 1 (2015): 3–22; and Rapoport, "Utopian Visions and Real Estate Dreams."
138. Datta, "India's Ecocity?"
139. Tim Bunnell and Diganta Das, "Urban Pulse—A Geography of Serial Seduction: Urban Policy Transfer from Kuala Lumpur to Hyderabad," *Urban Geography* 31, no. 3 (2010): 277–284; Datta, "A 100 Smart Cities, a 100 Utopias"; Datta, "New Urban Utopias of Postcolonial India"; and R. I. G. Hollands, "Will the Real Smart City Please Stand Up?," *City* 12, no. 3 (2008): 303–320.
140. Acuto, "High-Rise Dubai"; Datta, "A 100 Smart Cities, a 100 Utopias"; Datta, "New Urban Utopias of Postcolonial India"; and Murray, "Waterfall City (Johannesburg)."
141. Keller Easterling, "Zone," in *Urban Transformation*, ed. Ilka Ruby and Andreas Ruby (Berlin: Ruby, 2008), 30–45.
142. Koch, "Urban 'Utopias' "; and Laszezkowski, "Shrek Meets the President."
143. Easterling, "The Corporate City Is the Zone," 75.
144. Easterling, "The Corporate City Is the Zone"; and Easterling, "Zone."
145. Luís Carvalho, "Smart Cities from Scratch? A Socio-Technical Perspective," *Cambridge Journal of Regions, Economy and Society* 8, no. 1 (2015): 43–60; Cugurullo, "How to Build a Sandcastle"; Federico Cugurullo, "Urban Eco-Modernisation and the Policy Context of New Eco-City Projects: Where Masdar City Fails and Why," *Urban Studies* 53, no. 11 (2016): 2417–2433; Lawrence Crot, "Planning for Sustainability in Non-democratic Polities: The Case of Masdar City," *Urban Studies* 50, no. 13 (2013): 2809–2825; and Boris Brorman Jensen, "Masdar City: A Critical Retrospection," in *Under Construction: Logics of Urbanism in the Gulf Region*, ed. Steffan Wippel, Katrin Bromber, Christian Steiner, and Birget Krawweitz (Burlington, VT: Ashgate, 2014), 45–54.
146. Herbert and Murray, "Building New Cities from Scratch"; Murray, "Waterfall City (Johannesburg)"; and Hogan and Houston, "Corporate Cities."
147. Michael Batty, "Editorial: Smart Cities, Big Data," *Environment and Planning B* 39, no. 2 (2012): 191–193; Anthony Townsend, *Smart Cities: Big Data, Civic Hackers, and the Quest for a New Utopia* (New York: Norton, 2013); and Ola Söderström, Till Paasche, and Francisco Klauser, "Smart Cities as Corporate Storytelling," City 18, no. 3 (2014): 307–320.
148. Rugkhapan and Murray, "Songdo IBD (International Business District)."
149. Datta, "Introduction: Fast Cities in an Urban Age," 2–3, 11–12.
150. Murray, " 'City Doubles': Re-Urbanism in Africa"; and Koch, "Urban 'Utopias.' "
151. I-Chun Catherine Chang and Eric Sheppard, "China's Eco-Cities as Variegated Urban Sustainability: Dongtan Eco-City and Chongming Eco-Island," *Journal of Urban Technology* 20, no. 1 (2013): 57–75; Cugurullo, "How to Build a Sandcastle"; Datta, "India's Ecocity?"; Chigon Kim, "Place Promotion and Symbolic Characterization of New Songdo City, South Korea," *Cities* 27, no. 1 (2010): 13–19; Jun In Kim, "Making Cities Global"; and Sofia Shwayri, "A Model Korean Ubiquitous Eco-City? The Politics of Making Songdo," *Journal of Urban Technology* 20, no. 1 (2013): 39–55.
152. Easterling, "The Corporate City Is the Zone," 76. For a critique of this idea of borderless globalization and stateless cities, see Göran Therborn, "End of a Paradigm: The Current Crisis and the Idea of Stateless Cities," *Environment & Planning A* 43, no. 2 (2011): 272–285.
153. Bach, "Modernity and the Urban Imagination in Economic Zones"; Barry, "Technological Zones"; Easterling, "The Corporate City Is the Zone"; and Easterling, "Zone."

154. Easterling, "The Corporate City Is the Zone"; and Easterling, "Zone."
155. Ananya Roy, "Why India Cannot Plan Its Cities: Informality, Insurgence, and the Idiom of Urbanization," *Planning Theory* 8, no. 1 (2009): 76–87; and Shatkin, "Planning Privatopolis."
156. Bach, "Modernity and the Urban Imagination in Economic Zones"; Easterling, "The Corporate City Is the Zone"; and Easterling, "Zone."
157. Koolhaas, "The Generic City," 1248.
158. Marc Augé, *Non-Places: Introduction to an Anthropology of Supermodernity* (London: Verso, 1995), 86.
159. Ravi Sundaram, "Recycling Modernity: Pirate Electronic Cultures in India," *Third Text* 13, no. 47 (1999):59–65.
160. See Abidin Kusno, "The Post-Colonial Unconscious: Observing Mega-Imagistic Urban Projects in Asia," in *The Emerging Asian City: Concomitant Urbanites and Urbanisms*, ed. Vinayak Bharne (New York: Routledge, 2013), 168–178; and Rashmi Varma, "Provincializing the Global City: From Bombay to Mumbai," *Social Text* 81 (2004): 65–89.
161. George Katodrytis, "Metropolitan Dubai and the Rise of Architectural Fantasy," *Bidoun Magazine* 4 (2005), http://www.radicalurbantheory.com/misc/dubai.html, 42–43.
162. Davide Ponzini, "Large-Scale Development Projects and Star Architecture in the Absence of Democratic Politics: The Case of Abu Dhabi, UAE," *Cities* 28 (2011): 251–259; Agatino Rizzo, "Metro Doha," *Cities* 31 (2013): 533–543; Mark Jackson and Veronica della Dora, "Dreams So Big Only the Sea Can Hold Them: Man-Made Islands as Anxious Spaces, Cultural Icons, and Travelling Visions," *Environment and Planning A* 41, no. 9 (2009): 2086–2104; and Mohsen Mohammadzadeh, "Neo-Liberal Planning in Practice: Urban Development with/Without a Plan (A Post-Structural Investigation of Dubai's Development)," paper presented in Track 9 (Spatial Policies and Land Use Planning) at the Third World Planning Schools Congress, Perth, Western Australia, July 4–8, 2011.
163. See Gupta, "Futures, Fakes and Discourses"; and Mike Davis, "Sand, Fear, and Money in Dubai."
164. Davide Ponzini, "Branded Megaprojects and Fading Urban Structure in Contemporary Cities," in *Urban Megaprojects: A Worldwide View*, ed. Gerardo del Cerro Santamaría (Bingley, UK: Emerald, 2013), 107–129.
165. Kim, "Making Cities Global."
166. David Harvey, *The Condition of Postmodernity: An Enquiry Into the Origins of Cultural Change* (Malden, MA: Blackwell, 1989), 12.
167. Harvey, *The Condition of Postmodernity*, 8–10.
168. Vivien Green, "Utopia/Dystopia," *American Art* 25, no. 2 (2011): 1–7.
169. Ricky Burdett, "Designing Urban Democracy: Mapping Scales of Urban Identity," *Public Culture* 25, no. 2 (2013): 349–367; and Cassim Shepard, "Montage Urbanism: Essence, Fragment, Increment," *Public Culture* 25, no. 2 (2013): 230–231.
170. Christina Schwenkel, "Post-Socialist Affect: Ruination and Reconstruction of the Nation in Urban Vietnam," *Cultural Anthropology* 28, no. 2 (2013): 256.
171. Ivonne Audirac, "Urban Shrinkage Amid Fast Metropolitan Growth (Two Faces of Contemporary Urbanism)," in *The Future of Shrinking Cities: Problems, Patterns and Strategies of Urban Transformation in a Global Context*, ed. Katrina Pallagst et al. (Berkeley, CA: Institute of Urban and Regional Development, Center for Global Metropolitan Studies, and the Shrinking Cities International Research Network Monograph Series, 2009), 76, 77.
172. Audirac, "Urban Shrinkage Amid Fast Metropolitan Growth," 76–77.
173. Carl Honore, *In Praise of Slowness: Challenging the Cult of Speed* (New York: Harper Collins, 2004).
174. See, for example, Marco Clausen, "Prinzessinnengarten," in *Make_Shift: The Expanded Field of Critical Spatial Practice* (Berlin: TU Berlin, Institute for Architecture, 2012), 11–12.
175. Edward Relph, *Place and Placelessness* (London: Pion, 1976); and Augé, *Non-Places*, 96–98. See also Scherr, "The Synthetic City."

176. Marco Gandelsonas, "Slow Infrastructure," in *Fast-Forward Urbanism: Rethinking Architecture's Engagement with the City*, ed. Dana Cuff and Roger Sherman (New York: Princeton Architectural Press, 2011), 123.
177. Paul Knox, "Creating Ordinary Places: Slow Cities in a Fast World," *Journal of Urban Design* 10, no. 1 (2005): 4–5.
178. Jaimee Semmens and Claire Freeman, "The Value of Cittaslow as an Approach to Local Sustainable Development: A New Zealand Perspective," *International Planning Studies*, 17, no. 4 (2012): 353–375; and Sarah Pink and Lisa Servon, "Sensory Global Towns: An Experiential Approach to the Growth of the Slow City Movement," *Environment and Planning A* 45, no. 2 (2013): 451–466.
179. Sarah Pink, "Urban Social Movements and Small Places: Slow Cities as Sites of Activism," *City* 13 (2009): 451–465.
180. Knox, "Creating Ordinary Places," 6.
181. Pink and Servon, "Sensory Global Towns," 454.
182. Jana Carp, "The Town's Abuzz: Collaborative Opportunities for Environmental Professionals in the Slow City Movement," *Environmental Practice* 14, no. 2 (2012): 130–142; and Pink and Servon, "Sensory Global Towns."
183. Semmens and Freeman, "The Value of Cittaslow."
184. Knox, "Creating Ordinary Places," 6.
185. Knox, "Creating Ordinary Places," 7.
186. Pink and Servon, "Sensory Global Towns," 465.
187. Tomlinson, *The Culture of Speed*, 147.
188. Pink and Servon, "Sensory Global Towns," 455; Heike Mayer and Paul Knox, "Slow Cities: Sustainable Places in a Fast World," *Journal of Urban Affairs* 28, no. 4 (2006): 321; and Wendy Parkins and Geoffrey Craig, *Slow Living* (New York: Berg, 2006).
189. Mayer and Knox, "Slow Cities," 321.
190. Heike Mayer and Paul Knox, "Small-Town Sustainability: Prospects in the Second Modernity," *European Planning Studies* 18, no. 10 (2010): 1553.
191. Mayer and Knox, "Small-Town Sustainability," 1563.
192. Nicolai Ouroussoff, "A Building Forms a Bridge Between a University's Past and Future," *New York Times*, February 9, 2011.

CONCLUSION: URBAN FUTURES

1. Marshall Berman, *All That Is Solid Melts Into Air* (New York: Simon & Shuster, 1982), 15.
2. Nasser Abourahme, "Contours of the Neoliberal City: Fragmentation, Frontier Geographies and the New Circularity," *Occupied London No. 4*, May 23, 2009, https://baierle.me/2009/05/23/contours-of-the-neoliberal-city-fragmentation-frontier-geographies-and-the-new-circularity/.
3. Max Page, *The Creative Destruction of Manhattan, 1900–1940* (Chicago: University of Chicago Press, 1999), 1. See Dana Cuff, *The Provisional City: Los Angeles Stories of Architecture and Urbanism* (Cambridge, MA: MIT Press, 2001); and Thomas Bender, *The Unfinished City: New York and the Metropolitan Idea* (New York: New York University Press, 2002).
4. Page, *The Creative Destruction of Manhattan*, 3.
5. Paul Chatterton, "The Urban Impossible: A Eulogy for the Unfinished City," *City* 14, no. 3 (2010): 234–244.
6. Nigel Thrift, " 'Not a Straight Line but a Curve', or, Cities Are Not Mirrors of Modernity," in *City Visions*, ed. Duncan Bell and Azzedine Haddour (London: Longman, 2000), 234; Cuff, *The Provisional City*, 5, 120; and Bender, *The Unfinished City*, xiv–xv.

7. Thrift, " 'Not a Straight Line but a Curve,' " 233.
8. See Dilip Parameshwar Gaonkar, "On Alternative Modernities," in *Alternative Modernities*, ed. Dilip Parameshwar Gaonkar (Durham, NC: Duke University Press, 2001), 1–23; David Frisby, *Cityscapes of Modernity: Critical Explorations* (Cambridge: Polity, 2001), 1–5; and Timothy Mitchell, "The Stage of Modernity," in *Questions of Modernity*, ed. Timothy Mitchell (Minneapolis: University of Minnesota Press, 2000), 7.
9. Mitchell, "The Stage of Modernity," 7–9.
10. Austin Zeiderman, "Cities of the Future? Megacities and the Space/Time of Urban Modernity," *Critical Planning* 15 (2008): 23–39.
11. Austin Zeiderman, Sobia Ahmad Kaker, Jonathan Silver, and Astrid Wood, "Uncertainty and Urban Life," *Public Culture* 27, no. 2 (2015): 281–304. See also AbdouMaliq Simone, "Cities of Uncertainty, Jakarta, the Urban Majority, and Inventive Political Technologies," *Theory, Culture and Society* 30, no. 7–8 (2013): 243–263.
12. See Zeiderman et al., "Uncertainty and Urban Life," 282.
13. Georg Simmel, "The Metropolis and Mental Life [1903]," in *Classic Essays on the Culture of Cities*, ed. Richard Sennett (New York: Appleton-Century-Crofts, 1969), 47–60.
14. Simmel, "The Metropolis and Mental Life," 53.
15. Zeiderman et al., "Uncertainty and Urban Life," 282.
16. Ash Amin and Nigel Thrift, *Cities: Reimagining the Urban* (Cambridge: Polity, 2002), 1; and Neil Brenner, "Theses on Urbanization," *Public Culture* 25, no. 1 (2013): 86.
17. Brenner, "Theses on Urbanization," 85.
18. Colin McFarlane, "Assemblage and Critical Urbanism," *City* 15, no. 2 (2011): 204–224; Neil Brenner, David Madden, and David Wachsmuth, "Assemblage Urbanism and the Challenges of Critical Urban Theory," *City* 15, no. 2 (2011): 225–240.
19. Neil Brenner and Christian Schmid, "The 'Urban Age' in Question," *International Journal of Urban and Regional Research* 38, no. 3 (2014): 731–755; Ananya Roy, "The Twenty-First-Century Metropolis: New Geographies of Theory," *Regional Studies* 43, no. 6 (2009): 819–830; Allen Scott and Michael Storper, "The Nature of Cities: The Scope and Limits of Urban Theory," *International Journal of Urban and Regional Research* 39, no. 1 (2015): 1–15; and Jennifer Robinson, *Ordinary Cities: Between Modernity and Development* (New York: Routledge, 2006).
20. Scott and Storper, "The Nature of Cities," 1.
21. Zeiderman et al., "Uncertainty and Urban Life," 283.
22. Jeremy Walker and Melinda Cooper, "Genealogies of Resilience: From Systems Ecology to the Political Economy of Crisis Adaptation," *Security Dialogue* 42, no. 2 (2011): 143–160; and Andrew Lakoff, "Preparing for the Next Emergency," *Public Culture* 19, no. 2 (2007): 247–271.
23. Jon Coaffee, David Murakami Wood, and Peter Rogers, *The Everyday Resilience of the City: How Cities Respond to Terrorism and Disaster* (New York: Palgrave Macmillan, 2009); Stephen Graham, ed., *Disrupted Cities: When Infrastructure Fails* (New York: Routledge, 2010); Stephen Graham, *Cities Under Siege: The New Military Urbanism* (New York: Verso, 2011); Martin Coward, *Urbicide: the Politics of Urban Destruction* (New York: Routledge, 2008); and Daanish Mustafa, "The Production of an Urban Hazardscape in Pakistan: Modernity, Vulnerability, and the Range of Choice," *Annals of the Association of American Geographers* 95, no. 3 (2005): 556–586.
24. Jon Coaffee, "Risk, Resilience, and Environmentally Sustainable Cities," *Energy Policy* 36, no. 12 (2008): 4633–4638; Jack Ahern, "From Fail-Safe to Safe-to-Fail: Sustainability and Resilience in the New Urban World," *Landscape and Urban Planning* 100, no. 4 (2011): 341–343; and Will Medd and Simon Marvin, "From the Politics of Urgency to the Governance of Preparedness: A Research Agenda on Urban Vulnerability," *Journal of Contingencies and Crisis Management* 13, no. 2 (2005): 44–49.
25. Zeiderman et al., "Uncertainty and Urban Life," 283, 298.

26. Ash Amin, "Surviving the Turbulent Future," *Environment and Planning D* 31, no. 1 (2013): 140–156. For a one-sided (and wrongheaded) critique of "too many urban theories," see Michael Storper and Alan Scott, "Current Debates in Urban Theory: A Critical Assessment," *Urban Studies* 53, no. 6 (2016): 1114–1136.
27. Zeiderman et al., "Uncertainty and Urban Life," 299–300.
28. Trevor Hogan and Julian Potter, "Big City Blues," *Thesis Eleven* 121, no. 1 (2014): 3–8.
29. Neil Smith, "New Globalism, New Urbanism: Gentrification as Global Urban Strategy," *Antipode* 34, no. 3 (2002): 436.
30. Ananya Roy, "The 21st-Century Metropolis: New Geographies of Theory," *Regional Studies* 43, no. 6 (2009): 820; and Edgar Pieterse, *City Futures: Confronting the Crisis of Urban Development* (London: Zed, 2008).
31. See Jennifer Robinson, "Global and World Cities: A View from Off the Map," *International Journal of Urban and Regional Research* 26, no. 3 (2002): 531–554.
32. Idalina Baptista, "How Portugal Became an 'Unplanned Country': A Critique of Scholarship on Portuguese Urban Development and Planning," *International Journal of Urban and Regional Research* 36, no. 5 (2012): 1076.
33. Baptista, "How Portugal Became an 'Unplanned Country.' "
34. Stuart Elden, "Thinking Territory Historically," *Geopolitics* 15, no. 4 (2010): 757–761.
35. See Jung In Kim, "Making Cities Global: The New City Development of Songdo, Yujiapu and Lingang," *Planning Perspectives* 29, no. 3 (2014): 331.
36. Russell Weaver and Chris Holtcamp, "Geographical Approaches to Understanding Urban Decline: From Evolutionary Theory to Political Economy . . . and Back?," *Geography Compass* 9, no. 5 (2015): 286–302.
37. Kimon Krenz, "Capturing Patterns of Shrinkage and Growth in Post-Industrial Regions: A Comparative Study of the Ruhr Valley and Leipzig-Halle," *Proceedings of the 10th International Space Syntax Symposium* 72 (July 13–17, 2015): 1–18. See also Russell Weaver, Sharmistha Bagchi-Sen, Jason Knight, and Amy Frazier, *Shrinking Cities: Understanding Urban Decline in the United States* (New York: Routledge, 2017).
38. Katrin Großman, Marco Bontje, Annegret Haase, and Vlad Mykhnenko, "Shrinking Cities: Notes for the Further Research Agenda," *Cities* 35 (2013): 221–225; and John Schilling and Alan Mallach, *Cities in Transition: A Guide for Practicing Planners* (Washington, DC: American Planning Association Press, 2012).
39. Diana Reckien and Christina Martinez-Fernandez, "Why Do Cities Shrink?," *European Planning Studies* 19, no. 8 (2011): 1375.
40. Robert Beauregard, "Urban Population Loss in Historical Perspective: United States, 1820–2000," *Environment and Planning A* 41, no. 3 (2009): 514–528; and Robert Beauregard, "Shrinking Cities in the United States in Historical Perspective: A Research Note," in *The Future of Shrinking Cities: Problems, Patterns and Strategies of Urban Transformation in a Global Context*, ed. Karina Pallagst et al. (Berkeley, CA: Institute of Urban and Regional Development, Center for Global Metropolitan Studies, and the Shrinking Cities International Research Network Monograph Series, 2009), 61–68.
41. Justin Hollander, Karina Pallagst, Terry Schwarz, and Frank J. Popper, "Planning Shrinking Cities," *Progress in Planning* 72 (2009): 223–232; Margaret Dewar and June Thomas, eds., *The City After Abandonment* (Philadelphia: University of Pennsylvania Press, 2012); and Karina Pallagst, Thorsten Wiechmann, and Cristina Martinez-Fernandez, eds., *Shrinking Cities: An International Perspective* (New York: Routledge, 2014).
42. Benedict Anderson, *The Spectre of Comparisons: Nationalism, Southeast Asia, and the World* (New York: Verso, 1998), 1–2.
43. See Peng Cheah, "Grounds of Comparison," in *Grounds of Comparison: Around the Work of Benedict Anderson*, ed. Pheng Cheah and Jonathan Culler (New York: Routledge, 2003), 13.

44. Anderson, *The Spectre of Comparisons*, 2.
45. Timothy Mitchell, "The Stage of Modernity," in *Questions of Modernity*, ed. Timothy Mitchell (Minneapolis: University of Minnesota Press, 2000), 24, 26.
46. Gwendolyn Wright, "Building Global Modernisms," *Grey Room* 7 (2002): 127.
47. See Arjun Appadurai, *Modernity at Large: Cultural Dimensions of Globalization* (Minneapolis: University of Minnesota Press, 1996).
48. Wright, "Building Global Modernisms," 128.
49. For example, see Allen Scott, "Emerging Cities of the Third Wave," *City* 15, no. 3–4 (2011): 289–321.
50. Ozan Karaman, *Remaking Space for Globalization: Dispossession Through Urban Renewal in Istanbul* (PhD diss., University of Minnesota, 2010), 36.
51. Joshua Akers, "Emerging Market City," *Environment and Planning A* 47, no. 9 (2015): 1842–1858; and Joshua Akers, "Making Markets: Think Tanks Legislation and Private Property in Detroit," *Urban Geography* 34, no. 8 (2013): 1070–1095.
52. Seth Schindler, "Detroit After Bankruptcy: A Case of Degrowth Machine Politics," *Urban Studies* 53, no. 4 (2016): 824, 831; Mattias Bernt, "Partnerships for Demolition: The Governance of Urban Renewal in East Germany's Shrinking Cities," *International Journal of Urban and Regional Research* 33, no. 3 (2009): 754–769; and Joan Martinez-Alier, "Socially Sustainable Economic De-growth," *Development and Change* 40, no. 6 (2009): 1099–1119.
53. David Scobey, *Empire City: The Making and Meaning of the New York City Landscape* (Philadelphia: Temple University Press, 2003); Cuff, *The Provisional City*, 20–22; and Page, *The Creative Destruction of Manhattan*, 1–5.
54. See Partha Chatterjee, *The Politics of the Governed: Reflections on Popular Politics in Most of the World* (New York: Columbia University Press, 2001), 6–8.
55. See Achille Mbembé and Janet Roitman, "Figures of the Subject in Times of Crisis," *Public Culture* 7, no. 2 (1995): 323–352.
56. Paul Knox, "Creating Ordinary Places: Slow Cities in a Fast World," *Journal of Urban Design* 10, no. 1 (2005): 1–11.
57. Nicolas Bourriaud, "Altermodern," in *Altermodern: Tate Triennial*, ed. Nicolas Bourriaud (London: Tate, 2009).
58. Keith Moxey, "Is Modernity Multiple?," *Revista de História da Arte* 10 (2012): 50–57.
59. See Alessandro Gubitosi, "Fast Track Cities—Dubai's Time Machine," in *Instant Cities: Emergent Trends in Architecture and Urbanism in the Arab World: the Third International Conference of the Center for the Study of Architecture in the Arab Region*, ed. Amer Moustafa, Jamal Al-Qawasmi, and Kevin Mitchell (Sharaj, UAE: CSAAR Press, 2008), 89–102.
60. Max Rousseau, *"Re-imaging the City* Centre for the Middle Classes: Regeneration, Gentrification and Symbolic Policies in 'Loser Cities,' " *International Journal of Urban and Regional Research* 33, no. 3 (2009): 772.
61. Dieter Rink, Annegret Haase, Katrin Grossmann, Chris Couch, and Matthew Cocks, "From Long-Term Shrinkage to Re-growth? The Urban Development Trajectories of Liverpool and Leipzig," *Built Environment* 38, no. 2 (2012): 162–178.
62. Tim Rieniets, "Shrinking Cities: Causes and Effects of Urban Population Losses in the Twentieth Century," *Nature and Culture* 4, no. 3 (2009): 231–254; and Cristina Martinez-Fernandez, Ivonne Audirac, Sylvie Fol, and Emmanuéle Cunningham Sabot, "Shrinking Cities: Urban Challenges of Globalization," *International Journal of Urban and Regional Research* 36, no. 2 (2012): 213–225.
63. For a wider view, see the collection of essays in Pallagst et al., *The Future of Shrinking Cities*.

BIBLIOGRAPHY

BOOKS AND ARTICLES

Abbott, Andrew. "Los Angeles and the Chicago School: A Comment on Michael Dear." *City and Community* 1, no. 1 (2002): 33–38.

Abu-Lughod, Janet. *Rabat: Urban Apartheid in Morocco*. Princeton, NJ: Princeton University Press, 1980.

——. *New York, Los Angeles, Chicago: America's Global Cities*. Minneapolis: University of Minnesota Press, 1999.

Acuto, Michele. "High-Rise Dubai Urban Entrepreneurialism and the Technology of Symbolic Power." *Cities* 27, no. 4 (2010): 272–284.

Acuto, Michele, Cecelia Dinardi, and Colin Marx. "Transcending (In)formal Urbanism." *Urban Studies* 56, no. 3 (2019): 475–487.

Adey, Peter. "Air/Atmospheres of the Megacity." *Theory, Culture & Society* 30, no. 7–8 (2013): 291–308.

Adham, Khaled. "Rediscovering the Island: Doha's Urbanity from Pearls to Spectacle." In *The Evolving Arab City: Tradition, Modernity and Urban Development*, ed. Yasser Elsheshtawy, 218–256. London: Routledge, 2008.

Adranovich, Greg, Matthew Burbank, and Charles Heying. "Olympic Cities: Lessons Learned from Mega-event Politics." *Journal of Urban Affairs* 23, no. 2 (2001): 113–132.

Agamben, Giorgio. *Homo Sacer: Sovereign Power and Bare Life*. Trans. Daniel Heller-Roazen. Palo Alto, CA: Stanford University Press, 1998.

Aguilar, Adrián, and Peter Ward. "Globalization, Regional Development, and Mega-City Expansion in Latin America: Analyzing Mexico City's Peri-Urban Hinterland." *Cities* 20, no. 1 (2003): 3–21.

Aguilera, Thomas, and Alan Smart. "Squatting, North, South and Turnabout: A Dialogue Comparing Illegal Housing Research." In *Public Goods Versus Economic Interests: Global Perspectives on the History of Squatting*, ed. Freia Anders and Alexander Sedlmaier, 29–55. New York: Routledge, 2017.

Ahern, Jack. "From Fail-Safe to Safe-to-Fail: Sustainability and Resilience in the New Urban World." *Landscape and Urban Planning* 100, no. 4 (2011): 341–343.

Akers, Joshua. "Making Markets: Think Tank Legislation and Private Property in Detroit." *Urban Geography* 34, no. 8 (2013), 1070–1095.

——. "Emerging Market City." *Environment and Planning A* 47, no. 9 (2015): 1842–1858.

Albecker, Marie-Fleur. "The Effects of Globalization in the First Suburbs of Paris: From Decline to Revival?" *Berkeley Planning Journal* 23, no. 1 (2010): 102–131.

Albecker, Marie-Fleur, and Sylvie Fol. "The Restructuring of Declining Suburbs in the Paris Region." In *Shrinking Cities: International Perspectives and Policy Implications*, ed. Karina Pallagst, Cristina Martinez-Fernandez, and Thorsten Wiechmann, 78–98. New York: Routledge, 2014.

Alderson, Arthur, and Jason Beckfield. "Power and Position in the World City System." *American Journal of Sociology* 109, no. 4 (2004): 811–851.

Alexander, Anthony. *Britain's New Towns: From Garden Cities to Sustainable Communities.* New York: Routledge, 2009.

Alexander, Jeffrey C. "Robust Utopias and Civil Repairs." *International Sociology* 16, no. 4 (2001): 579–591.

Algier, Michel. "Between War and the City: Towards an Urban Ethnography of Refugee Camps." *Ethnography* 3, no. 3 (2002): 317–341.

Allen, John. "Ambient Power: Berlin's Potsdamer Platz and the Seductive Logic of Public Spaces." *Urban Studies* 43, no. 2 (2006): 441–455.

Allweil, Yael. "Shrinking Cities: Like a Slow-Motion Katrina." *Places* 9, no. 1 (2007): 191–195.

Allwinkle, Sam, and Peter Cruickshank. "Creating Smarter Cities: An Overview." *Journal of Urban Technology* 18, no. 2 (2011): 1–16.

Alofsin, Anthony. "Broadacre City: Nemesis." *American Art* 25, no. 2 (2011): 21–25.

AlSayyad, Nezar. "Global Norms and Urban Forms in the Age of Tourism: Manufacturing Heritage, Consuming Tradition." In *Consuming Tradition, Manufacturing Heritage: Global Norms and Urban Forms in the Age of Tourism*, ed. Nezar AlSayyad, 1–33. London: Routledge, 2001.

——. "Hybrid Culture/Hybrid Urbanism: Pandora's Box of the 'Third Place.'" In *Hybrid Urbanism*, ed. Nezar AlSayyad, 1–20. Westport, CT: Praeger, 2001.

——. "Urban Informality as a 'New' Way of Life." In *Urban Informality: Transnational Perspectives from the Middle East, Latin America, and South Asia*, ed. Ananya Roy and Nezar AlSayyad, 7–30. Lanham, MD: Lexington, 2004.

AlSayyad, Nezar, and Ananya Roy. "Urban Informality: Crossing Borders." In *Urban Informality: Transnational Perspectives from the Middle East, Latin America, and South Asia*, ed. Ananya Roy and Nezar AlSayyad, 1–6. Lanham, MD: Lexington, 2004.

——. "Medieval Modernity: On Citizenship and Urbanism in a Global Era." *Space and Polity* 10, no. 1 (2006): 1–20.

Altrock, Uwe. "Conceptualising Informality: Some Thoughts on the Way Towards Generalization." In *Urban Informalities: Reflections on the Formal and Informal*, ed. Colin McFarlane and Michael Waibel, 171–194. London: Routledge, 2016.

Altshuler, Alan, and David Luberoff. *Mega-Projects: The Changing Politics of Urban Public Investment.* Washington, DC: Brookings Institution Press, 2003.

Amin, Ash. "Spatialities of Globalization." *Environment & Planning A* 34, no. 3 (2002): 385–399.

——. "The Good City." *Urban Studies* 43, no. 5–6 (2006): 1009–1023.

——. "Surviving the Turbulent Future." *Environment and Planning D* 31, no. 1 (2013): 140–156.

——. "Telescopic Urbanism and the Poor." *City* 17, no. 4 (2013): 476–492.

——. "The Urban Condition: A Challenge to Social Science." *Public Culture* 25, no. 2 (2013): 201–208.

Amin, Ash, and Stephen Graham. "The Ordinary City." *Transactions of the Institute of British Geographers* 22, no. 4 (1997): 411–429.

Amin, Ash, and Nigel Thrift. "Neo-Marshallian Nodes in Global Networks." *International Journal of Urban and Regional Research* 16, no. 4 (1992): 571–587.

——. *Cities: Reimagining the Urban.* Cambridge: Polity, 2002.

——. *Seeing Like a City.* Cambridge: Polity, 2016.

Anderson, Benedict. *The Spectre of Comparisons: Nationalism, Southeast Asia, and the World.* New York: Verso, 1998.

Anderson, Michelle Wilde. "Dissolving Cities." *Yale Law Journal* 121 (2011): 1364–1436.

——. "The New Minimal Cities." *Yale Law Journal* 123 (2014): 1118–1227.

Angelini, Alessandro. "Book Review of *Spaces of Global Cultures*." *Future Anterior* 2, no. 1 (2005): 67–71.
Angelo, Hillary, and Kian Goh. "Out in Space: Difference and Abstraction in Planetary Urbanization." *International Journal of Urban and Regional Research* 45, 4 (2021): 732–744.
Angelo, Hillary, and David Wachsmuth. "Urbanizing Urban Political Ecology: A Critique of Methodological Cityism." *International Journal of Urban and Regional Research* 39, no. 1 (2015): 16–27.
——. "Why Does Everyone Think Cities Can Save the Planet?" *Urban Studies* 57, no. 11 (2020): 2201–2221.
Angin, Timir. "Jakarta Leftover Spaces." *Inter-Asia Cultural Studies* 12, no. 4 (2011): 568–583.
Angotti, Tom. "Apocalyptic Anti-Urbanism: Mike Davis and His Planet of Slums." *International Journal of Urban and Regional Research* 30, no. 4 (2006): 961–967.
——. *The New Century of the Metropolis: Urban Enclaves and Orientalism*. New York: Routledge, 2013.
Anholt, Simon. *Competitive Identity: The New Brand Management for Nations, Cities and Regions*. New York: Palgrave Macmillan, 2007.
Anjaria, Jonathan Shapiro. *The Slow Boil: Street Food, Rights, and Public Space in Mumbai*. Stanford, CA: Stanford University Press, 2016.
Appadurai, Arjun. *Modernity at Large: Cultural Dimensions of Globalization*. Minneapolis: University of Minnesota Press, 1996.
Appel, Hannah. "Offshore Work: Oil, Modularity, and the How of Capitalism in Equatorial Guinea." *American Ethnologist* 39, no. 4 (2012): 692–709.
Arabindoo, Pushpa. "Beyond the Return of the 'Slum'." *City* 15, no. 6 (2011): 631–635.
——. "Rhetoric of the Slum: Rethinking Urban Poverty." *City* 15, no. 6 (2011): 636–646.
Archer, Kevin. "The Limits to the Imagineered City: Sociospatial Polarization in Orlando." *Economic Geography* 73, no. 3 (1997): 322–336.
Arkaraprasertkul, Non. "Power, Politics, and the Making of Shanghai." *Journal of Planning History* 9, no. 4 (2010): 232–259.
Arthur, W. Brian. "Increasing Returns and the New World of Business." *Harvard Business Review* (July–August 1996): 100–109.
Assaad, Ragui. "Formal and Informal Institutions in the Labor Market, with Applications to the Construction Sector in Egypt." *World Development* 21, no. 6 (1993): 922–939.
Atkinson, Glen, and Ted Oleson. "Urban Sprawl as a Path Dependent Process." *Journal of Economic Issues* 30, no. 2 (1996): 609–615.
Atkinson, Rowland. "Domestication by Cappuccino or a Revenge on Urban Space? Control and Empowerment in the Management of Public Spaces." *Urban Studies* 40, no. 9 (2003): 1829–1843.
Atkinson, Rowland, and Gary Bridge. "Introduction." In *Gentrification in a Global Context: The New Urban Colonialism*, ed. Rowland Atkinson and Gary Bridge, 1–17. New York: Routledge, 2005.
Audirac, Ivonne. "Shrinking Cities in Latin America: An Oxymoron?" In *Shrinking Cities: A Global Perspective*, ed. Harry W. Richardson and Chang Woon Nam, 28–46. New York: Routledge, 2014.
——. "Introduction: Shrinking Cities from Marginal to Mainstream: Views from North America and Europe." *Cities* 75 (2018): 1–5.
——. "Shrinking Cities: An Unfit Term for American Urban Policy?" *Cities* 75 (2018): 12–19.
Audirac, Ivonne, and Jesús Arroyo Alejandre. "Introduction." In *Shrinking Cities South/North*, ed. Ivonne Audirac and Jesuś Arroyo Alejandre, 9–36. Tallahassee: Florida State University, 2010.
Audirac, Ivonne, Emmanuèle Cunningham Sabot, Sylvie Fol, and Sergio Torres Moraes. "Declining Suburbs in Europe and Latin America." *International Journal of Urban and Regional Research* 36, no. 2 (2012): 226–244.
Audirac, Ivonne, Sylvie Fol, and Cristina Martinez-Fernandez. "Shrinking Cities in a Time of Crisis." *Berkeley Planning Journal* 23, no. 1 (2010): 51–57.
Augé, Marc. *Non-Places: Introduction to an Anthropology of Supermodernity*. Trans. John Howe. New York: Verso, 1995.

Auyero, Javier. "Taking Bourdieu to the Shantytown." *International Journal of Urban and Regional Research* 45, no. 1 (2021): 176–185.
Ay, Deniz. "Review of Cities of the Global South Reader." *Journal of Planning Education and Research* 38, no. 3 (2018): 380–381.
Bach, Jonathan. "Modernity and Urban Imagination in Economic Zones." *Theory, Culture & Society* 28, no. 5 (2011): 98–122.
Bagaeen, Samer. "Brand Dubai: The Instant City; or the Instantly Recognizable City." *International Planning Studies* 12, no. 1 (2007): 173–197.
Balakrisnan, Melodena. "Dubai—A Star in the East: A Case Study in Strategic Destination Branding." *Journal of Place Management and Development* 1, no. 1 (2008): 62–91.
Balbo, Marcelo. "Beyond the City of Developing Countries: The New Urban Order of the 'Emerging City.'" *Planning Theory* 13, no. 3 (2014): 269–287.
Banerjee, Tridib. "The Future of Public Space: Beyond Invented Streets and Reinvented Places." *Journal of the American Planning Association* 67, no. 1 (2001): 9–24.
Baptista, Idalina. "How Portugal Became an 'Unplanned Country': A Critique of Scholarship on Portuguese Urban Development and Planning." *International Journal of Urban and Regional Research* 36, no. 5 (2012): 1076–1092.
——. "Practices of Exception in Urban Governance: Reconfiguring Power Inside the State." *Urban Studies* 50, no. 1 (2013): 39–54.
——. "Electricity Services Always in the Making: Informality and the Work of Infrastructure Maintenance and Repair in an African City." *Urban Studies* 56, no. 3 (2019): 510–525.
Barnett, Jonathan, *City Design: Modernist, Traditional, Green and Systems Perspectives.* 2nd ed. New York: Routledge, 2016.
Barry, Andrew. "Technological Zones." *European Journal of Social Theory* 9, no. 2 (2006): 239–253.
Barthel, Pierre-Arnaud. "Arab Mega-Projects: Between the Dubai Effect, Global Crisis, Social Mobilization and a Sustainable Shift." *Built Environment* 36, no. 2 (2010): 133–145.
Barthel, Pierre-Arnaud, and Sabine Planel. "Tangier-Med and Casa-Marina, Prestige Projects in Morocco: New Capitalist Frameworks and Local Context." *Built Environment* 36, no. 2 (2010): 176–191.
Bartley, Brendan, and Kasey Treadwell Shine. "Competitive City: Governance and the Changing Dynamics of Urban Regeneration in Dublin." In *The Globalized City: Economic Restructuring and Social Polarization in European Cities*, ed. Frank Moulaert, Arantxa Rodriguez, and Erik Swyngedouw, 145–166. Oxford: Oxford University Press, 2003.
Bassens, David, and Michiel van Meeteren. "World Cities Under Conditions of Financialized Globalization: Towards an Augmented World City Hypothesis." *Progress in Human Geography* 39, no. 6 (2015): 752–775.
Bassett, Keith. "Partnerships, Business Elites and Urban Politics: New Forms of Governance in an English City?" *Urban Studies* 33, no. 3 (1996): 539–555.
——. "Urban Cultural Strategies and Urban Regeneration: A Case Study and Critique." *Environment and Planning A* 25, no. 12 (1993): 1773–1788.
Batchelor, Peter. "The Origin of the Garden City Concept of Urban Form." *Journal of the Society of Architectural Historians* 28, no. 3 (1969): 184–200.
Batty, Michael. "Editorial: Smart Cities, Big Data." *Environment and Planning B* 39, no. 2 (2012): 191–193.
Batunova, Elena, and Maria Gunko. "Urban Shrinkage: An Unspoken Challenge of Spatial Planning in Russian Small and Medium-Sized Cities." *European Planning Studies* 26, no. 8 (2018): 1580–1597.
Bauman, Zygmunt. *Wasted Lives: Modernity and Its Outcasts.* Cambridge: Polity, 2004.
——. "Seeking Shelter in Pandora's Box." *City* 9, no. 2 (2005): 161–168.
Bayat, Asef. "Un-Civil Society: The Politics of the Informal People," *Third World Quarterly* 18, no. 1 (1997): 53–72.
——. "From 'Dangerous Classes' to 'Quiet Rebels': Politics of the Urban Subaltern in the Global South," *International Sociology* 15, no. 3 (2000): 269–278.

——. "Radical Religion and the Habitus of the Dispossessed: Does Islamic Militancy Have an Urban Ecology?" *International Journal of Urban and Regional Research* 31, no. 3 (2007): 579–590.

——. *Life as Politics: How Ordinary People Changed the Middle East.* Stanford, CA: Stanford University Press, 2010.

Bayliss, Darrin. "The Rise of the Creative City: Culture and Creativity in Copenhagen." *European Planning Studies* 15, no. 7 (2007): 889–903.

Beall, Jo. "Cities, Terrorism, and Development." *Journal of International Development* 18, no. 1 (2006): 105–120.

Beauregard, Robert A. "Representing Urban Decline: Postwar Cities as Narrative Objects." *Urban Affairs Review* 29, no. 2 (1993): 187–202.

——. "Federal Policy and Postwar Urban Decline: A Case of Government Complicity?" *Housing Policy Debate* 12, no. 1 (2001): 129–151.

——. "Aberrant Cities: Urban Population Loss in the United States, 1820–1930," *Urban Geography* 24, no. 8 (2003): 672–690.

——. "City of Superlatives." *City & Community* 2, no. 3 (2003): 183–199.

——. *Voices of Decline: The Postwar Fate of U.S. Cities.* 2nd ed. New York: Routledge, 2003.

——. "The Radical Break in Late Twentieth-Century Urbanization." *Area* 38, no. 2 (2006): 218–220.

——. *When America Became Suburban.* Minneapolis: University of Minnesota Press, 2006.

——. "Urban Population Loss in Historical Perspective: United States, 1820–2000." *Environment and Planning A* 41, no. 3 (2009): 514–528.

——. "Growth and Depopulation in the United States." In *Rebuilding America's Legacy Cities: New Directions for the Industrial Heartland*, ed. Allan Mallach, 1–24. New York: American Assembly, Columbia University, 2012.

——. "Radical Uniqueness and the Flight from Urban Theory." In *The City, Revisited: Urban Theory from Chicago, Los Angeles, and New York*, ed. Dennis Judd and Dick Simpson, 186–206. Minneapolis: University of Minnesota Press, 2013.

Beauregard, Robert A., and Anne Haila. "The Unavoidable Continuities of the City." In *Globalizing Cities: A New Spatial Order?*, ed. Peter Marcuse and Ronald van Kempen, 22–36. Oxford: Blackwell, 2000.

Beaverstock, Jonathan, Richard Smith, and Peter Taylor. "A Roster of World Cities." *Cities* 16, no. 6 (1999): 445–458.

——. "World-City Network: A New Metageography?" *Annals of the Association of American Geographers* 90, no. 1 (2004): 123–134.

Begg, Iain. "Cities and Competitiveness." *Urban Studies* 36, no. 5–6 (1999): 795–809.

——. "'Investability': The Key to Competitive Regions and Cities?" *Regional Studies* 36, no. 2 (2002): 187–193.

Bender, Thomas. *The Unfinished City: New York and the Metropolitan Idea.* New York: New York University Press, 2002.

Benjamin, Solomon. "Occupancy Urbanism: Radicalizing Politics and Economy Beyond Policy and Programs." *International Journal of Urban and Regional Research* 32, no. 3 (2008): 719–729.

Benjamin, Walter. *The Arcades Project.* Trans. Howard Eiland and Kevin McLaughlin. Cambridge, MA: Belknap, 1999.

——. *Illuminations.* London: Pimlico, 1999.

Bentley, George, Priscilla McCutcheon, Robert Cromley, and Dean Hanink. "Race, Class, Unemployment, and Housing Vacancies in Detroit: An Empirical Analysis." *Urban Geography* 37, no. 5 (2016): 785–800.

Beriatos, Elias, and Aspa Gospodini. "Glocalizing Urban Landscapes: Athens and the 2004 Olympics." *Cities* 21, no. 3 (2004): 187–202.

Berman, Marshall, *All That Is Solid Melts Into Air.* New York: Simon & Schuster, 1982.

Bernt, Matthias. "Partnerships for Demolition: The Governance of Urban Renewal in East Germany's Shrinking Cities." *International Journal of Urban and Regional Research* 33, no. 3 (2009): 754–769.

——. "The Limits of Shrinkage: Conceptual Pitfalls and Alternatives in the Discussion of Urban Population Loss." *International Journal of Urban and Regional Research* 40, no. 2 (2016): 441–450.

Bernt, Mattias, Annegret Haase, Katrin Großmann, Matthew Cocks, Chris Couch, Caterina Cortese, and Robert Krzysztofik. "How Does(n't) Urban Shrinkage Get Onto the Agenda? Experiences from Leipzig, Liverpool, Genoa and Bytom." *International Journal of Urban and Regional Research* 38, no. 5 (2014): 1749–1766.

Bezmez, Dikmen. "The Politics of Urban Waterfront Regeneration: The Case of Haliç (the Golden Horn), Istanbul." *International Journal of Urban and Regional Research* 32, no. 4 (2008): 815–840.

Bhan, Gautam. "The Real Lives of Urban Fantasies." *Environment and Urbanization* 26, no. 1 (2014): 232–235.

Bianchini, Franco, Jon Dawson, and Richard Evans. "Flagship Projects in Urban Regeneration." In *Rebuilding the City: Property-Led Urban Regeneration*, ed. Patsey Healey, Simin Davoudi, Mo O'Toole, David Usher, and Solmaz Tavsanoglu, 245–255. London: Spon, 199.

Biehl, João. *Vita: Life in a Zone of Social Abandonment.* Berkeley: University of California, 2005.

Bigon, Liora. *A History of Urban Planning in Two West African Colonial Capitals: Residential Segregation in British Lagos and French Dakar (1850–1930).* Lewiston, NY: Mellen, 2009.

——. "'Garden City' in the Tropics? French Dakar in Comparative Perspective." *Journal of Historical Geography* 38, no. 1 (2012): 35–44.

——. "Garden Cities in Colonial Africa: A Note on Historiography." *Planning Perspectives* 28, no. 3 (2013): 477–485.

Birch, Eugenie, and Susan Wachter. "World Urbanization: The Critical Issue of the Twenty-First Century." In *Global Urbanization*, ed. Eugenie Birch and Susan Wachter, 3–23. Philadelphia: University of Pennsylvania Press, 2011.

Bishop, Ryan, John Phillips, and Wei Yeo, eds. *Postcolonial Urbanism: Southeast Asian Cities and Global Processes.* New York: Routledge, 2003.

Boddy, Martin, and Michael Parkinson, eds. *City Matters: Competitiveness, Cohesion, and Urban Governance.* Bristol, UK: Policy Press, University of Bristol, 2004.

Bogaert, Koenraad. "New State Space Formation in Morocco: The Example of the Bouregreg Valley." *Urban Studies* 49, no. 2 (2012): 255–270.

——. *Globalized Authoritarianism: Megaprojects, Slums, and Class Relations in Urban Morocco.* Minneapolis: University of Minnesota Press, 2018.

Boland, Philip. "Unpacking the Theory–Policy Interface of Local Economic Development: An Analysis of Cardiff and Liverpool." *Urban Studies* 4, no. 5/6 (2007): 1019–1039.

Bolleter, Julian. "Charting a Changing Waterfront: A Review of Key Schemes for Perth's Foreshore." *Journal of Urban Design* 19, no. 5 (2014): 569–592.

Bontje, Marco. "Facing the Challenge of Shrinking Cities in East Germany: The Case of Leipzig." *Geojournal* 61 (2004): 13–21.

Bontje, Marco, and Sako Musterd. "Understanding Shrinkage in European Regions." *Built Environment* 38, no. 2 (2012): 153–161.

Booth, Douglas. "Regional Long Waves and Urban Policy." *Urban Studies* 24, no. 6 (1987): 447–459.

Booth, Philip. "Partnerships and Networks: The Governance of Urban Regeneration in Britain." *Journal of Housing and the Built Environment* 20, no. 3 (2005): 257–269.

Borer, Michael Ian. "The Location of Culture: The Urban Culturalist Perspective." *City & Community* 5, no. 2 (2006): 173–197.

Bourdieu, Pierre. *The State Nobility: Elite Schools in the Field of Power.* Trans. Lauretta C. Clough. Stanford, CA: Stanford University Press, 1996.

Bourne, Larry. "On Schools of Thought, Comparative Research, and Inclusiveness: A Commentary." *Urban Geography* 29, no. 2 (2008): 177–186.

Bourriaud, Nicolas. "Altermodern." In *Altermodern: Tate Triennial*, ed. Nicolas Bourriaud. London: Tate, 2009.

Boutang, Yann Moulier. *Cognitive Capitalism*. Cambridge: Polity, 2012.

Boyer, M. Christine. "Cities for Sale: Merchandising History at South Street Seaport." In *Variations on a Theme Park: The New American City and the End of Public Space*, ed. Michael Sorkin, 181–204. New York: Hill and Wang, 1992.

——. "The City of Illusion: New York's Public Places." In *The Restless Urban Landscape*, ed. Paul Knox, 111–126. Englewood Cliffs, NJ: Prentice Hall, 1993.

——. "The Great Frame-Up: Fantastic Appearances in Contemporary Spatial Politics." In *Spatial Practices*, ed. Helen Liggett and David Perry, 81–109. London: Sage, 1995.

——. *The City of Collective Memory: Its Historical Imagery and Architectural Entertainments*. Cambridge, MA: MIT Press, 1996.

——. "Twice Told Stories: The Double Erasure of Times Square." In *The Unknown City: Contesting Architecture and Social Space*, ed. Iain Borden, Joe Kerr, Jane Rendell, with Alicia Pivaro, 30–53. Cambridge, MA: MIT Press, 2001.

Boyle, Mark, Christopher McWilliams, and Gareth Rice. "The Spatialities of Actually Existing Neoliberalism in Glasgow, 1977 to Present." *Geografiska Annaler: Series B, Human Geography* 90, no. 4 (2008): 313–325.

Brand, Peter, and Julio Dávila. "Mobility Innovation at the Urban Margins." *City* 15, no. 6 (2011): 647–661.

Brandstetter, Benno, Thilo Lang, and Anne Pfeifer. "Umgang mit der schrumpfenden Stadt—ein Debattenüberblick." *Berliner Debatte Initial* 16, no. 6 (2005): 55–68.

Breckenfeld, Gurney. "Coping with City Shrinkage." *Civil Engineering* 48, no. 11 (1978): 112–113.

Breen, Ann, and Dick Rigby. *Waterfronts: Cities Reclaim Their Edge*. New York: McGraw Hill, 1993.

——. *The New Waterfront: A Worldwide Success Story*. London: Thames & Hudson, 1996.

Breman, Jan. "Slumlands." *New Left Review* 40 (2006): 141–148.

——. "Myth of the Global Safety Net." *New Left Review* 59 (2009): 29–38.

Brennan, Christiana, Darlenne Hackler, and Christopher Hoene. "Demographic Change in Small Cities, 1990 to 2000." *Urban Affairs Review* 40, no. 3 (2005): 342–361.

Brenner, Neil. "Stereotypes, Archetypes, and Prototypes: Three Uses of Superlatives in Contemporary Urban Studies." *City & Community* 2, no. 3 (2003): 205–216.

——. "Theses on Urbanism." *Public Culture* 25, no. 1 (2013): 85–114.

——, ed. *Implosions/Explosions: Towards a Study of Planetary Urbanization*. Berlin: Jovis, 2014.

——. "Introduction: Urban Theory Without an Outside." In *Implosions/Explosions: Towards a Theory of Planetary Urbanization*, ed. Neil Brenner, 14–30. Berlin: Jovis, 2014.

——. "Debating Planetary Urbanization: For an Engaged Pluralism." *Environment & Planning D* 36, no. 3 (2018): 570–590.

Brenner, Neil, David Madden, and David Wachsmuth. "Assemblage Urbanism and the Challenges of Critical Urban Theory." *City* 15, no. 2 (2011): 225–240.

Brenner, Neil, and Christian Schmid. "The 'Urban Age' in Question." *International Journal of Urban and Regional Research* 38, no. 3 (2014): 731–755.

——. "Towards a New Epistemology of the Urban?" *City* 19, no. 2–3 (2015): 151–182.

Brenner, Neil, and Nik Theodore. "Cities and the Geographies of 'Actually Existing Neoliberalism'." In *Spaces of Neoliberalism: Urban Restructuring in North America and Western Europe*, ed. Neil Brenner and Nik Theodore, 2–32. Oxford: Wiley-Blackwell, 2002.

——. "Preface: From the 'New Localism' to the Spaces of Neoliberalism." *Antipode* 34, no. 3 (2002): 341–347.

——. "Neoliberalism and the Urban Condition." *City* 9, no. 1 (2005): 101–107.

Brillembourg, Carlos. "The New Slum Urbanism of Caracas, Invasions and Settlements, Colonialism, Democracy, Capitalism and Devil Worship." *AD: Architectural Design* 74, no. 2 (2004): 77–81.

Britton, Stephen. "Tourism, Capital, and Place: Towards a Critical Geography of Tourism." *Environment and Planning D* 9, no. 4 (1991): 451–478.

Bromley, Ray. "The Informal Sector: Why Is It Worth Discussing?" *World Development* 6 (1978): 1033–1039.

Browns, Caru. "Shrinkage Happens . . . in Small Towns Too! Responding to De-population and Loss of Place in Susquehanna River Towns." *Urban Design International* 18, no. 1 (2013): 61–77.

Bruekner, Jan. "Urban Sprawl: Diagnosis and Remedies." *International Regional Science Review* 23, no. 2 (2000): 160–171.

Broudehoux, Anne-Marie. "Image Making, City Marketing and the Aesthetization of Social Inequality in Rio de Janeiro." In *Consuming Tradition, Manufacturing Heritage: Global Norms and Urban Forms in the Age of Tourism*, ed. Nezar AlSayyad, 273–297. New York: Routledge, 2001.

Buček, Ján. "Slovakia." In *Urban Issues and Urban Policies in the New EU Countries*, ed. Ronald Van Kempen, Marcel Vermeulen, and Ad Baan, 79–108. Aldershot, UK: Ashgate, 2005.

Buček, Ján, and Branislav Bleha. "Urban Shrinkage as a Challenge to Local Development Planning in Slovakia." *Morovian Geographical Reports* 21, no. 1 (2013): 5–15.

Buck, Nick, Ian Gordon, Alan Harding, and Ivan Turok. "Conclusion: Moving Beyond the Conventional Wisdom." In *Changing Cities: Rethinking Urban Competitiveness, Cohesion and Governance*, ed. Ian Gordon, Nick Buck, Alan Harding, and Ivan Turok, 265–282. New York: Palgrave Macmillan, 2005.

Buckley, Michelle. "From Kerala to Dubai and Back Again: Construction Migrants and the Global Economic Crisis." *Geoforum* 43, no. 2 (2012): 250–259.

Buckley, Michelle, and Kendra Strauss. "With, Against and Beyond Lefebvre: Planetary Urbanization and Epistemic Plurality." *Environment and Planning D: Society and Space* 34, no. 4 (2016): 617–636.

Buck-Morss, Susan. "The City as Dreamworld and Catastrophe." *October* 73 (1995): 3–26.

Buder, Stanley. "Ebenezer Howard: The Genesis of a Town Planning Movement." *Journal of the American Institute of Planners* 35, no. 6 (1969): 390–398.

——. *Visionaries and Planners: The Garden City Movement and the Modern Community.* Oxford: Oxford University Press, 1990.

Buhnik, Sophie. "From Shrinking Cities to Toshi no Shukushō: Identifying Patterns of Urban Shrinkage in the Osaka Metropolitan Area." *Berkeley Planning Journal* 23, no. 1 (2010): 132–155.

——. "The Uneven Impacts of Demographic Decline in a Japanese Metropolis: A Three-Scale Approach to Urban Shrinkage Patterns in the Osaka Metropolitan Area." In *Globalization and New Intra-Urban Dynamics in Asian Cities*, ed. Natacha Aveline-Dubach, Sue-Ching Jou, and Hsin-Huang Michael Hsiao, 199–238. Taipei: National Taiwan University Press, 2014.

——. "The Dynamics of Urban Degrowth in Japanese Metropolitan Areas: What Are the Outcomes of Urban Recentralisation Strategies?" *Town Planning Review* 88, no. 1 (2017): 79–92.

Bunker, Robert, and John Sullivan. "Integrating Feral Cities and Third Phase Cartels/Third Generation Gangs Research: The Rise of Criminal (Narco) City Networks and BlackFor." *Small Wars & Insurgencies* 22, no. 5 (2011): 764–786.

Bunnell, Tim. "Antecedent Cities and Inter-referencing Effects: Learning from and Extending Beyond Critiques of Neoliberalisation." *Urban Studies* 52, no. 11 (2015): 1983–2000.

——. "Smart City Returns." *Dialogues in Human Geography* 5, no. 1 (2015): 45–48.

Bunnell, Tim, and Diganta Das. "Urban Pulse—A Geography of Serial Seduction: Urban Policy Transfer from Kuala Lumpur to Hyderabad." *Urban Geography* 31, no. 3 (2010): 277–284.

Bunnell, Tim, Daniel Goh, Chee-Kien Lai, and Choon-Piew Pow. "Introduction: Global Urban Frontiers? Asian Cities in Theory, Practice and Imagination." *Urban Studies* 49, no. 13 (2012): 2785–2793.

Bunnell, Tim, and Andrew Harris. "Re-Viewing Informality: Perspectives from Urban Asia." *International Development Planning Review* 34, no. 4 (2012): 339–348.

Bunnell, Tim, and Anant Marinanti. "Practising Urban and Regional Research Beyond Metrocentricity." *International Journal of Urban and Regional Research* 34, no. 2 (2010): 415–420.

Bunting, Trudi, and Piere Filion. "Uneven Cities: Addressing Rising Inequality in the Twenty-First Century." *Canadian Geographer* 45, no. 1 (2001): 126–141.

Burdett, Ricky. "Designing Urban Democracy: Mapping Scales of Urban Identity." *Public Culture* 25, no. 2 (2013): 349–367.

Burdett, Ricky, and Philipp Rode. "Living in the Urban Age." In *Living in the Endless City*, ed. Ricky Burdett and Deyan Sudjic, 8–43. London: Phaidon, 2011.

Burgers, Jack, and Sako Musterd. "Understanding Urban Inequality: A Model Based on Existing Theories and an Empirical Illustration." *International Journal of Urban and Regional Research* 26, no. 2 (2002): 403–413.

Burgess, Patricia. "City Planning and the Planning of Cities: The Recent Historiography." *Journal of Planning Literature* 7, no. 4 (1993): 314–327.

Butler, Tim. "Re-urbanizing London Docklands: Gentrification, Suburbanization or New Urbanism?" *International Journal of Urban and Regional Research* 31, no. 4 (2007): 759–781.

Buzar, Stefan, Philip Ogden, and Ray Hall. "Households Matter: The Quiet Demography of Urban Transformation." *Progress in Human Geography* 29, no. 4 (2005): 413–436.

Buzar, Stefan, Philip Ogden, Ray Hall, Annegret Haase, Sigrun Kabisch, and Annett Steinführer. "Splintering Urban Populations: Emergent Landscapes of Reurbanisation in Four European Cities." *Urban Studies* 44, no. 4 (2007): 651–677.

Cabral, João, and Berta Rato. "Urban Development for Competitiveness and Cohesion: The Expo 98 Urban Project in Lisbon." In *The Globalized City: Economic Restructuring and Social Polarization in European Cities*, ed. Frank Moulaert, Arantxa Rodríguez, and Erik Swyngedouw, 209–228. Oxford: Oxford University Press, 2003.

Caldeira, Teresa. "Worlds Set Apart." In *Living in the Endless City*, ed. Ricky Burdett and Deyan Sudjic, 168–175. London: Phaidon, 2011.

Cameron, Catherine. "The Marketing of Tradition: The Value of Culture in American Life." *City & Society* 1, no. 2 (1987): 162–174.

——. "The Marketing of Heritage: From the Western World to the Global Stage." *City and Society* 20, no. 2 (2008): 160–168.

Campanella, Thomas. *The Concrete Dragon: China's Urban Revolution and What It Means for the World.* New York: Princeton Architectural Press, 2008.

Campbell, Tim, and Alana Campbell. "Emerging Disease Burdens and the Poor in Cities of the Developing World." *Journal of Urban Health* 84, no. 1 (2007): 54–64.

Caro, Robert. *The Power Broker: Robert Moses and the Fall of New York.* New York: Knopf, 1974.

Carp, Jana. "The Town's Abuzz: Collaborative Opportunities for Environmental Professionals in the Slow City Movement." *Environmental Practice* 14, no. 2 (2012): 130–142.

Carr, Matt. "Slouching Towards Dystopia: The New Military Futurism." *Race & Class* 51, no. 3 (2010): 13–32.

Carvalho, Luís. "Smart Cities from Scratch? A Socio-Technical Perspective." *Cambridge Journal of Regions, Economy and Society* 8, no. 1 (2015): 43–60.

Castells, Manuel. *The Informational City.* London: Blackwell, 1989.

——. *The Rise of the Network Society.* London: Blackwell, 1996.

——. *The Rise of the Network Society.* 2nd ed. New York: Blackwell, 2000.

Catungal, John Paul, Deborah Leslie, and Yvonne Hii. "Geographies of Displacement in the Creative City: The Case of Liberty Village, Toronto." *Urban Studies* 46, no. 5–6 (2009): 1095–1114.

Ceballos, Sara González. "The Role of the Guggenheim Museum in the Development of Urban Entrepreneurial Practices in Bilbao." *International Journal of Iberian Studies* 16, no. 3 (2004): 177–186.

Chang, I-Chun Catherine, and Eric Sheppard. "China's Eco-Cities as Variegated Urban Sustainability: Dongtan Eco-City and Chongming Eco-Island." *Journal of Urban Technology* 20, no. 1 (2013): 57–75.

Chang, T. C. "Theming Cities, Taming Places: Insights from Singapore. "*Geografiska Annaler: Series B, Human Geography* 82, no. 1 (2000): 35–54.

Chapin, Timothy. "From Growth Controls, to Comprehensive Planning, to Smart Growth: Planning's Emerging Fourth Wave." *Journal of the American Planning Association* 78, no. 1 (2012): 5–15.

Chatterjee, Partha. *The Politics of the Governed: Reflections on Popular Politics in Most of the World.* New York: Columbia University Press, 2001.

——. "The Rights of the Governed." *Identity, Culture and Politics* 3, no. 2 (2002): 51–72.

Chatterton, Paul. "Will the Real Creative City Please Stand Up?" *City* 4, no. 3 (2000): 390–397.

——. "The Urban Impossible: A Eulogy for the Unfinished City." *City* 14, no. 3 (2010): 234–244.

Cheah, Pheng. "Grounds of Comparison." In *Grounds of Comparison: Around the Work of Benedict Anderson*, ed. Pheng Cheah and Jonathan Culler, 1–21. New York: Routledge, 2003.

Chen, Martha Alter. "Rethinking the Informal Economy: Linkages with the Formal Economy and Formal Regulatory Environment." In *Linking the Formal and Informal Economy: Concepts and Politics*, ed. Basudeb Guha-Khasnobis, Ravi Kanbur, and Elinor Ostrom, 75–92. New York: Oxford University Press, 2007.

Cheong, Kee-Cheok. "Charter Cities: An Idea Whose Time Has Come or Should Have Gone?" *Malaysian Journal of Economic Studies* 47, no. 2 (2010): 165–168.

Cheshire, Paul. "Resurgent Cities, Urban Myths and Policy Hubris." *Urban Studies* 43, no. 8 (2006): 1231–1246.

Chibber, Vivek. "Reviving the Developmental State? The Myth of the National Bourgeoisie." In *The Empire Reloaded: The Socialist Register*, ed. Leo Panitch and Colin Leys, 144–165. New York: Monthly Review Press, 2005.

Chien, Shiuh-Shen, and Max Woodworth. "China's Urban Speed Machine: The Politics of Speed and Time in a Period of Rapid Urban Growth." *International Journal of Urban and Regional Research* 42, no. 4 (2018): 723–737.

Chin, Dong-Hung. "A Cluster of the Four Coal Mining Cities in Korea from a Global Perspective: How Sid the People Overcome a Crisis After a Massive Closure of Mines?" In *Shrinking Cities: International Perspectives and Policy Implications*, ed. Karina Pallagst, Thorsten Wiechmann, and Cristina Martinez-Fernandez, 186–204. New York: Routledge, 2014.

Chiodelli, Francesco, and Erez Tzfadia. "The Multifaceted Relation Between Formal Institutions and the Production of Informal Urban Spaces: An Editorial Introduction." *Geography Research Forum* 36 (2016): 1–14.

Christopherson, Susan. "The Fortress City: Privatized Spaces, Consumer Citizenship." In *Post-Fordism: A Reader*, ed. Ash Amin, 409–427. Oxford: Blackwell, 1994.

Cirtautas, Matas. "Urban Sprawl of Major Cities in the Baltic States." *Architecture and Urban Planning* 7 (2013): 72–79.

Clapson, Mark. *Invincible Green Suburbs, Brave New Towns: Social Change and Urban Dispersal in Post-War England.* Manchester, UK: Manchester University Press, 1998.

Clark, Terry Nichols, Richard Lloyd, Kenneth Wong, and Pushpam Jain. "Amenities Drive Urban Growth." *Journal of Urban Affairs* 24, no. 5 (2002): 493–515.

——, ed. *The City as an Entertainment Machine.* Rev. ed. Lanham, MD: Lexington, 2011.

Clausen, Marco. "Prinzessinnengarten." In *Make_Shift: The Expanded Field of Critical Spatial Practice*, 11–12. Berlin: TU Berlin, Institute for Architecture, 2012.

Coaffee, Jon. "Recasting the 'Ring of Steel': Designing Out Terrorism in the City of London?" In *Cities, War and Terrorism: Towards an Urban Geopolitics*, ed. Stephen Graham, 276–296. Oxford: Blackwell, 2004.

——. "Rings of Steel, Rings of Concrete and Rings of Confidence: Designing Out Terrorism in Central London Pre and Post September 11th." *International Journal of Urban and Regional Research* 28, no. 1 (2004): 201–211.

——. "Urban Renaissance in the Age of Terrorism: Revanchism, Automated Social Control or the End of Reflection?" *International Journal of Urban and Regional Research* 29, no. 2 (2005): 447–454.

——. "Risk, Resilience, and Environmentally Sustainable Cities." *Energy Policy* 36, no. 12 (2008): 4633–4638.

Coaffee, Jon, David Murakami Wood, and Peter Rogers. *The Everyday Resilience of the City: How Cities Respond to Terrorism and Disaster.* New York: Palgrave Macmillan, 2009.

Cocks, Mathew, and Chris Couch. "The Governance of a Shrinking City: Housing Renewal in the Liverpool Conurbation, UK." *International Planning Studies* 17, no. 3 (2012): 277–301.

Cohen, Erik. "Authenticity and Commodification in Tourism." *Annals of Tourism Research* 15, no. 3 (1988): 371–386.

——. "Contemporary Tourism—Trends and Challenges: Sustainable Authenticity or Contrived Post-Modernity?" In *Change in Tourism: People, Places, Processes*, ed. Richard Butler and Douglas Pearce, 12–29. New York: Routledge, 1995.

Cohen, Michael. "The Hypothesis of Urban Convergence: Are Cities in the North and South Becoming More Alike in an Age of Globalization?" In *Preparing for the Urban Future: Global Pressures and Local Forces*, ed. Michael Cohen, Blair Ruble, Joseph Tulchin, and Allison Garland, 25–38. Washington, DC: Woodrow Wilson Center Press, 1996.

Coles, Tim. "Urban Tourism, Place Promotion and Economic Restructuring: The Case of Post-Socialist Leipzig." *Tourism Geographies* 5, no. 2 (2003): 190–219.

Collins, John. "Lusaka: Urban Planning in a British Colony, 1931–1964." In *Shaping an Urban World*, ed. G. E. Cherry, 227–241. London: Mansell, 1980.

Colomb, Claire. *Staging the New Berlin: Place Marketing and the Politics of Urban Reinvention Post-1980.* New York: Routledge, 2012.

Comaroff, Jean, and John Comaroff. *Theory from the South: Or, How Euro-America Is Evolving Toward Africa.* New York: Paradigm, 2012.

Connell, Raewyn. *Southern Theory: The Global Dynamics of Knowledge in Social Science.* Sydney: Allen & Unwin, 2007.

——. "Using Southern Theory: Decolonizing Social Thought in Theory, Research and Application." *Planning Theory* 13, no. 2 (2014): 210–223.

Connolly, Creighton. "Urban Political Ecology Beyond Methodological Cityism." *International Journal of Urban and Regional Research* 39, no. 1 (2015): 63–75.

Constantinescu, Ilinca Păun. "Shrinking Cities in Romania: Former Mining Cities in Valea Jiului." *Built Environment* 38, no. 2 (2012): 214–228.

Cornelissen, Scarlett. "Crafting Legacies: The Changing Political Economy of Global Sport and the 2010 FIFA World Cup™." *Politikon* 34, no. 3 (2007): 241–259.

——. "Scripting the Nation: Sport, Mega-Events and State-Building in Post-Apartheid South Africa." *Sport in Society* 11, no. 4 (2008): 481–493.

Cornelissen, Scarlett, and Kamilla Swart. "The 2010 World Cup as a Political Construct: The Challenge of Making Good on an African Promise." *Sociological Review* 54, no. 2 (2006): 108–123.

Côté-Roy, Laurence, and Sarah Moser. " 'Does Africa Not Deserve Shiny New Cities?' The Power of Seductive Rhetoric Around New Cities in Africa." *African Studies* 56, no. 12 (2019): 2391–2407.

Couch, Chris, Jay Karecha, Henning Nuissl, and Dieter Rink. "Decline and Sprawl: An Evolving Type of Urban Development—Observed in Liverpool and Leipzig." *European Planning Studies* 13, no. 1 (2005): 117–136.

Couch, Chris, and Matthew Cocks. "Housing Vacancy and the Shrinking City: Trends and Policies in the UK and the City of Liverpool." *Housing Studies* 28, no. 3 (2013): 499–519.

Coward, Martin, *Urbicide: The Politics of Urban Destruction.* New York: Routledge, 2008.

Crang, Mike. "Grounded Speculations on Theories of the City and the City of Theory." *International Journal of Urban and Regional Research* 25, no. 3 (2001): 665–669.

——. "The Calculus of Speed: Accelerated Worlds, Worlds of Acceleration." *Time & Society* 19, no. 3 (2010): 404–410.

Crawford, Margaret. *Building the Workingman's Paradise: The Design of American Company Towns.* New York: Verso, 1995.

Crilley, Donald. "Architecture as Advertising: Constructing the Image of Redevelopment." In *Selling Places: The City as Cultural Capital, Past and Present*, ed. Gerry Kearns and Chris Philo, 233–234. Oxford: Pergamon, 1993.

Crossa, Veronica. "Reading for Difference on the Street: De-homogenising Street Vending in Mexico City." *Urban Studies* 53, no. 2 (2016): 287–301.

Crot, Lawrence. "Planning for Sustainability in Non-democratic Polities: The Case of Masdar City." *Urban Studies* 50, no. 13 (2013): 2809–2825.

Cruz, Teddy. "Tijuana Case Study: Tactics of Invasion: Manufactured Sites." *Architectural Design* 75, no. 5 (2005): 32–37.

Cuff, Dana. *Provisional City: Los Angeles Stories of Architecture and Urbanism.* Cambridge, MA: MIT Press, 2001.

Cugurullo, Federico. "The Business of Utopia: Estidama and the Road to the Sustainable City." *Utopian Studies* 24, no. 1 (2013): 66–88.

——. "How to Build a Sandcastle: An Analysis of the Genesis and Development of Masdar City." *Journal of Urban Technology* 20, no. 1 (2013): 23–37.

——. "Urban Eco-Modernisation and the Policy Context of New Eco-City Projects: Where Masdar City Fails and Why." *Urban Studies* 53, no. 11 (2016): 2417–2433.

Cunningham, Allen. "The Modern City Revisited—Envoi." In *The Modern City Revisited*, ed. Thomas Deckker, 247–251. London: Spon, 2000.

Cunningham, David. "Slumming It: Mike Davis's Grand Narrative of Urban Revolution." *Radical Philosophy* 142 (2007): 8–18.

Cunningham Sabot, Emmanuèle, and Sylvie Fol. "Schrumpfende Stäedte in Westeuropa: Fallstudien aus Frankreich und Grossbritannien." *Berliner Debatte Initial* 18, no. 1 (2007): 22–35.

Currid, Elizabeth. "New York as a Global Creative Hub: A Competitive Analysis of Four Theories on World Cities." *Economic Development Quarterly* 204 (2006): 330–350.

——. "How Art and Culture Happen in New York: Implications for Urban Economic Development." *Journal of the American Planning Association* 73, no. 4 (2007): 454–467.

——. *The Warhol Economy: How Fashion, Art, and Music Drive New York City.* Princeton, NJ: Princeton University Press, 2007.

Currid, Elizabeth, and Sarah Williams. "The Geography of Buzz: Art, Culture and the Social Milieu in New York and Los Angeles." *Journal of Economic Geography* 10, no. 3 (2010): 423–451.

Currid-Halkett, Elizabeth, and Allen J. Scott. "The Geography of Celebrity and Glamour: Reflections on Economy, Culture, and Desire in the City." *City, Culture and Society* 4 (2013): 2–11.

Cuthbert, Alexander. "Alphaville and Masdar: The Future of Urban Space and Form." In *Emergent Urbanism: Urban Planning & Design in Times of Structural and Systematic Change*, ed. Tigran Haas and Krister Olsson, 2nd ed., 9–18. New York: Routledge, 2016.

Dale, Britt. "An Institutionalist Approach to Local Restructuring: The Case of Four Norwegian Mining Towns." *European Urban and Regional Studies* 9, no. 1 (2002): 5–20.

Das, Raju. "David Harvey's Theory of Uneven Geographical Development: A Marxist Critique." *Capital & Class* 41, no. 3 (2017): 511–536.

Datta, Ayona. *The Illegal City: Space, Law and Gender in a Delhi Squatter Settlement.* London: Ashgate, 2012.

——. "India's Ecocity? Environment, Urbanisation, and Mobility in the Making of Lavasa." *Environment and Planning C* 30, no. 6 (2012): 982–996.

——. "New Urban Utopias of Postcolonial India: 'Entrepreneurial Urbanization' in Dholera Smart City, Gujarat." *Dialogues in Human Geography* 5, no. 1 (2015): 3–22.

——. "A 100 Smart Cities, a 100 Utopias." *Dialogues in Human Geography* 5, no. 1 (2015): 49–53.

——. "Introduction: Fast Cities in an Urban Age." In *Mega-Urbanization in the Global South: Fast Cities and New Urban Utopias of the Postcolonial State*, ed. Ayona Datta and Abdul Shaban, 1–28. New York: Routledge, 2017.

Davidson, Christopher. *Dubai: The Vulnerability of Success*. London: Hurst, 2008.

Davies, Katherine. "The Architecture of Appearance: Arendt's Feminism and Guatemala's Private City." *Arendt Studies* 4 (2020): 53–82. https://doi.org/10.5840/arendtstudies20209825.

Davis, Mike. *City of Quartz: Excavating the Future of Los Angeles*. New York: Vintage, 1990.

——. "Fortress Los Angeles: The Militarization of Urban Space." In *Variations on a Theme Park*, ed. Michael Sorkin, 154–180. New York: Noonday, 1992.

——. *Dead Cities and Other Tales*. New York: New Press, 2002.

——. "Planet of Slums: Urban Involution and the Informal Proletariat." *New Left Review* 26 (2004): 5–34.

——. "The Urbanization of Empire: Megacities and the Laws of Chaos." *Social Text* 22, no. 4 (2004): 9–15.

——. "Fear and Money in Dubai." *New Left Review* 41 (2006): 47–68.

——. *Planet of Slums*. New York: Verso, 2006.

Davoudi, Simin. "Polycentricity in European Spatial Planning: From an Analytical Tool to a Normative Agenda." *European Planning Studies* 11, no. 8 (2003): 979–990.

Dawson, Ashley, and Brent Hayes Edwards. "Introduction: Global Cities of the South." *Social Text* 22, no. 4 (2004): 1–7.

Dear, Michael, and Steven Flusty. "Postmodern Urbanism." *Annals of the Association of American Geographers* 88, no. 1 (1998): 50–72.

de Boeck, Filip, and Marie-Françoise Plissard. *Kinshasa: Tales of the Invisible City*. Antwerp: Ludion, 2006.

De Frantz, Monika. "From Cultural Regeneration to Discursive Governance: Constructing the Flagship of the 'Museumsquarter Vienna' as a Plural Symbol of Change." *International Journal of Urban and Regional Research* 29, no. 1 (2005): 50–66.

Degen, Monica. "Fighting for the Global Catwalk: Formalizing Public Life in Castlefield (Manchester) and Diluting Public Life in el Raval (Barcelona)." *International Journal of Urban and Regional Research* 27, no. 4 (2003): 867–880.

——. *Sensing Cities: Regenerating Urban Life in Barcelona and Manchester*. New York: Routledge 2008.

Degen, Monica, Caitlin De Silvey, and Gillian Rose. "Experiencing Visualities in Designed Urban Environments: Learning from Milton Keynes." *Environment & Planning A* 40, no. 8 (2008): 1901–1920.

Deitrick, Sabrina, and Robert Beauregard. "From Front-Runner to Also-Ran: The Transformation of a Once Dominant Industrial Region: Pennsylvania, USA." In *The Rise of the Rustbelt: Revitalizing Older Industrial Regions*, ed. Philip Cooke, 52–71. London: UCL Press, 1995.

de Kloet, Jeroen, and Lena Scheen. "Pudong: The Shanzhai Global City." *European Journal of Cultural Studies* 16, no. 6 (2013): 692–709.

Dellheim, Charles. "The Creation of Company Culture: Cadburys, 1861–1931." *American Historical Review* 92, no. 1 (1987): 13–44.

Dempsey, Michael Cameron. *Castles in the Sand: An Urban Planner in Abu Dhabi*. Jefferson, NC: MacFarland, 2014.

Derickson, Kate. "Masters of the Universe." *Environment and Planning D: Society and Space* 36, no. 3 (2018): 356–362.

Derudder, Ben. "The Mismatch Between Concepts and Evidence in the Study of a Global Urban Network." In *Cities in Globalization: Practices, Policies, and Theories*, ed. Peter Taylor, Ben Derudder, Pieter Saey, and Frank Witlox, 261–275. New York: Routledge, 2007.

Desan, Mathieu Hikaru. "Bankrupted Detroit." *Thesis Eleven* 121, no. 1 (2014): 122–130.

Desan, Mathieu Hikaru, and George Steinmetz. "The Spontaneous Sociology of Detroit's Hyper-Crisis." In *Reinventing Detroit: The Politics of Possibility*, ed. Michael Peter Smith and L. Owen Kirkpatrick, 15–36. New Brunswick, NJ: Transaction, 2015.

Desfor, Gene, and John Jørgensen. "Flexible Urban Governance: The Case of Copenhagen's Recent Waterfront Development." *European Planning Studies* 12, no. 4 (2004): 479–496.

de Sousa Santos, Boanventura. *Toward a New Common Sense: Law, Science and Politics in the Paradigmatic Transition*. New York: Routledge, 1995.

Devlin, Ryan. " 'An Area That Governs Itself': Informality, Uncertainty and the Management of Street Vending in New York City." *Planning Theory* 10, no. 1 (2011): 53–65.

Dewar, Margaret, and June Manning Thomas, eds. *The City After Abandonment*. Philadelphia: University of Pennsylvania Press, 2013.

——. "Introduction." In *The City After Abandonment*, ed. Margaret Dewar and June Manning Thomas, 1–14. Philadelphia: University of Pennsylvania Press, 2013.

Dewar, Margaret, and Matt Weber. "City Abandonment." In *The Oxford Handbook of Urban Planning*, ed. Rachel Weber and Randall Crane, 563–586. New York: Oxford University Press, 2011.

Dick, Howard, and Peter Rimmer. "Beyond the Third World City: The New Urban Geography of South-East Asia." *Urban Studies* 35, no. 12 (1998): 2303–2321.

Diep, Martina, and Stephanie Drabble. "Living with Difference? The 'Cosmopolitan City' and Urban Reimaging in Manchester, UK." *Urban Studies* 43, no. 10 (2006): 1687–1714.

Dinius, Oliver, and Angela Vergara, eds. *Company Towns in the Americas: Landscape, Power, and Working-Class Communities*. Athens, GA: University of Georgia Press, 2011.

D'Monte, Darryl. "A Matter of People." In *Living in the Endless City*, ed. Ricky Burdett and Deyan Sudjic, 94–101. London: Phaidon, 2011.

Dobraszczyk, Paul. *The Dead City: Urban Ruins and the Spectacle of Decay*. London: Taurus, 2017.

Domosh, Mona. "The Symbolism of the Skyscraper: Case Studies of New York's First Tall Buildings." *Journal of Urban History* 14, no. 3 (1988): 321–345.

Donald, Stephanie Hemelryk, Eleonore Kofman, and Catherine Kevin, eds. *Branding Cities: Cosmopolitanism, Parochialism, and Social Change*. New York: Routledge, 2008.

Donegan, Mary, and Nichola Lowe. "Inequality in the Creative City: Is There Still a Place for 'Old-Fashioned' Institutions?" *Economic Development Quarterly* 22, no. 1 (2008): 46–62.

Doshi, Sapana. "The Politics of the Evicted: Redevelopment, Subjectivity and Difference in Mumbai's Slum Frontier." *Antipode* 45, no. 3 (2013): 1–22.

Doucet, Brian. "Variations of the Entrepreneurial City: Goals, Roles and Visions in Rotterdam's *Kop van Zuid* and the Glasgow Harbour Megaprojects." *International Journal of Urban and Regional Research* 37, no. 6 (2013): 2035–2051.

Doucet, Brian, Ronald van Kempen, and Jan van Weesep. " 'We're a Rich City with Poor People': Municipal Strategies of New-Build Gentrification in Rotterdam and Glasgow." *Environment and Planning A* 43, no. 6 (2011): 1438–1454.

Douglass, Mike. "Mega-Urban Regions and World City Formation: Globalization, the Economic Crisis and Urban Policy Issues in Pacific Asia." *Urban Studies* 37, no. 12 (2000): 2315–2335.

Douglass, Mike, and Liling Huang. "Globalizing the City in Southeast Asia: Utopia on the Urban Edge—the Case of Phu My Hung, Saigon." *International Journal of Asia-Pacific Studies* 3, no. 2 (2007): 1–42.

Dovey, Kim. "Uprooting Critical Urbanism." *City* 15, no. 3–4 (2011): 347–354.

——. "Informal Urbanism and Complex Adaptive Assemblage." *International Development Planning Review* 34, no. 4 (2012): 349–367.

——. "Incremental Urbanisms: The Emergence of Informal Settlements." In *Emergent Urbanism: Urban Planning & Design in Times of Structural and Systemic Change*, ed. Kigran Haas and Krister Olsson, 45–54. New York: Routledge, 2014.

Dovey, Kim, and Ronald King. "Forms of Informality: Morphology and Visibility of Informal Settlements." *Built Environment* 37, no. 1 (2011): 11–29.

Dovey, Kim, and Ross King. "Informal Urbanism and the Taste for Slums." *Tourism Geographies* 14, no. 2 (2012): 275–293.

Downey, Davia, and Laura Reese. "Sudden Versus Slow Death of Cities: New Orleans and Detroit." *Du Bois Review: Social Science Research on Race* 14, no. 1 (2017): 219–243.

Downs, Anthony. "The Challenge of Our Declining Big Cities." *Housing Policy Debate* 8 (1997): 359–408.

——. "Some Realities About Sprawl and Urban Decline." *Housing Policy Debate* 10, no. 4 (1999): 955–974.

Drakakis-Smith, David. *Third World Cities.* London: Routledge, 2000.

Drozdz, Martine, Manuel Appert, and Andrew Harris. "High-Rise Urbanism in Contemporary Europe." *Built Environment* 43, no. 4 (2017): 469–480.

Dubeaux, Sarah, and Emmanuèle Cunningham Sabot. "Maximizing the Potential of Vacant Spaces Within Shrinking Cities, a German Approach." *Cities* 75 (2018): 6–11.

Duffy, Hazel. *Competitive Cities: Succeeding in the Global Economy.* 2nd ed. London: Spon, 2004.

Dunn, Bill. "Against Neoliberalism as a Concept." *Capital & Class* 41, no. 3 (2017): 435–454.

Dyson, Peter. "Slum Tourism: Representing and Interpreting 'Reality' in Dharavi, Mumbai." *Tourism Geographies* 14, no. 2 (2012): 254–274.

Easterling, Keller. "The Corporate City Is the Zone." In *Visionary Power: Producing the Contemporary City,* ed. Christine de Baan, Joachim Declerck, and Veronique Patteeuw, 75–85. (Rotterdam: NAi, 2007.

——. "The Wrong Story." *Perspecta* 41 (2008): 74–82.

——. "Zone." In *Urban Transformation,* ed. Ilka Ruby and Andreas Ruby, 30–45. Berlin: Ruby, 2008.

——. *Subtraction.* Berlin: Sternberg, 2014.

Eaton, Ruth. *Ideal Cities: Utopianism and the (Un)Built Environment.* London: Thames & Hudson, 2002.

Ebel, Robert. "Urban Decline in the World's Developed Economies: An Examination of the Trends," *Research in Urban Economics* 5 (1985): 1–19.

Eco, Umberto. *Travels in Hyperreality.* San Diego: Harcourt Brace Jovanovich, 1983.

Edensor, Tim. "Waste Matter: The Debris of Industrial Ruins and the Disordering of the Material World." *Journal of Material Culture* 10, no. 3 (2000): 311–332.

——. "The Ghosts of Industrial Ruins: Ordering and Disordering Memory in Excessive Space." *Environment and Planning D* 23, no. 6 (2005): 830–849.

——. *Industrial Waste: Space, Aesthetics and Materiality.* London: Berg, 2005.

Edensor, Tim, and Mark Jayne. "Afterword: A World of Cities." In *Urban Theory Beyond the West: A World of Cities,* ed. Tim Edensor and Mark Jayne, 329–331. New York: Routledge, 2012.

——. "Introduction: Urban Theory Beyond the West." In *Urban Theory Beyond the West: A World of Cities,* ed. Tim Edensor and Mark Jayne, 1–27. New York: Routledge, 2012.

Edgington, David. "Restructuring Japan's Rustbelt: The Case of Muroran, Hokkaido, 1985–2010." *Urban Affairs Review* 49, no. 4 (2012): 475–524.

Eeckhout, Bart. "The 'Disneyfication' of Times Square: Back to the Future?" In *Critical Perspectives on Urban Redevelopment: Volume 6,* ed. Kevin Fox Gotham, 379–428. New York: Elsevier, 2001.

Ehrenfuecht, Renia, and Marla Nelson. "Recovery in a Shrinking City: Challenges to Right-Sizing Post-Katrina New Orleans." In *The City After Abandonment,* ed. Margaret Dewar and June Manning Thomas, 133–150. Philadelphia: University of Pennsylvania Press, 2013.

Ehrkamp, Patricia. "Internationalizing Urban Theory: Toward Collaboration." *Urban Geography* 32, no. 8 (2011): 1122–1128.

Ehrlich, Kornelia, Agnes Kriszan, and Thilo Lang. "Urban Development in Central and Eastern Europe—Between Peripheralization and Centralization?" *disP—The Planning Review* 48, no. 2 (2012): 77–92.

Eisinger, Peter. "The Politics of Bread and Circuses: Building the City for the Visitor Class." *Urban Affairs Review* 35, no. 3 (2000): 316–333.

Elden, Stuart. "Thinking Territory Historically." *Geopolitics* 15, no. 4 (2010): 757–761.

El Khoury, Ann. "Pluralizing Global Urbanism: Replacing the God-Trick with Goddess Tactics." *Dialogues in Human Geography* 6, no. 3 (2016): 291–295.

Ellin, Nan. *Postmodern Urbanism.* Rev. ed. New York: Princeton Architectural Press, 1999.

Elsheshtawy, Yasser. "Redrawing Boundaries: Dubai, an Emerging Global City." In *Planning Middle Eastern Cities: An Urban Kaleidoscope in a Globalizing World*, ed. Yasser Elsheshtawy, 169–199. London: Routledge, 2004.

——. "Cities of Sand and Fog: Abu Dhabi's Global Ambitions." In *The Evolving Arab City: Tradition, Modernity and Urban Development*, ed. Yasser Elsheshtawy, 258–304. New York: Routledge, 2008.

——. "Navigating the Spectacle: Landscapes of Consumption in Dubai." *Architectural Theory Review* 13, no. 2 (2008): 164–187.

——. "Transitory Sites: Mapping Dubai's 'Forgotten' Urban Spaces." *International Journal of Urban and Regional Research* 32, no. 4 (2008): 668–688.

——. *Dubai: Behind an Urban Spectacle*. New York: Routledge, 2010.

Enwezor, Okwui. "Terminal Modernity: Rem Koolhaas' Discourse on Entropy." In *What Is OMA: Considering Rem Koolhaas and the Office of Metropolitan Architecture*, ed. V. Patteeuw, 116. Rotterdam: NAi, 2003.

Erbacher, Eric, and Sina Nitzsche. "Performing the Double Rupture: Kraftklub, Popular Music and Post-Socialist Urban Identity in Chemnitz, Germany." *International Journal of Cultural Studies* 20, no. 4 (2017): 437–455.

Etherington, David, and Martin Jones. "City-Regions: New Geographies of Uneven Development and Inequality." *Regional Studies* 43, no. 2 (2009): 247–265.

Evans, Graeme. "Hard-Branding the Cultural City—From Prado to Prada." *International Journal of Urban and Regional Research* 27, no. 2 (2003): 417–440.

——. "Cultural Industry Quarters: From Pre-Industrial to Post-Industrial Production." In *City of Quarters: Urban Villages in the Contemporary City*, ed. Duncan Bell and Mark Jayne, 71–92. Aldershot, UK: Ashgate, 2004.

——. "Measure for Measure: Evaluating the Evidence of Culture's Contribution to Regeneration." *Urban Studies* 42, no. 5–6 (2005): 959–983.

——. "Branding the City of Culture—The Death of City Planning?" In *Culture, Urbanism and Planning*, ed. Javier Monclús and Manuel Guardia, 197–214. Aldershot, UK: Ashgate, 2006.

——. "Tourism, Creativity and the City." In *Tourism, Creativity and Development*, ed. Greg Richards and Julie Wilson, 57–72. London: Routledge, 2007.

——. "Creative Cities, Creative Spaces and Urban Policy." *Urban Studies* 46, no. 5–6 (2009): 1003–1040.

——. "From Cultural Quarters to Creative Clusters—Creative Spaces in the New City Economy." In *The Sustainability and Development of Cultural Quarters: International Perspectives*, ed. Mattias Legner, 32–59. Stockholm: Institute of Urban History, 2009.

Evans, Peter. *Livable Cities?* Berkeley: University of California Press, 2002.

Evenson, Norma. *Chandigarh*. Berkeley: University of California Press, 1966.

Fainstein, Susan. "Mega-Projects in New York, London and Amsterdam." *International Journal of Urban and Regional Research* 32, no. 4 (2008): 867–880.

Fainstein, Susan, and Dennis Judd. "Cities as Places to Play." In *The Tourist City*, ed. Dennis Judd and Susan Fainstein, 261–272. New Haven, CT: Yale University Press, 1999.

——, eds. *The Tourist City*. New Haven, CT: Yale University Press, 1999.

Fainstein, Susan, Lily Hoffman, and Dennis Judd. "Introduction." In *Cities and Visitors: Regulating People, Markets, and City Space*, ed. Lily Hoffman, Susan Fainstein, and Dennis Judd, 1–20. New York: Blackwell, 2003.

Fanon, Franz. *The Wretched of the Earth*. New York: Grove, 1993.

Fawaz, Mona. "An Unusual Clique of City-Makers: Social Networks in the Production of a Neighborhood in Beirut (1950–75)." *International Journal of Urban and Regional Research* 32, no. 3 (2008): 565–585.

Ferenćuhová, Slavomíra. "Urban Theory Beyond the 'East/West Divide'? Cities and Urban Research in Postsocialist Europe." In *Urban Theory Beyond the West: A World of Cities*, ed. Tim Edensor and Mark Jayne, 65–74. New York: Routledge, 2012.

Ferguson, James. "Seeing Like an Oil Company: Space, Security, and Global Capital in Neoliberal Africa." *American Anthropologist* 107, no. 3 (2005): 377–382.

——. *Global Shadows: Africa in the Neoliberal World.* Durham, NC: Duke University Press, 2006.

Ferman, Louis, Stuart Henry, and Michelle Hoyman. "Issues and Prospects for the Study of Informal Economies: Concepts, Research Strategies, and Policy." *Annals of the American Academy of Political and Social Science* 493 (1987): 154–172.

Fernandez, Rosa Flores. "Physical and Spatial Characteristics of Slum Territories Vulnerable to Natural Disasters." *Les Cahiers de l'Afrique de l'Est* 44 (2011): 5–22.

Fernandez-Galiano, Luis. "Spectacle and Its Discontents; or, the Joys of Architainment." In *Commodification and Spectacle in Architecture: A Harvard Design Magazine Reader*, ed. William Saunders, 1–7. Minneapolis: University of Minnesota Press, 2005.

Fernandez-Kelly, Patricia, and Anna Garcia. "Informalization at the Core: Hispanic Women, Homework, and the Advanced Capitalist State." In *The Informal Economy: Studies in Advanced and Less Developed Countries*, ed. Alejandro Portes, Manuel Castells, and Laura Benton, 247–264. Baltimore: Johns Hopkins University Press, 1989.

Ferrari, Ed, and Jonathan Roberts. "Liverpool—Changing Urban Form." In *Shrinking Cities: Working Papers—Manchester/Liverpool II*, ed. Philipp Oswalt, 130–136. Berlin: Office Philipp Oswalt, 2004.

Ferreira, Sanette, and Gustav Visser. "Creating an African Riviera: Revisiting the Impact of the Victoria and Alfred Waterfront Development in Cape Town." *Urban Forum* 18, no. 3 (2007): 227–246.

Ferreira da Silva, Denise. "Towards a Critique of the Socio-Logos of Justice: The Analytics of Raciality and the Production of Universality." *Social Identities* 7, no. 3 (2001): 421–454.

Ferreri, Mara. "The Seductions of Temporary Urbanism." *Ephemera* 15, no. 1 (2015): 181–191.

Fischer, Brodwyn. "Introduction." In *Cities from Scratch: Poverty and Informality in Urban Latin America*, ed. Brodwyn Fischer, Bryan McCann, and Javier Auyero, 1–8. Durham, NC: Duke University Press, 2014.

Fishman, Robert. *Urban Utopias in the Twentieth Century: Ebenezer Howard, Frank Lloyd Wright, Le Corbusier.* Cambridge, MA: MIT Press, 1982.

——. *Bourgeois Utopias: The Rise and Fall of Suburbia.* New York: Basic Books, 1987.

Florentin, Daniel. "The 'Perforated City': Leipzig's Model of Urban Shrinkage Management." *Berkeley Planning Journal* 23, no. 1 (2010): 83–101.

Florida, Richard. The Rise of the Creative Class. New York: Basic Books, 2002.

——. "Cities and the Creative Class." *City & Community* 2, no. 1 (2003): 3–19.

——. *Cities and the Creative Class.* New York: Routledge, 2004.

Flüchter, Winfried. "Megapolises and Rural Peripheries: Shrinking Cities in Japan." In *Shrinking Cities: Volume 1, International Research*, ed. Philipp Oswalt, 83–95. Ostfildern-Ruit, Germany: Hatje Cantz, 2005.

——. "Shrinking Cities in Japan: Between Megalopolises and Rural Peripheries." *Electronic Journal of Contemporary Japanese Studies* (2008): 3–8.

——. "Urbanisation, City, and City System in Japan Between Development and Shrinking: Coping with Shrinking Cities in Times of Demographic Change." In *Urban Spaces in Japan: Cultural and Social Perspectives*, ed. Christoph Brumann and Evelyn Schulz, 15–36. London: Routledge, 2012.

Flusty, Steven. "The Banality of Interdiction: Surveillance, Control and the Displacement of Diversity." *International Journal of Urban and Regional Research* 25, no. 3 (2001): 658–664.

Flyvbjerg, Bent. "Design by Deception: The Politics of Megaproject Approval." *Harvard Design Magazine* 22 (Spring–Summer 2005): 50–59.

——. "Machiavellian Megaprojects." *Antipode* 37, no. 1 (2005): 18–22.

Flyvbjerg, Bent, Nils Bruzelius, and Werner Rothengatter. *Megaprojects and Risk: An Anatomy of Ambition.* Cambridge: Cambridge University Press, 2003.

Fol, Sylvie. "Urban Shrinkage and Socio-Spatial Disparities: Are the Remedies Worse Than the Disease?" *Built Environment* 38, no. 2 (2012): 259–275.

Fol, Sylvie, and Emmanuèle Cunningham Sabot. "Urban Decline and Shrinking Cities: A Critical Assessment of Approaches to Urban Shrinkage." *Annales de Géographie* 119 (2010): 359–383.

Ford, Larry. "World Cities and Global Change: Observations on Monumentality in Urban Design." *Eurasian Geography and Economics* 49, no. 3 (2008): 237–262.

Fox, Sean. "Urbanization as a Global Historical Process: Theory and Evidence from Sub-Saharan Africa." *Population and Development Review* 38, no. 2 (2012): 285–310.

Franz, Peter. "Suburbanization and the Clash of Urban Regimes: Developmental Problems of East Germany Cities in a Free Market Environment." *European Urban and Regional Studies* 7, no. 2 (2000): 135–146.

Fraser, James. "Globalization, Development and Ordinary Cities: A Review Essay." *Journal of World-Systems Research* 12, no. 1 (2006): 189–197.

Frebzel, Fabian, Ko Koens, Malke Steinbrink, and Christian Rogerson. "Slum Tourism: State of the Art." *Tourism Review International* 18, no. 4 (2015): 237–252.

Freiman, Christopher. "Cosmopolitanism Cithin Borders: On Behalf of Charter Cities." *Journal of Applied Philosophy* 30, no. 1 (2013): 40–52.

Friedman, John. "The Good City: Defense of Utopian Thinking." *International Journal of Urban and Regional Research* 24, no. 2 (2000): 460–472.

Friedman, Michael, Jacob Bustad, and David Andrews. "Feeding the Downtown Monster: (Re)developing Baltimore's 'Tourist Bubble'." *City, Culture and Society* 3, no. 3 (2012): 209–218.

Friedmann, John. "The World-City Hypothesis." *Development and Change* 17, no. 1 (1986): 69–83.

——. "Where We Stand: A Decade of World City Research." In *World Cities in a World System*, ed. Paul Knox and Peter Taylor, 21–47. Cambridge: Cambridge University Press, 1995.

Friedmann, John, and Mike Douglass. *Cities for Citizens: Planning and the Rise of Civil Society in a Global Age.* New York: Wiley, 1998.

Friedmann, John, and Goetz Wolff. "World City Formation: An Agenda for Research and Action." *International Journal of Urban and Regional Research* 6, no. 3 (2002): 309–344.

Friedrichs, Jürgen. "A Theory of Urban Decline: Economy, Demography and Political Elites." *Urban Studies* 30, no. 6 (1993): 907–917.

Frisby, David. *Cityscapes of Modernity: Critical Explorations.* Cambridge: Polity, 2001.

Frisch, Thomas. "Glimpses of Another World: The *Favela* as a Tourist Attraction." *Tourism Geographies* 14, no. 2 (2012): 320–338.

Fritsche, Miriam, Marcel Langner, Hadia Köhler, Anke Ruckes, Daniela Schüler, Betka Zakirova, Katharina Appel, Valeska Contardo, Ellen Diermayer, Mathias Hofmann, Christoph Kulemeyer, Peter Meffert, and Janneke Westermann. "Shrinking Cities: A New Challenge for Research in Urban Ecology." In *Shrinking Cities: Effects on Urban Ecology and Challenges for Urban Development*, ed. Marcell Langner and Wilfried Endlicher, 17–34. Frankfurt: Peter Lang, 2007.

Fuhrich, Manfred, and Robert Kaltenbrunner. "Der Osten—jetzt auch im Westen? Stadtumbau-West und Stadtumbau Ost—zwei ungleiche Geschwister." *Berliner Debatte Initial* 16, no. 6 (2005): 41–54.

Fuji, Yasuyuki. "Shrinkage in Japan." In *Shrinking Cities: Volume 1, International Research*, ed. Philipp Oswalt, 96–100. Ostfildern-Ruit, Germany: Hatje Cantz, 2006.

Gadanho, Pedro. "Mirroring Uneven Growth: A Speculation on Tomorrow's Cities Today." In *Uneven Growth: Tactical Urbanisms for Expanding Megacities*, ed. Pedro Gadanho, 14–25. New York: Museum of Modern Art, 2015.

Gaffikin, Frank, and David Perry. "The Contemporary Urban Condition: Understanding the Globalizing City as Informal, Contested, and Anchored." *Urban Affairs Review* 48, no. 5 (2012): 701–730.

Galland, Daniel, and Carsten Hansen. "The Roles of Planning in Waterfront Redevelopment: From Plan-Led and Market-Driven Styles to Hybrid Planning?" *Planning Practice and Research* 27, no. 2 (2012): 203–225.

Ganapati, Sukumar. "The Paradox of Shrinking Cities in India." In *Shrinking Cities: A Global Perspective*, ed. Harry W. Richardson and Chang Woon Nam, 169–181. New York: Routledge, 2014.

Gandelsonas, Marco. "Slow Infrastructure." In *Fast-Forward Urbanism: Rethinking Architecture's Engagement with the City*, ed. Dana Cuff and Roger Sherman, 122–131. New York: Princeton Architectural Press, 2011.

Gandy, Matthew. "Learning from Lagos." *New Left Review* 33 (2005): 36–52.

Gaonkar, Dilip Parameshwar. "On Alternative Modernities." In *Alternative Modernities*, ed. Dilip Parameshwar Gaonkar, 1–23. Durham, NC: Duke University Press, 2001.

Garner, John, ed. *The Model Company Town: Urban Design Through Private Enterprise in Nineteenth-Century New England*. Amherst: University of Massachusetts Press, 1984.

——, ed. *The Company Town: Architecture and Society in the Early Industrial Age*. Oxford: Oxford University Press, 1992.

Gaskell, Martin. "Model Industrial Villages in S. Yorkshire/N. Derbyshire and the Early Town Planning Movement." *Town Planning Review* 50, no. 4 (1979): 437–458.

Gberie, Lansana. "Review Article: Africa—The Troubled Continent." *African Affairs* 104, no. 415 (2005): 337–342.

Gellert, Paul, and Barbara Lynch. "Mega-Projects as Displacements." *International Social Science Journal* 55, no. 175 (2003): 15–25.

Gherner, Asher. *Rule by Aesthetics: World-Class City Making in Delhi*. New York: Oxford University Press, 2015.

Ghirardo, Diane. *Architecture After Modernism*. London: Thames & Hudson, 1996.

Gibson, Chris. "Economic Geography, to What Ends? From Privilege to Progressive Performances of Expertise." *Environment and Planning A: Economy and Space* 51, no. 3 (2019): 805–813.

Gilbert, Alan. "The Return of the Slum: Does Language Matter?" *International Journal of Urban and Regional Research* 31, no. 4 (2007): 697–713.

——. "Slums, Tenants and Home-Ownership: On Blindness to the Obvious." *International Development Planning Review* 30, no. 2 (2008): i–x.

——. "Extreme Thinking About Slums and Slum Dwellers: A Critique." *SAIS Review* 29, no. 1 (2009): 35–48.

——. "Epilogue." *City* 15, no. 6 (2011): 722–726.

Gillespie, Tom. "From Quiet to Bold Encroachment: Contesting Dispossession in Accra's Informal Sector." *Urban Geography* 38, no. 7 (2017): 974–992.

Gillham, Oliver. *The Limitless City: A Primer on the Urban Sprawl Debate*. London: Island, 1992.

Gilman, Theodore. *No Miracles Here: Fighting Urban Decline in Japan and the United States*. Albany: State University of New York Press, 2001.

Gilmour, Tony. *Sustaining Heritage: Giving the Past a Future*. Sydney: Sydney University Press, 2007.

Giorgi, Gabriel. "Improper Selves: Cultures of Precarity." *Social Text* 31, no. 2 (115) (2013): 69–81.

Glaeser, Edward. *Triumph of the City: How Our Greatest Invention Makes Us Richer, Smarter, Greener, Healthier, and Happier*. New York: Penguin, 2011.

Glaeser, Edward, and William Kerr. "What Makes a City Entrepreneurial?" *Harvard Kennedy School Policy Briefs*, February 2010, 1–4.

Glasze, Georg, Chris Webster, and Klaus Franz, eds. *Private Cities: Global and Local Perspectives*. London: Routledge, 2006.

Gleeson, Brendan. "Critical Commentary. The Urban Age: Paradox and Prospect." *Urban Studies* 49, no. 5 (2012): 931–943.

Glock, Birgit, and Hartmut Häußermann. "New Trends in Urban Development and Public Policy in Eastern Germany: Dealing with the Vacant Housing Problem at the Local Level." *International Journal of Urban and Regional Research* 28, no. 4 (2004): 919–929.

Goodfellow, Tom. "Urban Fortunes and Skeleton Cityscapes: Real Estate and Late Urbanization in Kigali and Addis Ababa." *International Journal of Urban and Regional Research* 41, no. 5 (2017): 786–803.

Gollin, Douglas, Remi Jedwan, and Dietrich Vollrath. "Urbanization With and Without Industrialization." *Journal of Economic Growth* 21, no. 1 (2016): 35–70.

Golubchikov, Oleg, Anna Badyina, and Alla Makhrova. "The Hybrid Spatialities of Transition: Capitalism, Legacy and Uneven Urban Economic Restructuring." *Urban Studies* 51, no. 4 (2014): 617–633.

Gomez, María. "Reflective Images: The Case of Urban Regeneration in Glasgow and Bilbao." *International Journal of Urban and Regional Research* 22, no. 1 (1998): 106–121.

Gomez, María, and Sara Gonzalez. "A Reply to Beatriz Plaza's 'The Guggenheim-Bilbao Museum Effect.' " *International Journal of Urban and Regional Research* 25, no. 4 (2001): 898–900.

Gonzalez, Carmen. "Squatters, Pirates, and Entrepreneurs: Is Informality the Solution to the Urban Housing Crisis?" *University of Miami Inter-American Law Review* 40, no. 2 (2008): 239–259.

González de la Rocha, Mercedes. "The Myth of Survival." *Development & Change* 38, no. 1 (2007): 45–66.

Goodman, Allen. "Central Cities and Housing Supply: Growth and Decline in US Cities." *Journal of Housing Economics* 14, no. 4 (2005): 315–335.

Goonewardena, Kanishka. "Planetary Urbanization and Totality." *Environment and Planning D: Society and Space* 36, no. 3 (2018): 456–473.

Gordon, Colin. *Mapping Decline: St. Louis and the Fate of the American City.* Philadelphia: University of Pennsylvania Press, 2008.

Gordon, David. "Planning, Design and Managing Change in Urban Waterfront Redevelopment." *Town Planning Review* 67, no. 3 (1996): 261–290.

Gordon, Ian, and Nick Buck. "Introduction: Cities in the New Conventional Wisdom." In *Changing Cities: Rethinking Urban Competitiveness, Cohesion and Governance*, ed. Nick Buck, Ian Gordon, Alan Harding, and Ivan Turok (New York: Palgrave Macmillan, 2005): 1–21.

Gordon, Peter, and Harry Richardson. "Beyond Polycentricity: The Dispersed Metropolis, Los Angeles, 1970–1990." *Journal of the American Planning Association* 62, no. 3 (1996): 289–295.

Gospodini, Aspa. "Urban Waterfront Redevelopment in Greek Cities: A Framework for Redesigning Space." *Cities* 18, no. 5 (2001): 285–295.

——. "European Cities in Competition and the New 'Uses' of Urban Design." *Journal of Urban Design* 7, no. 1 (2002): 59–73.

Goss, Jon. "Placing the Market and Marketing the Place: Tourist Advertising of the Hawaiian Islands, 1972–1992." *Environment and Planning D* 11, no. 6 (1993): 663–698.

——. "Disquiet on the Waterfront: Reflections on Nostalgia and Utopia in the Urban Archetypes of Festival Market Places." *Urban Geography* 17, no. 3 (1996): 221–247.

Gotham, Kevin Fox. "Marketing Mardi Gras: Commodification, Spectacle and the Political Economy of Tourism in New Orleans." *Urban Studies* 39, no. 10 (2002): 1735–1756.

——. "Theorizing Urban Spectacles: Festivals, Tourism, and the Transformation of Urban Space." *City* 9, no. 2 (2005): 225–246.

——. "Tourism from Above and Below: Globalization, Localization and New Orleans's Mardi Gras." *International Journal of Urban and Regional Research* 29, no. 2 (2005): 309–326.

——. *Authentic New Orleans: Tourism, Culture, and Race in the Big Easy.* New York: New York University Press, 2007.

——. "(Re)Branding the Big Easy: Tourism Rebuilding in Post-Katrina New Orleans." *Urban Affairs Review* 42, no. 6 (2007): 823–850.

Gottdiener, Mark. *Theming of America: Dreams, Visions, and Commercial Spaces.* Boulder, CO: Westview, 1997.

——, ed. *New Forms of Consumption: Consumers, Culture, and Commodification.* New York: Rowman & Littlefield, 2000.

Graham, Stephen. "Constructing Premium Network Spaces: Reflections on Infrastructure Networks and Contemporary Urban Development," *International Journal of Urban and Regional Research* 24, no. 1 (2000): 183–200.

——, ed. *Disrupted Cities: When Infrastructure Fails.* New York: Routledge, 2010.

——. *Cities Under Siege: The New Military Urbanism.* New York: Verso, 2011.

——. "Vanity and Violence: On the Politics of Skyscrapers." *City* 20, no. 5 (2016): 755–771.

——. *Vertical: The City from Satellites to Bunkers.* New York: Verso, 2016.

Graham, Stephen, and Lucy Hewitt. "Getting Off the Ground: On the Politics of Urban Verticality." *Progress in Human Geography* 37 (2013): 72–92.

Graham, Stephen, and Simon Marvin. *Splintering Urbanism: Networked Infrastructures, Technological Mobilities and the Urban Condition.* New York: Routledge, 2001.

Grant, Richard, and Jan Nijman. "Globalization and the Corporate Geography of Cities in the Less-Developed World." *Annals of the Association of American Geographers* 92, no. 2 (2002): 320–340.

Green, Hardy. *The Company Town: The Industrial Edens and Satanic Mills That Shaped the American Economy.* New York: Basic Books, 2010.

Green, Vivien. "Utopia/Dystopia." *American Art* 25, no. 2 (2011): 1–7.

Greenberg, Miriam. "Branding Cities: A Social History of the Urban Lifestyle Magazine." *Urban Affairs Review* 36, no. 2 (2000): 228–263.

——. *Branding New York: How a City in Crisis Was Sold to the World.* New York: Routledge, 2008.

Gregory, Derek. "A Real Differentiation and Post-Modern Human Geography." In *Horizons in Human Geography*, ed. Derek Gregory and Rex Walford, 67–96. New York: Palgrave, 1989.

——. "The Black Flag: Guantánamo Bay and the Space of Exception." *Geografiska Annaler: Series B, Human Geography* 88, no. 4 (2006): 405–427.

Griffiths, Ron. "Making Sameness: Place Marketing and the New Urban Entrepreneurialism." In *Cities, Economic Competition and Urban Policy*, ed. Nick Oatley, 41–57. London: Paul Chapman, 1998.

Großmann, Katrin, Marco Bontje, Annegret Haase, and Vlad Mykhnenko. "Shrinking Cities: Notes for the Further Research Agenda." *Cities* 35 (2013): 221–225.

Großmann, Katrin, Annegret Haase, Dieter Rink, and Annett Steinführer. "Urban Shrinkage in East Central Europe? Benefits and Limits of a Cross-National Transfer of Research Approaches." In *Declining Cities/Developing Cities: Polish and German Perspectives*, ed. Marek Nowak and Michał Nowosielski, 77–99. Poznan, Poland: Instytut Zachodni, 2008.

Grodach, Carl. "Museums as Urban Catalysts: The Role of Urban Design in Flagship Cultural Development." *Journal of Urban Design* 13, no. 2 (2008): 195–212.

Grodach, Carl, and Anastasia Loukaitou-Sideris. "Cultural Development Strategies and Urban Revitalization: A Survey of US Cities." *International Journal of Cultural Policy* 13, no. 4 (2007): 349–370.

Growth, Jacqueline, and Eric Corijn. "Reclaiming Urbanity: Indeterminate Spaces, Informal Actors and Urban Agenda Setting." *Urban Studies* 42, no. 3 (2005): 503–520.

Gubitosi, Alessandro. "Fast Track Cities—Dubai's Time Machine." In *Instant Cities: Emergent Trends in Architecture and Urbanism in the Arab World: The Third International Conference of the Center for the Study of Architecture in the Arab Region*, ed. Amer Moustafa, Jamal Al-Qawasmi, and Kevin Mitchell, 89–102. Sharaj, UAE: CSAAR Press, 2008.

Gugler, Josef. "Over-urbanization Reconsidered." In *Cities in the Developing World: Issues, Theory and Policy*, ed. Josef Gugler, 114–123. Oxford: Oxford University Press, 1997.

——. "World Cities in Poor Countries: Conclusions from Case Studies of the Principal Regional and Global Players." *International Journal of Urban and Regional Research* 27, no. 3 (2003): 707–712.

Gunton, Thomas. "Megaprojects and Regional Development: Pathologies in Project Planning." *Regional Studies* 37, no. 5 (2003): 505–519.

Gupta, Akhil, and James Ferguson. "Spatializing States: Toward an Ethnography of Neo-Liberal Governmentality." *American Ethnologist* 29, no. 4 (2000): 981–1002.

Gupta, Pamila. "Futures, Fakes and Discourses of the Gigantic and Miniature in 'The World' Islands, Dubai." *Island Studies Journal* 10, no. 2 (2015): 181–196.

Haartsen, Tiadla, and Viktor Venhorst. "Planning for Decline: Anticipating on Population Decline in the Netherlands." *Tijdschrift Voor Economische En Sociale Geografie* 101, no. 2 (2010): 218–227.

Haase, Annegret, Alexandra Athanasopoulou, and Dieter Rink. "Urban Shrinkage as an Emerging Concern for European Policymaking." *European Urban and Regional Studies* 23, no. 1 (2016): 103–107.

Haase, Annegret, Matthias Bernt, Katrin Großmann, Vlad Mykhnenko, and Dieter Rink. "Varieties of Shrinkage in European Cities." *European Urban and Regional Studies* 23, no. 1 (2016): 86–102.

Haase, Annegret, Anja Nelle, Alan Mallach. "Representing Urban Shrinkage: The Importance of Discourse as a Frame for Understanding Conditions and Policy." *Cities* 69 (2017): 95–101.

Haase, Annegret, Dieter Rink, Katrin Grossmann, Matthias Bernt, and Vlad Mykhnenko. "Conceptualizing Urban Shrinkage." *Environment and Planning A* 46, no. 7 (2014): 1519–1534.

Haase, Annegret, Antonin Vaishar, and Grzegorz Węcławowicz. "The Post-Socialist Condition and Beyond: Framing and Explaining Urban Change in East Central Europe." In *Residential Change and Demographic Challenge: The Inner City of East Central Europe in the 21st Century*, ed. Annegret Haase, Annett Steinführer, Sigrun Kabisch, Katrin Großmann, and Ray Hall, 63–83. Burlington, VT: Ashgate, 2011.

Haase, Dagmar, Sven Lautenbach, and Ralf Seppelt. "Modeling and Simulating Residential Mobility in a Shrinking City Using an Agent-Based Approach." *Environmental Modeling and Software* 25, no. 10 (2010): 1225–1240.

Hackenbroch, Kirsten, and Shahadat Hossain. " 'The Organised Encroachment of the Powerful'—Everyday Practices of Public Space and Water Supply in Dhaka, Bangladesh." *Planning Theory & Practice* 13, no. 3 (2012): 397–420.

Hackworth, Jason. *The Neoliberal City: Governance, Ideology, and Development in American Urbanism.* Ithaca, NY: Cornell University Press, 2007.

——. "Rightsizing as Spatial Austerity in the American Rust Belt." *Environment and Planning A* 47, no. 4 (2015): 766–782.

——. "Why There Is No Detroit in Canada." *Urban Geography* 37, no. 2 (2016): 272–295.

——. *Manufacturing Decline: How Racism and the Conservative Movement Crush the Rust Belt.* New York: Columbia University Press, 2019.

Haines, Chad. "Cracks in the Façade: Landscapes of Hope and Desire in Dubai." In *Worlding Cities: Asian Experiments and the Art of Being Global*, ed. Ananya Roy and Ailwa Ong, 160–181. Malden, MA: Wiley Blackwell, 2011.

Hajer, Maarten. "The Generic City." *Theory, Culture & Society* 16, no. 4 (1999): 137–144.

Hall, C. M. "Urban Entrepreneurship, Corporate Interests and Sports Mega-Events: The Thin Policies of Competitiveness Within the Hard Outcomes of Neoliberalism." *Sociological Review* 54, no. 2 (2006): 59–70.

Hall, Peter. *Cities in Civilization: Culture, Innovation, and Urban Order.* London: Weindenfield & Nicolson, 1998.

——. *Cities of Tomorrow: An Intellectual History of Urban Planning and Design in the Twentieth Century.* Oxford: Blackwell, 2002.

——. "Balancing European Territory: The Challenge of the Post-Liberal Era." In *Parallel Patterns of Shrinking Cities and Urban Growth: Spatial Planning for Sustainable City Regions and Rural Areas*, ed. Robin Ganser and Rocky Piro, 27–44. Burlington, VT: Ashgate, 2012.

Hall, Peter, and Kathy Pain. "From Metropolis to Polyopolis." In *The Polycentric Metropolis: Learning from Mega-City Regions in Europe*, ed. Peter Hall and Kathy Pain, 3–18. New York: Earthscan, 2006.

Hall, Tim, and Phil Hubbard. "The Entrepreneurial City: New Urban Politics, New Urban Geography?" *Progress in Human Geography* 20, no. 2 (1996): 153–174.

Halpern, Orit, Jesse LeCavalier, Nerea Calvillo, and Wolfgang Pietsch. "Test Bed Urbanism." *Public Culture* 25, no. 2 [70] (2013): 272–306.

Hamer, Andrew. "Economic Impacts of Third World Mega-Cities: Is Size the Issue?" In *Mega-City Growth and the Future*, ed. Roland Fuchs, Ellen Brennan, Joseph Chamie, Fuchen Lo, and Joha Uitto, 172–191. Tokyo: United Nations University Press, 1994.

Hamnett, Chris. "Social Polarisation in Global Cities: Theory and Evidence." *Urban Studies* 31, no. 3 (1994): 410–424.

——. *Unequal City: London in the Global Arena.* New York: Routledge, 2003.

Han, Heejin. "Singapore, a Garden City: Authoritarian Environmentalism in a Developmental State." *Journal of Environment & Development* 26, no. 1 (2017): 3–24.
Hanlon, Bernadette. "The Decline of Older, Inner Suburbs in Metropolitan America." *Housing Policy Debate* 19, no. 3 (2008): 423–456.
——. "A Typology of Inner-Ring Suburbs: Class, Race, and Ethnicity in U.S. Suburbia." *City & Community* 8, no. 3 (2009): 221–246.
Hannemann, Christine. "The Industrial City as a Shrinking City and the Special Case of Flint, MI." In *Industrial Cities: History and Future*, ed. Clemens Zimmerman, 142–164. Frankfurt-on-Main: Campus Verlag, 2013.
Hannigan, John. *Fantasy City: Pleasure and Profit in the Postmodern Metropolis.* New York: Routledge, 1998.
Hansen, Karen Tranberg, and Mariken Vaa, eds. *Reconsidering Informality: Perspectives from Urban Africa.* Uppsula, Sweden: Nordiska Afrikainstitutet, 2004.
Hansen, Susan, Carolyn Ban, and Leonard Huggins. "Explaining the Brain Drain from Older Industrial Regions: The Pittsburgh Region." *Economic Development Quarterly* 17, no. 2 (2003): 132–147.
Harding, Jeremy. "It Migrates to Them." *London Review of Books* 29, no. 5 (March 8, 2007): 25–27.
Hardy, Dennis. "Dubai—Planning Miracle or Desert Mirage?" *Town and Country Planning* 77, no. 3 (2008): 143.
Harker, Christopher. "Theorizing the Urban from the 'South'?" *City* 15, no. 1 (2011): 120–122.
Harris, Andrew. "The Metonymic Urbanism of Twenty-First-Century Mumbai." *Urban Studies* 49, no. 13 (2012): 2955–2973.
——. "Vertical Urbanisms: Opening Up Geographies of the Three-Dimensional City." *Progress in Human Geography* 39, no. 5 (2015): 601–620.
Harris, Richard, and Charlotte Vorms. "Introduction." In *What's in a Name? Talking About Urban Peripheries*, ed. Richard Harris and Charlotte Vorms, 3–35. Toronto: University of Toronto Press, 2017.
Hart, Keith. "Informal Income Opportunities and Urban Employment in Ghana." *Modern African Studies* 11, no. 1 (1973): 61–89.
Hartt, Maxwell, and Jason Hackworth. "Shrinking Cities, Shrinking Households, or Both?" *International Journal of Urban and Regional Research* 44, no. 6 (2020): 1083–1095.
Hartt, Maxwell Douglas. "How Cities Shrink: Complex Pathways to Population Decline." *Cities* 75 (2018): 38–49.
Harvey, David. *The Condition of Postmodernity: An Enquiry Into the Origins of Cultural Change.* Cambridge, MA: Blackwell, 1989.
——. "From Managerialism to Entrepreneurialism: The Transformation in Urban Governance in Late Capitalism." *Geografiska Annaler: Series B, Human Geography* 71, no. 1 (1989): 3–17.
——. *The Urban Experience.* New York: Blackwell, 1989.
——. *Spaces of Hope.* Berkeley: University of California Press, 2000.
——. "Neo-Liberalism as Creative Destruction." *Geografiska Annaler: Series B, Human Geography* 88, no. 2 (2006): 145–158.
——. *Spaces of Global Capitalism: A Theory of Uneven Geographical Development.* New York: Verso, 2006.
——. "Neoliberalism and the City." *Studies in Social Justice* 1, no. 1 (2007): 1–13.
Hashim, Alamira Reem Bani, Clara Irazábal, and Greta Byrum. "The Scheherazade Syndrome: Fiction and Fact in Dubai's Quest to Become a Global City." *Architectural Theory Review* 15, no. 2 (2010): 210–231.
Hatherley, Owen. "Comparing Capitals." *New Left Review* 105 (2017): 107–132.
Häußermann, Hartmut, and Katja Simons. "Facing Fiscal Crisis: Urban Flagship Projects in Berlin." In *The Globalized City: Economic Restructuring and Social Polarization in European Cities*, ed. Frank Moulaert, Arantxa Rodriguez, and Erik Swyngedouw, 107–124. Oxford: Oxford University Press, 2003.
Häußermann, Hartmut, and Walter Siebel. *Neue Urbanität.* Frankfurt: Suhrkamp, 1987.
——. "Die Schrumpfende Stadt und die Stadtsoziologie" [The shrinking city and urban sociology]. In *Soziologische Stadtforschung*, ed. Jürgen Friedrichs, 78–94. Opladen, Germany: Westdeutscher Verlag, 1998.

Healey, Patsy. "The Universal and the Contingent: Some Reflections on the Transnational Flow of Planning Ideas and Practices." *Planning Theory* 11, no. 2 (2012): 188–207.

Hebblethwaite, Rupert. "The Municipal Housing Programme in Sheffield Before 1914." *Architectural History* 30 (1987): 143–179.

Hentschel, Christine. "City Ghosts: The Haunted Struggles for Downtown Durban and Berlin *Neukölln*." In *Locating Right to the City in the Global South*, ed. Tony Samara, Shenjing He, and Guo Chen, 195–217. New York: Routledge, 2013.

——. "Postcolonizing Berlin and the Fabrication of the Urban." *International Journal of Urban and Regional Research* 39, no. 1 (2015): 79–91.

Herbert, Claire, and Martin J. Murray. "Building New Cities from Scratch: Privatized Urbanism and the Spatial Restructuring of Johannesburg After Apartheid." *International Journal of Urban and Regional Research* 39, no. 3 (2015): 471–494.

Herfert, Günter. "Berlin: The Metropolitan Periphery Between Boom and Shrinkage." *Revue European Spatial Research Policy* 13, no. 2 (2006): 5–22.

Hetherington, Kevin. "Manchester's Urbis: Urban Regeneration, Museums and Symbolic Economies." *Cultural Studies* 21, no. 4–5 (2007): 630–649.

High, Steven. *Industrial Sunset: The Making of North America's Rust Belt, 1969–1984*. Toronto: University of Toronto Press, 2003.

Hill, Richard Child. "Cities and Nested Hierarchies." *International Social Science Journal* 56, no. 181 (2004): 373–384.

Hiller, Harry. "Mega-Events, Urban Boosterism and Growth Strategies: An Analysis of the Objectives and Legitimations of the Cape Town 2004 Olympic Bid." *International Journal of Urban and Regional Research* 24, no. 2 (2000): 449–458.

Hoekveld, Josje. "Time–Space Relations and the Differences Between Shrinking Regions." *Built Environment* 38, no. 2 (2012): 179–195.

——. "Understanding Spatial Differentiation in Urban Decline Levels, *European Planning Studies* 22, no. 2 (2014): 362–382.

——. "Spatial Differentiation of Population Development in a Declining Region: The Case of Saarland." *Geografiska Annaler: Series B, Human Geography* 97, no. 1 (2015): 47–68.

Hoffman, Danny. *Monrovia Modern: Urban Form and Political Imagination in Liberia*. Durham, NC: Duke University Press, 2017.

Hogan, Trevor, and Christopher Houston. "Corporate Cities: Urban Gateways of Gated Communities Against the City: The Case of Lippo, Jakarta." In *Critical Reflections on Cities in Southeast Asia*, ed. Tim Bunnell, Lisa Drummond, and K. C. Ho, 243–264. Singapore: Times Academic, 2002.

Hogan, Trevor, and Julian Potter. "Big City Blues." *Thesis Eleven* 121, no. 1 (2014): 3–8.

Holcomb, Briavel. "Revisioning Place: De- and Re-constructing the Image of the Industrial City." In *Selling Places: The City as Cultural Capital, Past and Present*, ed. Gerry Kearns and Chris Philo, 133–144. New York: Pergamon, 1993.

——. "Marketing Cities for Tourism." In *The Tourist City*, ed. Susan Fainstein and Dennis Judd, 54. New Haven, CT: Yale University Press, 1999.

Hollander, Justin. *Polluted and Dangerous: America's Worst Abandoned Properties and What Can Be Done About Them*. Burlington: University of Vermont Press, 2009.

——. "Moving Toward a Shrinking Cities Metric: Analyzing Land Use Changes Associated with Depopulation in Flint, Michigan." *Cityscape: A Journal of Policy Development and Research* 12, no. 1 (2010): 133–151.

——. "Can a City Successfully Shrink? Evidence from Survey Data on Neighborhood Quality." *Urban Affairs Review* 47, no. 1 (2011): 129–141.

——. *Sunburnt Cities: The Great Recession, Depopulation, and Urban Planning in the American Sunbelt*. New York: Routledge, 2011.

Hollander, Justin, Karina Pallagst, Terry Schwarz, and Frank J. Popper. "Planning Shrinking Cities." *Progress in Planning* 72 (2009): 223–232.

Hollands, R. I. G. "Will the Real Smart City Please Stand Up?" *City* 12, no. 3 (2008): 303–320.

Hollands, Robert, and Paul Chatterton. "Producing Nightlife in the New Urban Entertainment Economy: Corporatization, Branding, and Market Segmentation." *International Journal of Urban and Regional Research* 27, no. 2 (2003): 361–385.

Holston, James. *The Modernist City: An Anthropological Critique of Brasilia.* Chicago: University of Chicago Press, 1989.

——. *Insurgent Citizenship: Disjunctions of Democracy and Modernity in Brazil.* Princeton, NJ: Princeton University Press, 2009.

——. "Insurgent Citizenship in an Era of Global Urban Peripheries." *City & Society* 21, no. 2 (2009): 245–267.

Holston, James, and Teresa Caldeira. "Urban Peripheries and the Invention of Citizenship." *Harvard Design Magazine* 28 (2008): 18–23.

Home, Robert. "Town Planning and Garden Cities in the British Colonial Empire, 1910–1940." *Planning Perspectives* 5, no. 1 (1990): 23–37.

——. *Of Planting and Planning: The Making of British Colonial Cities.* London: Spon, 1997.

Honore, Carl. *In Praise of Slowness: Challenging the Cult of Speed.* New York: Harper Collins, 2004.

Hospers, Gert-Jan. "Coping with Shrinkage in Europe's Cities and Towns." *Urban Design International* 18, no. 1 (2013): 78–89.

Hospers, Gert-Jan, and Cees-Jan Pen. "A View on Creative Cities Beyond the Hype." *Creativity and Innovation Management* 17, no. 4 (2008): 259–270.

Howard, Ebenezer. *The Garden Cities of To-morrow.* London: Swan Sonnenschein, 1902.

Howes, David. "Introduction: Commodities and Cultural Borders." In *Cross-Cultural Consumption: Global Markets, Local Realities*, ed. David Howes, 1–16. New York: Routledge, 1996.

Hoyle, Brian. "Global and Local Change on the Port-City Waterfront." *Geographical Review* 90, no. 3 (2000): 395–417.

Hoyler, Michael, and John Harrison. "Global Cities Research and Urban Theory Making." *Environment and Planning A* 49, no. 12 (2017): 2853–2858.

Hoyt, Lorene. "Do Business Improvement District Organizations Make a Difference? Crime in and Around Commercial Areas in Philadelphia." *Journal of Planning Education and Research* 25, no. 2 (2005): 185–199.

——. "Planning Through Compulsory Commercial Clubs: Business Improvement Districts." *Economic Affairs* 25, no. 4 (2005): 24–27.

Huat, Chua Beng. "Singapore as Model: Planning Innovations, Knowledge Experts." In *Worlding Cities: Asian Experiments and the Art of Being Global*, ed. Ananya Roy and Aiwha Ong, 29–54. Walden, MA: Wiley-Blackwell, 2011.

Hubbard, Phil. "Urban Design and Local Economic Development: A Case Study in Birmingham." *Cities* 12, no. 4 (1995): 243–251.

——. "Urban Design and City Regeneration: Social Representations and Entrepreneurial Landscapes." *Urban Studies* 33, no. 8 (1996): 1441–1461.

——. *City.* London: Routledge, 2006.

Hubbard, Phil, and Keith Lilley. "Pacemaking the Modern City: The Urban Politics of Speed and Slowness." *Environment and Planning D* 22, no. 2 (2004): 273–294.

Huchzermeyer, Marie. "Troubling Continuities: Use and Utility of the Term 'Slum.'" In *The Routledge Handbook on Cities of the Global South*, ed. Susan Parnell and Sophie Oldfield, 86–97. New York: Routledge, 2014.

Hummel, Daniel. "Right-Sizing Cities in the United States: Defining Its Strategies." *Journal of Urban Affairs* 37, no. 4 (2015): 397–409.

Huxtable, Ada Louise. *The Unreal America: Architecture and Illusion.* New York: New Press, 1997.

Huyssen, Andreas. "Authentic Ruins: Products of Modernity." In *Ruins of Modernity*, ed. Julia Hell and Andreas Schönle, 17–28. Durham, NC: Duke University Press, 2010.

Irving, Allan. "The Modern/Postmodern Divide in Urban Planning." *University of Toronto Quarterly* 62, no. 4 (1993): 474–488.

Isin, Engin. "Historical Sociology of the City." In *Handbook of Historical Sociology*, ed. Gerard Delanty and Engin Isin, 312–325. London: Sage, 2002.

——. "City.State: Critique of Scalar Thought." *Citizenship Studies* 11, no. 2 (2007): 211–228.

Jackson, Kenneth. *Crabgrass Frontier: The Suburbanization of the United States.* New York: Oxford University Press, 1985.

Jackson, Mark, and Veronica della Dora. "Dreams So Big Only the Sea Can Hold Them: Man-Made Islands as Anxious Spaces, Cultural Icons, and Travelling Visions." *Environment and Planning A 41*, no. 9 (2009): 2086–2104.

Jackson, Peter. "Domesticating the Street." In *Images of the Street: Planning, Identity and Control in Public Space*, ed. Nicholas Fyfe, 176–191. London: Routledge, 1998.

Jacobs, Jane M. "A Geography of Big Things." *Cultural Geography* 13, no. 1 (2006): 1–27.

——. "Commentary: Comparing Comparative Urbanism." *Urban Geography* 33, no. 6 (2012): 904–914.

Jacobs, Keith. "Waterfront Redevelopment: A Critical Discourse Analysis of the Policy-Making Process Within the Chatham Maritime Project." *Urban Studies* 41, no. 4 (2004): 817–832.

Jameson, Frederic. *Postmodernism: or, The Cultural Logic of Late Capitalism.* New York: Verso, 1984.

Jayne, Mark, and Kevin Ward. "A Twenty-First Century Introduction to Urban Theory." In *Urban Theory: New Critical Perspectives*, ed. Mark Jayne and Kevin Ward, 1–18. London: Routledge, 2016.

Jencks, Christopher. *What Is Postmodernism?* New York: St. Martin's, 1986.

Jensen, Boris Brorman. "Masdar City: A Critical Retrospection." In *Under Construction: Logics of Urbanism in the Gulf Region*, ed. Steffan Wippel, Katrin Bromber, Christian Steiner, and Birget Krawweitz, 45–54. Burlington, VT: Ashgate, 2014.

Jessen, Johann. "Conceptualizing Shrinking Cities—A Challenge for Planning Theory." In *Parallel Patterns of Shrinking Cities and Urban Growth: Spatial Planning for Sustainable City Regions and Rural Areas*, ed. Robin Ganser and Rocky Piro, 45–58. Burlington, VT: Ashgate, 2012.

Jessop, Bob. "The Entrepreneurial City: Re-imaging Localities, Redesigning Economic Governance, or Restructuring Capital?" In *Transforming Cities: Contested Governance and New Spatial Divisions*, ed. Nick Jewson and Susanne MacGregor, 28–41. London: Routledge, 1997.

——. "The Narrative of Enterprise and the Enterprise of Narrative." In *The Entrepreneurial City: Geographies of Politics, Regime, and Representation*, ed. Tim Hall and Phil Hubbard, 77–99. Chichester, NY: Wiley, 1998.

Jones, Andrew. "Issues in Waterfront Regeneration: More Sobering Thoughts—A UK Perspective." *Planning Practice & Research* 13, no. 4 (1998): 433–442.

——. "The 'Global City' Misconceived: The Myth of 'Global Management' in Transnational Service Firms." *Geoforum* 33, no. 3 (2002): 335–350.

Jones, Gareth A. "Slumming About." *City* 15, no. 6 (2011): 696–708.

Jones, Gareth A., and Romola Sanyal. "Spectacle and Suffering: The Mumbai Slum as a Worlded Space." *Geoforum* 65 (2015): 431–439.

Jones, Roy, and Brian Shaw. "Development, Postmodernism and the Postcolonial City in Southeast Asia." *Australian Development Studies Network: Development Bulletin* 45 (1998): 20–22.

Judd, Dennis, ed. *The Infrastructure of Play: Building the Tourist City.* Armonk, NY: Sharp, 2004.

Julier, Guy. "Urban Designscapes and the Production of Aesthetic Consent." *Urban Studies* 42, no. 5–6 (2005): 869–887.

Kabisch, Nadja, Dagmar Haase, and Annegret Haase. "Urban Population Development in Europe, 1991–2008: The Examples of Poland and the UK." *International Journal of Urban and Regional Research* 36, no. 6 (2012): 1326–1348.

Kalia, Ravi. *Chandigarh: The Making of an Indian City.* New York: Oxford University Press, 2002.

Kaminer, Tahl, Miguel Robles-Durán, and Heidi Sohn. "Introduction: Urban Asymmetries." In *Urban Asymmetries: Studies and Projects on Neoliberal Urbanization*, ed. Tahl Kaminer, Miguel Robles-Durán, and Heidi Sohn, 10–20. Rotterdam: 011 Publishers, 2011.

——, eds. *Urban Asymmetries: Studies and Projects on Neoliberal Urbanization.* Rotterdam: 011 Publishers, 2011.

Kanai, Juan Miguel. "Buenos Aires Beyond (Homo)Sexualized Urban Entrepreneurialism: Queer Geographies of Tango." *Antipode* 47, no. 3 (2015): 652–670.

——. "On the Peripheries of Planetary Urbanization: Globalizing Manaus and Its Expanding Impact." *Environment and Planning D: Society and Space* 32, no. 6 (2014): 1071–1087.

——. "Buenos Aires, the Capital of Tango: Tourism, Redevelopment and the Cultural Politics of Neoliberal Urbanism." *Urban Geography* 35, no. 8 (2014): 1111–1117.

Kanai, Miguel, and Iliana Ortega-Alcázar. "The Prospects for Culture-Led Urban Regeneration in Latin America: Cases from Mexico City and Buenos Aires." *International Journal of Urban and Regional Research* 33, no. 2 (2009): 483–501.

Kanna, Ahmed. "Dubai in a Jagged World." *Middle East Report* 243 (2007): 22–29.

——. "Flexible Citizenship in Dubai: Neoliberal Subjectivity in the Emerging 'City-Corporation.' " *Cultural Anthropology* 25, no. 1 (2010): 100–129.

Kantor, Paul, H. V. Savitch, and Serena Vicari Haddock. "The Political Economy of Urban Regimes: A Comparative Perspective." *Urban Affairs Review* 32, no. 3 (1997): 348–377.

Kapp, Paul Hardin, and Paul Armstrong, eds. *SynergiCity: Reinventing the Postindustrial City.* Champaign: University of Illinois Press, 2012.

Kavaratzis, Michalis. "From City Marketing to City Branding: Towards a Theoretical Framework for Developing City Brands." *Place Branding* 1, no. 1 (2004): 58–73.

Kavaratzis, Michalis, and Gregory Ashworth. "City Branding: An Effective Assertion of Identity or a Transitory Marketing Trick?" *Tijdschrift voor Economische en Sociale Geografie* 96, no. 5 (2005): 506–514.

Keil, Roger. "The Empty Shell of the Planetary: Re-rooting the Urban in the Experience of the Urbanites." *Urban Geography* 39, no. 10 (2018): 1589–1602.

——. "Extended Urbanization, 'Disjunct Fragments' and Global Suburbanisms." *Environment and Planning D: Society and Space* 3, no. 2 (2018): 494–511.

Kenny, Judith. "Making Milwaukee Famous: Cultural Capital, Urban Image and the Politics of Place." *Urban Geography* 16 (1995): 440–458.

Keyder, Caglar. "Globalization and Social Exclusion in Istanbul." *International Journal of Urban and Regional Research* 29, no. 1 (2005): 124–134.

Kilian, Darryll, and Belinda Dodson. "Between the Devil and the Deep Blue Sea: Functional Conflicts in Cape Town's Victoria and Alfred Waterfront." *Geoforum* 27, no. 4 (1996): 495–507.

Kim, Chigon. "Place Promotion and Symbolic Characterization of New Songdo City, South Korea." *Cities* 27, no. 1 (2010): 13–19.

Kim, Jung In. "Making Cities Global: The New City Development of Songdo, Yujiapu and Lingang." *Planning Perspectives* 29, no. 3 (2014): 329–356.

Kinder, Kimberley. "Guerrilla-Style Defensive Architecture in Detroit: A Self-Provisional Strategy in a Neoliberal Space of Disinvestment." *International Journal of Urban and Regional Research* 38, no. 5 (2014): 1767–1784.

——. *DIY Detroit: Making Do in a City Without Services.* Minneapolis: University of Minnesota Press, 2016.

King, Anthony. *Global Cities: Post-Imperialism and the Internationalization of London.* London: Routledge and Kegan Paul, 1990.

——. "Worlds in the City: Manhattan Transfer and the Ascendance of Spectacular Space." *Planning Perspectives* 11, no. 2 (1996): 87–114.

——. "(Post)Colonial Geographies: Material and Symbolic." *Historical Geography* 27, no. 1 (1999): 99–118.

Kinney, Rebecca. *Beautiful Wasteland: The Rise of Detroit as America's Postindustrial Frontier.* Minneapolis: University of Minnesota Press, 2016.

Kirshenblatt-Gimblett, Barbara. *Destination Culture: Tourism, Museums, and Heritage.* Berkeley: University of California Press, 1998.

Klingmann, Anna. *Brandscapes: Architecture in the Experience Economy.* Cambridge, MA: MIT Press, 2007.

Knox, Paul. "Creating Ordinary Places: Slow Cities in a Fast World." *Journal of Urban Design* 10, no. 1 (2005): 1–11.

——. *Metroburbia, USA.* New Brunswick, NJ: Rutgers University Press, 2008.

——. *Cities and Design.* New York: Routledge, 2011.

Knox, Paul, and Sallie Marston. *Human Geography: Places and Regions in Global Context.* Upper Saddle River, NJ: Prentice-Hall, 2000.

Knox, Paul, and Heike Mayer. *Small Towns Sustainability.* Basel, Switzerland: Birkhauser, 2009.

Knox, Paul, and Kathy Pain. "Globalization, Neoliberalism and International Homogeneity in Architecture and Urban Development." *Informationen zur Raumentwicklung* 5, no. 6 (2010): 417–428.

Koch, Natalie. "The Monumental and the Miniature: Imagining 'Modernity' in Astana." *Social and Cultural Geography* 11, no. 8 (2010): 769–787.

——. "Urban 'Utopias': The Disney Stigma and Discourses of 'False Modernity'." Environment and Planning A 44, no. 10 (2012): 2445–2462.

——. "Why Not a World City? Astana, Ankara, and Geopolitical Scripts in Urban Networks." *Urban Geography* 34, no. 1 (2013): 109–130.

——. "The Violence of Spectacle: Statist Schemes to Green the Desert and Constructing Astana and Ashgabat as Urban Oases." *Social and Cultural Geography* 16, no. 6 (2015): 675–697.

—— *The Geopolitics of Spectacle: Space, Synecdoche, and the New Capitals of Asia.* Ithaca, NY: Cornell University Press, 2018.

Koch, Regan, and Alan Latham. "On the Hard Work of Domesticating a Public Space." *Urban Studies* 50, no. 1 (2013): 6–21.

Kohn, Margaret. *Brave New Neighborhoods: The Privatization of Public Space.* New York: Routledge, 2004.

Koolhaas, Rem. "What Ever Happened to Urbanism?" In *S,M,L,XL*, by Rem Koolhaas and Bruce Mau, 959–971. New York: Monacelli, 1995.

——. "The Generic City." In *S,M,L,X*, by Rem Koolhaas and Bruce Mau, 1239–1264. New York: Monacelli, 1995.

Koolhaas, Rem, and Harvard Project on the City. "Lagos." In *Mutations*, ed. Francine Fort and Michel Jacques, 650–720. Barcelona: Actar, 2001.

Koolhaas, Rem, and AMO. *The Gulf.* Baden, Switzerland: Lars Muller, 2007.

Koolhaas, Rem, and Bruce Mau. *S,M,L,XL.* New York: Monacelli, 1995.

Koonings, Kess, and Dirk Kruijt. "Fractured Cities, Second-Class Citizenship, and Urban Violence." In *Fractured Cities: Social Exclusion, Urban Violence and Contested Spaces in Latin America*, ed. Kess Koonings and Dirk Kruijt, 7–22. London: Zed, 2007.

——, eds. *Megacities: The Politics of Urban Exclusion and Violence in the Global South.* London: Zed, 2010.

Köppen, Bernhard. "The Production of a New Eurasian Capital on the Kazakh Steppe: Architecture, Urban Design, and Identity in Astana." *Nationalities Papers: The Journal of Nationalism and Ethnicity* 41, no. 4 (2013): 590–605.

Koselleck, Reinhart. *Futures Past: On the Semantics of Historical Time.* Trans. Keith Tribe. Cambridge, MA: MIT Press, 1985.

Koster, Martijn, and Monique Nuijten. "Coproducing Urban Space: Rethinking the Formal/Informal Dichotomy." *Singapore Journal of Tropical Geography* 37, no. 3 (2016): 282–294.

Kotilainen, Juha, Ilkka Eisto, and Eero Vatanen. "Uncovering Mechanisms for Resilience: Strategies to Counter Shrinkage in a Peripheral City in Finland." *European Planning Studies* 23, no. 1 (2015): 53–68.

Kraas, Frauke, Surinder Aggarwal, Martin Coy, and Günter Mertins, eds. *Megacities: Our Global Urban Future.* New York: Springer, 2012.

Kracauer, Siegfried. "Ahasuerus, or the Riddle of Time." In *History: The Last Things Before the Last*, 139–163. New York: Oxford University Press, 1969.

Krane, Jim. *Dubai: The Story of the World's Fastest City.* London: Atlantic, 2009.

Krätke, Stefan. " 'Creative Cities' and the Rise of the Dealer Class: A Critique of Richard Florida's Approach to Urban Theory." *International Journal of Urban and Regional Research* 34, no. 4 (2010): 835–853.

——. "How Manufacturing Industries Connect Cities Across the World: Extending Research on 'Multiple Globalizations'. " *Global Networks* 14, no. 2 (2013): 121–147.

——. "Cities in Contemporary Capitalism." *International Journal of Urban and Regional Research* 38, no. 5 (2014): 1660–1677.

Kreibich, Volker. "The Mode of Informal Urbanisation: Reconciling Social and Statutory Regulation in Urban Land Management." In *Urban Informalities: Reflections on the Formal and Informal*, ed. Colin McFarlane and Michael Waibel, 149–170. London: Routledge, 2016.

Kruijt, Dirk, and Kees Koonings. "The Rise of Megacities and the Urbanization of Informality, Exclusion, and Violence." In *Mega-Cities: The Politics of Urban Exclusion and Violence in the Global South*, ed. Kees Koonings and Dirk Kruijt, 8–26. London: Zed, 2009.

Kudva, Neema. "The Everyday and the Episodic: The Spatial and Political Impacts of Urban Informality." *Environment and Planning A* 41, no. 7 (2009): 1614–1628.

Kühn, Manfred, Matthias Bernt, and Laura Colini. "Power, Politics and Peripheralization: Two Eastern German Cities." *European Urban and Regional Studies* 24, no. 3 (2017): 258–273.

Kumar, Krishan. "Aspects of the Western Utopian Tradition." *History of the Human Sciences* 16, no. 1 (2003): 63–77.

Kusno, Abidin. *Behind the Postcolonial.* London: Routledge, 2000.

——. "Runaway City: Jakarta Bay, the Pioneer and the Last Frontier." *Inter-Asia Cultural Studies* 12, no. 4 (2011): 513–531.

——. "The Post-Colonial Unconscious: Observing Mega-Imagistic Urban Projects in Asia." In *The Emerging Asian City: Concomitant Urbanites and Urbanisms*, ed. Vinayak Bharne, 168–178. New York: Routledge, 2013.

Kusno, Abidin, Melani Budianta, and Hilmar Farid. "Editorial Introduction: Runaway City/Leftover Spaces." *Inter-Asia Cultural Studies* 12, no. 4 (2011): 473–481.

Labbé, Danielle, and Julie-Ann Boudreau. "Understanding the Causes of Urban Fragmentation in Hanoi: The Case of New Urban Areas." *International Development Planning Review* 33, no. 3 (2011): 273–291.

Lakoff, Andrew. "Preparing for the Next Emergency." *Public Culture* 19, no. 2 (2007): 247–271.

Landry, Charles. *The Art of City Making.* London: Earthscan, 2006.

——. *The Creative City: A Toolkit for Urban Innovation.* London: Earthscan, 2000.

Landry, Charles, and Franco Bianchini. *The Creative City.* London: Demos Comedia, 1995.

Lang, Robert, and Paul Knox. "The New Metropolis: Rethinking Megalopolis." *Regional Studies* 43, no. 6 (2009): 789–802.

Lang, Thilo. "Shrinkage, Metropolization and Peripheralization in East Germany." *European Planning Studies* 20, no. 10 (2012): 1747–1754.

Langner, Marcel, and Wilfried Endlicher, eds. *Shrinking Cities: Effects on Urban Ecology and Challenges for Urban Development.* Frankfurt: Peter Lang, 2007.

La Porta, Rafael, and Andrei Shleifer. "Informality and Development." *Journal of Economic Perspectives* 28, no. 3 (2014): 109–126.

Laszezkowski, Mateusz. "Shrek Meets the President: Magical Authoritarianism in a 'Fairy Tale' City." In *Kazakhstan in the Making: Legitimacy, Symbols, and Social Changes*, ed. Marlene Laruelle, 63–88. New York: Lexington, 2016.

Law, Christopher. "Urban Tourism and Its Contribution to Economic Regeneration." *Urban Studies* 29, no. 3–4 (1992): 599–618.

Lawhon, Mary, Jonathan Silver, Henrik Ernstson, and Joseph Pierce. "Unlearning (Un)located Ideas in the Provincialization of Urban Theory." *Regional Studies* 50, no. 9 (2016): 1611–1622.

Leary, Michael Edema. "A Lefebvrian Analysis of the Production of Glorious, Gruesome Public Space in Manchester." *Progress in Planning* 85 (2013): 1–52.

Lee, Jae Seung, Sehyung Won, and Saehoon Kim. "Describing Changes in the Built Environment of Shrinking Cities: Case Study of Incheon, South Korea." *Journal of Urban Planning and Development* 142, no. 2 (2016).

Lee, Zack. "Eco-Cities as an Assemblage of Worlding Practices." *International Journal of Built Environment and Sustainability* 2, no. 3 (2015): 183–191.

Lees, Loretta. "The Ambivalence of Diversity and the Politics of Urban Renaissance: The Case of Growth in Downtown Portland, Maine." *International Journal of Urban and Regional Research* 27, no. 3 (2003): 613–634.

——. "The Geography of Gentrification: Thinking Through Comparative Urbanism." *Progress in Human Geography* 36, no. 2 (2012): 155–171.

Lees, Loretta, Hyun Bang Shin, and Ernesto Lopez-Morales. *Urban Futures: Planetary Gentrification.* Cambridge: Polity, 2016.

Lees, Loretta, Tom Slater, and Elvin Wyly. *Gentrification.* New York: Routledge, 2008.

Leetmaa, Kadri, Agnes Kriszan, Mari Nuga, and Joachim Burdack. "Strategies to Cope with Shrinkage in the Lower End of the Urban Hierarchy in Estonia and Central Germany." *European Planning Studies* 23, no. 1 (2015): 147–165.

Lehrer, Ute. "Willing the Global City: Berlin's Cultural Strategies of Interurban Competition After 1989." In *The Global Cities Reader*, ed. Neil Brenner and Roger Keil, 332–338. New York: Routledge, 2006.

Lehrer, Ute, and Jennifer Laidy. "Old Mega-Projects Newly Packaged? Waterfront Redevelopment in Toronto." *International Journal of Urban and Regional Research* 32, no. 4 (2008): 786–803.

Leigh, Nancey Green, and Nathanael Hoelzel. "Smart Growth's Blind Side." *Journal of the American Planning Association* 78, no. 1 (2012): 87–103.

Leitner, Helga, and Eric Sheppard. "Provincializing Critical Urban Theory." *International Journal of Urban and Regional Research* 40, no. 1 (2016): 228–235.

Lejano, Raul, and Corinna Del Bianco. "The Logic of Informality: Pattern and Process in a São Paulo Favela." *Geoforum* 91 (2018): 195–205.

Leon, Joshua. "Global Cities at Any Cost: Resisting Municipal Merchantilism." *City* 21, no. 1 (2017): 6–24.

Leontidou, Lila. "Alternatives to Modernism in (Southern) Urban Theory: Exploring In-Between Spaces." *International Journal of Urban and Regional Research* 30, no. 2 (1996): 178–195.

Lerner, Isaac. "Instant City: City as Recreation/Re-creation." In *Instant Cities: Emergent Trends in Architecture and Urbanism in the Arab World: The Third International Conference of the Center for the Study of Architecture in the Arab Region*, ed. Amer Moustafa, Jamal Al-Qawasmi, and Kevin Mitchell, 33–44. Sharaj, UAE: CSAAR Press, 2008.

Lever, William, and Ivan Turok. "Competitive Cities: Introduction to the Review." *Urban Studies* 36, no. 5–6 (1999): 791–793.

Ley, David. "Transnational Spaces and Everyday Lives." *Transactions of the Institute of British Geographers*, n.s., 29, no. 2 (2004): 151–164.

Lieberherr-Gardiol, Françoise. "Slums Forever? Globalisation and Its Consequences." *European Journal of Development Research* 18, no. 2 (2006): 275–283.

Liebmann, Heike, and Thomas Kuder. "Pathways and Strategies of Urban Regeneration—Deindustrialized Cities in Eastern Germany." *European Planning Studies* 20, no. 7 (2012): 1155–1172.

Lim, H. "Cultural Strategies for Revitalizing the City: A Review and Evaluation." *Regional Studies* 27, no. 6 (1993): 589–595.

Lin, George. "Chinese Urbanism in Question: State, Society, and the Reproduction of Urban Spaces." *Urban Geography* 28 (2007): 7–29.

Linden, Eugene. "The Exploding Cities of the Developing World." *Foreign Affairs* 75, no. 1 (1996): 52–65.

Lintz, Gerd, Bernhart Müller, and Karl Schmude. "The Future of Industrial Cities and Regions in Central and Eastern Europe." *Geoforum* 38, no. 3 (2007): 512–519.

Liu, Cathy. "From Los Angeles to Shanghai: Testing the Applicability of Five Urban Paradigms." *International Journal of Urban and Regional Research* 36, no. 6 (2012): 1127–1145.

Liu, Xingjian, and Ben Derudder. "Two-Mode Networks and the Interlocking World City Network Model: A Reply to Neal." *Geographical Analysis* 44, no. 2 (2012): 171–173.

Lloyd, Richard. *Neo-Bohemia: Art and Commerce in the Post-Industrial City.* New York: Routledge, 2006.

Lloyd, Richard, and Terry Nichols Clark. "The City as Entertainment Machine." In *Critical Perspectives on Urban Redevelopment: Volume 6*, ed. Kevin Fox Gotham, 357–378. New York: Elsevier, 2001.

Lo, Fu-chen, and Peter Marcotullio. "Globalisation and Urban Transformations in the Asia-Pacific Region: A Review." *Urban Studies* 37, no. 1 (2000): 77–111.

Loftman, Patrick, and Brendan Nevin. "Prestige Projects and Urban Regeneration in the 1980s and 1990s: A Review of the Benefits and Limitations." *Planning Practice and Research* 10, no. 3–4 (1995): 299–315.

——. "Going for Growth: Prestige Projects in Three British Cities." *Urban Studies* 33, no. 6 (1996): 991–1019.

Lomnitz, Larissa. *Networks and Marginality: Life in a Mexican Shantytown.* New York: Academic, 1977.

Longoria, Rafael, and Susan Rogers. "Exodus Within an Expanding City: The Case of Houston's Historic African-American Communities." *Urban Design International* 18, no. 1 (2012): 24–42.

Lötsher, Leinhard. "Shrinking East German Cities?" *Geographia Polonica* 78, no. 1 (2005): 79–98.

Loukaitou-Sideris, Anastasia, and Tridip Banerjee. *Urban Design Downtown: Poetics and Politics of Form.* Berkeley: University of California Press, 1998.

Low, Setha. "How Private Interests Take Over Public Space." In *The Politics of Public Space*, ed. Setha Low and Neil Smith, 81–104. New York: Routledge, 2006.

Luescher, Andreas, and Sujata Shetty. "An Introductory Review to the Special Issue: Shrinking Cities and Towns: Challenge and Responses." *Urban Design International* 18, no. 1 (2013): 1–5.

Lynch, Casey. " 'Vote with Your Feet': Neoliberalism, the Democratic Nation-State, and Utopian Enclave Libertarianism." *Political Geography* 59 (2017): 82–91.

Ma, Lawrence, and Fulong Wu. "Restructuring the Chinese City: Diverse Processes and Reconstituted Spaces." In *Restructuring the Chinese City: Changing Society, Economy, and Space*, ed. Lawrence Ma and Fulong Wu, 10. New York: Routledge, 2005.

Mabin, Alan. "Comparative Segregation: The Origins of the Group Areas Act and Its Planning Apparatuses." *Journal of Southern African Studies* 18, no. 2 (1992): 405–429.

——. "Grounding Southern City Theory in Time and Place." In *The Routledge Handbook on Cities of the Global South*, ed. Susan Parnell and Sophie Oldfield, 21–36. London: Routledge, 2014.

Mabin, Alan, and Dan Smith. "Reconstructing South Africa's Cities? The Making of Urban Planning 1900–2000." *Planning Perspectives* 12, no. 2 (1997): 193–223.

MacCannell, Dean. "Staged Authenticity: Arrangements of Social Space in Tourist Settings." *American Journal of Sociology* 79, no. 3 (1973): 589–603.

——. *The Tourist: A New Theory of the Leisure Class.* London: Macmillan, 1976.

——. *Empty Meeting Grounds.* New York: Routledge, 1992.

MacClaran, Andrew. "Masters of Space: The Property Development Sector." In *Making Space: Property Development and Urban Planning*, ed. Andrew MacClaran, 2nd ed., 7–62. New York: Routledge, 2014.

MacClaran, Andrew, and Pauline McGuirk. "Planning the City." In *Making Space: Property Development and Urban Planning*, ed. Andrew MacClaran, 2nd ed., 63–94. New York: Routledge, 2014.

Mace, Alan, Peter Hall, and Nick Gallert. "New East Manchester: Urban Renaissance or Urban Opportunism?" *European Planning Studies* 15, no. 1 (2007): 51–65.

MacLeod, Gordon. "From Urban Entrepreneurialism to a 'Revanchist City'? On the Spatial Injustices of Glasgow's Renaissance." *Antipode* 34, no. 3 (2002): 602–623.

——. "Urban Politics Reconsidered: Growth Machine to Post-Democratic City?" *Urban Studies* 48, no. 12 (2011): 2629–2660.

MacLeod, Gordon, and Kevin Ward. "Spaces of Utopia and Dystopia: Landscaping the Contemporary City." *Geografiska Annaler: Series B, Human Geography* 84, no. 3–4 (2002): 153–170.

Madden, David. "Revisiting the End of Public Space: Assembling the Public in an Urban Park." *City & Community* 9, no. 2 (2010): 187–207.

Maes, Marieke, Maarten Loopmans, and Christian Kesteloot. "Urban Shrinkage and Everyday Life in Post-Socialist Cities: Living with Diversity in Hrušov, Ostrava, Czech Republic." *Built Environment* 38, no. 2 (2012): 229–243.

Mah, Alice. *Industrial Ruination, Community, and Place: Landscapes and Legacies of Urban Decline.* Toronto: University of Toronto Press, 2012.

Mahmud, Tayyab. "Slums, Slumdogs, and Resistance." *American University Journal of* Gender Social Policy *and* Law 18, no. 3 (2010): 685–710.

——. " 'Surplus Humanity' and the Margins of Legality: Slums, Slumdogs, and Accumulation by Dispossession." *Chapman Law Review* 4, no. 1 (2010): 1–73.

Malaquais, Dominique. "Anti-Teleology: Re-mapping the Imag(in)ed City." In *The African Cities Reader II: Mobilities and Fixtures*, ed. Ntone Edjabe and Edgar Pieterse, 7–23. Cape Town: African Centre for Cities, 2010.

Mallach, Alan. "What We Talk About When We Talk About Shrinking Cities: The Ambiguity of Discourse and Policy Response in the United States." *Cities* 69 (2017): 109–115.

Mallach, Alan, Annegret Haase, and Keiro Hattori. "The Shrinking City in Comparative Perspective: Contrasting Dynamics and Responses to Urban Shrinkage." *Cities* 69 (2017): 102–108.

Malone, Patrick, ed. *City, Capital and Water.* London: Routledge, 1996.

Marcinkoski, Christopher. *The City That Never Was: Reconsidering the Speculative Nature of Contemporary Urbanization.* New York: Princeton Architectural Press, 2016.

Marcuse, Peter, and Ronald van Kempen. "Conclusion: A Changed Spatial Order." In *Globalizing Cities: A New Spatial Order?*, ed. Peter Marcuse and Ronald van Kempen, 249–275. London: Blackwell, 2000.

——, eds. *Globalizing Cities: A New Spatial Order?* London: Blackwell, 2000.

——. "Introduction." In *Globalizing Cities: a New Spatial Order?*, ed. Peter Marcuse and Ronald van Kempen, 1–21. London: Blackwell, 2000.

Marlow, Susan, Scott Taylor, and Amanda Thompson. "Informality and Formality in Medium-Sized Companies: Contestation and Synchronization." *British Journal of Management* 21, no. 4 (2010): 954–966.

Marr, Stephen. "Worlding and Wilding: Lagos and Detroit as Global Cities." *Race & Class* 57, no. 4 (2016): 3–21.

Marsh, Ben. "Continuity and Decline in the Anthracite Towns of Pennsylvania." *Annals of the Association of American Geographers* 77, no. 3 (1987): 337–352.

Marshall, Richard. "Contemporary Urban Space-Making at the Water's Edge." In *Waterfronts in Post-Industrial Cities*, ed. Richard Marshall, 3–14. New York: Spon, 2001.

——. *Emerging Urbanity: Global Urban Projects in the Asia Pacific Rim.* New York: Spon, 2003.

Martinez, Javier. "Selling Avant-Garde: How Antwerp Became a Fashion Capital (1990–2002)." *Urban Studies* 44, no. 2 (2007): 2449–2464.

Martinez, Richardo, Tim Bunnell, and Michelle Acuto. "Productive Tensions? The 'City' Across Geographies of Planetary Urbanization and the Urban Age." *Urban Geography*, October 20, 2020, https://doi.org/10.1080/02723638.2020.1835128.

Martinez-Alier, Joan. "Socially Sustainable Economic De-growth." *Development and Change* 40, no. 6 (2009): 1099–1119.

Martinez-Fernandez, Cristina. "De-industrialization and De-innovation in Shrinking Cities." In *Shrinking Cities South/North*, ed. Ivonne Audirac and Jesuś Arroyo Alejandre, 51–68. Tallahassee: Florida State University, 2010.

Martinez-Fernandez, Cristina, Ivonne Audirac, Sylvie Fol, and Emmanuèle Cunningham Sabot. "Shrinking Cities: Urban Challenges of Globalization." *International Journal of Urban and Regional Research* 36, no. 2 (2012): 213–225.

Martinez-Fernandez, Cristina, Tamara Weyman, Sylvie Fol, Ivonne Audirac, Emmanuèle Cunningham Sabot, Thorsten Wiechmann, and Hiroshi Yahagi. "Shrinking Cities in Australia, Japan, Europe and the USA: From a Global Process to Local Policy Responses." *Progress in Planning* 105 (2016): 1–48.

Martinez-Fernandez, Cristina, Chung-Tong Wu, Laura Schatz, Nobuhisa Taira, and José Vargas-Hernandez. "The Shrinking Mining City: Urban Dynamics and Contested Territory." *International Journal of Urban and Regional Research* 36, no. 2 (2012): 245–260.

Marvin, Simon, and Steven Graham. *Splintering Urbanism: Networked Infrastructures, Technological Mobilities and the Urban Condition*. New York: Routledge, 2001.

Marx, Benjamin, Thomas Stoker, and Tavneet Suri. "The Economics of Slums in the Developing World." *Journal of Economic Perspectives* 27, no. 4 (2013): 187–210.

Marx, Colin, and Emily Kelling. "Knowing Urban Informalities." *Urban Studies* 56, no. 3 (2019): 494–509.

Massey, Doreen. "Cities in the World." In *City Worlds*, ed. Doreen Massey, John Allen, and Steve Pile, 93–150. London: Routledge, 1999.

——. *For Space*. Thousand Oaks, CA: Sage, 2005.

Mattssen, Helena. "Staging a Milieu." In *Foucault, Biopolitics, and Governmentality*, ed. Jakob Nilsson and Sven-Olov Wallenstein, 123–132. Stockholm: Södertörn Philosophical Studies, 2013.

Mayer, Heike, and Paul Knox. "Slow Cities: Sustainable Places in a Fast World." *Journal of Urban Affairs* 28, no. 4 (2006): 321–334.

——. "Small-Town Sustainability: Prospects in the Second Modernity." *European Planning Studies* 18, no. 10 (2010): 1545–1565.

Mayne, Alan. *Slums: The History of a Global Injustice*. London: Reaktion, 2017.

Mazlish, Bruce. "A Tale of Two Enclosures: Self and Society as a Setting for Utopias." *Theory, Culture & Society* 20, no. 1 (2003): 43–60.

Mbaye, Jenny, and Cecilia Dinardi. "Ins and Outs of the Cultural Polis: Informality, Culture and Governance in the Global South." *Urban Studies* 56, no. 3 (2009): 578–593.

Mbembé, Achille, and Sarah Nuttall. "Writing the World from an African Metropolis." *Public Culture* 16, no. 3 (2004): 347–372.

Mbembé, Achille, and Janet Roitman. "Figures of the Subject in Times of Crisis." *Public Culture* 7, no. 2 (1995): 323–352.

McCann, Eugene. "Collaborative Visioning or Urban Planning as Therapy? The Politics of Public–Private Policy Making." *Professional Geographer* 53, no. 2 (2001): 207–218.

——. "The Urban as an Object of Study in Global Cities Literatures: Representational Practices and Conceptions of Place and Scale." In *Geographies of Power: Placing Scale*, ed. Andrew Herod and Melissa Wright, 61–84. (Cambridge, MA: Blackwell, 2002.

——. "Urban Political Economy Beyond the 'Global City'." *Urban Studies* 41, no. 12 (2004): 2315–2334.

——. "Urban Policy Mobilities and Global Circuits of Knowledge: Toward a Research Agenda." *Annals of the Association of American Geographers 101*, no. 1 (2011): 107–130.

McCann, Eugene, and Kevin Ward. "Introduction. Urban Assemblages: Territories, Relations, Practices, and Power." In *Mobile Urbanism*, ed. Eugene McCann and Kevin Ward, xiii–xxxv. Minneapolis: University of Minnesota Press, 2011.

McCarthy, John. "Regeneration of Cultural Quarters: Public Art for Place Image or Place Identity?" *Journal of Urban Design* 11, no. 2 (2006): 243–262.

McCarthy, Mary. *Venice Observed*. New York: Harcourt Brace, 1963.

McFarlane, Colin. "Sanitation in Mumbai's Informal Settlements: State, 'Slum,' and Infrastructure." *Environment and Planning A* 40, no. 1 (2008): 88–107.

——. "Urban Shadows: Materiality, the 'Southern City' and Urban Theory." *Geography Compass* 2, no. 2 (2008): 340–358.

——. "Thinking with and Beyond the Informal–Formal Relation in Urban Thought." *Urban Studies* 56, no. 3 (2009): 620–623.

——. "The Comparative City: Knowledge, Learning, Urbanism." *International Journal of Urban and Regional Research* 34, no. 4 (2010): 725–742.

——. "Assemblage and Critical Urbanism." *City* 15, no. 2 (2011): 204–224.

——. "The City as Assemblage: Dwelling and Urban Space." *Environment and Planning D: Society and Space* 29, no. 4 (2011): 649–671.

——. "The Entrepreneurial Slum: Civil Society, Mobility and the Co-production of Urban Development." *Urban Studies* 49, no. 13 (2012): 2795–2816.

——. "Rethinking Informality: Politics, Crisis, and the City." *Planning Theory and Practice* 13, no. 1 (2012): 89–108.

McFarlane, Colin, and Michael Waibel. "Introduction: The Formal–Informal Divide in Context." In *Urban Informalities: Reflections on the Formal and Informal*, ed. Colin McFarlane and Michael Waibel, 1–12. New York: Routledge, 2012.

McGee, T. G. *Urbanization in the Third World: Explorations in Search of a Theory.* London: Bell, 1971.

McGovern, Stephen. "Evolving Visions of Waterfront Development in Postindustrial Philadelphia: The Formative Role of Elite Ideologies." *Journal of Planning History* 7, no. 4 (2008): 295–326.

McGuigan, Jim. "Neo-Liberalism, Culture and Policy." *International Journal of Cultural Policy* 11, no. 3 (2005): 229–241.

McLean, Heather. "In Praise of Chaotic Research Pathways: A Feminist Response to Planetary Urbanization." *Environment and Planning D: Society and Space* 36, no. 3 (2018): 547–555.

McLennan, Gregor. "Postcolonial Critique: The Necessity of Sociology." In *Postcolonial Sociology*, ed. Julian Go, 119–144. Bingley, UK: Emerald, 2013.

McNeill, Donald. "Skyscraper Geography." *Progress in Human Geography* 29, no. 1 (2005): 41–55.

——. *Global Cities and Urban Theory.* London: Sage, 2017.

Meacham, Standish. *Regaining Paradise: Englishness and the Early Garden City Movement.* New Haven, CT: Yale University Press, 1999.

Meagher, Kate. "Crisis, Informalization and the Urban Informal Sector in Sub-Saharan Africa." *Development and Change* 26, no. 2 (1995): 259–284.

——. "The Tangled Web of Associational Life: Urban Governance and the Politics of Popular Livelihoods in Nigeria." *Urban Forum* 21 (2010): 299–313.

Medd, Will, and Simon Marvin. "From the Politics of Urgency to the Governance of Preparedness: A Research Agenda on Urban Vulnerability." *Journal of Contingencies and Crisis Management* 13, no. 2 (2005): 44–49.

Meegan, Richard, Patricia Kennett, Gerwyn Jones, and Jacqui Croft. "Global Economic Crisis, Austerity and Neoliberal Urban Governance in England." *Cambridge Journal of Regions, Economy and Society* 7, no. 1 (2014): 137–153.

Mehrotra, Rahul. "Negotiating the Static and Kinetic Cities: The Emergent Urbanism of Mumbai." In *Other Cities, Other Worlds: Urban Imaginaries in a Globalizing* Age, ed. Andreas Huyysen, 205–218. Durham, NC: Duke University Press, 2008.

——. "Forward." In *Rethinking the Informal City: Critical Perspectives from Latin America*, ed. Filipe Hernández, Peter Kellett, and Lea Allen, xiii–xiv. New York: Berghahn, 2010.

Mele, Christopher. *Selling the Lower East Side: Culture, Real Estate, and Resistance in New York City.* Minneapolis: University of Minnesota Press, 2000.

Melo, João Marcelo. "Aesthetics and Ethics in City of God: Content Fails, Form Talks." *Third Text* 18, no. 5 (2004): 475–481.

Merrifield, Andy. "The Urban Question Under Planetary Urbanization." *International Journal of Urban and Regional Research* 37, no. 3 (2013): 909–922.

——. "Planetary Urbanisation: *Une affaire de perception*." *Urban Geography* 39, no. 10 (2018): 1603–1607.

Meschkank, Julia. "Investigations Into Slum Tourism in Mumbai: Poverty Tourism and the Tensions Between Different Constructions of Reality." *GeoJournal* 76, no. 1 (2011): 47–62.

Methan, Kevin. *Tourism in Global Society: Place, Culture, and Consumption.* New York: Palgrave, 2001.

Mikulić, Davorka, and Lidija Petrić. "Can Culture and Tourism Be the Foothold of Urban Regeneration? A Croatian Case Study." *Tourism* 62, no. 4 (2014): 377–395.

Miles, Malcolm. "Interruptions: Testing the Rhetoric of Culturally Led Urban Development." *Urban Studies* 42, no. 5–6 (2005): 889–911.

Millington, Nate. "Post-Industrial Imaginaries: Nature, Representation and Ruin in Detroit, Michigan." *International Journal of Urban and Regional Research* 37, no. 1 (2013): 279–296.

Millsbaugh, Martin. "Waterfronts as Catalysts for City Renewal." In *Waterfronts in Post-Industrial Cities*, ed. Richard Marshall, 74–85. New York: Spon, 2001.

Minca, Claudio. "The Return of the Camp." *Progress in Human Geography* 29 (2004): 405–412.

Minton, Anna. *The Privatisation of Public Space: What Kind of World Are We Building?* London: Royal Institution of Chartered Surveyors, 2006.

Misztal, Barbara. *Informality: Social Theory and Contemporary Practice.* New York: Routledge, 2000.

Mitchell, Don. "The End of Public Space? People's Park, Definitions of the Public Democracy." *Annals of the Association of American Geographers* 85, no. 1 (1995): 108–133.

——. "The S.U.V. Model of Citizenship: Floating Bubbles, Buffer Zones and the Rise of the Purely Atomic Individual." *Political Geography* 24, no. 1 (2004): 77–100.

Mitchell, Jerry. "Business Improvement Districts and the 'New' Revitalization of Downtown." *Economic Development Quarterly* 15, no. 2 (2001): 115–123.

Mitchell, Katharyne. "Transnationalism, Neo-Liberalism, and the Rise of the Shadow State." *Economy & Society* 30, no. 2 (2001): 165–189.

Mitchell, Timothy. "Fixing the Economy." *Cultural Studies* 12, no. 1 (1998): 80–101.

——. "The Stage of Modernity." In *Questions of Modernity*, ed. Timothy Mitchell, 1–34. Minneapolis: University of Minnesota Press, 2000.

——. "Dreamland." In *Evil Paradises: Dreamworlds of Neoliberalism*, ed. Mike Davis and Dan Monk, 1–33. New York: New Press, 2007.

——. "The Properties of Markets." In *Do Economists Make Markets? On the Performativity of Economics*, ed. Donald MacKenzie, Fabian Muniesa, and Lucia Siu, 244–275. Princeton, NJ: Princeton University Press, 2007.

Molotch, Harvey. "Place in Product." *International Journal of Urban and Regional Research* 26, no. 4 (2002): 665–688.

Morckelm, Victoria. "Why the Flint, Michigan, USA Water Crisis is an Urban Planning Failure." *Cities* 62 (2017): 23–27.

Morley, David, and Kevin Robins. *Spaces of Identity: Global Media, Electronic Landscapes and Cultural Boundaries.* New York: Routledge, 1995.

Moroni, Stefano, and David Emanuel Andersson. "Introduction: Private Enterprise and the Future of Urban Planning." In *Cities and Private Planning: Property Rights, Entrepreneurship and Transaction Costs*, ed. David Emanuel Andersson and Stefano Moroni, 1–16. Northampton, MA: Edward Elgar, 2014.

Moser, Sarah, Marian Swain, and Mohammed Alkhabbaz. "King Abdullah Economic City: Engineering Saudi Arabia's Post-Oil Future." *Cities* 45 (2015): 71–80.

Moss, Timothy. "Utilities, Land-Use Change and Urban Development: Brownfield Sites as 'Cold-Spots' of Infrastructure Networks in Berlin." *Environment and Planning A* 35, no. 3 (2003): 511–529.

——. "'Cold Spots' of Urban Infrastructure: 'Shrinking' Process in Eastern Germany and the Modern Infrastructure Ideal." *International Journal of Urban and Regional Research* 32, no. 2 (2007): 436–451.

Moudon, Anne. "Proof of Goodness: A Substantive Basis for New Urbanism." *Places* 13, no. 2 (2000): 38–43.

Mould, Oli. "A Limitless Urban Theory? A Response to Scott and Storper's 'The Nature of Cities: The Scope and Limits of Urban Theory.' " *International Journal of Urban and Regional Research* 39, no. 1 (2015): 157–163.

Moxey, Keith. "Is Modernity Multiple?" *Revista de História da Arte* 10 (2012): 50–57.

Mozingo, Louise. *Pastoral Capitalism: A History of Corporate Suburban Landscapes*. Cambridge, MA: MIT Press, 2011.

Müller, Bernhard, and Stefan Siedentop. "Growth and Shrinkage in Germany: Trends, Perspectives and Challenges for Spatial Planning and Development." *German Journal of Urban Studies* 44, no. 1 (2004): 14–32.

Mullins, Patrick. "Tourism Urbanization." *International Journal of Urban and Regional Research* 15, no. 3 (1991): 326–342.

Mumford, Louis. *The Story of Utopia*. New York: Boni & Liveright, 1922.

Murray, Martin J. "Revitalization and Displacement in the Inner City." In *Taming the Disorderly City: The Spatial Landscape of Johannesburg After Apartheid*, 189–224. Ithaca, NY: Cornell University Press, 2008.

——. "Fire and Ice: Unnatural Disasters and the Disposable Urban Poor in Post-Apartheid Johannesburg." *International Journal of Urban and Regional Research* 33, no. 1 (2009): 165–192.

——. "Afterword: Re-engaging with Transnational Urbanism." In *Locating Right to the City in the Global South*, ed. Tony Samara, Shenjing He, and Guo Chen, 285–310. New York: Routledge, 2013.

——. "The Quandary of Post-Public Space: New Urbanism, Melrose Arch and the Rebuilding of Johannesburg After Apartheid." *Journal of Urban Design* 18, no. 1 (2013): 119–144.

——. "'City Doubles': Re-urbanism in Africa." In *Cities and Inequalities in a Global and Neoliberal World*, ed. Faranak Miraftab, David Wilson, and Kenneth Salo, 92–109. New York: Routledge, 2015.

——. "Waterfall City (Johannesburg): Privatized Urbanism in Extremis." *Environment & Planning A* 47, no. 3 (2015): 503–520.

——. "Large-Scale, Master-Planned Redevelopment Projects in Urbanizing Africa: Frictionless Utopias for the Contemporary Urban Age." In *Mega-Urbanization in the Global South: Fast Cities and New Urban Utopias in the Postcolonial State*, ed. Ayona Datta and Abdul Shaban, 31–53. New York: Routledge, 2017.

——. *The Urbanism of Exception: Global City Building in the Twenty-First Century*. New York: Cambridge University Press, 2017.

——. "Cities on a Grand Scale: Instant Urbanism at the Start of the Twenty-First Century." In *Routledge Handbook on Urban Spaces*, ed. David Wilson, Byron Miller, Kevin Ward, and Andrew Jonas, 184–196. New York: Routledge, 2018.

——. "Ruination and Rejuvenation: Rethinking Growth and Decline Through an Inverted Telescope." *International Journal of Urban and Regional Research* 45, no. 2 (2021): 348–362.

Mustafa, Daanish. "The Production of an Urban Hazardscape in Pakistan: Modernity, Vulnerability, and the Range of Choice." *Annals of the Association of American Geographers* 95, no. 3 (2005): 556–586.

Musterd, Sako, and Alan Murie. "Making Cities Competitive: Debates and Challenges." In *Making Competitive Cities*, ed. Sako Musterd and Alan Murie, 3–16. Oxford: Wiley-Blackwell, 2010.

——. "The Idea of the Creative or Knowledge-Based City." In *Making Competitive Cities*, ed. Sako Musterd and Alan Murie, 17–32. Oxford: Wiley-Blackwell, 2010.

Myers, Garth Andrew. *Verandahs of Power: Colonialism and Space in Urban Africa*. Syracuse, NY: Syracuse University Press, 2003.

——. "Remaking the Edges: Surveillance and Flows in Sub-Saharan Africa's New Suburbs." In *The Design of Frontier Spaces: Control and Ambiguity*, ed. Carolyn Loeb and Andreas Luescher, 45–64. Burlington, VT: Ashgate, 2015.

Myers, Garth, and Martin J. Murray. "Introduction: Situating Contemporary Cities in Africa." In *Cities in Contemporary Africa*, ed. Martin J. Murray and Garth Myers, 1–26. New York: Palgrave Macmillan, 2008.

Mykhnenko, Vlad, and Ivan Turok. "East European Cities: Patterns of Growth and Decline, 1960–2005." *International Planning Studies* 13, no. 4 (2008): 311–342.

Nasr, Joe, and Mercedes Volait. "Introduction: Transporting Planning." In *Urbanism: Imported or Exported?*, ed. Joe Nasr and Mercedes Volait xi–xxxviii. Chichester, UK: Wiley-Academy, 2003.

Nastar, Maryam. "The Quest to Become a World City: Implications for Access to Water." *Cities* 41 (2014): 1–9.

Neal, Zackery. "Structural Determinism in the Interlocking World City Network." *Geographical Analysis* 44, no. 2 (2012): 162–170.

Németh, Jeremy. "Defining a Public: The Management of Privately Owned Public Space." *Urban Studies* 46, no. 1 (2009): 2463–2490.

——. "Control in the Commons: How Public Is Public Space?" *Urban Affairs Review* 48, no. 6 (2012): 814–838.

Németh, Jeremy, and Justin Hollander. "Security Zones and New York City's Shrinking Public Space." *International Journal of Urban and Regional Research* 34, no. 1 (2010): 20–34.

Németh, Jeremy, and Stephan Schmidt. "Toward a Methodology for Measuring the Security of Publicly Accessible Spaces." *Journal of the American Planning Association* 73, no. 3 (2007): 283–297.

Neuwirth, Robert. *Shadow Cities: A Billion Squatters, a New World.* New York: Routledge, 2005.

Nijman, Jan. "Comparative Urbanism—An Introduction." *Urban Geography* 28, no. 1 (2007): 1–6.

——. "A Study of Space Inside Mumbai's Slums." *Tijdschrift voor Economische en Sociale Geografie* 101, no. 1 (2010): 4–17.

Norcliffe, Glen, Keith Bassett, and Tony Hoare. "The Emergence of Postmodernism on the Urban Waterfront: Geographical Perspectives on Changing Relationships." *Journal of Transport Geography* 4, no. 2 (1996): 123–134.

Nordlund, Carl. "A Critical Comment on the Taylor Approach for Measuring World City Interlock Linkages." *Geographical Analysis* 36, no. 3 (2004): 290–296.

Norris, Andrew. "The Exemplary Exception: Philosophical and Political Decisions in Giorgio Agamben's *Homo Sacer*." In *Politics, Metaphysics and Death: Essays on Giorgio Agamben's Homo Sacer*, ed. Andrew Norris, 262–283. Durham, NC: Duke University Press, 2005.

Norton, Richard. "Feral Cities." *Naval War College Review* 56, no. 4 (2003): 97–106.

——. "Feral Cities: Problems Today, Battlefields Tomorrow?" *Marine Corps University Journal* 1, no. 1 (2010): 50–77.

Novy, Johannes, and Deike Peters. "Railway Station Mega-Projects as Public Controversies: The Case of Stuttgart 21." *Built Environment* 38, no. 1 (2012): 128–145.

Nowak, Marek, and Michał Nowosielski, eds. *Declining Cities/Developing Cities: Polish and German Perspectives.* Poznán: Instytut Zachodny, 2008.

Nuijten, Monique, and Gerhard Anders. "Corruption and the Secret of Law: An Introduction." In *Corruption and the Secret of Law: A Legal Anthropological Perspective*, ed. Monique Nuijten and Gerhard Anders, 2nd ed., 1–26. New York: Routledge, 2017.

Nuissl, Henning, and Dieter Rink. "The 'Production' of Urban Sprawl: Urban Sprawl in Eastern Germany as a Phenomenon of Post-Socialist Transformation." *Cities* 22, no. 2 (2005): 123–134.

Nuttall, Sarah, and Achille Mbembé. "A Blasé Attitude: A Response to Michael Watts." *Public Culture* 17, no. 1 (2005): 193–202.

Oakley, Susan. "Working Port or Lifestyle Port? A Preliminary Analysis of the Port Adelaide Waterfront Redevelopment." *Geographical Research* 43, no. 3 (2005): 319–326.

Obarrio, Juan. "Symposium Theory from the South." *Johannesburg Salon* 5 (2012): 5–9.

Olmedo, Clara, and Martin J. Murray. "The Formalization of Informal/Precarious Labor in Contemporary Argentina." *International Sociology* 17, no. 2 (2002): 421–443.

Olsen, Aksel. "Shrinking Cities: Fuzzy Concept or Useful Framework?" *Berkeley Planning Journal* 26, no. 1 (2013): 107–132.

Ondřej, Šerý, Svobodová Hana, Šilhan Zdeněk, and Szczyrba Zdeněk. "Shrinking of Cities in the Czech Republic and Its Reflection on Society: Case Study of Karviná City." *Geographica Pannonica* 22, no. 1 (2018): 68–80.

Ong, Aihwa. *Neoliberalism as Exception: Mutations in Citizenship and Sovereignty.* Durham, NC: Duke University Press, 2006.

——. "Hyperbuilding: Spectacle, Speculation, and the Hyperspace of Sovereignty." In *Worlding Cities: Asian Experiments and the Art of Being Global*, ed. Ananya Roy and Aihwa Ong, 205–226. Malden, MA: Wiley-Blackwell, 2011.

——. "Introduction: Worlding Cities, or the Art of being Global." In *Worlding Cities: Asian Experiments and the Art of Being Global*, ed. Ananya Roy and Aihwa Ong, 1–26. Malden, MA: Wiley-Blackwell, 2011.

Ortiz-Moya, Fernando, and Nieves Moreno. "The Incredible Shrinking Japan: Cinematic Representations of Urban Decline." *City* 20, no. 6 (2017): 880–903.

Orueta, Fernando Diaz, and Susan Fainstein. "The New Mega-Projects: Genesis and Impacts." *International Journal of Urban and Regional Research* 32, no. 4 (2008): 759–767.

Osbaldiston, Nick. "Consuming Space Slowly: Reflections on Authenticity, Place and the Self." In *Culture of the Slow: Social Deceleration in an Accelerated World*, ed. Nick Osbaldiston, 71–93. New York: Palgrave Macmillan, 2013.

Osborne, Stephen. *Public–Private Partnerships: Theory and Practice in International Perspective.* London: Routledge, 2000.

Oswalt, Philipp. "The Ephemeral." In *Urban Ecology: Detroit and Beyond*, ed. Kyong Park, 115–120. Hong Kong: Map Book Publishers, 2005.

——, ed. *Shrinking Cities: Volume 1, International Research.* Ostfildern-Ruit, Germany: Hatje Cantz, 2006.

——, ed. *Shrinking Cities: Volume 2, Interventions.* Ostfildern, Germany: Hatje Cantz, 2006.

Oswalt, Philipp, and Tim Rieniets, eds. *Atlas of Shrinking Cities.* Ostfildern, Germany: Hatje Cantz, 2006.

——. "Introduction." In *Atlas of Shrinking Cities*, 6–7. Ostfildern, Germany: Hatje Crantz, 2006.

Oswin, Natalie. "Planetary Urbanization: A View from Outside." *Environment and Planning D: Society and Space*, 36, no. 3 (2018): 540–546.

Pacione, Michael. "City Profile Dubai." *Cities* 22, no. 3 (2005): 255–265.

Page, Max. *The Creative Destruction of Manhattan, 1900–1940.* Chicago: University of Chicago Press, 1999.

Paling, Willem. "Planning a Future for Phnom Penh: Mega Projects, Aid Dependence and Disjointed Governance." *Urban Studies* 49, no. 13 (2012): 2889–2912.

Pallagst, Karina. "Das Ende der Wachstumsmaschine." *Berliner Debatte Initial* 18, no. 1 (2007): 4–13.

——. *Growth Management in the U.S.: Between Theory and Practice.* Aldershot, UK: Ashgate, 2007.

——. "Shrinking Cities: Planning Challenges from an International Perspective." In *Cities Growing Smaller*, ed. Steve Rugare and Terry Schwarz, 6–16. Cleveland: Kent State University and Cleveland Urban Design Collaborative, 2008.

——. "Shrinking Cities in the United States of America: Three Cases, Three Planning Stories." In *The Future of Shrinking Cities: Problems, Patterns and Strategies of Urban Transformation in a Global Context*, ed. Karina Pallagst, Jasmin Aber, Ivonne Audirac, Emmanuèle Cunningham Sabot, Sylvie Fol, Cristina MartinezFernandez, Moraes Sergio, Helen Mulligan, José VargasHernández, Thorsten Wiechmann, Chung-Tong Wu, and Jessica Rich, 81–88. Berkeley: Institute of Urban and Regional Development, Center for Global Metropolitan Studies, and the Shrinking Cities International Research Network, 2009.

Pallagst, Karina, Jasmin Aber, Ivonne Audirac, Emmanuèle Cunningham Sabot, Sylvie Fol, Cristina Martinez-Fernandez, Moraes Sergio, Helen Mulligan, José Vargas-Hernández, Thorsten Wiechmann, Chung-Tong Wu, and Jessica Rich, eds. *The Future of Shrinking Cities: Problems, Patterns and Strategies of Urban Transformation in a Global Context.* Berkeley: Center for Global Metropolitan Studies and the Institute of Urban and Regional Development and the Shrinking Cities International Research Network, 2009.

Pallagst, Karina, Cristina Martinez-Fernandez, and Thorsten Wiechmann. "Introduction." In *Shrinking Cities: International Perspectives and Policy Implications*, ed. Karina Pallagst, Thorsten Wiechmann, and Cristina Martinez-Fernandez, 1–13. New York: Routledge, 2014.

Pallagst, Karina, and Thorsten Wiechmann. "Shrinking Smart—Städtische Schrumpfungsprozesse in den USA." In *Jahrbuch Stadtregion 2004/05, Schwerpunkt: Schrumpfende Staedte*, ed. Norbert Gestring et al., 105–127. Wiesbaden, Germany: V. S. Verlag fuer Sozialwissenschaften, 2005.

Pallagst, Karina, Thorsten Wiechmann, and Cristina Martinez-Fernandez, eds. *Shrinking Cities: International Perspectives and Policy Implications*. New York: Routledge, 2014.

Park, Robert, Ernest Burgess, Roderick McKenzie. *The City*. Chicago: University of Chicago Press, 1925.

Parkins, Wendy, and Geoffrey Craig. *Slow Living*. New York: Berg, 2006.

Parnell, Susan, and Sophie Oldfield. " 'From the South.' " In *The Routledge Handbook on Cities of the Global South*, ed. Susan Parnell and Sophie Oldfield, 1–4. New York: Routledge, 2014.

Parnell, Susan, and Jennifer Robinson. "(Re)theorizing Cities from the Global South: Looking Beyond Neoliberalism." *Urban Geography* 33, no. 4 (2012): 593–617.

Patel, Ronak, and Thomas Burke. "Urbanization—An Emerging Humanitarian Disaster." *New England Journal of Medicine* 361 (2009): 741–743.

Peake, Linda. "On Feminism and Feminist Allies in Urban Geography." *Urban Geography* 37, no. 6 (2016): 830–838.

——. "The Twenty-First Century Quest for Feminism and the Global Urban." *International Journal of Urban and Regional Research* 40, no. 1 (2016): 219–227.

Peake, Linda, Darren Patrick, Rajyashree Reddy, Gökbörü Sarp Tanyildiz, Sue Ruddick, and Roza Tchoukaleyska. "Placing Planetary Urbanization in Other Fields of Vision." *Environment and Planning D: Society and Space* 36, no. 3 (2018): 374–386.

Pearson, Leonie, Peter Newton, and Peter Roberts, eds. *Resilient Sustainable Cities: A Future*. New York: Routledge, 2014.

Peck, Jamie. "Struggling with the Creative Class." *International Journal of Urban and Regional Research* 24, no. 4 (2005): 740–770.

——. "Review of Jason Hackworth, The Neoliberal City." *Annals of the Association of American Geographers* 97, no. 4 (2007): 806–809.

——. "Creative Moments: Working Culture, Through Municipal Socialism and Neoliberal Urbanism." In *Mobile Urbanism: Cities and Policymaking in the Global Age*, ed. Eugene McCann and Kevin Ward, 41–70. Minneapolis: University of Minnesota Press, 2011.

——. "Austerity Urbanism." *City* 16, no. 6 (2012): 626–655.

——. "Entrepreneurial Urbanism: Between Uncommon Sense and Dull Compulsion." *Geografiska Annaler: Series B, Human Geography* 96, no. 4 (2014): 396–401.

——. "Cities Beyond Compare?" *Regional Studies* 49, no. 1 (2015): 160–182.

——. "Economic Rationality Meets Celebrity Urbanology: Exploring Edward Glaeser's City." *International Journal of Urban and Regional Research* 40, no. 1 (2016): 1–30.

——. "Transatlantic City, Part 2: Late Entrepreneurialism." *Urban Studies* 54, no. 2 (2017): 327–363.

Peck, Jamie, and Nik Theodore. "Mobilizing Policy: Models, Methods, and Mutations." *Geoforum* 41, no. 2 (2010): 169–174.

Peck, Jamie, Nik Theodore, and Neil Brenner. "Neoliberal Urbanism Redux?" *International Journal of Urban Regional Research* 37, no. 3 (2013): 1091–1099.

Peck, Jamie, and Adam Tickell. "Neoliberalizing Space." *Antipode* 34, no. 3 (2002): 380–404.

Percival, Tom, and Peter Waley. "Articulating Intra-Asian Urbanism: The Production of Satellite Cities in Phnom Penh." *Urban Studies* 49, no. 13 (2012): 2872–2888.

Perlman, Janice. *The Myth of Marginality*. Berkeley: University of California Press, 1976.

——. "Six Misconceptions About Squatter Settlements." *Development* 4 (1986): 40–44.

——. *Favela: Four Decades of Living on the Edge in Rio de Janeiro*. New York: Oxford University Press, 2010.

Peters, Deike. "The Renaissance of Inner-City Railway Station Areas: A Key Element in the Contemporary Dynamics of Urban Restructuring." *Critical Planning* 15, no. 1 (2009): 162–185.

Petti, Alessandro. "Dubai Offshore Urbanism." In *Heterotopia and the City*, ed. Michiel Dehaene and Lieven De Cauter, 287–296. London: Routledge, 2008.

Pieterse, Edgar. *City Futures: Confronting the Crisis of Urban Development.* London: Zed, 2008.

——. "African Cities: Grasping the Unknowable: Coming to Grips with African Urbanisms." *Social Dynamics* 37, no. 1 (2011): 5–23.

——. "Grasping the Unknowable: Coming to Grips with African Urbanisms." In *Rogue Urbanism: Emergent African Cities*, ed. Edgar Pieterse and AbdouMaliq Simone, 19–36. Cape Town: Jacana, 2013.

Pinder, David. "In Defence of Utopian Urbanism: Imaging Cities After the End of Utopia." *Geografiska Annaler: Series B, Human Geography* 84, no. 3–4 (2002): 229–241.

Pink, Sarah. "Urban Social Movements and Small Places: Slow Cities as Sites of Activism." *City* 13 (2009): 451–465.

Pink, Sarah, and Lisa Servon. "Sensory Global Towns: An Experiential Approach to the Growth of the Slow City Movement." *Environment and Planning A* 45, no. 2 (2013): 451–466.

Piot, Charles. *Nostalgia for the Future: West Africa After the Cold War.* Chicago: University of Chicago Press, 2010.

Pithouse, Richard. "Review of Mike Davis in *Planet of Slums*." *Journal of Asian and African Studies* 43 (2008): 567–574.

Platt, Harold. *Shock Cities: The Environmental Transformation and Reform of Manchester and Chicago.* Chicago: University of Chicago Press, 2005.

Plaza, Beatriz. "The Guggenheim-Bilbao Museum Effect: A Reply to María Gomez." *International Journal of Urban and Regional Research* 23, no. 3 (1999): 589–592.

——. "Evaluating the Influence of a Large Cultural Artifact in the Attraction of Tourism: The Guggenheim-Bilbao Museum Case." *Urban Affairs Review* 36, no. 2 (2000): 264–274.

——. "The Return on Investment of the Guggenheim Museum, Bilbao." *International Journal of Urban and Regional Research* 30, no. 2 (2006): 452–467.

Ponzini, Davide. "Large-Scale Development Projects and Star Architecture in the Absence of Democratic Politics: The Case of Abu Dhabi, UAE." *Cities 28* (2011): 251–259.

——. "Branded Megaprojects and Fading Urban Structure in Contemporary Cities." In *Urban Megaprojects: A Worldwide View*, ed. Gerardo del Cerro Santamaría, 107–129. Bingley, UK: Emerald, 2013.

Ponzini, Davide, and Ugo Rossi. "Becoming a Creative City: The Entrepreneurial Mayor, Network Politics and the Promise of an Urban Renaissance." *Urban Studies* 47, no. 5 (2010): 1037–1057.

Popper, Deborah, and Frank Popper. "Small Can Be Beautiful: Coming to Terms with Decline." *Planning* 68, no. 7 (2002): 20–23.

Porter, Libby. "Informality, the Commons and the Paradoxes for Planning: Concepts and Debates for Informality and Planning." *Planning Theory & Practice* 12 (2011): 115–153.

Portes, Alejandro. "The Informal Economy and Its Paradoxes." In *The Handbook of Economic Sociology*, ed. Neil Smelser and Richard Swedberg, 426–450. Princeton, NJ: Princeton University Press, 1994.

Potts, Deborah. "Shanties, Slums, Breeze Blocks and Bricks." *City* 15, no. 6 (2011): 709–721.

Pratt, Andy. "Creative Cities: The Cultural Industries and the Creative Class." *Geografiska Annaler: Series B, Human Geography* 90, no. 2 (2008): 107–117.

——. "Formality as Exception." *Urban Studies* 56, no. 3 (2019): 612–615.

Prouty, Richard. "Buying Generic: The Generic City in Dubai." *London Consortium Static* 8 (2009): 1–8.

Pruijt, Hans. "Culture Wars, Revanchism, Moral Panics, and the Creative City: A Reconstruction of a Decline of Tolerant Public Policy: The Case of Dutch Anti-Squatting Legislation." *Urban Studies* 50, no. 6 (2013): 1114–1129.

Pukhopadhya, Partha, Marie-Hélène Zérah, and Eric Denis. "Subaltern Urbanization: Indian Insights for Urban Theory." *International Journal of Urban and Regional Research* 44, no. 4 (2020): 582–598.

Quinn, Bernadette. "Arts Festivals and the City." *Urban Studies* 42, no. 5–6 (2005): 927–943.
Rakodi, Carole. "Global Forces, Urban Change, and Urban Management in Africa." In *The Urban Challenge in Africa: Growth and Management of its Largest Cities*, ed. Carole Rakodi, 17–73. Tokyo: United Nations University Press, 1997.
——. "Order and Disorder in African Cities: Governance, Politics, and Land Development Processes." In *Under Siege: Four African Cities: Freetown, Johannesburg, Kinshasa, Lagos*, ed. Okwui Enwezor et al., 45–80. Ostfildern-Ruit, Germany: Hatje Cantz, 2002.
Ramos, Stephen. *Dubai Amplified: The Engineering of a Port Geography*. New York: Routledge, 2010.
Randall, James, and R. Geoff Ironside. "Communities on the Edge: An Economic Geography of Resource-Dependent Communities in Canada." *Canadian Geographer* 40, no. 1 (1996): 17–35.
Rao, Vyjayanthi. "Slum as Theory." *International Journal of Urban and Regional Research* 30, no. 1 (2006): 225–232.
——. "Proximate Distances: The Phenomenology of Density in Mumbai." *Built Environment* 33, no. 2 (2007): 227–248.
——. "Slum as Theory." *Editoriale Lotus* 143 (2010): 10–17.
——. "Slum as Theory: Mega-Cities and Urban Models." In *The Sage Handbook of Architectural Theory*, ed. C. Greig Crysler, Stephen Cairns, and Hilde Heynen, 671–686. Thousand Oaks, CA: Sage, 2012.
Rapoport, Elizabeth. "Utopian Visions and Real Estate Dreams: The Eco-City Past, Present and Future." *Geography Compass* 8, no. 2 (2014): 137–149.
Rappaport, Amos. "Spontaneous Settlements as Vernacular Design." In *Spontaneous Shelter: International Perspectives and Prospects*, ed. Carl Patton, 51–77. Philadelphia: Temple University Press, 1988.
Ratiu, Dan Eugen. "Creative Cities and/or Sustainable Cities: Discourses and Practices." *City, Culture and Society* 4 (2013): 125–135.
Read, Stephen. "The Form of the Future." In *Future City*, ed. Stephen Read, Jürgen Rosemann, and Job van Eldijk, 3–17. Londonk: Spon, 2005.
Reckien, Diana, and Cristina Martinez-Fernandez. "Why Do Cities Shrink?" *European Planning Studies* 19, no. 8 (2011): 1375–1397.
Reddy, Rajyashree. "The Urban Under Erasure: Towards a Postcolonial Critique of Planetary Urbanization." *Environment and Planning D: Society and Space* 36, no. 3 (2018): 529–539.
Reese, Laura. "Economic Versus Natural Disasters: If Detroit Had a Hurricane." *Economic Development Quarterly* 20, no. 3 (2006): 219–231.
Reher, David. "The Demographic Transition Revisited as a Global Process." *Population, Space and Place* 10, no. 1 (2004): 19–41.
Rehfeld, Dieter. "Disintegration and Reintegration of Production Clusters in the Ruhr Area." In *The Rise of the Rustbelt: Revitalizing Older Industrial Regions*, ed. Philip Cooke, 85–102. London: UCL Press, 1995.
Reis, José, Elisabete Silva, and Paulo Pinho. "Measuring Space: A Review of Spatial Metrics for Urban Growth and Shrinkage." In *The Routledge Handbook of Planning Research Methods*, ed. Elisabete Silva, Patsey Healey, Neil Harris, and Pieter Van den Broeck, 279–292. New York: Routledge, 2014.
——. "Spatial Metrics to Study Urban Patterns in Growing and Shrinking Cities." *Urban Geography* 37, no. 2 (2016): 246–271.
Reisz, Todd. "Making Dubai: A Process in Crisis." [In "Post-Traumatic Urbanism" Special Issue] *Architectural Design* 80, no. 5 (2010): 38–43.
Relph, Edward. *Place and Placelessness*. London: Pion, 1976.
Richards, Greg, and Julie Wilson. "The Impact of Cultural Events on City Image: Rotterdam, Cultural Capital of Europe 2001." *Urban Studies* 41, no. 10 (2004): 1931–1951.
Rieniets, Tim. "Shrinking Cities: Causes and Effects of Urban Population Losses in the Twentieth Century." *Nature & Culture* 4 (2009): 231–254.
Rimmer, Peter, and Howard Dick. *The City in Southeast Asia: Patterns, Processes and Policy*. Singapore: Nation University of Singapore Press, 2009.

Rink, Dieter. "Wilderness: The Nature of Urban Shrinkage? The Debate on Urban Restructuring and Restoration in Eastern Germany." *Nature and Culture* 4, no. 3 (2009): 275–292.

Rink, Dieter, Annegret Haase, Katrin Grossmann, Chris Couch, and Matthew Cocks. "From Long-Term Shrinkage to Re-growth? The Urban Development Trajectories of Liverpool and Leipzig." *Built Environment* 38, no. 2 (2012): 162–178.

Rink, Dieter, and Sigrun Kabisch. "Introduction: The Ecology of Shrinkage." *Nature and Culture* 4, no. 3 (2009): 223–230.

Ritzer, George. *The Globalization of Nothing.* New York: Pine Forge, 2004.

Rizzo, Agatino. "Metro Doha." *Cities* 31 (2013): 533–543.

Roberts, Bryan. "Globalization and Latin American Cities." *International Journal of Urban and Regional Research* 29, no. 1 (2006): 110–123.

Roberts, Bryan, Ruth Finnegan, and Duncan Gallie. "Introduction." In *New Approaches to Economic Life*, ed. Bryan Roberts, Ruth Finnegan, and Duncan Gallie, 8–12. Manchester, UK: Manchester University Press, 1985.

Robinson, Jennifer. "Global and World Cities: A View from Off the Map." *International Journal of Urban and Regional Research* 26, no. 3 (2002): 531–554.

——. "Postcolonialising Geography: Tactics and Pitfalls." *Singapore Journal of Tropical Geography* 24, no. 3 (2003): 273–289.

——. "In the Tracks of Comparative Urbanism: Difference, Urban Modernity and the Primitive." *Urban Geography* 25, no. 8 (2004): 709–723.

——. "Urban Geography: World Cities, or a World of Cities." *Progress in Human Geography* 29, no. 6 (2005): 757–765.

——. *Ordinary Cities: Between Modernity and Development.* New York: Routledge, 2006.

——. "Developing Ordinary Cities: City Visioning Processes in Durban and Johannesburg." *Environment and Planning A* 40, no. 1 (2008): 74–87.

——. "Cities in a World of Cities: The Comparative Gesture." *International Journal of Urban and Regional Research* 35, no. 1 (2010): 1–23.

——. "Living in Dystopia: Past, Present and Future." In *Noir Urbanisms*, ed. Gyan Prakash, 218–240. Princeton, NJ: Princeton University Press, 2010.

——. "Comparisons: Cosmopolitan or Colonial?" *Singapore Journal of Tropical Geography* 32, no. 2 (2011): 125–140.

——. "The Spaces of Circulating Knowledge: City Strategies and Global Urban Governmentality." In *Mobile Urbanism: Cities and Policy-Making in the Global Age*, ed. Eugene McCann and Kevin Ward, 15–40. Minneapolis: University of Minnesota Press, 2011.

——. "'Arriving at' Urban Policies/the Urban: Traces of Elsewhere in Making Urban Futures." In *Critical Mobilities*, ed. Ola Söderström et al., 1–28. London: Routledge, 2013.

——. "The Urban Now: Theorizing Cities Beyond the New." *European Journal of Cultural Studies* 16, no. 6 (2013): 659–677.

——. "Thinking Cities Through Elsewhere: Comparative Tactics for a More Global Urban Studies." *Progress in Human Geography* 40, no. 1 (2015): 3–29.

——. "Comparative Urbanism: New Geographies and Cultures of Theorizing the Urban." *International Journal of Urban and Regional Research* 40, no. 1 (2016): 187–197.

——. "Theorizing the Global Urban with 'Global and World Cities Research': Beyond Cities and Synechdoche." *Dialogues in Human Geography* 6, no. 3 (2016): 268–272.

Robinson, Jennifer, and Ananya Roy. "Debate on Global Urbanisms and the Nature of Urban Theory." *International Journal of Urban and Regional Research* 40, no. 1 (2016): 181–186.

Rodgers, Dennis. "The State as a Gang: Conceptualising the Governmentality of Violence in Contemporary Nicaragua." *Critique of Anthropology* 26, no. 3 (2006): 315–330.

——. "Slum Wars of the 21st Century: Gangs, Mano Dura, and the New Urban Geography of Conflict in Central America." *Development and Change* 40, no. 5 (2009): 949–976.

——. "An Illness Called Managua: 'Extraordinary' Urbanization and 'Mal-development' in Nicaragua." In *Urban Theory Beyond the West*, ed. Tim Edensor and Mark Jayne, 121–135. New York: Routledge, 2012.

Rodriguez, Arantxa, and Elena Martinez. "Restructuring Cities: Miracles and Mirages in Urban Revitalization in Bilbao." In *The Globalized City: Economic Restructuring and Social Polarization in European Cities*, ed. Frank Moulaert, Arantxa Rodriguez, and Erik Swyngedouw, 181–208. Oxford: Oxford University Press, 2003.

Rogers, Dallas, and Chris Gibson. "Unsolicited Urbanism: Development Monopolies, Regulatory-Technical Fixes and Planning-as-Deal-Making." *Environment and Planning A: Society and Space* 50, no. 3 (2020): 1–20.

Rothman, Hal, and Mike Davis. "Introduction: The Many Faces of Las Vegas." In *The Grit Beneath the Glitter: Tales from the Real Las Vegas*, ed. Hal Rothman and Mike Davis, 1–14. Berkeley: University of California Press, 2002.

Rousseau, Max. "Re-imaging the City Centre for the Middle Classes: Regeneration, Gentrification and Symbolic Policies in 'Loser Cities'." *International Journal of Urban and Regional Research* 33, no. 3 (2009): 770–788.

Roy, Ananya. "The Gentleman's City: Urban Informality in the Calcutta of New Communism." In *Urban Informality: Transnational Perspectives from the Middle East, Latin America, and South Asia*, ed. Ananya Roy and Nezar AlSayyad, 147–170. Lanham, MD: Lexington, 2004.

——. "Urban Informality: Toward an Epistemology of Planning." *Journal of the American Planning Association* 71, no. 2 (2005): 147–158.

——. "The 21st-Century Metropolis: New Geographies of Theory." *Regional Studies* 43, no. 6 (2009): 819–830.

——. "Why India Cannot Plan Its Cities: Informality, Insurgence and the Idiom of Urbanization." *Planning Theory* 8, no. 1 (2009): 76–87.

——. "Conclusion: Postcolonial Urbanism: Speed, Hysteria, Mass Dreams." In *Worlding Cities: Asian Experiments and the Art of Being Global*, ed. Ananya Roy and Aihwa Ong, 307–335. Malden, MA: Wiley-Blackwell, 2011.

——. "Slumdog Cities: Rethinking Subaltern Urbanism." *International Journal of Urban and Regional Research* 35, no. 4 (2011): 223–238.

——. "Worlding the South: Towards a Post-Colonial Urban Theory." In *The Routledge Handbook on Cities of the Global South*, ed. Sue Parnell and Sophie Oldfield, 9–20. New York: Routledge, 2014.

——. "Who's Afraid of Postcolonial Theory?" *International Journal of Urban and Regional Research* 40, no. 1 (2015): 200–209.

——. "What Is Urban About Critical Urban Theory?" *Urban Geography* 37, no. 6 (2016): 810–823.

Ruby, Ilka, and Andreas Ruby. "Forward." In *Urban Transformation*, ed. Ilka Ruby and Andreas Ruby, 10–13. Berlin: Ruby, 2008.

Ruddick, Sue, Linda Peake, Gökbörü Tanyildiz, and Darren Patrick. "Planetary Urbanization: An Urban Theory for Our Time?" *Environment and Planning D: Society and Space* 36, no. 3 (2018): 387–404.

Rugkhapan, Napong, and Martin J. Murray. "Songdo IBD (International Business District): Experimental Prototype for the City of Tomorrow?" *International Planning Studies* 24, no. 3–4 (2019): 272–292.

Rundell, John. "Imagining Cities, Others: Strangers, Contingency and Fear." *Thesis Eleven* 121, no. 1 (2014): 9–22.

Rutheiser, Charles. *Imagineering Atlanta: The Politics of Place in the City of Dreams*. New York: Verso, 1996.

Ryan, Brent. "The Restructuring of Detroit: City Block Form Change in a Shrinking City, 1900–2000." *Urban Design International* 13 (2008): 156–168.

——. *Design After Decline: How America Rebuilds Shrinking Cities*. Philadelphia: University of Pennsylvania Press, 2012.

Rybczynski, Witold, and Peter Linneman. "How to Save Our Shrinking Cities." *Public Interest* 135 (1999): 30–44.

Ryberg-Webster, Stephanie, and J. Rosie Tighe. "The Legacy of Legacy Cities." In *Legacy Cities: Continuity and Change Amid Decline and Revival*, ed. Stephanie Ryberg-Webster and J. Rosie Tighe, 3–17. Pittsburgh: University of Pittsburgh Press, 2019.

Sack, Robert David. *Place, Modernity, and the Consumer's World: A Relational Framework for Geographical Analysis*. Baltimore: Johns Hopkins University Press, 1992.

Sagar, Rahul. "Are Charter Cities Legitimate?" *Journal of Political Philosophy* 24, no. 4 (2016): 509–529.

Salmon, Scott. "Imagineering the Inner City? Landscapes of Pleasure and the Commodification of Cultural Spectacle in the Postmodern City." In *Popular Culture: Production and Consumption*, ed. C. Lee Harrington and Denise Bielby, 106–120. Malden, MA: Blackwell, 2001.

Sánchez-Moral, Simón, Ricardo Méndez, and José Prada-Trigo. "Resurgent Cities: Local Strategies and Institutional Networks to Counteract Shrinkage in Avilés (Spain)." *European Planning Studies* 23, no. 1 (2015): 33–52.

Sandercock, Leonie, and Kim Dovey. "Pleasure, Politics and the 'Public Interest': Melbourne's Riverscape Revitalization." *Journal of the American Planning Association* 68, no. 2 (2002): 151–164.

Sanyal, Romola. "Slum Tours as Politics: Global Urbanism and Representations of Poverty." *International Political Sociology* 9, no. 1 (2015): 93–96.

Sapotichne, Joshua. "Rhetorical Strategy in Stadium Development Politics." *City, Culture and Society* 3, no. 1 (2012): 169–180.

Sassen, Saskia. "The Urban Complex in the World Economy." *International Social Science Journal*, 139 (1994): 42–63.

——. *Globalization and Its Discontents: Essays on the New Mobility of People and Money*. New York: New Press, 1999.

——, ed. *Cities and Their Cross-Border Networks*. New York: Routledge, 2000.

——. "New Frontiers Facing Urban Sociology at the Millennium." *British Journal of Sociology* 51, no. 1 (2000): 143–159.

——. *Global Cities: New York, London, Tokyo*. 2nd ed. Princeton, NJ: Princeton University Press, 2001.

——. "Reading the City in a Global Digital Age: Between Topographical Representation and Spatialized Power Projects." In *Future City*, ed. Stephen Read, Jürgen Rosemann and Job van Eldijk, 145–155. London: Spon, 2005.

——. "Introduction: Deciphering the Global." In *Deciphering the Global, Its Scales, Spaces, and Subjects*, ed. Saskia Sassen, 1–18. New York: Routledge, 2007.

——. "Cityness." In *Urban Transformation*, ed. Ilka Ruby and Andreas Ruby, 84–87. Berlin: Ruby, 2008.

——. *Cities in a World Economy*. 4th ed. Thousand Oaks, CA: Pine Forge, 2011.

——. "Does the City Have Speech?" *Public Culture* 25, no. 2 (2013): 209–221.

——. *Expulsions: Brutality and Complexity in the Global Economy*. Cambridge, MA: Belknap, 2014.

Sassen, Saskia, and Frank Roost. "The City: Strategic Site for the Global Entertainment Industry." In *The Tourist City*, ed. Susan Fainstein and Dennis Judd, 143–154. New Haven, CT: Yale University Press, 1999.

Savitch, H. V., and Paul Kantor. *Cities in the International Marketplace: The Political Economy of Urban Development in North America and Western Europe*. Princeton, NJ: Princeton University Press, 2002.

Sawyer, Lindsay. "Piecemeal Urbanisation at the Peripheries of Lagos." *African Studies* 7, no. 2 (2014): 271–289.

Schatz, Edward. "What Capital Cities Say About State and Nation Building." *Nationalism and Ethnic Politics* 9, no. 4 (2004): 111–140.

Scherr, Richard. "The Synthetic City: Excursions Into the Real–Not Real." *Places* 18, no. 2 (2008): 6–15.

Schilling, John, and Alan Mallach. *Cities in Transition: A Guide for Practicing Planners*. Washington, DC: American Planning Association Press, 2012.

Schilling, Joseph, and John Logan. "Greening the Rust Belt: A Green Infrastructure Model for Rightsizing America's Shrinking Cities." *Journal of the American Planning Association* 74, no. 4 (2008): 451–466.

Schindler, Seth. "Governing the Twenty-First Century Metropolis and Transforming Territory." *Territory, Politics, Governance* 3, no. 1 (2015): 7–26.

——. "Detroit After Bankruptcy: A Case of Degrowth Machine Politics." *Urban Studies* 53, no. 4 (2016): 818–836.

——. "Towards a Paradigm of Southern Urbanism." *City* 21, no. 1 (2017): 47–64.

Schmid, Christian. "Networks, Borders, and Difference: Towards a Theory of the Urban." In *Implosions/ Explosions: Towards a Study of Planetary Urbanization*, ed. Neil Brenner, 67–79. Berlin: Jovis, 2014.

——. "Journeys Through Planetary Urbanization: Decentering Perspectives on the Urban." *Environment and Planning D: Society and Space* 36, no. 3 (2018): 591–610.

Schmidt, Stephan. "Sprawl Without Growth in Eastern Germany." *Urban Geography* 32, no. 1 (2011): 105–128.

Schmidt, Volker. "Multiple Modernities or Varieties of Modernity?" *Current Sociology* 54, no. 1 (2006): 77–97.

Schwartz, Amy Ellen, and Ingrid Gould Ellen. "No Easy Answers: Cautionary Notes for Competitive Cities." *Brookings Review* 18, no. 3 (2000): 44–47.

Schwartz, Vanessa. *Spectacular Realities: Early Mass Culture in Fin-de-Siecle Paris.* Berkeley: University of California Press, 1998.

Schwenkel, Christina. "Post-Socialist Affect: Ruination and Reconstruction of the Nation in Urban Vietnam." *Cultural Anthropology* 28, no. 2 (2013): 252–277.

Sclar, Elliott, Pietro Garau, and Gabriella Carolini. "The 21st Century Health Challenge of Slums and Cities." *Lancet* 365, no. 9462 (2005): 901–903.

Scobey, David. *Empire City: The Making and Meaning of the New York City Landscape.* Philadelphia: Temple University Press, 2003.

Scott, Allen J. "The Cultural Economy of Cities." *International Journal of Urban and Regional Research* 21, no. 2 (1997): 323–339.

——. *The Cultural Economy of Cities: Essays on the Geography of Image Producing Industries.* London: Sage, 2000.

——. "Cultural-Products Industries and Urban Economic Development: Prospects for Growth and Market Contestation in a Global Context." *Urban Affairs Review* 39, no. 4 (2004): 461–490.

——. "Creative Cities: Conceptual Issues and Policy Questions." *Journal of Urban Affairs* 28, no. 1 (2006): 1–17.

——. "Capitalism and Urbanization in a New Key? The Cognitive–Cultural Dimension." *Social Forces* 85 (2007): 1465–1482.

——. *Social Economy of the Metropolis: Cognitive–Cultural Capitalism and the Global Resurgence of Cities.* Oxford: Oxford University Press, 2008.

——. "Human Capital Resources and Requirements Across the Metropolitan Hierarchy of the United States." *Journal of Economic Geography* 9 (2009): 207–226.

——. "Cultural Economy and the Creative Field of the City." *Geografiska Annaler: Series B, Human Geography* 92 (2010): 115–130.

——. "Emerging Cities of the Third Wave." *City* 15, no. 3–4 (2011): 289–321.

Scott, Allen J., John Agnew, Edward W. Soja, and Michael Storper. "Global City-Regions." In *Global City-Regions: Trends, Theory, Policy*, ed. Allan J. Scott, 11–30. Oxford: Oxford University Press, 2002.

Scott, Allen J., and Edward W. Soja. "Los Angeles: Capital of the Late 20th Century." *Environment and Planning D* 4, no. 3 (1986): 249–254.

Scott, Allen J., and Michael Storper. "The Nature of Cities: The Scope and Limits of Urban Theory." *International Journal of Urban and Regional Research* 39, no. 1 (2015): 1–16.

Scott, James C. *Weapons of the Weak: Everyday Forms of Peasant Resistance.* New Haven, CT: Yale University Press 1985.

——. "The High-Modernist City: An Experiment and a Critique." In *Seeing Like a State: How Certain Schemes to Improve the Human Condition Have Failed*, 103–146. New Haven, CT: Yale University Press, 1998.

Seabrook, Jeremy. *In the Cities of the South: Some Scenes from a Developing World.* New York: Verso, 1996.

Semmens, Jaimee, and Claire Freeman. "The Value of Cittaslow as an Approach to Local Sustainable Development: A New Zealand Perspective." *International Planning Studies*, 17, no. 4 (2012): 353–375.

Seo, J.-K. "Re-urbanisation in Regenerated Areas of Manchester and Glasgow." *Cities* 19, no. 2 (2002): 113–121.

Shahid, Yusuf, and Kaoru Nabeshima. "Creative Industries in East Asia." *Cities* 22, no. 2 (2005): 109–122.

Shatkin, Gavin. " 'Fourth World' Cities in the Global Economy: The Case of Phnom Penh, Cambodia." *International Journal of Urban and Regional Research* 22, no. 3 (1998): 378–393.

——. "Global Cities of the South: Emerging Perspectives on Growth and Inequality." *Cities* 24, no. 1 (2007): 1–15.

——. "The City and the Bottom Line: Urban Megaprojects and the Privatization of Planning in Southeast Asia." *Environment and Planning A* 40, no. 2 (2008): 383–401.

——. "Planning Privatopolis: Representation and Contestation in the Development of Urban Integrated Mega-Projects." In *Worlding Cities: Asian Experiments and the Art of Being Global*, ed. Ananya Roy and Ailwa Ong, 77–97. Malden, MA: Wiley-Blackwell, 2011.

——. "Contesting the Indian City: Global Visions and the Politics of the Local." *International Journal of Urban and Regional Research* 38, no. 1 (2014): 1–13.

——. *Cities for Profit: The Real Estate Turn in Asia's Urban Politics.* Ithaca, NY: Cornell University Press, 2017.

Shatkin, Gavin, and Sanjeev Vidyarthi. "Introduction: Contesting the Indian City: Global Visions and the Politics of the Local." In *Contesting the Indian City: Global Visions and the Politics of the Local*, ed. Gavin Shatkin, 1–38. Malden, MA: Wiley-Blackwell, 2014.

Shaw, Annapurna. "Town Planning in Postcolonial India, 1947–1965: Chandigarh Re-examined." *Urban Geography* 30, no. 8 (2009): 857–878.

Shaw, Stephen, Susan Bagwell, and Joanna Karmowska. "Ethnoscapes as Spectacle: Reimaging Multicultural Districts as New Destinations for Leisure and Tourism Consumption." *Urban Studies* 41, no. 10 (2004): 1983–2000.

Sheller, Mimi, and John Urry, eds. *Tourism Mobilities: Places to Play, Places in Play.* New York: Routledge, 2004.

Shepard, Cassim. "Montage Urbanism: Essence, Fragment, Increment." *Public Culture* 25, no. 2 (2013): 223–232.

Sheppard, Eric, Helga Leitner, and Anant Maringanti. "Provincializing Global Urbanism: A Manifesto." *Urban Geography* 34, no. 7 (2013): 893–900.

Sheppard, Eric, and Richa Nagar. "From East–West to North–South." *Antipode* 36, no. 4 (2004): 557–563.

Shields, Rob. "The Urban Question as Cargo Cult: Opportunities for a New Urban Pedagogy." *International Journal of Urban and Regional Research* 32, no. 3 (2008): 712–718.

Short, John Rennie, Lisa Benton, William Luce, and Judith Walton. "The Reconstruction of the Image of a Postindustrial City." *Annals of the Association of American Geographers* 83, no. 2 (1993): 207–224.

Short, John Rennie, Bernadette Hanlon, and Thomas Vicino. "The Decline of Inner Suburbs: The New Suburban Gothic in the United States." *Geography Compass* 1, no. 3 (2007): 641–656.

Shoshkes, Ellen. "East–West: Interactions Between the United States and Japan and Their Effect on Utopian Realism." *Journal of Planning History* 3, no. 3 (2004): 215–240.

Shwayri, Sofia. "A Model Korean Ubiquitous Eco-City? The Politics of Making Songdo." *Journal of Urban Technology* 20, no. 1 (2013): 39–55.

Sidaway, James. "Enclave Space: A New Metageography of Development?" *Area* 39, no. 3 (2007): 331–339.

Sidaway, James, Tim Bunnell, and Brenda Yeoh. "Geography and Postcolonialism." *Singapore Journal of Tropical Geography*, 24, no. 3 (2003): 269–272.

Siedentop, Stefan, and Stefan Fina. "Who Sprawls Most? Exploring the Patterns of Urban Growth Across 26 European Countries." *Environment and Planning A* 44, no. 11 (2012): 2765–2784.

Siemiatycki, Matti. "The Making of a Mega Project in the Neoliberal City: The Case of Mass Rapid Transit Infrastructure Investment in Vancouver, Canada." *City* 9, no. 1 (2005): 67–83.

Sigler, Thomas. "Relational Cities: Doha, Panama City, and Dubai as 21st Century Entrepôts." *Urban Geography* 34, no. 5 (2013): 612–633.

Silver, Jonathan. "Incremental Infrastructures: Material Improvisation and Social Collaboration Across Post-Colonial Accra." *Urban Geography* 35, no. 6 (2014): 788–804.

Silverman, Robert Mark, Li Yin, and Kelly Patterson. "Dawn of the Dead City: An Exploratory Analysis of Vacant Addresses in Buffalo, NY 2008–2010." *Journal of Urban Affairs* 35, no. 2 (2013): 131–152.

Simmel, Georg. "The Metropolis and Mental Life [1903]." In *Classic Essays on the Culture of Cities*, ed. Richard Sennett, 47–60. New York: Appleton-Century-Crofts, 1969.

Simmie, James. "Trading Places: Competitive Cities in the Global Economy." *European Planning Studies* 10, no. 2 (2002): 201–214.

Simmie, James, and Peter Wood. "Innovation and Competitive Cities in the Global Economy: Introduction to the Special Issue." *European Planning Studies* 10, no. 2 (2002): 149–151.

Simon, David. "Situating Slums." *City* 15, no. 6 (2011): 674–685.

Simone, AbdouMaliq. "Between Ghetto and Globe: Remaking Urban Life in Africa." In *Associational Life in African Cities: Popular Responses to the Urban Crisis*, ed. Arne Tostensen, Inge Tvedten, and Mariken Vaa, 46–63. Stockholm: Nordiska Afrikainstitutet, 2001.

——. "On the Worlding of African Cities." *African Studies Review* 44, no. 2 (2001): 15–42.

——. "Straddling the Divides: Remaking Associational Life in the Informal City." *International Journal of Urban and Regional Research* 25, no. 1 (2001): 102–117.

——. "The Visible and Invisible: Remaking Cities in Africa." In *Under Siege: Four African Cities: Freetown, Johannesburg, Kinshasa, Lagos*, ed. Okwui Enwezor et al., 23–44. Ostfildern-Ruit, Germany: Hatje Cantz, 2002.

——. "Resource of Intersection: Remaking Social Collaboration in Urban Africa." *Canadian Journal of African Studies* 37, no. 2–3 (2003): 513–538.

——. *For the City Yet to Come: Changing African Life in Four Cities.* Durham, NC: Duke University Press, 2004.

——. "People as Infrastructure: Intersecting Fragments in Johannesburg." *Public Culture* 16, no. 3 (2004): 407–429.

——. "Emergency Democracy and the 'Governing Composite'." *Social Text* 95, no. 26 (2) (2008): 13–33.

——. *City Life from Jakarta to Dakar: Movements at the Crossroads.* New York: Routledge, 2010.

——. "The Ineligible Majority: Urbanizing the Postcolony in Africa and Southeast Asia." *Geoforum* 42 (2011): 266–270.

——. "The Urbanity of Movement: Dynamic Frontiers in Contemporary Africa." *Journal of Planning Education and Research* 31, no. 4 (2011): 379–391.

——. "Cities of Uncertainty: Jakarta, the Urban Majority, and Inventive Political Technologies." *Theory, Culture and Society* 30 (2013): 243–263.

——. "It's Just the City After All!" *International Journal of Urban and Regional Research* 39, no. 1 (2015): 210–218.

——. "The Urban Poor and Their Ambivalent Exceptionalities: Some Notes from Jakarta." *Current Anthropology* 56, no. S11 (2015): 515–523.

——. "What You See Is Not Always What You Know: Struggles Against Re-containment and the Capacities to Remake Urban Life in Jakarta's Majority World." *South East Asia Research* 23, no. 2 (2015): 227–244.

——. "(Non)Urban Humans: Questions for a Research Agenda (the Work the Urban *Could* Do)." *International Journal of Urban and Regional Research* 44, no. 4 (2020): 755–767.

Simone, AbdouMaliq, and Vyjayanthi Rao. "Securing the Majority: Living Through Uncertainty in Jakarta." *International Journal of Urban and Regional Research* 36, no. 2 (2012): 315–335.

Simpson, Tim. "Tourist Utopias: Biopolitics and the Genealogy of the Post–World Tourist City." *Current Issues in Tourism* 19, no. 1 (2016): 27–59.

Simson, Alan. "The Post-Romantic Landscape of Telford New Town." *Landscape and Urban Planning* 52, no. 2–3 (2001): 189–197.

Sklair, Leslie. "Iconic Architecture and Capitalist Globalization." *City* 10, no. 1 (2006).

——. "Commentary: From the Consumerist/Oppressive City to the Functional/Emancipatory City." *Urban Studies* 46, no. 12 (2009): 2703–2711.

Smith, Andrew. " 'Borrowing' Public Space to Stage Major Events: The Greenwich Park Controversy." *Urban Studies* 51, no. 2 (2014): 247–263.

Smith, David. *Third World Cities in Global Perspective: The Political Economy of Uneven Urbanization.* Boulder, CO: Westview, 1996.

Smith, David, and Michael Timberlake. "Hierarchies of Dominance Among World Cities: A Network Approach." In *Global Networks, Linked Cities*, ed. Saskia Sassen, 117–141. New York: Routledge, 2002.

Smith, Michael Peter. *Transnational Urbanism: Locating Globalization.* Malden, MA: Blackwell, 2001.

Smith, Neil. *The New Urban Frontier: Gentrification and the Revanchist City.* London: Routledge, 1996.

——. "The Satanic Geographies of Globalization: Uneven Development in the 1990s." *Public Culture* 10, no. 1 (1997): 169–187.

——. "New Globalism, New Urbanism: Gentrification as Global Urban Strategy." *Antipode* 34, no. 3 (2002): 427–450.

Smith, Neil, Paul Caris, and Elvin Wyly. "The 'Camden Syndrome' and the Menace of Suburban Decline: Residential Disinvestment and the Discontents in Camden County, New Jersey." *Urban Affairs Review* 36, no. 4 (2001): 497–531.

Smith, Richard G. "The Ordinary City Trap." *Environment and Planning A* 45, no. 10 (2013): 2290–2304.

——. "Beyond the Global City Concept and the Myth of 'Command and Control'." *International Journal of Urban and Regional Research* 38, no. 1 (2014): 98–115.

Smyth, Herbert. *Marketing the City: The Role of Flagship Developments in Urban Regeneration.* London: Spon, 1994.

Snowden, Frank. *Naples in the Time of Cholera.* New York: Cambridge University Press, 1995.

Söderström, Ola, Till Paasche, and Francisco Klauser. "Smart Cities as Corporate Storytelling." *City* 18, no. 3 (2014): 307–320.

Soja, Edward W. "It All Comes Together in Los Angeles." In *Postmodern Geographies: The Reassertion of Space in Critical Social Theory*, 190–221. New York: Verso, 1989.

——. "Inside Exopolis: Scenes from Orange County." In *Variations on a Theme Park*, ed. Michael Sorkin, 94–122. New York: Hill & Wang, 1992.

——. *Postmetropolis: Critical Studies of Cities and Regions.* New York: Wiley-Blackwell, 2000.

Soja, Edward W., and Miguel Kanai. "The Urbanization of the World." In *The Endless City*, ed. Ricky Burdett and Deyan Sudjic, 54–68. London: Phaidon, 2007.

——. "The Urbanization of the World." In *Implosions/Explosions: Towards a Study of Planetary Urbanization*, ed. Neil Brenner, 142–159. Berlin: Jovis, 2014.

Sorensen, Andre, and Junichiro Okata, eds. *Megacities: Urban Form, Governance, and Sustainability.* New York: Springer, 2011.

Sorensen, Anthony, and Richard Day. "Libertarian Planning." *Town Planning Review* 52, no. 4 (1981): 390–402.

Sousa, Sílvia, and Paulo Pinho. "Shrinkage in Portuguese National Policy and Regional Spatial Plans: Concern or Unspoken Word?" *Journal of Spatial and Organizational Dynamics* 2, no. 4 (2014): 260–275.

——. "Planning for Shrinkage: Paradox or Paradigm." *European Planning Studies* 23, no. 1 (2015): 12–32.

Stansel, Dean. "Why Some Cities Are Growing and Others Are Shrinking." *Cato Journal* 31, no. 2 (2011): 285–303.

Steinführer, Annett, Adam Bierzynski, Katrin Großmann, Annegret Haase, and Sigrun Kabish. Nadja Kabisch, "Population Decline in Polish and Czech Cities During Post-Socialism? Looking Behind the Official Statistics." *Urban Studies* 47, no. 11 (2010): 2325–2346.

Steinführer, Annett, and Annegret Haase. "Demographic Change as Future Challenge for Cities in East Central Europe." *Geografiska Annaler: Series B, Human Geography* 89, no. 2 (2007): 183–195.

Sternberg, Ernest. "What Makes Buildings Catalytic? How Cultural Facilities Can Be Designed to Spur Surrounding Development." *Journal of Architectural and Planning Research* 19, no. 1 (2002): 30–43.

Stevens, Quintin, and Kim Dovey. "Appropriating the Spectacle: Play and Politics in a Leisure Landscape." *Journal of Urban Design* 9, no. 3 (2004): 351–365.

Stillwaggon, Eileen. *Stunted Lives, Stagnant Economies: Poverty, Disease and Underdevelopment.* New Brunswick, NJ: Rutgers University Press, 1998.

Stoker, Gerry. "Public–Private Partnerships and Urban Governance." In *Partnerships in Urban Governance: European and American Experience*, ed. Jon Pierre, 34–51. London: Macmillan, 1998.

Stokes, Charles. "A Theory of Slums." *Land Economics* 38, no. 3 (1962): 187–197.

Storper, Michael. "Governing the Large Metropolis." *Territory, Politics, Governance* 2, no. 2 (2014): 115–134.

Storper, Michael, and Allen J. Scott. "Current Debates in Urban Theory: A Critical Assessment." *Urban Studies* 53, no. 16 (2016): 1114–1136.

Stren, Richard. "Local Development and Social Diversity in the Developing World: New Challenges for Globalizing City-Regions." In *Global City-Regions: Trends, Theory, Policy*, ed. Allen J. Scott, 193–213. New York: Oxford University Press, 2001.

Streule, Monika, Ozan Karaman, Lindsay Sawyer, and Christian Schmid. "Popular Urbanization: Conceptualizing Urbanization Processes Beyond Informality." *International Journal of Urban and Regional Research* 44, no. 4 (2020): 652–672.

Strom, Elizabeth. "Let's Put on a Show: Performing Arts and Urban Revitalization in Newark, New Jersey." *Journal of Urban Affairs* 21, no. 4 (1999): 423–435.

——. "From Pork to Porcelain: Cultural Institutions and Downtown Development." *Urban Affairs Review* 38, no. 1 (2002): 3–21.

Stryjakiewicz, Tadeusz, Przemysław Ciesiółka, and Emilia Jaroszewska. "Urban Shrinkage and the Post-Socialist Transformation: The Case of Poland." *Built Environment* 38, no. 2 (2012): 196–213.

Sudjic, Deyan. *The 100-Mile City.* San Diego: Harcourt Brace, 1993.

——. *The Edifice Complex: How the Rich and Powerful—and Their Architects—Shape the World.* New York: Penguin, 2005.

Sugrue, Thomas. *The Origins of the Urban Crisis.* Princeton, NJ: Princeton University Press, 1996.

Sundaram, Ravi. "Recycling Modernity: Pirate Electronic Cultures in India." *Third Text* 13, no. 47 (1999): 59–65.

——. "Revisiting the Pirate Kingdom." *Third Text* 23, no. 3 (2009): 335–345.

——. *Pirate Modernity: Delhi's Media Urbanism.* London: Routledge, 2010.

Swyngedouw, Erik. "Exit 'Post'—The Making of 'Glocal' Urban Modernities." In *Future City*, ed. Stephen Read, Jürgen Rosemann, and Job van Eldijk, 125–144. London: Spon, 2005.

Swyngedouw, Erik, Frank Moulaert, and Arantxa Rodriguez. "Neoliberal Urbanization in Europe: Large-Scale Urban Development Projects and the New Urban Policy." *Antipode* 34, no. 3 (2002): 542–577.

——. "The Contradictions of Urbanizing Globalization." In *The Globalized City: Economic Restructuring and Social Polarization in European Cities*, ed. Frank Moulaert, Arantxa Rodriguez, Erik Swyngedouw (Oxford: Oxford University Press, 2005): 247–265.

——. " 'The World in a Grain of Sand': Large-scale Urban Development Projects and the Dynamics of 'Glocal' Transformations." In *The Globalized City: Economic Restructuring and Social Polarization in European Cities*, ed. Frank Moulaert, Arantxa Rodriguez, and Erik Swyngedouw, 9–28. Oxford: Oxford University Press, 2005.

Tabb, William. "The Wider Context of Austerity Urbanism." *City* 18, no. 2 (2014): 87–100.

Tan, Puay Yok, James Wang, and Angela Sia. "Perspectives on Five Decades of the Urban Greening of Singapore." *Cities* 32 (2013): 24–32.

Tanaka, Stefan. "History Without Chronology." *Public Culture* 28, no. 1 (2015): 161–186.

Tay, Jinna. "Creative Cities." In *Creative Industries*, ed. John Hartley, 220–232. Malden, MA: Blackwell, 2005.

Taylor, Ian. "European Ethnoscapes and Urban Redevelopment: The Return of Little Italy in 21st Century Manchester." *City* 4, no. 1 (2000): 27–42.

Taylor, Peter. "World Cities and Territorial States: The Rise and Fall of Their Mutuality." In *World Cities in a World System*, ed. Peter Taylor and Paul Knox, 48–62. Cambridge: Cambridge University Press, 1995.

——. "World Cities and Territorial States Under Conditions of Contemporary Globalisation." *Political Geography* 19, no. 1 (2000): 5–32.

——. "Specification of the World City Network." *Geographical Analysis* 33, no. 2 (2001): 181–194.

——. *World City Network: A Global Urban Analysis.* New York: Routledge, 2004.

——. "Leading World Cities: Empirical Evaluations of Urban Nodes in Multiple Networks." *Urban Studies* 42, no. 9 (2005): 1593–1608.

Taylor, Peter, Ben Derudder, Pieter Saey, and Frank Witlox. "Introduction: Cities in Globalization." In *Cities in Globalization: Practices, Policies, and Theories*, ed. Peter Taylor, Ben Derudder, Pieter Saey, and Frank Witlox, 13–18. New York: Routledge, 2007.

Teaford, Jon. *The Twentieth-Century American City.* Baltimore: Johns Hopkins University Press, 1986.

——. *Post-Suburbia: Government and Politics in the Edge Cities.* Baltimore: Johns Hopkins University Press, 1997.

Teo, Peggy. "The Limits of Imagineering: A Case Study of Penang." *International Journal of Urban and Regional Research* 27, no. 3 (2003): 545–563.

Therborn, Göran. "End of a Paradigm: The Current Crisis and the Idea of Stateless Cities." *Environment & Planning A* 43, no. 2 (2011): 272–285.

——. *Cities of Power: The Urban, the National, the Popular, the Global.* London: Verso, 2017.

Thieme, Tatiana. "The 'Hustle' Amongst Youth Entrepreneurs in Mathare's Informal Waste Economy." *Journal of Eastern African Studies* 7 (2013): 389–412.

——. "The Hustle Economy: Informality, Uncertainty and the Geographies of Getting By." *Progress in Human Geography* 42, no. 4 (2018): 529–548.

Thierstein, Alain, and Elisabeth Schein. "Emerging Cities on the Arabian Peninsula: Urban Space in the Knowledge Economy Context." *International Journal of Architectural Research* 2, no. 2 (2008): 178–195.

Thornbury, Robert. *The Changing Urban School.* London: Taylor & Francis, 1978.

Thrift, Nigel. " 'Not a Straight Line but a Curve,' or, Cities Are Not Mirrors of Modernity." In *City Visions*, ed. Duncan Bell and Azzedine Haddour, 233–263. London: Longman, 2000.

Tighe, J. Rosie, and Joanna Ganning. "The Divergent City: Unequal and Uneven Development in St. Louis." *Urban Geography* 36, no. 5 (2015): 654–673.

Tipps, Dean. "Modernization Theory and the Comparative Study of Societies: A Critical Perspective." *Comparative Studies in Society and History* 15, no. 2 (1973): 199–226.

Tomlinson, John. *The Culture of Speed: The Coming of Immediacy.* London: Sage, 2007.

Townsend, Anthony. *Smart Cities: Big Data, Civic Hackers, and the Quest for a New Utopia.* New York: Norton, 2013.

Tsikata, Dzodzi. "Informalization, the Informal Economy, and Urban Women's Livelihoods in Sub-Saharan Africa Since the 1990s." In *The Gendered Impacts of Liberalization: Towards "Embedded Liberalism"?*, ed. Shahra Razavi, 131–162. New York: Routledge, 2009.

Turner, Charles. "Mannheim's Utopia Today." *History of the Human Sciences* 16, no. 1 (2002): 27–47.

Turok, Ivan. "Cities, Regions, and Competitiveness." *Regional Studies* 38, no. 9 (2004): 1069–1083.

——. "Cities, Competition and Competitiveness: Identifying New Connections." In *Changing Cities: Rethinking Urban Competitiveness, Cohesion and Governance*, ed. Nick Buck, Ian Gordon, Alan Harding, and Ivan Turok, 25–43. New York: Palgrave Macmillan, 2005.

———. "The Distinctive City: Pitfalls in the Pursuit of Differential Advantage." *Environment and Planning A* 41 (2009): 13–30.

Turok, Ivan, and Vlad Mykhnenko. "The Trajectories of European Cities, 1960–2005." *Cities* 24, no. 3 (2007): 165–182.

Tyler, Anne. *The Accidental Tourist.* New York: Berkley, 1986.

UN-Habitat. "The Challenge of Slums: Global Report on Human Settlements 2003." *Management of Environmental Quality* 15, no. 3 (2004): 337–338.

Urry, John. *The Tourist Gaze: Leisure and Travel in Contemporary Societies.* Thousand Oaks, CA: Sage, 1990.

———. "The Power of Spectacle." In *Visionary Power: Producing the Contemporary City*, ed. Christine de Baan, Joachim Declerck, and Veronique Patteeuw, 131–141. Rotterdam: NAi, 2007.

Van Aalst, Irina, and Inez Boogaarts. "From Museum to Mass Entertainment: The Evolution of the Role of Museums in Cities." *European Urban and Regional Studies* 9, no. 3 (2002): 195–209.

Van de Kaa, Dirk. "Europe's Second Demographic Transition." *Population Bulletin* 42 (1987): 1–57.

van der Merve, Izak. "The Global Cities of Sub-Saharan African: Fact or Fiction?" *Urban Forum* 15, no. 1 (2004): 36–47.

van Meeteren, Michiel, David Bassens, Ben Derudder. "Doing Global Urban Studies: On the Need for Engaged Pluralism, Frame Switching, and Methodological Cross-Fertilization." *Dialogues in Human Geography* 6, no. 3 (2016): 296–301.

van Meeteren, Michiel, Ben Derudder, and David Bassens. "Can the Straw Man Speak? An Engagement with Postcolonial Critiques of 'Global Cities Research.'" *Dialogues in Human Geography* 6, no. 3 (2016): 247–267.

Vanolo, Alberto. "The Image of the Creative City: Some Reflections on Urban Branding in Turin." *Cities* 25, no. 6 (2008): 370–382.

———. "Smartmentality: The Smart City as Disciplinary Strategy." *Urban Studies* 51, no. 5 (2014): 883–898.

van Noorloos, Femke, and Marjan Kloosterboer. "Africa's New Cities: The Contested Future of Urbanization." *Urban Studies* 55, no. 6 (2018): 1223–1241.

Varley, Ann. "Feminist Perspectives on Urban Poverty." In *Rethinking Feminist Interventions Into the Urban*, ed. Linda Peake and Martina Rieker, 125–141. London: Routledge, 2013.

———. "Postcolonialising Informality?" *Environment & Planning D: Society and Space* 31, no. 1 (2013): 4–22.

Varma, Rashmi. "Provincializing the Global City: From Bombay to Mumbai." *Social Text* 81 (2004): 65–89.

Vasudevan, Alexander. "The Autonomous City: Towards a Critical Geography of Occupation." *Progress in Human Geography* 39, no. 3 (2015): 313–334.

———. "The Makeshift City: Towards a Global Geography of Squatting." *Progress in Human Geography* 39, no. 3 (2015): 338–359.

Velegrinis, Steven, and Richard Weller. "The 21st Garden City? The Metaphor of the Garden in Contemporary Singaporean Urbanism." *Journal of Landscape Architecture* 2, no. 2 (2007): 30–41.

Vengopal, Rajesh. "Neoliberalism as a Concept." *Economy and Society* 44, no. 2 (2015): 165–187.

Venturi, Robert, *Complexity and Contradiction in Architecture.* New York: Museum of Modern Art, 1996.

Venturi, Robert, Denise Scott Brown, and Steven Izenour. *Learning from Las Vegas: The Forgotten Symbolism of Architectural Form.* Cambridge, MA: MIT Press, 1972.

Vicario, Lorenzo, and P. Manuel Martínez Monje. "Another 'Guggenheim Effect'? The Generation of a Potentially Gentrifiable Neighbourhood in Bilbao." *Urban Studies* 40, no. 12 (2003): 2383–2400.

Wachsmuth, David. "City as Ideology: Reconciling the Explosion of the City Form with the Tenacity of the City Concept." *Environment & Planning D* 32, no. 1 (2014): 75–90.

Wacquant, Loïc. "Class, Ethnicity, and the State in the Making of Marginality: Revisiting Territories of Urban Relegation." In *Territories of Poverty: Rethinking North and South*, ed. Ananya Roy and Emma Shaw Crane, 247–259. Athens, GA: University of Georgia Press, 2015.

Walker, Jeremy, and Melinda Cooper. "Genealogies of Resilience: From Systems Ecology to the Political Economy of Crisis Adaptation." *Security Dialogue* 42, no. 2 (2011): 143–160.

Walker, Richard. "Why Cities? A Response." *International Journal of Urban and Regional Research* 40, no. 1 (2016): 164–180.

Wansborough, Matthew, and Andrea Mangeean. "The Role of Urban Design in Cultural Regeneration." *Journal of Urban Design* 5, no. 2 (2000): 181–197.

Ward, Kevin. "Entrepreneurial Urbanism, State Restructuring and Civilising 'New' East Manchester." *Area* 35, no. 2 (2003): 116–127.

——. "Entrepreneurial Urbanism and Business Improvement Districts in the State of Wisconsin: A Cosmopolitan Critique." *Annals of the Association of American Geographers*, 100, no. 5 (2010): 1177–1196.

——. "Towards a Relational Comparative Approach to the Study of Cities." *Progress in Human Geography* 34, no. 4 (2010): 471–487.

Ward, Stephen V. "Spatial Planning Approaches to Growth and Shrinkage: A Historical Perspective." In *Parallel Patterns of Shrinking Cities and Urban Growth: Spatial Planning for Sustainable Development of City Regions and Rural Areas*, ed. Robin Ganser and Rocky Piro, 7–26. Burlington, VT: Ashgate, 2012.

Watson, Vanessa. "Seeing from the South: Refocusing Urban Planning on the Globe's Central Urban Issues." *Urban Studies* 46, no. 11 (2009): 2259–2275.

——. "Planning and the 'Stubborn Realities' of Global South-East Cities: Some Emerging Ideas." *Planning Theory* 12, no. 1 (2013): 81–100.

——. "African Urban Fantasies: Dreams or Nightmares?" *Environment and Urbanization* 26, no. 1 (2014): 215–231.

Watts, Michael. "Development II: The Privatization of Everything?" *Progress in Human Geography* 18, no. 3 (1994): 371–384.

Weaver, Robert. "The Suburbanization of America or the Shrinking of the Cities." *Civil Rights Digest* 9, no. 3 (1977): 2–11.

Weaver, Russell, Sharmistha Bagchi-Sen, Jamie Knight, and Amy Frazier. *Shrinking Cities: Understanding Urban Decline in the United States.* New York: Routledge, 2017.

Weaver, Russell, and Chris Holtcamp. "Geographical Approaches to Understanding Urban Decline: From Evolutionary Theory to Political Economy . . . and Back?" *Geography Compass* 9, no. 5 (2015): 286–302.

Webber, Richard. "The Metropolitan Habitus: Its Manifestations, Locations, and Consumption Profiles." *Environment and Planning A* 39 (2007): 182–207.

Weber, Rachel. "Extracting Value from the City: Neoliberalism and Urban Redevelopment." *Antipode* 34, no. 3 (2002): 519–539.

——. "Selling City Futures: The Financialization of Urban Redevelopment Policy." *Economic Geography* 86, no. 3 (2010): 251–274.

——. *From Boom to Bubble: How Finance Built the New Chicago.* Chicago: University of Chicago Press, 2015.

——. *Why We Overbuild.* Chicago: University of Chicago Press, 2015.

Weinstein, Liza. "Mumbai's Development Mafias: Globalization, Organized Crime and Land Development." *International Journal of Urban and Regional Research* 32, no. 1 (2008): 22–39.

——. *The Durable Slum: Dharavi and the Right to Stay Put in Globalizing Mumbai.* Minneapolis: University of Minnesota Press, 2014.

Westlund, Hans. "Urban Futures in Planning, Policy and Regional Science: Are We Entering a Post-Urban World?" *Built Environment* 40, no. 4 (2014): 447–457.

Wettenhall, Roger. "The Rhetoric and Reality of Public–Private Partnerships." *Public Organization Review* 3, no. 1 (2003): 77–107.

While, Aidan, Andrew Jonas, and David Gibbs. "The Environment and the Entrepreneurial City: Searching for the Urban 'Sustainability Fix' in Manchester and Leeds." *International Journal of Urban and Regional Research* 28, no. 3 (2004): 549–569.

White, James. "Old Wine, Cracked Bottle?: Tokyo, Paris, and the Global City Hypothesis." *Urban Affairs Review* 33, no. 4 (1998): 451–477.

Wiechmann, Thorsten. "Errors Expected: Aligning Urban Strategy with Demographic Uncertainty in Shrinking Cities." *International Planning Studies* 13, no. 4 (2008): 431–446.
Wiechmann, Thorsten, and Marco Bontje. "Responding to Tough Times: Policy and Planning Strategies in Shrinking Cities." *European Planning Studies* 23, no. 1 (2015): 1–11.
Wiechmann, Thorsten, and Karina Pallagst. "Urban Shrinkage in Germany and the USA: A Comparison of Transformation Patterns and Local Strategies." *International Journal of Urban and Regional Research* 36, no. 2 (2012): 261–280.
Wiedmann, Florian, Ashraf Salama, and Alain Thierstein. "A Framework for Investigating Urban Qualities in Emerging Knowledge Economies: The Case of Doha." *Archnet-IJAR, International Journal of Architectural Research* 6, no. 1 (2012): 42–56.
Williams, Gwyndaf. "Rebuilding the Entrepreneurial City: The Master Planning Response to the Bombing of Manchester." *Environment and Planning B* 27, no. 4 (2000): 485–505.
Wilson, David. "Metaphors, Growth Coalition Discourses and Black Poverty Neighborhoods in a U.S. City." *Antipode* 28, no. 1 (1996): 72–96.
——. "Performative Neoliberal-Parasitic Economies and the Making of Political Realities: The Chicago Case." *International Journal of Urban and Regional Research* 35, no. 4 (2011): 691–711.
Wilson, David, Dean Beck, and Adrian Bailey. "Neoliberal-Parasitic Economies and Space Building: Chicago's Southwest Side." *Annals of the Association of American Geographers* 99, no. 2 (2009): 301–324.
Windelband, Wilhem. "History and Natural Science." *Theory and Psychology* 8, no. 1 (1894/1998): 5–22.
Wolch, Jennifer. "Radical Openness as Method in Urban Geography." *Urban Geography* 24, no. 8 (2003): 645–646.
Wolff, Manuel, Sylvie Fol, Hélène Roth, and Emmanuèle Cunningham Sabot. "Is Planning Needed? Shrinking Cities in the French Urban System." *Transnational Planning Review* 88, no. 1 (2017): 131–145.
Wolff, Manuel, and Thorsten Wiechmann. "Thorsten Wiechmann, "Urban Growth and Decline: Europe's Shrinking Cities in a Comparative Perspective 1990–2010." *European Urban and Regional Studies* 25, no. 2 (2018): 122–139.
Wong, C. M. L. "The Developmental State in Ecological Modernization and the Politics of Environmental Framings: The Case of Singapore and Implications for East Asia." *Nature and Culture* 7, no. 1 (2012): 95–119.
Wright, Gwendolyn, *The Politics of Design in French Colonial Urbanism.* Chicago: University of Chicago Press, 1991.
——. "Building Global Modernisms." *Grey Room* 7 (2002): 124–134.
Wright, Herbert. *Instant Cities.* London: Black Dog, 2008.
Yang, Dan. "Tourism Development and Transformation of Resource-Exhausted City: A Case Study of Wansheng District, Chongqing, China." *International Journal of Humanities and Social Science* 4, no. 10 (2014): 194–198.
Yeoh, Brenda. "Global/Globalizing Cities." *Progress in Human Geography* 23, no. 4 (1999): 607–616.
——. "Postcolonial Cities." *Progress in Human Geography* 25, no. 3 (2001): 456–468.
Yiftachel, Oren. "Essay: Re-engaging Planning Theory? Towards 'South-Eastern' Perspectives." *Planning Theory* 2006 5, no. 3 (2006): 211–222.
——. "(Un)Settling Colonial Presents." *Political Geography* 27, no. 3 (2008): 365–370.
——. "Critical Theory and Gray Space: Mobilization of the Colonized." *City* 13, no. 2/3 (2009): 246–263.
——. "Theoretical Notes on 'Gray Cities': The Coming of Urban Apartheid." *Planning Theory* 8, no. 1 (2009): 88–100.
Young, Iris Marilyn. *Justice and the Politics of Difference.* Princeton, NJ: Princeton University Press, 1990.
Yu, Chang, Martin de Jong, and Baodong Cheng. " Getting Depleted Resource-Based Cities Back on Their Feet Again—the Example of Yichun in China." *Journal of Cleaner Production* 134 (part A), no. 15 (2016): 42–50.

Yuen, Belinda. "Creating the Garden City: The Singapore Experience." *Urban Studies* 33, no. 6 (1996): 955–970.

Zakirova, Betka. "Shrinkage at the Urban Fringe: Crisis or Opportunity?" *Berkeley Planning Journal* 23, no. 1 (2010): 58–82.

Zeiderman, Austin. "Cities of the Future? Megacities and the Space/Time of Urban Modernity." *Critical Planning* 15 (2008): 23–39.

——. *Endangered City: The Politics of Security and Risk in Bogotá.* Durham, NC: Duke University Press, 2016.

——. "Beyond the Enclave of Urban Theory?" *International Journal of Urban and Regional Research* 42, no. 6 (2018): 1114–1126.

Zeiderman, Austin, Sobia Ahmad Kaker, Jonathan Silver, and Astrid Wood. "Uncertainty and Urban Life." *Public Culture* 27, no. 2 (2015): 281–304.

Zimmerman, Jeffrey. "From Brew Town to Cool Town: Neoliberalism and the Creative City Development Strategy in Milwaukee." *Cities* 25, no. 4 (2008): 230–242.

Zukin, Sharon. *Landscapes of Power: From Detroit to Disneyworld.* Berkeley: University of California Press, 1991.

——. *The Cultures of Cities.* Oxford: Blackwell, 1995.

——. "Cultural Strategies for Economic Development and the Hegemony of Vision." In *Urbanization of Injustice*, ed. Erik Swyngedouw and Andrew Merrifield, 223–243. London: Lawrence and Wishart, 1996.

——. "Urban Lifestyles: Diversity and Standardization in Spaces of Consumption." *Urban Studies* 35, no. 5/6 (1998): 825–839.

——. *Naked City: The Death and Life of Authentic Urban Places.* New York: Oxford University Press, 2010.

NEWSPAPERS, PERIODICALS, AND TELEVISION SOURCES

Alexander, Brian. "What America Is Losing As Its Small Towns Struggle: To Erode Small-Town Culture Is to Erode the Culture of the Nation." *Atlantic*, October 18, 2017.

Allen, Kate. "Shrinking Cities: Population Decline in the World's Rust-Belt Areas." *Financial Times*, June 16, 2017.

Anjaria, Jonathan Shapiro. "Street Hawkers and Public Space in Mumbai." *Economic and Political Weekly*, May 27, 2006, 2140–2146.

Anonymous. "Charter Cities: Unchartered Territory." *Economist*, October 6, 2012.

Anonymous. "Cities: The Doughnut Effect." *Economist*, January 17, 2002.

Anonymous. "Honduras Shrugged." *Economist*, December 10, 2011.

Anonymous. "Rustbelt Britain: The Urban Ghosts." *Economist*, October 12, 2013, 67.

Atkins, Mathieu. "Gangs of Karachi: Meet the Mobsters Who Run the Show in One of the World's Deadliest Cities." *Harpers* 331, no. 1984 (September 2015): 34–46.

Boudreau, John. "Cisco Systems Helps Build 'Cites-in-a-Box.' " *Washington Post*, June 9, 2010.

Chakrabortty, Aditya. "Paul Romer Is a Brilliant Economist—but His Idea for Charter Cities Is Bad." *Guardian*, July 17, 2010.

Dickey, Christopher. "Is Dubai's Party Over?" *Newsweek*, December 15, 2008.

Draper, Robert, and Pascal Maitre. "Kinshasa, Urban Pulse of the Congo." *National Geographic* 224, no. 4 (September 2013): 100–123.

Gecan, Michael. "On Borrowed Time: Urban Decline Moves to the Suburbs." *Boston Review*, March 8, 2008.

Hawkes, Steve. " 'Failing' Cities and Towns." *Telegraph*, October 11, 2013.

Hughes, Robert. *Trouble in Utopia*, episode 4, "The Shock of the New." BBC and Time-Life Films, 1980. DVD, New York: Ambrose Video, 2001.

Kaplan, Robert. "The Coming Anarchy: How Scarcity, Crime, Overpopulation, Tribalism, and Disease Are Rapidly Destroying the Social Fabric of Our Planet." *Atlantic Monthly* 273, no. 2 (February 1994): 44–76.

Kotkin, Joel, and Wendell Cox. "The World's Fastest Growing Megacities." *Forbes Magazine*, April 8, 2013.

Lindsay, Greg. "Cities-in-a-Box." *Slate*, November 26, 2011.

Maciag, Mike. "Are Cities That Lost Population Making a Comeback?" *Governing*, May 23, 2013.

Mason, Paul. "Slumlands—Filthy Secret of the Modern Mega-City." *New Statesman*, August 8, 2011.

Ouroussoff, Nicolai. "A Building Forms a Bridge Between a University's Past and Future." *New York Times*, February 9, 2011.

Overberg, Paul. "The Divide Between America's Prosperous Cities and Struggling Small Towns—in 20 Charts." *Wall Street Journal*, December 29, 2017.

Packer, George. "The Megacity: Decoding the Chaos of Lagos." *New Yorker*, November 13, 2006, 62–75.

Perraudin, Frances. "Ten of Top 12 Most Declining UK Cities Are in North of England—Report." *Guardian*, February 29, 2016.

Polgreen, Lydia. "In a Dream City, a Nightmare for the Common Man." *New York Times*, December 13, 2006.

Power, Matthew. "The Magic Mountain: Trickle-Down Economics in a Philippine Garbage Dump." *Harper's Magazine*, December 2006, 57–68.

Staff Reporter. "Britain's Decaying Towns: City Sicker." *Economist*, October 12, 2013.

Stohr, Kate. "Shrinking City Syndrome." *New York Times*, February 5, 2004, D8.

Swift, Andrew. "The FP Top 100 Global Thinkers." *Foreign Policy*, November 29, 2010.

Taitz, Laurice. "Shock and Awe." *Sunday Times Lifestyle* [Johannesburg], November 25, 2007, 12.

Webster, Paul, and Jacob Burke. "How the Rise of the Megacity Is Changing the Way We Live." *Observer*, January 21, 2012.

Yinan, Hu. "Tough Time for Towns Where Mines Run Dry." *China Daily*, April 16, 2010.

RESEARCH REPORTS AND WORKING PAPERS

Audirac, Ivonne. "Urban Shrinkage Amid Fast Metropolitan Growth (Two Faces of Contemporary Urbanism)." In *The Future of Shrinking Cities: Problems, Patterns and Strategies of Urban Transformation in a Global Context*, ed. Katrina Pallagst et al., 69–80. Berkeley, CA: Institute of Urban and Regional Development, Center for Global Metropolitan Studies, and the Shrinking Cities International Research Network Monograph Series, 2009.

Beauregard, Robert. "Shrinking Cities in the United States in Historical Perspective: A Research Note." In *The Future of Shrinking Cities: Problems, Patterns and Strategies of Urban Transformation in a Global Context*, ed. Karina Pallagst et al., 61–68. Berkeley, CA: Institute of Urban and Regional Development, Center for Global Metropolitan Studies, and the Shrinking Cities International Research Network Monograph Series, 2009.

Berglund, Eeva. "Troubled Landscapes of Change: Limits and Natures in Grassroots Urbanism." In *Dwelling in Political Landscapes: Contemporary Anthropological Perspectives*, ed. Ann Lounela, Eeva Berglund, and Timo Kallinen, 196–212. Helsinki: Finnish Literature Society, 2019.

Bunker, Robert J. "The Emergence of Feral and Criminal Cities: U.S. Military Implications in a Time of Austerity." Land Warfare Paper 99W. Arlington, VA: Association of the United States Army, April 2014.

Comaroff, Jean, and John Comaroff. "Theory from the South: A Rejoinder." In *The Johannesburg Salon: Volume Five*, ed. Achille Mbembe, 30–36. Johannesburg: Johannesburg Workshop in Theory and Criticism, 2012.

Cunningham Sabot, Emmanuèle, and Sylvie Fol. "Shrinking Cities in France and Great Britain: A Silent Process." In *The Future of Shrinking Cities: Problems, Patterns and Strategies of Urban Transformation in a Global Context*, ed. Karina Pallagst et al., 24–35. Berkeley, CA: Institute of Urban and Regional Development, Center for Global Metropolitan Studies, and the Shrinking Cities International Research Network Monograph Series, 2009.

Evans, Graeme. "From Cultural Quarters to Creative Clusters—Creative Spaces in the New City Economy." In *The Sustainability and Development of Cultural Quarters: International Perspectives*, ed. Mattias Legner, 32–59. Stockholm: Institute of Urban History, 2009.

Flusty, Steven. *Building Paranoia: The Proliferation of Interdictory Space and the Erosion of Spatial Justice.* Los Angeles: Los Angeles Forum for Architecture and Urban Design, 1994.

Hoyt, Lorene, and André Leroux. *Voices from Forgotten Cities: Innovative Revitalization Coalitions in America's Older Small Cities.* PolicyLink, Citizens' Housing and Planning Association, and MIT School of Architecture and Planning, 2007.

Mallach, Alan. *Facing the Urban Challenge: The Federal Government and America's Older Distressed Cities.* Metropolitan Policy Program at Brookings, May 2010.

Martinez-Fernandez, Cristina, Noya Antonella, and Weyman Tamara, eds. *Demographic Change and Local Development Shrinkage, Regeneration and Social Dynamics.* Paris: OECD, Local and Employment Development, 2012.

Mitchell, Timothy. "The Properties of Markets: Informal Housing and Capitalism's Mystery." Cultural Political Economy Working Paper Number 2. Lancaster, UK: Institute for Advanced Studies in Social and Management Studies, University of Lancaster, 2007.

Pike, Andy, Danny MacKinnon, Mike Coombes, Tony Champion, David Bradley, Andrew Cumbers, Liz Robson, and Colin Wymer. "Unequal Growth: Tackling Declining Cities." Report from the Joseph Roundtree Foundation. Newcastle, UK: Newcastle University, 2016.

Theodore, Georgeen. "Improve Your Lot." In *Cities Growing Smaller (Urban Infill, Vol. 1)*, ed. Karina Pallagst, 47–64. Cleveland: Kent State University and Cleveland Urban Design Collaborative, 2008.

Ubarevičienė, Rūta, and Maarten van Ham. "Population Decline in Lithuania: Who Lives in Declining Regions and Who Leaves?" Discussion Paper No. 10160. Bonn: Institute for the Study of Labor, 2016.

Warner, Sam Bass, and Lawrence Vale. "Introduction: Cities, Media, and Imaging." In *Imaging the City: Continuing Struggles and New Directions*, ed. Lawrence Vale and Sam Bass Warner, viii–xxiii. New Brunswick, NJ: Center for Urban Policy Research, 2001.

GOVERNMENT REPORTS

Bairoch, Paul. *Urban Unemployment in Developing Countries: The Nature of the Problem and Proposals for its Solution.* Geneva: International Labour Office, 1973.

Moreno, Eduardo López (principal author). *State of the World's Cities 2008/9: Harmonious Cities—UN-HABITAT.* London: Earthscan, 2008.

Rodas, Luis Guillermo Pineda. "The Making of a Free City: The Foundation of Laissez-Faire Capitalist Free Cities in Honduras in the Juncture of Globalization." Unpublished Paper. Department of Society and Globalization, Roskilde Universitet [Denmark], 2013.

UN-Habitat. *The State of the World's Cities.* London: Earthscan, 2006.

United Nations. *State of the World's Cities, 2002.* New York: United Nations Center for Human Settlement, 2002.

THESES AND DISSERTATIONS

Clement, Daniel. "The Spatial Injustice of Crisis-Driven Neoliberal Urban Restructuring in Detroit." PhD diss., University of Miami, 2013.

Karaman, Ozan. "Remaking Space for Globalization: Dispossession through Urban Renewal in Istanbul." PhD diss., University of Minnesota, 2010.

Puri, Siddharth. "Specifying the Generic: A Theoretical Unpacking of Rem Koolhaas' 'Generic City.' " PhD diss., University of Cincinnati, 2007.

INTERNET SOURCES

Abourahme, Nasser. "Contours of the Neoliberal City: Fragmentation, Frontier Geographies and the New Circularity." *Occupied London No. 4*, May 23, 2009. https://baierle.me/2009/05/23/contours-of-the-neoliberal-city-fragmentation-frontier-geographies-and-the-new-circularity/.

Aravamudan, Srinivas. "Surpassing the North: Can the Antipodean Avantgarde Trump Postcolonial Belatedness?" *Society for Cultural Anthropology*, February 25, 2012. https://culanth.org/fieldsights/surpassing-the-north-can-the-antipodean-avantgarde-trump-postcolonial-belatedness.

Axel-Lute, Miriam. "Small Is Beautiful—Again." *Shelterforce*, July 23, 2007. http://www.nhi.org/online/issues/150/smallisbeautiful.html.

Fuller, Brandon, and Paul Romer. "Cities from Scratch." City Journal 20, no. 4 (2010). http://city-journal.org/2010/20_4_charter-cities.html.

Katodrytis, George. "Metropolitan Dubai and the Rise of Architectural Fantasy." *Bidoun Magazine* 4 (2005). http://www.radicalurbantheory.com/misc/dubai.html.

Lindsay, Greg. "Cisco's Big Bet on New Songdo: Creating Cities from Scratch." *Fast Company*, February 1, 2010. https://www.fastcompany.com/1514547/ciscos-big-bet-new-songdo-creating-cities-scratch.

Lumumba, Jane. "Why Africa Should Be Wary of Its 'New Cities.' " *The Rockefeller Foundation's Informal City Dialogues*, February 5, 2013. http://nextcity.org/informalcity/entry/why-africa-should-be-wary-of-its-new-cities.

McEwen, Mitch. "Interview with Keller Easterling About Subtraction." *Archinect*, August 26, 2014. https://www.archinect.com/another/interview-with-keller-easterling-about-subtraction.

Oosterman, Arjen. "Editorial: Notes from the Tele-Present," January 29, 2013. In *Volume #34: City in a Box*, ed. Arjen Oosterman, Michelle Provoost and Paul Kroese. Amsterdam: Stichting Archis, 2013. archis.org/volume/notes-from-the-tele-present/.

Oswalt, Philipp, Project Shrinking Cities. "Hypotheses on Urban Shrinking in the 21st Century." Accessed September 23, 2013. http://www.shrinkingcities.com/hypothesen.0.html%3F&L=1.html.

Peck, Jamie. "Creativity Fix." *Eurozine*, June 28, 2007. https://www.eurozine.com/the-creativity-fix/.

Romer, Paul. "Why the World Needs Charter Cities." *TEDGlobal* 2009. http://www.ted.com/talks/paul_romer.html.

Ruiz-Goiriena, Romina. "Guatemala Builds Private City to Escape Crime." *Yahoo! News*, January 8, 2013. https://news.yahoo.com/guatemala-builds-private-city-escape-crime-180518632.html.

Samuels, Linda C. "Book Review: *The City That Never Was*, by Christopher Marcinkoski." *Journal of Architectural Education*, June 2016, 1–11. https://www.jaeonline.org/articles/review/city-never-was#/page1/.

Sherman, Roger. "Opinion: Risky Business." *Planetizen*, September 18, 2007. http://www.planetizen.com/node/27120.

Woolsey, Matt. "America's Fastest-Dying Towns." *Forbes*, December 9, 2008. https://www.forbes.com/2008/12/08/towns-ten-economy-forbeslife-cx_mw_1209dying_slide.html?sh=7e96672214f0.

Zumbrun, Joshua. "America's Fastest-Dying Cities." *Forbes*, August 5, 2008. https://www.forbes.com/2008/08/04/economy-ohio-michigan-biz_cx_jz_0805dying.html?sh=2b57f68b180d.

WORKSHOPS, PRESENTATIONS, AND CONFERENCE PROCEEDINGS

Audirac, Ivonne. "Urban Shrinkage amid Fast Metropolitan Growth (Two Faces of Contemporary Urbanism)." Presented to The Future of Shrinking Cities: Problems, Patterns and Strategies of Urban Transformation in a Global Context, Centre for Global Metropolitan Studies, IURD. Berkeley: University of California, 2007.

Gregory, Siobhan. "Detroit Is a Blank Slate: Metaphors in the Journalistic Discourse of Art and Entrepreneurship in the City of Detroit." *EPIC [Ethnographic Praxis Industry Conference] 2012 Proceedings* (2012): 217–233.

Krenz, Kimon. "Capturing Patterns of Shrinkage and Growth in Post-Industrial Regions: A Comparative Study of the Ruhr Valley and Leipzig-Halle." *Proceedings of the 10th International Space Syntax Symposium* 72 (July 13–17, 2015): 1–18.

Mohammadzadeh, Mohsen. "Neo-Liberal Planning in Practice: Urban Development with/Without a Plan (A Post-Structural Investigation of Dubai's Development)." Paper presented in Track 9 (Spatial Policies and Land Use Planning) at the Third World Planning Schools Congress, Perth, Western Australia, July 4–8, 2011.

Pieterse, Edgar. "Exploratory Notes on African Urbanism." Paper presented at the Third European Conference on African Studies, Leipzig, June 4–7, 2009. http://www.stellenboschheritage.co.za/wp-content/uploads/Exploratory-Notes-on-African-Urbanism.pdf.

Roy, Ananya, and Helga Leitner. Social Science Research Council Dissertation Proposal Development Fellowship Spring 2011 Workshop. http://webarchive.ssrc.org/DPDF/2011%20-%20Provincializing%20Global%20Urbanism.pdf.

Schwarz, Nina, and Dagmar Haase. "Urban Shrinkage: A Vicious Circle for Residents and Infrastructure? Coupling Agent-Based Models on Residential Location Choice and Urban Infrastructure Development." *Proceedings of the Fifth Biennial Conference of the International Environmental Modelling and Software Society* (*iEMSs*), Ottawa, July 5–8, 2010, 2:817–824.

INDEX

abandonment, in struggling cities, 107, 110–111, *114*
Accidental Tourist, The (Tyler), 83
actor-network theory, 33
advertising. *See* place marketing
Akers, Joshua, 207
Allen, John, 89
alpha cities, competition between, xv. *See also* world-class cities
Amin, Ash, 57
Anderson, Benedict, 205
Angelo, Hillary, 35
Anjaria, Jonathan Shapiro, 162–163
Arcades Project, The (Benjamin), 174–175
architecture: in city of spectacle, 77; cultural mining and, 74; of instant cities, 181; modernist, 76; place marketing and, 76; postmodernist, 74–76; signature, 76–78; of slums, 145–150; themed entertainment destinations and, 74; theming and, 238n37; in tourist-entertainment cities, 72–73. *See also* city building; spatial patterns
artwork, on urban shrinkage, 95, 100–101
Assaad, Ragui, 157
assemblage theory, 33
asymmetrical urbanism, xiv
Atkinson, Rowland, 47
Audirac, Ivonne, 91, 94, 262n198
Augé, Marc, 192
austerity urbanism, 96
Australia, urban shrinkage in, 107
Babbage, Charles, 192
Bauman, Zygmunt, 128, 184
Bayat, Asef, 150–151, 272n163
Beauregard, Robert, xxi, 33, 38, 111, 121, 253n58
Benjamin, Walter, 24–25, 174–175, 206
Berlin, Germany, 118
Berman, Marshall, 199
Bernt, Mattias, 122–123
Bilbao effect, 76, 78
Bourdieu, Pierre, 52–53
bourgeois utopias, 173
Bourriaud, Nicolas, 209
Boyer, Christine, 72
branding, 71, 76
brandscapes, 73
Brenner, Neil, xi, xxii, 3, 4, 34, 35, 201
Bridge, Gary, 47
Broadacre City, 172
Brown, Denise Scott, 241n76
built environment, 38
Bunnell, Tim, 6
bypass-implant urbanism, 169–170

Caldeira, Teresa, 151
capitalism: cognitive, 48; cognitive-cultural, 14; globalization impact on, 5–6; informality and, 157; megacities, social inequality and, xvi; real estate, 38; slums and, 143–144, 265n15; tourist-entertainment cities and, 73. *See also* neoliberalism
Charter Cities, 186–188

Chicago, Illinois, 25
circulation, fast-track urbanism and, 184
cities: in comparative urbanism, 37; corporate, 189–191; distressed, 8–9; entrepreneurial, 68–71; free, 187; generic, 45–46, 55, 191–192; informational, 68; master-planned, 188–189; networked, 6–8; new particularism and, 30; power and injustice in, 32–33; radical uniqueness of, 33–34; resource-depleted, urban shrinkage and, 112; rethinking meaning of, xxii–xxiii; "shock," megacities as, 42–43, 228n116; similarities and differences among, 37–38; time/space of, 206–209; types of, 10–11, *12–13*, 19, 34, 207–208; universalizing theorizations on, 29–32; in world of cities, 29. *See also specific types*
Cittaslow movement, 194–196
city boosterism, 26; instant cities and, 181; mega-events and, 84–85; place marketing and, 75–76; signature architecture for, 76; world-class cities and, 42
city building: of Dubai, UAE, *179*, 179–183; expansion and contraction in, 110; fast-track urbanism and, 2; generic cities and, 191–192; incompleteness of, xxi; informality and, 155; instant cities and new model of, 17, 168–170, 173–175; modernist agenda in, 17; monumentality in, 72; neoliberalism and, 6; postmodernist architecture in, 74–76; types of, 10–11, *12–13*, 34. *See also* fast-track urbanism; instant cities; megacities
city improvement districts, 89
"city-in-a-box," 190
city of spectacle, 72, 77
civic freedoms, xvii
Clark, Terry Nichols, 171
Cleveland, Ohio, 118
closure, in tourist-entertainment cities, 89–90
cognitive capitalism, 48
cognitive-cultural capitalism, 14
cohesion, public-private partnerships and, 247n178
Comaroff, Jean, 62
Comaroff, John, 62
company towns, 171
comparative urbanism, xx; cities in, 37; in urban studies, 37–38
competition: alpha cities and, xv; creative cities fostering, 48–49; struggling cities and, 91–92; tourist-entertainment cities and, 69; unsolicited urbanism and, 87; world-class cities and globalization causing, 26, 67–68, 91–92
competitive world-class cities. *See* world-class cities
continuous settlement, xxii
contraction. *See* urban shrinkage
conventional urban studies: conceptual blindspots of, 135–138; of Global North, 57; limitations of, 40–44; rethinking, xviii–xx, 3–5, 7–10, 25–26, 200, 202–205. *See also* urban studies
convergence thesis: creative cities and, 47–53; deficiencies of, 53–56; spatial patterns and, 55; universalizing theorizations and, 54–55; in urban studies, 45–47; world-class cities and, 45–47
corporate cities, 189–191
Crawford, Margaret, 171
creative cities: competition fostered by, 48–49; convergence thesis and, 47–53; economic growth and, 49–50; elements of, 51; endemic policy mobility and, 50; income inequality and, 53; interpretations of, 51–52; neoliberalism and, 51–52; smart growth and, 48, 68; struggling cities and formula of, 52–53; thesis of, 49–51; urban regeneration and, 52–53
creative class, 49–50
cultural mining, architecture and, 74
culture-led urban regeneration, 75–77, 80
Cunningham, Allen, 67–68
Cunningham Sabot, Emmanuèle, 106, 262n198

Davis, Mike, xvii, 15, 42, 131, 146, 180
"decorated sheds," 241n76
Degen, Monica, 70, 79
"degrowth machine," 207
deindustrialization, 15, 95–96; struggling cities from, 99, 106; urban shrinkage from, 113–114. *See also* struggling cities
Del Bianco, Corinna, 146
deregulation: informality and, 162; instant cities and, 180; tourist-entertainment cities, governance and, 86
designscapes, 72
deterritorialization, from globalization, 41–42
Detroit, Michigan, *100*, 101, *114*, 117, 122
developmentalism: promise of, 39; on slums, 140; struggling cities and, 24; in urban studies, attachment to, 38–40
Dick, Howard, 46
Dickey, Christopher, 179

Disneyfication, 71, 185
distressed cities, 8–9
diversity, Southern theory and, 60
Doha, Qatar, *176*
domestication of space, 90
Donegan, Mary, 53
"doughnut effect," of urban shrinkage, 117
Dovey, Kim, 80, 158
downtown revitalization, 67–70
Dubai, UAE, *179*, 179–183
"ducks," 241n76

Easterling, Keller, 91, 93
economic growth, creative cities and, 49–50
economic zones, special, 186–187
Economist, 107
employment: informality and, 155, 161; megacities and, 20, 128, 155, 161; shrinkage syndrome and, 95–96; urban shrinkage and, 116
endemic policy mobility, creative cities and, 50
Engels, Friedrich, 145
entrepreneurial cities: goals of, 69; managerialism and, 69; neoliberalism and, 71; place marketing and, 68–71; public-private partnerships and, 70–71; PUDs and, 70. *See also* tourist-entertainment cities
entrepreneurial urbanism, 189
essentialism, in urban studies, 40
ethnoscapes, 85
excessive urbanism, 42
exclusion, in tourist-entertainment cities, 89–90
experience economy, 77
experimental urbanism, 175–177
extended urbanization, xxii
extralegality, in slums, 143, 156

Fainstein, Susan, 68
fast-track urbanism, 2, 11; circulation and, 184; corporate cities and, 189–191; Dubai as example of, *179*, 179–183; experimentation of, 168, 175–177; features of, *13*, 16–17; generic cities and, 191–192; historical antecedents of, 171–174; instant cities and, 167–171; libertarianism and, 185–188; location of, 170; market-logic of, 170–171; master-planned cities and, 188–189; narratives of, 174–175; postmodern hyperspace of, 17; profitability of, 171; progress and, 184; sustainable urban development as opposite of, 194–196; temporal dimensions of, 183–185; trends and possibilities of, 192–193; utopianism of, 177–178. *See also* instant cities
Ferguson, James, 133, 136–137
festival marketplaces, 80
First World cities. *See* Global North; world-class cities
Fishman, Robert, 138, 173
Florida, Richard, 49–50
Flyvbjerg, Bent, 82
Fol, Sylvie, 106, 262n198
Forbes, 103
Fordist global production, 6
forgotten cities, 5. *See also* struggling cities
formality, informality and, dividing lines of, 152–154, 158, 161–162
fortress urbanism, 89
France, urban shrinkage in, 105–106
Fraser, James, 38–39
free cities, 187

Gandy, Matthew, 132, 134–135
Garden City movement, 172, 177
gated communities, 89
Gehry, Frank, 76
generalizations in urban studies, inherent dangers of, 24–26
generic cities: city building and, 191–192; globalization and, 45; global urbanism and, 55; identity of, 46. *See also* convergence thesis
gentrification, 38; urbanization and, 47; urban regeneration and, 79
Germany, urban shrinkage in, 101–102, 105, 111–112, 118–121, 255n101
Gibson, Chris, 86, 87
Gilbert, Allen, 141
Glaeser, Edward, xvi
global cities. *See* world-class cities
globalization: capitalism restructured through, 5–6; corporate cities and, 191; deterritorialization from, 41–42; generic cities and, 45, 192; income inequality and, 47; local governance tension with, 55–57; megacities and, 129–130; modernity and, 39; networked cities and, 6–8; reterritorialization from, 41–42; scale and reach of, 41; social inequality and, 6; urban shrinkage and, 113; world-class cities and competition from, 26, 67–68, 91–92
Global North: conventional urban studies of, 57; terminology of, 211n5

Global South: analytical imprecision of, 62; meanings of, 62; megacities of, 131; terminology of, 211n5; urban population growth in, xi–xii; urban studies from perspective of, xx, xxiii. *See also* Southern theory
global urbanism, 28; advantages and disadvantages of, xvi; analytic framework of, 18–20; demographic patterns shifting in, 126–127; divergent trajectories of, xviii–xx, 2–3, 11, 34, 37–38, 203–204, 209–210; future of, xiv–xvii; generic cities and, 55; inverted telescopes and, 205–206; open-ended approach to understanding, xxiv–xxv; patterns of, xii–xiii; planetary urbanization and, 18; progress and, 201; provincializing, 61; Southern theory and, 58; success and, 1–2; in 21st century, complexity of, 130–135
Gonzales, Carmen, 160
governance, urban: deregulation and, 86; through exception, 87–88; gray spaces, informality and, 163; public-private partnerships and, 86; tourist-entertainment cities, neoliberalism and, 85–88; unsolicited urbanism and, 87
Graham, Stephen, 57
gray spaces, informality and, 163
Great Britain, urban shrinkage in, 105–107
growth fetishism, 207
growth-*versus*-decline couplet, 38
Guggenheim Bilbao, 76, 78

Hall, Tim, 68
Halle, 101
hard-branding, 71
Hart, Keith, 152–153
Harvey, David, 69
Häußermann, Hartmut, 99
Healey, Patsy, xviii
hierarchies: conventional urban studies and, xviii, 4–5, 19, 136; megacities in, 136–137; nested, 55–56
Hill, Richard Child, 55
historic old towns, urban regeneration of, 75, 79
Hogan, Trevor, xv
Holston, James, 151, 162
Howard, Ebenezer, 172, 174
Hubbard, Phil, 68
Hughes, Robert, 174
idiographic approaches, 31
imagineering, 71
import substitution industrialization (ISI), 117
income inequality: creative cities and, 53; globalization and, 47; informality and, 154
industrialization, modernity and, 8
informality: ambiguity of, 157–160; capitalism and, 157; city building and, 155; complexity of, 138; concept of, 157–158; deregulation and, 162; employment and, 155, 161; formality and, dividing lines of, 152–154, 158, 161–162; gray spaces and, 163; income inequality and, 154; politics of, 155–156; as regulation, 160–163; rethinking/reframing, 154–157; slums and, 137–138, 154–155; as social organization, 158–159, 164–165; value and, 159–160
informal settlements, in slums, 145–150
informal urbanism, 163–165
informational city, 68
infrastructure, 38
injustice, in cities, 32–33
innovation: interpretation of, 51; megacities and, 129; in slums, 151–152; urbanism and, xvi. *See also* creative cities
instantaneous urbanism, 184
instant cities, 11; architecture of, 181; benefits of, 168; city boosterism and, 181; corporate cities and, 189–191; deregulation and, 180; Dubai as example of, *179*, 179–183; experimentation of, 168, 175–177; fast-track urbanism and, 167–171; features of, *13*, 16–17; generic cities and, 191–192; global spread of, 17–18; historical antecedents of, 171–174; libertarianism and, 185–188; location of, 170; market-logic of, 170–171; master-planned, 188–189; modernity and, 20, 169; as new city building model, 17, 168–170, 173–175; postmodern urbanism and, 169; profitability of, 171; spatial patterns of, 169–170; success and failures of, 176–177; sustainable urban development as opposite of, 194–196; temporal dimensions of, 183–185; time/space of, 208–209; trends and possibilities of, 192–193; utopianism of, 177–178. *See also* fast-track urbanism
inverted telescopes, global urbanism and, 205–206
ISI (import substitution industrialization), 117
Isin, Egin, xvii
Ivanovo, Russia, 101
Izenour, Steven, 241n76

Jameson, Frederic, 17
Japan, urban shrinkage in, 107
Jerde, Jon, 74
Judd, Dennis, 68
Julier, Guy, 72

Kantor, Paul, 55
Kaplan, Robert, 132, 133
Klingmann, Anna, 73
Koch, Natalie, 185
Koolhaas, Rem, 46, 94, 133, 178, 191, 268n51
Koselleck, Reinhart, 184
Kusno, Abidin, 56

Lagos, Nigeria, 132–135, 268n51
Las Vegas, Nevada, 180
Lee, Zack, 185
Lees, Loretta, 75–76
legacy cities, struggling cities as, 102
Leipzig, Germany, 101, 118
Leitner, Helga, xx
Lejano, Raul, 146
Leontidou, Lila, 60
libertarianism, instant cities and, 185–188
Lin, George, 55
Linden, Eugene, 133
Liverpool, England, 101, 106–107
Lloyd, Richard, 171
local governance, globalization tension with, 55–57
localization, 44; population decline and, 108; in tourist-entertainment cities, 84
Logan, John, 119
Los Angeles, California, 25
loser cities, xiii, 220n78. *See also* struggling cities
Lowe, Nichola, 53

Ma, Lawrence, 55
"Machiavellian formula," 82
Machine-Age City, 177
Madden, David, 90
makeshift housing, in slums, 148–149
managerialism, 69
Manchester, England, 101, 106–107
Marinanti, Anant, 6
marketing. *See* place marketing
Martinez-Fernandez, Cristina, 262n198
Massey, Doreen, 39, 131
master-planned cities, 188–189
McCarthy, Mary, 83
McFarlane, Colin, 131
Meagher, Kate, 163
megacities, 11; capitalism, social inequality and, xvi; contrasting images of, 128–129, 131–132; employment and, 20, 128, 155, 161; features of, *13*, 15–16, 127; formality/informality dividing lines in, 152–154, 158, 161–162; future of, 131, 133–135; globalization and, 129–130; of Global South, 131; growth of, 126; in hierarchies, 136–137; importance of, xv, xvi; informality as regulation in, 160–163; informality in, ambiguity of, 157–160; informality in, rethinking/reframing, 154–157; informal urbanism and, 163–165; innovation and, 129; modernity and, xv–xvi, xxv–xxvi; poverty and, xiv, 127; as runaway urbanism, 128, 130; as shadow cities, 128–129; as "shock cities," 42–43, 228n116; as slums, 128, 132, 138–140, 143–144; spatial patterns of slums in, 145–152; time/space of, 208; urbanism and future of, 16; urban population growth and, xii; in urban studies, 16, 24, 130–131, 135–138. *See also* slums
mega-events, city boosterism and, 84–85
"megaproject paradox," 82
megaprojects, speculative investment in, 80–82
Mehrotra, Rahul, 163–164
methodological cityism, 35
migration, urban shrinkage and, 114–115
mining, 112
Minton, Anna, 247n178
Mitchell, Timothy, 156, 206
mixed-use redevelopment, 79
modernist architecture, 76
modernity: city building agenda of, 17; in flux, 199; globalization and, 39; industrialization and, 8; instant cities and, 20, 169; megacities and, xv–xvi, xxv–xxvi; promise of, 39; Southern theory and, 58; underdevelopment and, 166; in urban studies, attachment to, 38–40
monocentric urbanism, xxii
monumentality, in city building, 72
Moss, Timothy, 117
Mould, Oli, 32
Mumbai, India, *152*, 162–163
museums, 76–77

Nairobi, Kenya, *134*, *139*, *144*
natural disasters, urban shrinkage and, 120

neoliberalism: city building and, 6; creative cities and, 51–52; entrepreneurial cities and, 71; of speculative investment in megaprojects, 81; terminology of, 86; tourist-entertainment cities and governance under, 85–88; unsolicited urbanism and, 87. *See also* capitalism
nested hierarchies, 55–56
networked cities, 6–8
Neue Urbanität (Häußermann and Siebel), 99
Neuwirth, Robert, 131
new particularism, 30
New Town initiative, 173
New York City, New York, 25, *70*
New Yorker, 132
Nijman, Jan, 38
nomothetic approaches, 31
noncompetitive city. *See* struggling cities
"nowhere places," 223n16

Oldfield, Sophie, 59
Ong, Aihwa, 183
ordinary cities, in urban studies, 9–10, 25, 40, 43
Orientalism, 62
Oswalt, Philipp, 98, 100–101, 119
otherism, Southern theory and, 61
Ouroussoff, Nicolai, 196
overbuilding, in tourist-entertainment cities, 82
overlay zones, 77

Packer, George, 132
Paris, 25
Parnell, Susan, 59
Paseo Cayala, Guatemala, 187
Peck, Jamie, xx, 9, 48, 62, 86
Perlman, Janice, 160
perverse urbanism, 42
Philadelphia, Pennsylvania, 104
Pieterse, Edgar, 15, 138, 203
Piot, Charles, 175
place marketing: architecture and, 76; city boosterism and, 75–76; designscapes and, 72; entrepreneurial city and, 68–71; public-private partnerships and, 70–71; of Salford Quays on the Manchester Ship Canal, 73; of tourist-entertainment cities, 71–73, 83–84; urban regeneration and, 75–76
planetary urbanization, 206; critiques of, 35–36; extended urbanization and, xxii; global urbanism and, 18; hypothesis grounding, 36–37; methodological cityism and, 35; processes of, 35; in urban studies, 28–29, 34–37
Planet of Slums (Davis), 131
planned unit developments (PUDs), 70
Platt, Harold, 228n116
political autonomy, xvii
population decline: localization and, 108; as new normal, 105; struggling cities and, 102–104; systemic shock from, 104–105. *See also* urban shrinkage
population growth. *See* urban population growth
population shrinkage. *See* urban shrinkage
postcolonial theory, 33; Southern theory roots in, 57–58, 60, 62
post-Fordist global production, 6
postindustrial cities in decline. *See* struggling cities
postmodern hyperspace, of fast-track urbanism, 17
postmodernist architecture, 74–76
postmodern urbanism: instant cities and, 169; world-class cities and, 14
post-public space, 88–90
post-structuralist philosophy, in urban studies, 31–32
Potter, Julian, xv
poverty: megacities and, xiv, 127; urban population growth and, xii, xiv. *See also* slums
power, in cities, 32–33
Power, Matthew, 133
Pratt, Andy, 164
progress: fast-track urbanism and, 184; global urbanism and, 201; optimism and, 183; sustainable urban development and, xxiv
progressive urbanism, 49
property-led urban regeneration, 72–73
pseudo-public space, 88–90
public-private partnerships: cohesion and, 247n178; deregulation and, 86; entrepreneurial cities and, 70–71; governance and, 86
public space, post-, 88–90
PUDs. *See* planned unit developments

radical uniqueness, of cities, 33–34
Rao, Vyjayanthi, 142
real estate capitalism, 38
recycled urbanism, 192
regeneration. *See* urban regeneration
Regeneration East (Stadtumbau Ost) Program, 100
regulation, informality as, in megacities, 160–163
Relph, Edward, 74

resource-depleted cities, urban shrinkage and, 112
reterritorialization, from globalization, 41–42
Rieniets, Tim, 119
Riis, Jacob, 145
Rimmer, Peter, 46
Rio de Janeiro, Brazil, 160
Rise of the Creative Class, The (Florida), 50
Robinson, Jennifer, xiii–xiv, xxv, 3, 7, 9, 23–24, 40, 224n24
Rogers, Dallas, 87
Rogers, Dennis, 37
Romer, Paul, 186
Rosseau, Max, 220n78
Rothman, Hal, 180
Roy, Ananya, xviii, 9, 15, 61, 142, 158–159, 161–162, 203
rule of law, in slums, 143, 156
runaway urbanism, 42, 128, 130

Said, Edward, 62
Salford Quays on the Manchester Ship Canal, 73
São Paulo, Brazil, *150*
Sassen, Saskia, xv, xxi, xxv
satellite cities. *See* fast-track urbanism; instant cities
Savitch, Hank, 55
AlSayyad, Nezar, 56, 159
Schilling, Joseph, 119
Schmid, Christian, xi, xxii, 4, 34, 36
scientific naturalism, 32
SCIRN (Shrinking Cities International Research Network), 101–102
Scotland, urban shrinkage in, 106
Scott, Allen, 30–32
Scott, James, 272n163
scripted spaces, 73
shadow cities, megacities as, 128–129
Shatkin, Gavin, 55–56, 169–170
Sheppard, Eric, xx
Sherman, Roger, 77
Shields, Rob, 27
Shiuh-Shen Chien, 11
"shock cities," megacities as, 42–43, 228n116
shrinkage identity, 111–112
shrinkage syndrome, 95–97
shrinking cities. *See* struggling cities; urban shrinkage
Shrinking Cities International Research Network (SCIRN), 101–102
Shrinking Cities Project, 100–101
Siebel, Walter, 99
signature architecture, 76–78
Simmel, Georg, 201
Simon, David, 142
Simone, AbdouMaliq, 44, 149
Sinclair, Iain, 80–81
Singapore, 181–182
Sklair, Leslie, 73
Slow City movement, 194–196
Slow Food movement, 195
slow urbanism, 194
slums, xvii, 42; capitalism and, 143–144, 265n15; characteristics of, 147; in conventional urban studies, 137; definitions of, 142; developmentalism on, 140; global population living in, 143; housing in, 145–150; informality and, 137–138, 154–155; innovation in, 151–152; megacities as, 128, 132, 138–140, 143–144; process and outcomes of, 146; rule of law and extralegality in, 143, 156; spatial patterns of, 140–141, 145–152; stereotypes of, 141; subaltern urbanism and, 142–143; terminology of, 138, 141–142; theoretical construction of, 140–143; tours of, 129; victimology and, 150–151
smart growth, creative cities and, 48, 68
Smith, Neil, 47
social inequality, 47, 140; globalization and, 6; megacities, capitalism and, xvi
social services, privatization of, 43
soft urbanism, xxv
Soja, Edward, 25
Soto, Hernando de, 156
Sousa Santos, Bonaventura de, xxvi
Southern theory, 56; analytical imprecision of, 62; deficiencies of, 60–64; diversity and, 60; global urbanism and, 58; modernity and, 58; otherism and, 61; postcolonial theory roots of, 57–58, 60, 62; on struggling cities, 59–60; subaltern theory roots of, 60; on world-class cities, 59
southern urbanism, 59
South Korea, urban shrinkage in, 107
spatial patterns, 38; convergence thesis and, 55; domestication of space and, 90; of instant cities, 169–170; of slums, 140–141, 145–152; theming and, 238n37; of urbanization, xxii; urban shrinkage and, 109, 122; in urban studies, 29, 45; in world-class cities, 67–68. *See also* architecture
special economic zones, 186–187

spectacle-ization, 242n86
speculative investment, in urban regeneration, 80–82
speculative urbanism, 177
Speirs Wharf Canalside, Glasgow, 81
sprawl. *See* megacities
sprawling megacities of hypergrowth. *See* megacities
Stadtumbau Ost (Regeneration East) Program, 100
staging, 71; of differences, 206; of tourist-entertainment city, 75–76
Stevens, Quintin, 80
Stillwaggon, Eileen, 146
St. Louis, Missouri, 117
Storper, Michael, 30–32
struggling cities: abandonment in, 107, 110–111, *114*; causes and consequences of, 112–116, 204–205; causes of, 94; characteristics of, 92–93; competition and, 91–92; creative cities formula for, 52–53; from deindustrialization, 99, 106; deindustrialization and, 15; developmentalism and, 24; faulty premise of, 93; features of, *12–13*, 14–15; historical and conceptual origins of, 99–102; as legacy cities, 102; market-led solutions to, 123–124; migration and, 114–115; mythical construction of, 110–111; natural disasters and, 120; patterns of, 116–118; population decline and, 102–104; post-public space in, 88–90; rapid spread of, 102; SCIRN and, 101–102; scope and breadth of, 102–108, 122; shrinkage identity of, 111–112; shrinkage syndrome and, 95–97; Shrinking Cities Project and, 100–101; Southern theory on, 59–60; Stadtumbau Ost (Regeneration East) Program on, 100; stagnation and contraction patterns in, 97–99; from suburbanization, 99, 116; terminology of, 108–109, 119–120; time/space of, 208; as tourist-entertainment cities, 69, 78–80; urban decline and shrinkage in, 125; as urbanism outliers, 25; urbanization without growth in, 42; urban planning and, 98–99; urban regeneration of, 78–80, 102–103; urban studies on, 23–24, 40–41, 44; useful frameworks for, urban shrinkage and, 118–124; world-class cities compared to, 10–11; world-class cities competing with, 19. *See also* urban shrinkage
subaltern urbanism, 57, 60, 142–143
suburbanization, struggling cities from, 99, 116
surplus urbanism, 42
sustainable urban development, xi; authenticity of, 193; Cittaslow movement and, 194–196; progress and, xxiv

themed entertainment destinations, 74
theming, 71, 74, 238n37
"third space," 56
Third World. *See* Global South; Southern theory
Thrift, Nigel, 199
tourist-entertainment cities: appearance over substance in, 72; architecture in, 72–73; capitalism and, 73; closure and exclusion in, 89–90; competition and, 69; deregulation and, 86; designscapes and, 72; ethnoscapes in, 85; goals for, 68–69; localization in, 84; museums in, 76–77; neoliberalism and governance in, 85–88; New York City as, *70*; overbuilding in, 82; overlay zones in, 77; place marketing of, 71–73, 83–84; post-public space in, 88–90; PUDs and, 70; risks of, 77; signature architecture in, 76–78; speculative investment in, 80–82; staging of, 75–76; struggling cities as, 69, 78–80; themed entertainment destinations in, 74; unsolicited urbanism and, 87
tourist urbanism, 83–85
Tyler, Anne, 83

uncertainty, urbanism and, 200–201
underdevelopment, modernity and, 166
uniqueness, radical, 33–34
United States, urban shrinkage in, 103, 108, 112, 115–117
universalizing theorizations, in urban studies, 29–32, 45, 54–55, 202
unsolicited urbanism, 87
Urban Age, xi
urban contraction. *See* urban shrinkage
urban decline, 110–111, 121, 125. *See also* struggling cities
urban governance. *See* governance, urban
urban integrated megaprojects, 170
urbanism: austerity, 96; bypass-implant, 169–170; comparative, xx, 37–38; entrepreneurial, 189; of exception, 29; excessive, 42; experimental, 175–177; fortress, 89; informal, 163–165; innovation and, xvi; instantaneous, 184; megacities and future of, 16; monocentric, xxii;

perverse, 42; postmodern, 14, 169; progressive, 49; recycled, 192; runaway, 42, 128, 130; slow, 194; soft, xxv; southern, 59; speculative, 177; struggling cities as outliers of, 25; subaltern, 57, 60, 142–143; surplus, 42; tourist, 83–85; uncertainty and, 200–201; unsolicited, 87; world-class cities as examples of, 24–25. *See also* fast-track urbanism; global urbanism
urbanization: extended, xxii; gentrification and, 47; without growth, 42; new particularism and, 30; patterns of, 34; planetary, xxii, 28–29, 34–37, 206; population growth forecasts and, 126; spatial patterns of, xxii; tourist, 83–85; universalizing theorizations on, 29–32
urban planning, struggling cities and, 98–99
urban population growth: accelerating rate of, xii; in Global South, xi–xii; megacities and, xii; poverty and, xii, xiv; urbanization and forecasts for, 126
urban regeneration: creative cities and, 52–53; culture-led, 75–77, 80; downtown revitalization and, 70; gentrification and, 79; of historic old towns, 75, 79; market-driven approaches to, 69; megaprojects of, 80; overbuilding in, 82; place marketing and, 75–76; property-led, 72–73; signature architecture for, 76; speculative investment in, 80–82; of struggling cities, 78–80, 102–103. *See also* tourist-entertainment cities
urban shrinkage, 10–11, 14, 253n58; artwork on, 95, 100–101; in Australia, 107; causes and consequences of, 112–116, 204–205; characteristics of, 94; from deindustrialization, 113–114; "doughnut effect" of, 117; employment and, 116; in France, 105–106; in Germany, 101–102, 105, 111–112, 118–121, 255n101; global examples of, xiii; globalization and, 113; in Great Britain, 105–107; in Japan, 107; market-led solutions to, 123–124; migration and, 114–115; mythical construction of, 110–111; natural disasters and, 120; parallel processes of, 210; patterns of, 116–118; resource-depleted cities and, 112; scope of, 122; in Scotland, 106; sequence of, 121–122; shrinkage identity and, 111–112; shrinkage syndrome and, 95–97; in South Korea, 107; spatial patterns and, 109, 122; suburbanization and, 99, 116; terminology of, 119–120, 262n198; in United States, 103, 108, 112, 115–117; urban decline compared to, 125; use and meaning of, 108–109; as useful framework, 118–124. *See also* struggling cities
urban studies: comparative urbanism in, 37–38; conventional, conceptual blindspots of, 135–138; conventional, of Global North, 57; conventional, rethinking, xviii–xx, 3–5, 7–10, 25–26, 200, 202–205; convergence thesis deficiencies and, 53–56; convergence thesis in, 45–47; creative cities and convergence in, 47–53; deconstructing ill-fitting universalisms, 32–34; developmentalism in, attachment to, 38–40; essentialism in, 40; formality/informality dividing lines in, 152–154; generalizations in, inherent dangers of, 24–26; from Global South perspective, xx, xxiii; limitations of conventional, 40–44; megacities in, 16, 24, 130–131, 135–138; modernity in, attachment to, 38–40; new particularism and, 30; nomothetic-idiographic divide in, 31; ordinary cities in, 9–10, 25, 40, 43; planetary urbanization in, 28–29, 34–37; polarizing tendencies in, 28–32; post-structuralist philosophy in, 31–32; radical uniqueness in, 33–34; rethinking, 56–57, 61–62; Southern theory and, 56–64; spatial patterns in, 29, 45; on struggling cities, 23–24, 40–41, 44; terminology in, 28–29; universalizing theorizations on, 29–32, 45, 54–55, 202; urbanization without growth in, 42; on world-class cities, 23, 26–28; world-class cities in, 9
urban transformation: contradictory processes of, 2; divergent trajectories of, 10–18; multidirectional patterns of, xxiv, 5–8; universalizing theories of, 2–3
urban wastelands. *See* struggling cities
Urry, John, 242n86
utopianism, 171–173; of instant cities, 177–178

Vale, Lawrence, 67
value, informality and, 159–160
Varley, Ann, 62, 63
Venice Observed (McCarthy), 83
Venturi, Robert, 74, 241n76
victimology, slums and, 150–151

Wachsmuth, David, 35
Walker, Mabel, 253n58
Warner, Sam Bass, 67

waterfront redevelopment, 79, 81
Watson, Vanessa, xx
Wiechmann, Thorsten, 103, 124
Windelband, Wilhelm, 31
Wolff, Manuel, 103, 124
Woodworth, Max, 11
world city hypothesis, 1, 6
world city networks, 6
world-class cities, 5; city boosterism and, 42; cognitive-cultural capitalism and, 14; convergence thesis and, 45–47; creative cities thesis and, 49–51; features of, *12–13*, 14; framework of, benefits of, 26–27; framework of, limitations of, 27–28; globalization causing competition among, 26, 67–68, 91–92; normative idealism of, 47–48; postmodern urbanism and, 14; ranking, 28; as regulating fiction, 224n24; Southern theory on, 59; spatial patterns in, 67–68; struggling cities compared to, 10–11; struggling cities competing with, 19; thesis of, 1, 6; time/space of, 207–208; as urbanism examples, 24–25; in urban studies, 9; urban studies on, 23, 26–28. *See also* tourist-entertainment cities
worlding cities, 28
Wright, Frank Lloyd, 172
Wu, Fulong, 55

Yiftachel, Oren, 163, 220n63

Zeiderman, Austin, 28, 29, 133, 134
Zukin, Sharon, 69, 78, 79

GPSR Authorized Representative: Easy Access System Europe, Mustamäe tee 50, 10621 Tallinn, Estonia, gpsr.requests@easproject.com

www.ingramcontent.com/pod-product-compliance
Ingram Content Group UK Ltd.
Pitfield, Milton Keynes, MK11 3LW, UK
UKHW040331190325
456107UK00006B/173

* 9 7 8 0 2 3 1 2 0 4 0 7 1 *